# DYNAMIC PROBABILISTIC SYSTEMS

# DYNAMIC PROBABILISTIC SYSTEMS

## Volume I: Markov Models

RONALD A. HOWARD

*Stanford University*

DOVER PUBLICATIONS, INC.
Mineola, New York

*Bibliographical Note*

This Dover edition, first published in 2007, is an unabridged republication of the work originally published by John Wiley & Sons, Inc., New York, in 1971. The author has written a new Preface to the Dover Edition.

*International Standard Book Number*
*ISBN-13: 978-0-486-45870-0*
*ISBN-10: 0-486-45870-9*

Manufactured in the United States of America
Dover Publications, Inc., 31 East 2nd Street, Mineola, N.Y. 11501

To Polly,
my trapping state.

# PREFACE TO THE DOVER EDITION

When these volumes were published over three decades ago, a major difficulty in their application was assigning the necessary transition probabilities and holding-time distributions. The original motivation 50 years ago was a rare business application where the relevant information was available. The story of this application appears in:

"Comments on the Origin and Application of Markov Decision Processes," Operations Research, Vol. 50, No. 1, January-February 2002, pp. 100-102

With the development of modern data acquisition and processing systems in many current enterprises, data paucity has been changed to data abundance, if not data overload. Decision methods that could only rarely be implemented are now widely economically feasible. The republication of this work will bring these methods to those who can use them.

RONALD A. HOWARD
Stanford University
February, 2007

# PREFACE

Uncertainty, complexity, and dynamism have been continuing challenges to man's understanding and control of his physical environment. In the development of logical structures to describe these phenomena, the model originated by A. A. Markov stands as a major accomplishment. Where previous contributors had modeled uncertainty as a sequence of independent trials, Markov saw the advantage of introducing dependence of each trial on the result of its predecessor. While it is tempting to consider even more complex dependency of the present trial on the results of past trials, such temptation usually leads to results that are both analytically and computationally intractable.

Consequently, Markov models represent the first outpost in the domain of dependent models that is powerful both in capturing the essence of many dependent systems observed in practice and in producing the analytical and computational results necessary to gain insight into their behavior. The purpose of this book is to equip the reader to formulate, analyze, and evaluate simple and advanced Markov models of systems that range from genetics to space engineering to marketing. This book should be viewed not as a collection of techniques (although many will value it as such), but rather as a consistent application of the fundamental principles of probability and linear system theory.

It is often said that good ideas are simple; the Markov process is no exception. In fact, there is no problem considered in this book that cannot be made clear to a child. The device we use to make such expositions simple is a pond covered with lily pads among which a frog may jump. Although his jumps may be random, the frog never falls into the water. Furthermore, he may occupy the same pad after successive jumps. We can use the lily pond analogy to clarify any discussion in the text.

The book is an integrated work published in two volumes. The first volume, Chapters 1–9, treats the basic Markov process and its variants; the second, Chapters 10–15, semi-Markov and decision processes. Although the reader who is already familiar with Markov process terminology will understand the design of the book from the Contents, it should be helpful to all readers to discuss each chapter using the lily pond analogy.

Chapter 1 introduces the basic Markov model. The number of pads in the lily pond is assumed to be finite. The Markov assumption states that the probability of the frog's making his next jump to any pad depends only on the pad he currently occupies and not on how he reached his present pad. The discussion then considers the question, given the starting pad and the fact that $n$ jumps have been made without our observing the pond, what probability must we assign to the frog's occupying each pad in the pond. Graphical and transform methods are used to find these probabilities and to illustrate the typical behavior of a variety of processes.

Chapter 2 is a diversion to present the theory of linear systems and graphical methods for their analysis. Chapter 3 shows how to apply these powerful techniques to answering the problems of Chapter 1 with respect to the lily pond.

Chapter 4 considers the case where certain pads, once occupied, cannot be left and investigates the statistics for the number of jumps required before such trapping occurs. Chapter 5 treats two types of statistics: First, how many jumps will be made onto each pad in a given number of total jumps and, second, how many jumps will be required to reach one pad from another pad for the first time.

Chapter 6 finds the statistics of jumps from one set of pads to another set of pads. It allows identifying certain jumps as being of special interest and then finding the statistics of these jumps. Furthermore, it considers the implications of the information provided by these jumps for probability assignments on the frog's location.

Chapter 7 focuses on the case where we observe only one pad and where the frog's returns to that pad can be governed by an arbitrary probability distribution. We explore the statistics of these returns from several points of view. This chapter also discusses for the first time the possibility that the pond may have an infinite number of pads and illustrates the analytic modifications necessary to treat this situation.

Chapter 8 investigates the possibility of having several frogs jumping in the pond at the same time. It expands the discussion to the case where at certain times frogs may be added to each pad in the pond in varying numbers. Finally, it treats the case where the frogs may breed to produce new generations of jumpers.

Chapter 9 considers the case where the probability of a jump from one pad to another may be different on successive jumps. For example, the frog may be getting tired.

Chapter 10 introduces another dimension to the discussion. We explicitly consider the time between jumps as a multiple of some time unit and let this be a random variable that depends on the jump made by the frog. We can now consider the probability of the frog's occupying each pad given both the number of jumps he has made and the total time units he has been jumping. All previous questions about the frog's behavior are re-examined in this generalized form.

Chapter 11 allows the time between jumps to be a continuous random variable and extends all previous discussions to this case. Chapter 12 considers the continuous-time Markov process, the case in which the time that the frog has occupied the same pad has no bearing on the amount of time he will continue to occupy it. It also discusses the infinite-pad lily pond once again.

Chapter 13 concerns the payment of rewards on the basis of the frog's jumping. He may earn a reward for making a particular jump or from occupying a pad. We develop and illustrate expressions for the expected total reward the frog will earn in a finite and infinite time with and without discounting.

Chapter 14 presents another diversion, a discussion of the optimization procedure of dynamic programming, the natural way to formulate optimization problems for Markov processes.

Chapter 15 combines the material of the last several chapters to develop procedures for optimizing the expected reward generated by the frog. Control is effected by altering the probabilistic structure and reward structure of his jumping.

Since the level of interest in frog-jumping is relatively low, the examples throughout are drawn from a variety of fields: consumer purchasing, taxicab operation, inventory control, rabbit reproduction, coin-tossing, gambling, family-name extinction, search, car rental, machine repair, depreciation, production, action-timing, reliability, reservation policy, machine maintenance and replacement, network traversal, project scheduling, space exploration, and success in business. The applicability of the models to still other areas is also made evident.

The background suggested for the reader if he is to achieve greatest benefit from the book is a foundation in calculus, probability theory, and matrix theory. The emphasis is on the fundamental rather than the esoteric in these areas, so preparation should not be a major stumbling block. The book may be read with ease by anyone with the basic background whether his field be engineering, management, psychology, or whatever. Specialized material from any field used in examples is presented as required. Consequently, the professional reading independently to expand his knowledge of Markov models should be able to proceed rapidly and effectively.

This book developed from courses taught at M.I.T. and Stanford University over a decade. Although the advanced undergraduate can master the subject with no difficulty, the course has always been offered at the graduate level. The material is developed naturally for course presentation. To cover the whole book typically requires the entire academic year. The two volumes serve well as texts for successive one-semester courses. If a two-quarter version is desired, then Chapters 8 and 9 can be placed in the optional category.

One feature of the development that is becoming increasingly advantageous to exploit is the suitability of examples and problems for solution by computer, particularly the time-shared variety. Virtually every chapter contains material appropriate for computer demonstration, such as simulation of Markov process behavior, or solution for statistics via difference equations. While such integration with computer methods is not necessary, it does provide an opportunity for curriculum unification.

Some of the material presented here has appeared in more primitive form elsewhere. For example, Chapter 2 is a highly revised form of "System Analysis of Linear Models," *Multistage Inventory Models and Techniques*, eds., Scarf, Shelley, Guilford, 143–184, Stanford University Press, 1963; Chapter 13 is an expanded version of "Dynamic Programming," *Management Science*, Vol. 12, No. 5, 317–348, January 1966. Much of the development in the second volume is based on "System Analysis of Semi-Markov Processes," *Trans. IEEE Prof. Group on Mil. Elec.*, Vol. MIL–8, No. 2, 114–124, April 1964, and on "Semi-Markovian Decision Processes," *Proceedings of the 34th Session International Statistical Institute*, 625–652, Ottawa, Canada, August 1963.

Finally, some credit where credit is due. Much of the original work on this book was performed during a visiting year at Stanford University under the sponsorship of what has become the Department of Engineering-Economic Systems. Most of the final touches were supplied during a visiting year as Ford Research Professor at the Stanford Graduate School of Business. The book contains research results that grew from supported research. Most of the early research was sponsored by the Office of Naval Research; more recently the research was supported by the National Science Foundation. To all these organizations I express my appreciation.

The question of personal acknowledgments is a difficult one for any author. Only he knows the debt he owes to so many over a period of several years. By selecting a few for special mention, he accepts the risk that the important contributions of others will not receive sufficient notice. Accepting that risk, let me reveal how this work is the result of many contributors.

Several individuals have provided me with advice, criticism, and suggestions on the development of the manuscript. In the early versions, I benefited greatly from my contact with Edward S. Silver and Jerome D. Herniter. Richard D. Smallwood has provided a continuing stream of suggestions that have materially improved the manuscript. Herbert F. Ayres and W. Howard Cook have each made several suggestions that clarified the text. Finally, I must credit the Herculean task of my friend James E. Matheson who read through the completed manuscript in detail with no hope of reward but the promise of a feast in a fancy New York restaurant. Having fulfilled that promise, I am happy to report that he has found and corrected all mistakes in the manuscript, thus relieving me of the necessity of making that boring statement that any idiocies in the book are the full responsibility of the author.

I believe that there is no more important part of a textbook than its problems, and so I asked Richard D. Smallwood to collate and augment the problem sets we have used in courses over the years. He has done a thorough and imaginative job and has demonstrated that the challenging need not be humorless. Many of the problems were originally composed by teaching assistants; age and evolutionary change have made their authorship obscure. However, in many I discern the deft touch of Edward A. Silver who has a special knack for making a student smile and sweat at the same time.

Any book that would illustrate complex Markov models will require extensive electronic computation. I have been fortunate in having the services of individuals who have unusual flair and ability in this area, namely, Claude Dieudonné, Richard D. Smallwood, Paul Schweitzer, Leif Tronstad, and Alain Viguier. Their contribution ranged from solution of numerical examples to the development of programming systems that treated large classes of models: The reader is the beneficiary of their ingenuity.

A challenging part of the manuscript was the translation of Markov's paper. Here I needed someone who could translate classic Russian into English—I found him in George Petelin. Mr. Petelin's English translation provided the basis for the final version that appears in the book; for its accuracy I personally bear full responsibility.

Obviously a project of this magnitude is a secretarial nightmare. Fortunately, I have found myself over the years in the hands of a succession of ladies whose secretarial skills were exceeded only by their intelligence and charm. The first was Mrs. Nobuko McNeill who personally supervised the typing of over one thousand pages of Gregg shorthand directly into finished manuscript. The second was Mrs. Edna Tweet who prepared the final chapter of Volume II. The third was Mrs. Louise Goodrich who suffered through the endless revision, galley reading, and proof checking necessary to bring this project to a successful conclusion.

Therefore, let me express to all these fine people my appreciation and thanks for helping me tell my story about the Markov process and what became of it.

Palo Alto, California
January, 1971

RONALD A. HOWARD

# CONTENTS

# DYNAMIC PROBABILISTIC SYSTEMS

# 1 THE BASIC MARKOV PROCESS

## 1.1 INTRODUCTION

The Markov process is a probabilistic model useful in analyzing complex systems. Central to the theory of Markov process models are the concepts of state and state transition.

### The State Concept

To explain "state" briefly, let us first realize that the present situation in a physical system can usually be specified by giving the values of a number of variables that describe the system. For example, a chemical system can often be specified by the values of temperature, pressure, and volume; the instantaneous description of a spacecraft would include its position in spatial coordinates, its mass, and its velocity. Such critical variables of a system are called "state" variables. For the chemical process the state variables are temperature, pressure, and volume; for the spacecraft they are position, mass, and velocity. When the values of all state variables of a system are known, we can say that its state has been specified. The state of a system thus represents all we need to know to describe the system at any instant.

### The Transition Concept

In the course of time a system passes from state to state and thus exhibits dynamic behavior. In the chemical system such changes are caused by the application of heat, an increase in volume, etc. The spacecraft would experience a change of state if it burned fuel or even when it simply coasted under the earth's gravitational influence. Such changes of state are called "state transitions" or simply, "transitions."

The most general "state—transition" model would allow states described by continuous variables and transitions that could occur at any time. We shall not be that general. We shall consider systems that may occupy only a finite number of

**Figure 1.1.1** The lily pond.

states, or in some special situations, a countably infinite number of states. In this first chapter we shall not concern ourselves with the time interval separating transitions, but rather with the sequence of states occupied by the system. Later we shall consider the time between transitions as a random process in its own right.

### The Lily Pond

I have always found it useful to think of a Markov process as describing the behavior of a frog in a lily pond, perhaps like the one pictured in Figure 1.1.1. The frog must always sit on a pad; he never swims in the water. Occasionally, the frog jumps in the air and lands on a lily pad—sometimes the one he had been sitting on, sometimes a different one. For the moment we are interested not in the time pattern of his jumping, but in the location of the frog after successive jumps.

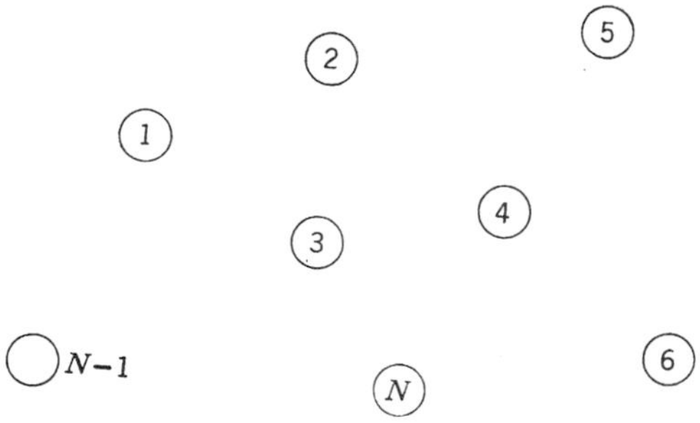


**Figure 1.1.2** The pond abstracted.

To make our discussion specific, we consider the pond to have a finite number of lily pads, $N$, numbered from 1 to $N$. Figure 1.1.2 is a diagrammatic representation of the pond. If we use $n$ to denote the number of jumps made by the frog, and $s(n)$ to denote the number of the pad occupied after the $n$th jump, then the sequence

$$s(0), s(1), s(2), \ldots, s(n), s(n+1) \tag{1.1.1}$$

specifies the frog's entire itinerary. We call such a sequence the trajectory of the process.

We now have a physical realization of a "state—transition" process. The state of the process is the number of the pad occupied by the frog. The quantity $s(n)$ represents the state of the process after its $n$th transition, which for the present we shall simply call the state of the process at time $n$. In an $N$-state process, $s(n)$ may take on any integer value $1, 2, \ldots, N$ at any integer time $n = 0, 1, 2, \ldots$ that shows the number of transitions made by the process.

As we observe the sequence of states occupied by the frog, i.e., the process trajectory, we may find that it is random. The statistical behavior of this process could be specified if the probability

$$\mathscr{P}\{s(n+1) = j \mid s(n) = i, s(n-1) = k, \ldots, s(0) = m\} \tag{1.1.2}$$

were known for all values of its arguments. This is the probability that each state will be occupied after the $n + 1$st transition given the entire trajectory or history of state occupancies through time $n$. If this conditional probability could be specified arbitrarily for every value of $n$, then the behavior of the process could be extremely complex. Indeed, specification of the probabilities would itself be a problem.

### The Markovian Assumption

The Markovian assumption greatly simplifies both the possible behavior of the process and the problem of specifying the process. The assumption is: Only the last state occupied by the process is relevant in determining its future behavior. With this assumption,

$$\mathscr{P}\{s(n+1) = j \mid s(n) = i, s(n-1) = k, \ldots, s(0) = m\}$$

becomes

$$\mathscr{P}\{s(n+1) = j \mid s(n) = i\}. \tag{1.1.3}$$

Thus, the probability of making a transition to each state of the process depends only on the state presently occupied. Equivalently, the future trajectory of a process depends only on its present state.

The Markovian assumption is very strong: There are few physical systems that we could expect to be so memoryless in a strict sense. Yet the Markov process is an extremely useful model for wide classes of systems ranging from genetics to inventory control.

No experiment can ever show the ultimate validity of the Markovian assumption; hence, no physical system can ever be classified absolutely as either Markovian or non-Markovian—the important question is whether the Markov model is useful. If the Markovian assumption can be justified, then the investigator can enjoy analytical and computational convenience not often found in complex models.

However, even if the history of the system before the last state occupied does influence future behavior, the Markovian assumption may still be satisfied by a change in the state structure. For example, suppose that the last two states occupied both influenced the transition to the next state. Then we could define a new process with $N^2$ states—each state in the new process would correspond to a pair of successive states in the old process. With this redefinition, the property of Equation 1.1.3 could be satisfied, but at the expense of considerable increase in computational complexity. Any dependence of future behavior on a finite amount of past history can, at least theoretically, be treated in the same way.

### Transition Probabilities

To define a Markov process we must specify for each state in the process and for each transition time the probability of making the next transition to each other state given these conditions. Thus the quantity

$$\mathscr{P}\{s(n+1)=j\,|\,s(n)=i\} \tag{1.1.4}$$

must be specified for all $1 \le i, j \le N$, and for $n = 0, 1, 2, \ldots$.

Our notation allows the possibility that the probability of making each transition may change as successive transitions are made. Although we shall want to consider such behavior in a later chapter, we shall at present restrict ourselves to the case where these probabilities stay constant for all transitions. We thus define the transition probability $p_{ij}$:

$$p_{ij} = \mathscr{P}\{s(n+1)=j\,|\,s(n)=i\} \qquad 1 \le i,j \le N,\ n = 0, 1, 2, \ldots. \tag{1.1.5}$$

The transition probability $p_{ij}$ is the probability that a process presently in state $i$ will occupy state $j$ after its next transition. Since each transition probability $p_{ij}$ is a probability, it must satisfy the requirement,

$$0 \le p_{ij} \le 1 \qquad 1 \le i,j \le N. \tag{1.1.6}$$

Notice that we allow the possibility of the same state's being occupied after a transition—the probabilities $p_{ii}$, $i = 1, 2, \ldots, N$ are not necessarily zero. Since the process must occupy one of its $N$ states after each transition,

$$\sum_{j=1}^{N} p_{ij} = 1 \qquad i = 1, 2, \ldots, N. \tag{1.1.7}$$

(*Remember*: The frog is not allowed to fall in the water.)

The $N^2$ transition probabilities that describe a Markov process are conveniently represented by an $N$ by $N$ transition probability matrix $P$ with elements $p_{ij}$,

$$P = \{p_{ij}\} = \begin{bmatrix} p_{11} & p_{12} & \cdots & p_{1N} \\ p_{21} & p_{22} & \cdots & p_{2N} \\ \vdots & & & \\ p_{N1} & p_{N2} & \cdots & p_{NN} \end{bmatrix}. \tag{1.1.8}$$

The entries in $P$ must satisfy the requirements imposed by Equations 1.1.6 and 1.1.7. A matrix whose elements cannot lie outside the range (0, 1) and whose rows sum to one is called a stochastic matrix; thus the transition probability matrix that defines a Markov process is a stochastic matrix. Because the rows of the transition probability matrix sum to one, only $N(N - 1)$ parameters are necessary to specify the probabilistic behavior of an $N$-state Markov process.

The transition probability matrix of a Markov process, and hence the process itself, can be graphically represented by a transition diagram—like that shown in Figure 1.1.3—formed of nodes and directed line segments called branches. Each node is numbered to represent one state of the process. A directed line segment or branch is drawn from each node $i$ to each node $j$ and labeled with the transition probability $p_{ij}$. In numerical examples we shall use the convention that only those

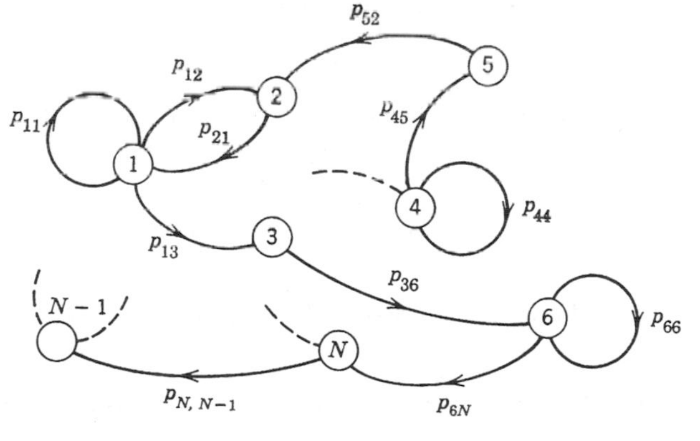


**Figure 1.1.3** The transition diagram.

directed line segments corresponding to non-zero transition probabilities will be shown in the transition diagram. Because the rows of the transition matrix sum to one, the numbers on all the branches leaving a node must also sum to one. We shall find that the transition diagram is the key not only to visualization of a Markov process, but also to powerful analytic techniques.

## 1.2 MARKOV PROCESS BEHAVIOR

Suppose we have reason to believe that the frog's activities can be described by a Markov process with a specified transition probability matrix. What questions might we then ask about his behavior?

### Multistep Transition Probabilities

An interesting question is this: What can we say about the position of the frog after $n$ jumps have been made if we know the pad on which he started? The answer to the question is the probability assignment that we would make on the frog's pad after $n$ jumps given that we knew the pad he originally occupied. Let $\phi_{ij}(n)$ be the probability that the process will occupy state $j$ at time $n$ given that it occupied state $i$ at time 0,

$$\phi_{ij}(n) = \mathscr{P}\{s(n) = j|s(0) = i\} \qquad 1 \leq i, j \leq N, n = 0, 1, 2, \ldots. \tag{1.2.1}$$

The quantity $\phi_{ij}(n)$ is called the $n$-step transition probability of the Markov process from state $i$ to state $j$. The behavior of $\phi_{ij}(n)$ for all values of $i$, $j$, and $N$ is the most important derived characteristic of a Markov process.

We can easily relate the multistep transition probabilities, such as $\phi_{ij}(n)$, to the transition probabilities $p_{ij}$. Suppose that the system started in state $i$. The probability that it is in state $j$ after $n + 1$ transitions is then

$$\mathscr{P}\{s(n + 1) = j|s(0) = i\}. \tag{1.2.2}$$

This probability may be written in terms of the joint probability that state $j$ is occupied at time $n + 1$ and state $k$ is occupied at time $n$ in the form

$$\mathscr{P}\{s(n + 1) = j|s(0) = i\} = \sum_{k=1}^{N} \mathscr{P}\{s(n + 1) = j, s(n) = k|s(0) = i\}. \tag{1.2.3}$$

From the definition of conditional probability we obtain

$$\mathscr{P}\{s(n + 1) = j|s(0) = i\} = \sum_{k=1}^{N} \mathscr{P}\{s(n) = k|s(0) = i\}$$
$$\times\ \mathscr{P}\{s(n + 1) = j|s(n) = k, s(0) = i\}. \tag{1.2.4}$$

If $n \geq 1$, then by the Markovian assumption that the future trajectory is influenced only by the present state,

$$\begin{aligned}\mathscr{P}\{s(n+1) = j \mid s(n) = k, s(0) = i\} \\ = \mathscr{P}\{s(n+1) = j \mid s(n) = k\} = p_{kj}.\end{aligned} \tag{1.2.5}$$

By using definition (1.2.1) Equation 1.2.4 can now be written

$$\phi_{ij}(n+1) = \sum_{k=1}^{N} \phi_{ik}(n)p_{kj}, \qquad n = 1, 2, 3, \ldots. \tag{1.2.6}$$

This equation relates the $\phi_{ij}$ after any transition to the $\phi_{ij}$ before that transition. For $n = 1$ it relates the 2-step transition probabilities $\phi_{ij}(2)$ to the 1-step transition probabilities $\phi_{ij}(1)$. But what are the 1-step transition probabilities? They are just the transition probabilities $p_{ij}$. Equation 1.2.6 will produce this result for $n = 0$ provided that $\phi_{ij}(0)$ is equal to $\delta_{ij}$, the Kronecker delta:

$$\phi_{ij}(0) = \delta_{ij} = \begin{cases} 1 & i = j \\ 0 & i \neq j \end{cases} \tag{1.2.7}$$

Such a value for $\phi_{ij}(0)$ is certainly required by the definition 1.2.1. Now we can write Equation 1.2.6 for all $n$, including $n = 0$, as

$$\begin{aligned}\phi_{ij}(n+1) &= \sum_{k=1}^{N} \phi_{ik}(n)p_{kj} \qquad n = 0, 1, 2, \ldots \\ \phi_{ij}(0) &= \delta_{ij} \qquad 1 \leq i, j \leq N.\end{aligned} \tag{1.2.8}$$

The multistep transition probabilities defined by Equations 1.2.8 satisfy the same requirements as do the transition probabilities. Thus,

$$0 \leq \phi_{ij}(n) \leq 1 \qquad 1 \leq i, j \leq N, n = 0, 1, 2, \ldots \tag{1.2.9}$$

and

$$\sum_{j=1}^{N} \phi_{ij}(n) = 1 \qquad i = 1, 2, \ldots, N, n = 0, 1, 2, \ldots. \tag{1.2.10}$$

### Matrix Formulation

Equations 1.2.8 provide a recursive method for calculating the multistep transition probabilities of a Markov process when its transition probability matrix $P$ is known. The $n$-step transition probabilities for such a process can be placed in an $N$ by $N$ matrix called the $n$-step transition probability matrix, $\Phi(n)$,

$$\Phi(n) = \{\phi_{ij}(n)\} = \begin{bmatrix} \phi_{11}(n) & \phi_{12}(n) & \cdots & \phi_{1N}(n) \\ \phi_{21}(n) & \phi_{22}(n) & \cdots & \phi_{2N}(n) \\ \vdots & & & \\ \phi_{N1}(n) & \phi_{N2}(n) & \cdots & \phi_{NN}(n) \end{bmatrix} \qquad n = 0, 1, 2, \ldots. \qquad (1.2.11)$$

Equations 1.2.9 and 1.2.10 show that $\Phi(n)$ is a stochastic matrix.

The matrix form of Equation 1.2.8 is

$$\begin{aligned} \Phi(n+1) &= \Phi(n)P \qquad n = 0, 1, 2, \ldots \\ \Phi(0) &= I \end{aligned} \qquad (1.2.12)$$

where $I$ is the identity matrix, the matrix whose elements are the Kronecker delta. Let us compute $\Phi(n)$ for successive $n$ by using Equation 1.2.12. We find

$$\begin{aligned} \Phi(0) &= I \\ \Phi(1) &= \Phi(0)P = IP = P \\ \Phi(2) &= \Phi(1)P = P^2 \\ \Phi(3) &= \Phi(2)P = P^3 \end{aligned} \qquad (1.2.13)$$

and in general

$$\Phi(n) = P^n \qquad n = 0, 1, 2, \ldots \qquad (1.2.14)$$

where we understand that $P^0 = I$.

Equation 1.2.14 is the fundamental equation for the basic Markov process. It says that the $n$-step transition probability matrix can be obtained by raising the transition probability matrix to the $n$th power. The rows of $\Phi(n)$ specify the probability distribution that will exist over the states of the process after $n$ transitions for each possible starting state. Thus, if we were given the transition probability matrix for the frog and his starting pad, we would have no difficulty, in principle, in calculating the probabilistic effect of any number of jumps that he might make. From now on we shall seldom mention the lily pond explicitly, but we can use it as a means of visualization whenever the need arises.

### A Marketing Example

Let us consider a Markov modeling example from the field of marketing. In a hypothetical market there are only two brands, $A$ and $B$. A typical consumer in this market buys brand $A$ with probability 0.8 if his last purchase was brand $A$ and with probability 0.3 if his last purchase was brand $B$. We shall model the system as a two-state Markov process; state 1 is the state occupied by the consumer if

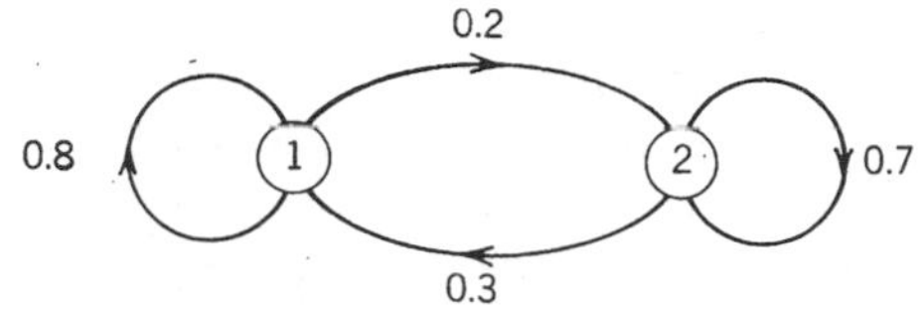


**Figure 1.2.1** Transition diagram for marketing example.

his last purchase was brand $A$, state 2 is the state occupied if his last purchase was brand $B$. The transition probability matrix for this process is

$$P = \begin{bmatrix} 0.8 & 0.2 \\ 0.3 & 0.7 \end{bmatrix}. \tag{1.2.15}$$

The corresponding transition diagram appears in Figure 1.2.1. We index the purchase occasions as $n = 0, 1, 2, \ldots$, where $n = 0$ corresponds to the purchase that initiates the purchase sequence.

***Multistep transition probabilities***

How does the probability of the consumer's purchasing each brand depend on the number of purchases he has made and on the brand he purchased initially? To answer this question, we compute the multistep transition probability matrices generated by raising the $P$ matrix to various powers. We find

$$\begin{aligned}
\Phi(0) &= I = \begin{bmatrix} 1 & 0 \\ 0 & 1 \end{bmatrix} \\
\Phi(1) &= P = \begin{bmatrix} 0.8 & 0.2 \\ 0.3 & 0.7 \end{bmatrix} \\
\Phi(2) &= P^2 = \begin{bmatrix} 0.7 & 0.3 \\ 0.45 & 0.55 \end{bmatrix} \\
\Phi(3) &= P^3 = \begin{bmatrix} 0.65 & 0.35 \\ 0.525 & 0.475 \end{bmatrix} \\
\Phi(4) &= P^4 = \begin{bmatrix} 0.625 & 0.375 \\ 0.5625 & 0.4375 \end{bmatrix} \\
\Phi(5) &= P^5 = \begin{bmatrix} 0.6125 & 0.3875 \\ 0.58125 & 0.41875 \end{bmatrix} \\
&\vdots
\end{aligned} \tag{1.2.16}$$

The probability that the consumer will buy brand $A$ on purchase occasion 2 is, therefore, $\phi_{11}(2) = 0.70$ if he was originally a brand $A$ user and $\phi_{21}(2) = 0.45$ if he was not. This example will allow us to show how computing the matrix $\Phi(n)$ by raising $P$ to the power $n$ is an efficient way to sum the probabilities of all relevant trajectories, and thus that $\phi_{ij}(n)$ is the sum of the probabilities of all trajectories of length $n$ that originate in state $i$ and terminate in state $j$.

Figure 1.2.2(a) shows the two possible trajectories relevant to the computation of $\phi_{11}(2)$. A trajectory of length 2 that starts in state 1 and ends in state 1 could have made its first transition from state 1 to either state but then would have to make its second transition to state 1. The sum of the probabilities of the two trajectories would then be

$$\phi_{11}(2) = p_{11}p_{11} + p_{12}p_{21} = (0.8)(0.8) + (0.2)(0.3) = 0.7. \qquad (1.2.17)$$

Similarly, Figure 1.2.2(b) shows the two trajectories that contribute to $\phi_{21}(2)$. A trajectory of length 2 that originates in state 2 and terminates in state 1 could have made its first transition from state 2 to either state; its second transition would have to be to state 1. The sum of the probabilities of the two possible trajectories is

$$\phi_{21}(2) = p_{21}p_{11} + p_{22}p_{21} = (0.3)(0.8) + (0.7)(0.3) = 0.45. \qquad (1.2.18)$$

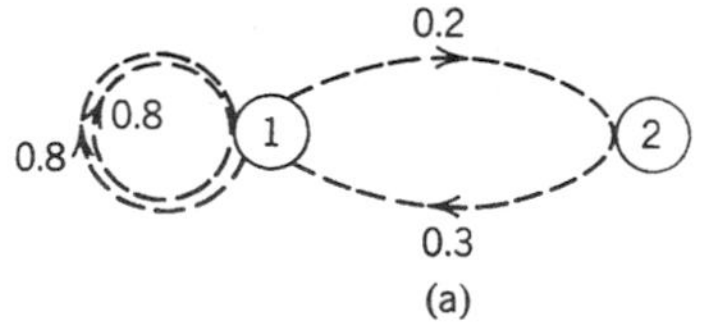


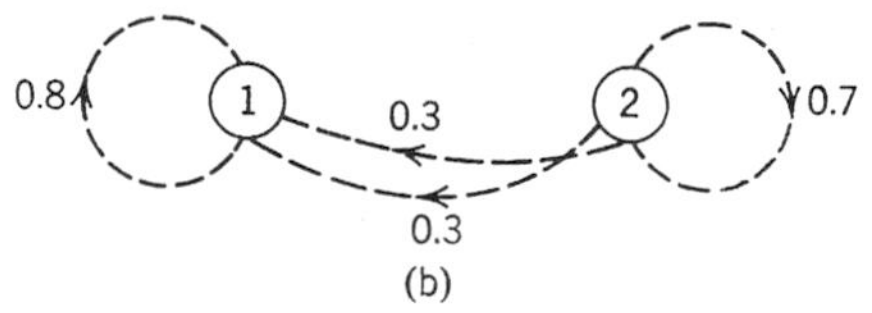


**Figure 1.2.2** Interpretation of $\Phi(n)$ as sum of trajectory probabilities. (a) Two possible trajectories for $1 \rightarrow 1$ in two transitions. (b) Two possible trajectories for $2 \rightarrow 1$ in two transitions.

We can readily verify that every multistep transition probability that was systematically produced in Equation 1.2.16 by matrix multiplication has a direct interpretation as the sum of the probabilities of all trajectories satisfying the multistep transition conditions.

***Limiting behavior***

The difference between the two rows of the $\Phi(n)$ matrices in Equation 1.2.16 seems to be decreasing as $n$ increases. As more and more purchases are made, the influence on the consumer's behavior of the brand originally used continually diminishes. The absolute value of the difference between the rows is the sequence 1, 0.5, 0.25, 0.125, 0.0625, . . . . This sequence approaches zero—when $n$ is very large, the influence of the first purchase will have vanished entirely.

What will be the entries in the two rows when they are identical? The 11 element of $\Phi(n)$ is given by the sequence 1, 0.8, 0.7, 0.65, 0.625, . . . ; this sequence is approaching the value 0.6. Therefore the value of the $\Phi(n)$ matrix when $n$ is very large is given by

$$\Phi(\infty) = \begin{bmatrix} 0.6 & 0.4 \\ 0.6 & 0.4 \end{bmatrix}. \tag{1.2.19}$$

We define

$$\Phi = \Phi(\infty) = P^{\infty} \tag{1.2.20}$$

and call the matrix $\Phi$ the limiting multistep transition probability matrix of the process. Its rows give the probability assignment over the states of the process that must exist for each starting state when the number of transitions of the process is allowed to increase beyond all bounds. In the present example we have found that the rows of $\Phi$ are identical; the ultimate probability assignment does not depend on where the process started. This behavior is not surprising in the marketing example—we would be amazed if the starting brand could maintain its influence into the indefinite future.

But do all Markov processes have this "forgetful" nature? The answer is that most do, but some don't. Separating the two classes will be one of our major goals.

## A Question of Interpretation

When we examine a sequence of multistep transition matrices like that in Equation 1.2.16, it is easy to fall into habits of thought and expression that may be misleading. For example, the sequence of values 1, 0.8, 0.7, 0.65, . . . etc., for $\phi_{11}(n)$ represents the probability that state 1 will be occupied after $n = 0, 1, 2, 3, \ldots$ transitions given that it was initially occupied. It is easy to speak of this sequence as demonstrating the transient behavior of a probability that approaches 0.6 when the number of transitions becomes very large. We thus find it natural to discuss the transient and limiting behavior of Markov process statistics.

However, there is one point we must constantly bear in mind. The process transitions are governed only by the transition probability matrix. If we ask the frog, "Are you in the steady state?", he doesn't know what we mean. He would reply that he selects the next state he will occupy given that he is now in state $i$ according to the probabilities recorded in the $i$th row of the transition probability matrix, and that beyond this, one transition is just like another to him.

*The transients that we observe in $\Phi(n)$ and other statistics are not process transients, but rather transients in our state of information and consequently probability assignments regarding the process.*

Thus the proper way to interpret the multistep transition probabilities is that these are the probabilities we must assign to the process's occupying each state after $n$ transitions if we observe its initial state, if we do not observe it in the interim, and if the operation of the process is to be consistent with its transition probability matrix. This probability assignment over states does exhibit transient behavior for $n = 0, 1, 2, 3, \ldots$ as we have seen and ultimately approaches for the example a limiting value that is independent of the starting state because starting information has become irrelevant in this example after many transitions. In fact, consistency with the transition probability matrix $P$ requires that we assign a 0.6 limiting probability to state 1. The next section will illuminate further the need for that particular assignment.

However, the fact that this limiting probability measures our knowledge about the process rather than the process itself is amply demonstrated by noting that if at any time we are able to observe the process, our probability assignment would change so as to assign probability 1 to occupancy of the state actually observed.

Now that we understand that we are studying the observer's information and not the process itself, we can often take liberties for the sake of brevity and speak of process transient or limiting behavior. However, we shall find it informative and necessary occasionally to make explicit the state-of-information interpretation of our results.

## 1.3 STATE PROBABILITIES

On many occasions we shall want to speak of the probability that a certain state is occupied after $n$ transitions without including in our notation the state in which the process was started. We shall call such a probability a state probability; the probability that state $i$ is occupied at time $n$ will be given the symbol $\pi_i(n)$ and defined by

$$\pi_i(n) = \mathscr{P}\{s(n) = i\} \qquad i = 1, 2, \ldots, N, n = 0, 1, 2, \ldots \tag{1.3.1}$$

The state probabilities $\pi_i(n)$ must sum to one over all states of the process:

$$\sum_{i=1}^{N} \pi_i(n) = 1 \qquad n = 0, 1, 2, \ldots. \tag{1.3.2}$$

When $n = 0$, the state probabilities $\pi_i(0)$ give us a way of describing a probability distribution over the starting states of the process rather than a specification of a particular state. It will not be unusual in our models to have the starting state for a Markov process generated by another random process.

### Calculation of State Probabilities

To see the equations that govern the $\pi_i(n)$, let us write the defining equation for the multistep transition probabilities, Equation 1.2.1,

$$\phi_{ij}(n) = \mathscr{P}\{s(n) = j \,|\, s(0) = i\}. \tag{1.3.3}$$

If we multiply both sides of this equation by $\mathscr{P}\{s(0) = i\}$ and sum over $i$ from 1 to $N$, we obtain

$$\sum_{i=1}^{N} \mathscr{P}\{s(0) = i\}\phi_{ij}(n) = \sum_{i=1}^{N} \mathscr{P}\{s(0) = i\}\mathscr{P}\{s(n) = j \,|\, s(0) = i\}$$

$$\sum_{i=1}^{N} \pi_i(0)\phi_{ij}(n) = \sum_{i=1}^{N} \mathscr{P}\{s(0) = i, s(n) = j\} = \mathscr{P}\{s(n) = j\}$$

and finally

$$\sum_{i=1}^{N} \pi_i(0)\phi_{ij}(n) = \pi_j(n)$$

or

$$\pi_j(n) = \sum_{i=1}^{N} \pi_i(0)\phi_{ij}(n) \qquad j = 1, 2, \ldots, N;\ n = 0, 1, 2, \ldots. \tag{1.3.4}$$

The state probabilities at time $n$ can be determined by multiplying the multistep transition probabilities by the probabilities of starting in each state and summing over all states.

The row vector formed by the $N$-state probabilities at time $n$ is called the state probability vector at time $n$ and is denoted by $\boldsymbol{\pi}(n)$,

$$\boldsymbol{\pi}(n) = [\pi_1(n), \pi_2(n), \ldots, \pi_N(n)] \qquad n = 0, 1, 2, \ldots. \tag{1.3.5}$$

With this definition, Equation 1.3.4 can be written as

$$\boldsymbol{\pi}(n) = \boldsymbol{\pi}(0)\Phi(n) \qquad n = 0, 1, 2, \ldots. \tag{1.3.6}$$

The state probability vector at time $n$ is thus the state probability vector at time 0 postmultiplied by the $n$-step transition probability matrix. From Equation 1.2.14 we then have

$$\pi(n) = \pi(0)P^n \qquad n = 0, 1, 2, \ldots. \tag{1.3.7}$$

Another important relation that shows how the $\pi(n)$ vectors are modified by the operation of the process is obtained by writing this equation for $\pi(n + 1)$. We find

$$\pi(n + 1) = \pi(0)P^{n+1} = \pi(0)P^nP$$

or

$$\pi(n + 1) = \pi(n)P \qquad n = 0, 1, 2, \ldots. \tag{1.3.8}$$

The state probability vector at any time can be calculated by postmultiplying the state probability vector at the preceding time by $P$.

***State probabilities of the marketing example***

We can illustrate the use of state probabilities in the marketing example of the last section. Suppose that the customer had a probability 0.4 of having purchased brand $A$ on purchase occasion 0 before the sequence of purchases $n = 1, 2, \ldots$ that we are considering in the example. The initial state probability vector for this customer would be

$$\pi(0) = [0.4 \quad 0.6]. \tag{1.3.9}$$

The probability that he will purchase brand $A$ on his second purchase occasion is then $\pi_1(2)$, the first element of the $\pi(2)$ vector. Equation 1.3.6 shows that $\pi(2)$ is given by

$$\pi(2) = \pi(0)\Phi(2). \tag{1.3.10}$$

The 2-step transition probability matrix for this example is found in Equation 1.2.16; we obtain

$$\pi(2) = [0.4 \quad 0.6]\begin{bmatrix} 0.7 & 0.3 \\ 0.45 & 0.55 \end{bmatrix} = [0.55 \quad 0.45]. \tag{1.3.11}$$

The probability that the customer will purchase brand $A$ on purchase occasion 2 is, therefore, 0.55.

## Asymptotic Behavior of State Probabilities

What happens to the state probability vector when the process is allowed to make a large number of transitions? Equation 1.3.7 shows that when $n$ is very large, the corresponding state probability vector $\pi(\infty)$ is given by

$$\pi(\infty) = \pi(0)P^\infty. \tag{1.3.12}$$

We shall call $\boldsymbol{\pi}(\infty)$ the limiting state probability vector of the process and give it the symbol $\boldsymbol{\pi}$. Then Equation 1.3.12 can be written in the form

$$\boldsymbol{\pi} = \boldsymbol{\pi}(0)\Phi \tag{1.3.13}$$

where $\Phi$ is the limiting multistep transition probability matrix defined in Equation 1.2.20. The limiting state probability vector is, therefore, the initial state probability vector postmultiplied by the limiting multistep transition probability matrix.

***Limiting state probabilities of monodesmic processes***

We still have not learned much about the form of the $\Phi$ matrix beyond the fact that it is a stochastic matrix. We shall soon understand not only its form, but how that form is achieved. However, let us anticipate one result. It is the rule rather than the exception that the rows of $\Phi$ are identical. For reasons that will become clear presently, we call a Markov process that has a $\Phi$ with equal rows a monodesmic process. We shall soon develop precise conditions for a process to be monodesmic. However, a sufficient condition for a process to be monodesmic is that the process be able to make a transition (have a non-zero transition probability) from any state to any other state. A necessary condition is that there exist only one subset of states (perhaps all the states of the process) that must be occupied after infinitely many transitions. Consequently, the marketing example we have discussed is a monodesmic process.

Equation 1.3.13 provides an important special result for monodesmic processes. First, we write this equation in terms of its elements,

$$\pi_j = \sum_{i=1}^{N} \pi_i(0)\phi_{ij} \qquad j = 1, 2, \ldots, N. \tag{1.3.14}$$

Because all rows of $\Phi$ are equal for a monodesmic process, each element $\phi_{ij}$ is equal to a value $\phi_j$ that depends only on the column index $j$. Then

$$\begin{aligned} \pi_j &= \sum_{i=1}^{N} \pi_i(0)\phi_j \\ &= \phi_j \sum_{i=1}^{N} \pi_i(0) \\ &= \phi_j. \end{aligned} \tag{1.3.15}$$

Thus a monodesmic Markov process has a limiting state probability vector $\boldsymbol{\pi}$ that is independent of the initial state probability vector $\boldsymbol{\pi}(0)$. The process approaches the same distribution of probability over its states regardless of where it started. Equation 1.3.15 shows that each row of $\Phi$ is just this unique limiting state probability vector $\boldsymbol{\pi}$.

***Direct solution for limiting state probabilities***

Fortunately, we do not need to raise $P$ to higher and higher power to find the vector $\boldsymbol{\pi}$. Equation 1.3.8 is the relation that successive state probability vectors must satisfy. If the state probability vector has attained its limiting value, $\boldsymbol{\pi}$, it must then satisfy the equation

$$\boldsymbol{\pi} = \boldsymbol{\pi} P. \tag{1.3.16}$$

Equation 1.3.16 implies the $N$ simultaneous equations

$$\pi_j = \sum_{i=1}^{N} \pi_i p_{ij} \qquad j = 1, 2, \ldots, N. \tag{1.3.17}$$

If we sum Equations 1.3.17 for all values of $j$, we obtain the identity $1 = 1$. Therefore the equations are linearly dependent, and we shall never be able to find a unique solution for the limiting state probabilities $\pi_j$ by solving them. We can see the arbitrariness in the solution of these equations directly by observing that the equations are invariant to the multiplication of the $\pi_j$'s by any constant. However, we can find a unique $\boldsymbol{\pi}$ vector for a monodesmic process by using Equations 1.3.17 in conjunction with the requirement that the state probabilities sum to one over all states,

$$\sum_{i=1}^{N} \pi_i = 1. \tag{1.3.18}$$

Equations 1.3.17 and 1.3.18 provide an important shortcut in the frequently encountered situation where we want to find only what the limiting state probabilities are rather than information about how they are approached. Happily, if we attempt to use this method in a Markov process that is not monodesmic, we obtain an indeterminate set of equations rather than an incorrect result. Thus the ability to find a unique solution for the linear simultaneous Equations 1.3.17 and 1.3.18 tests whether a process is monodesmic; we shall have more to say on this subject later.

**Limiting state probabilities of the marketing example.** Let us find the limiting state probability vector for the marketing example. The transition probability matrix for this example is

$$P = \begin{bmatrix} 0.8 & 0.2 \\ 0.3 & 0.7 \end{bmatrix}. \tag{1.3.19}$$

The general $\boldsymbol{\pi}$ for a two-state problem is

$$\boldsymbol{\pi} = [\pi_1 \quad \pi_2]. \tag{1.3.20}$$

Equation 1.3.16 then yields

$$\boldsymbol{\pi} = \boldsymbol{\pi} P$$

$$[\pi_1 \quad \pi_2] = [\pi_1 \quad \pi_2]\begin{bmatrix} 0.8 & 0.2 \\ 0.3 & 0.7 \end{bmatrix}$$

$$[\pi_1 \quad \pi_2] = [0.8\pi_1 + 0.3\pi_2 \quad 0.2\pi_1 + 0.7\pi_2]. \tag{1.3.21}$$

Equation 1.3.21 contains the two Equations 1.3.17,

$$\begin{aligned} \pi_1 &= 0.8\pi_1 + 0.3\pi_2 \\ \pi_2 &= 0.2\pi_1 + 0.7\pi_2 \end{aligned} \tag{1.3.22}$$

However, either of the two Equations 1.3.22 reduces to the same equation,

$$0.2\pi_1 = 0.3\pi_2. \tag{1.3.23}$$

We can obtain a solution for $\boldsymbol{\pi}$ only when we add the condition 1.3.18,

$$\pi_1 + \pi_2 = 1. \tag{1.3.24}$$

Equations 1.3.23 and 1.3.24 have the simultaneous solution

$$\pi_1 = 0.6, \qquad \pi_2 = 0.4 \tag{1.3.25}$$

or

$$\boldsymbol{\pi} = [0.6 \quad 0.4], \tag{1.3.26}$$

as we found before.

Thus we have obtained directly the rows of the matrix $\Phi$ in Equation 1.2.19. We shall find it a considerable advantage to be able to discover the limiting state probability vector $\boldsymbol{\pi}$ so simply.

### *Equilibrium of probabilistic flows*

Before we pass on to other considerations, it is worth commenting on a physical interpretation of the limiting state probabilities. It seems intuitive that in the steady state (where we make probability assignments based only on the knowledge of the transition probability matrix and the fact that many, many transitions have been made) the probability that the process will make its next transition out of state $j$ must equal the probability that it will make its next transition into state $j$. The probability that the next transition will carry the process out of state $j$ is just the probability $\pi_j$ that it is in state $j$ times the probability $p_{ji}$ of its making a transition to state $i$ (possibly $i = j$), summed over all destination states $i$. Similarly, the probability that the next transition will carry the process into state $j$ is just the sum over all states $i$ of the product of the probability $\pi_i$ that it occupies state $i$ and the transition probability $p_{ij}$. Therefore,

$$\sum_{i=1}^{N} \pi_j p_{ji} = \sum_{i=1}^{N} \pi_i p_{ij}$$

$$\pi_j \sum_{i=1}^{N} p_{ji} = \sum_{i=1}^{N} \pi_i p_{ij} \tag{1.3.27}$$

$$\pi_j = \sum_{i=1}^{N} \pi_i p_{ij}$$

and we have produced Equations 1.3.17.

This "equilibrium of probabilistic flows" argument allows us to write the equations governing limiting state probabilities directly by inspection of the transition diagram for the process. Thus Figure 1.2.1 immediately implies Equations 1.3.22. To determine the unique values of these quantities, we need only add the requirement that the sum of the limiting state probabilities be one.

A moment's reflection allows us to extend the "equilibrium of probabilistic flows" argument to obtain a more general result. If we consider some of the states of the process to belong to a subset $\mathscr{X}$, the probability that we assign in the steady state to a transition from some state of $\mathscr{X}$ to some state of the entire process must be the same as the probabilities we assign to a transition in the reverse direction. Thus,

$$\sum_{\substack{i=1,2,\ldots,N\\ j\in\mathscr{X}}} \pi_j p_{ji} = \sum_{j\in\mathscr{X}} \pi_j = \sum_{\substack{i=1,2,\ldots,N\\ j\in\mathscr{X}}} \pi_i p_{ij}. \tag{1.3.28}$$

Equation 1.3.27 then corresponds to the special case where the subset $\mathscr{X}$ contains only the state $j$; conversely, summing Equation 1.3.27 over all states $j$ that are members of $\mathscr{X}$ establishes the validity of Equation 1.3.28.

We obtain an even more important consequence of the flow argument if we partition the states of the process by dividing them between two mutually exclusive and collectively exhaustive sets $A$ and $B$. In this case, the probability we assign in the steady state to a transition across the partition must be the same in both directions,

$$\sum_{\substack{i\in A\\ j\in B}} \pi_j p_{ji} = \sum_{\substack{i\in A\\ j\in B}} \pi_i p_{ij}. \tag{1.3.29}$$

This property is particularly useful for writing the equations governing the limiting state probabilities by inspection of the transition diagram. For example, suppose we imagine a vertical partition to be constructed between nodes 1 and 2 in Figure 1.2.1. Equating steady-state probabilistic flows in both directions across this partition allows us to write Equation 1.3.23 immediately and hence establish that the ratio of $\pi_1$ to $\pi_2$ for the marketing example is 3:2. The normalization $\pi_1 + \pi_2 = 1$ then produces the limiting state probability vector $\boldsymbol{\pi} = [0.6 \quad 0.4]$.

## 1.4 THE TRANSIENT BEHAVIOR OF STATE PROBABILITIES—SHRINKAGE

Now that we have found the limiting form of the state probability vector, let us turn to the question of how this form is approached. We shall first develop a physical feeling for the convergence process and then investigate its mathematical characteristics.

### A Geometric Interpretation

The use of geometric analogies will be especially helpful in acquiring physical intuition for a Markov process. We realize that although the state probability vector for a two-state process has two elements, it is really described by either of them because of the requirement that the elements sum to one. Consequently, any state probability vector for a two-state process can be represented by a point on a line of unit length like that in Figure 1.4.1. The distance from the point to the right-hand end of the line is $\pi_1(n)$; the distance to the left-hand end of the line is $\pi_2(n)$. Then the end numbered 1 represents the vector [1 0], the initial state probability vector if the system starts in state 1; the point numbered 2 represents the vector [0 1], corresponding to a system starting in state 2. We call such a diagram a state probability diagram.

*The state probability diagram of the marketing example*

Let us draw a diagram of this type for our marketing example; Figure 1.4.2 shows the result. Suppose the customer was initially in state 1, $\boldsymbol{\pi}(0) = [1 \quad 0]$. The first row of $\Phi(1) = P$ in Equation 1.2.16 shows that his state probability vector after one purchase is $\boldsymbol{\pi}(1) = [0.8 \quad 0.2]$. The dotted line $n = 1$ in Figure 1.4.2 that begins at the point marked 1 shows the transformation that the state

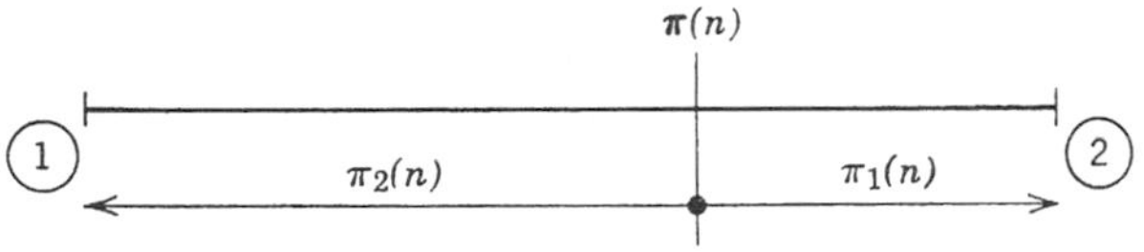


**Figure 1.4.1** A geometric interpretation of a two-state process.

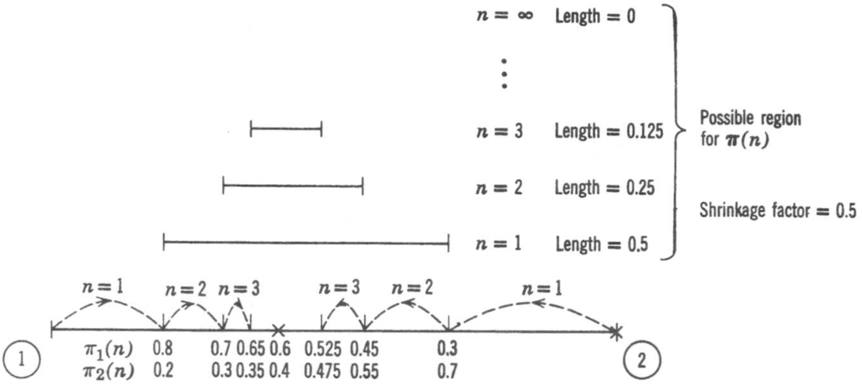


**Figure 1.4.2** State probability diagram for the marketing example.

probability vector experienced. If the customer started initially in state 2, $\boldsymbol{\pi}(0) = [0 \quad 1]$, then the second row of $\Phi(1) = P$ shows that his state probability vector after one purchase is $\boldsymbol{\pi}(1) = [0.3 \quad 0.7]$. This transformation of state probability vector as the result of one transition is indicated by the dotted line $n = 1$ that begins at the point marked 2.

Suppose that the customer had an initial state probability vector of the form $\boldsymbol{\pi}(0) = [a \quad 1 - a]$. Then according to Equation 1.3.6 his state probability vector after one transition would be given by

$$\begin{aligned}\boldsymbol{\pi}(1) &= \boldsymbol{\pi}(0)\Phi(1) = [a \quad 1 - a]\begin{bmatrix}0.8 & 0.2\\ 0.3 & 0.7\end{bmatrix}\\ &= [0.8a + 0.3(1 - a)\; 0.2a + 0.7(1 - a)].\end{aligned} \tag{1.4.1}$$

Since $0 \le a \le 1$, $\boldsymbol{\pi}(1)$ must lie in the region between $\boldsymbol{\pi}(1) = [0.8 \quad 0.2]$ and $\boldsymbol{\pi}(1) = [0.3 \quad 0.7]$ in Figure 1.4.2. In other words, regardless of the initial state probability vector, the state probability vector after one transition must lie in the region bounded by the vectors that correspond to the two rows of $\Phi(1)$. The possible region in which $\boldsymbol{\pi}(1)$ can lie is indicated by the line marked $n = 1$ drawn above the diagram; it has length 0.5.

If a customer whose initial state is state 1 makes two purchases, his $\boldsymbol{\pi}(2)$ is given by the first row of $\Phi(2) = P^2$ in Equation 1.2.16, $\boldsymbol{\pi}(2) = [0.7 \quad 0.3]$. The effect of this second purchase is shown in Figure 1.4.2 by the dotted line $n = 2$ that goes from the point $[0.8 \quad 0.2]$ to the point $[0.7 \quad 0.3]$. Similarly, if a customer starting in state 2 makes two purchases, the second row of $\Phi(2)$ shows us that his $\boldsymbol{\pi}(2) = [0.45 \quad 0.55]$. The dotted line $n = 2$ connecting $[0.3 \quad 0.7]$ with $[0.45 \quad 0.55]$ illustrates the effect of the second purchase. By the same argument as before, the state probability vector after two purchases is restricted to the region between [0.7 0.3] and [0.45 0.55] regardless of the initial state probability vector $\pi(0)$. The possible region for $\pi(2)$ is shown above the diagram; it has length 0.25.

We could continue the construction indefinitely. The possible region for $\pi(3)$ is bounded by the two rows of $\Phi(3) = P^3$ from Equation 1.2.16. Figure 1.4.2 shows that the length of this region is 0.125. The possible region for $\pi(n)$ continually decreases as $n$ increases. Ultimately it approaches the point $\pi = [0.6 \quad 0.4]$, the limiting state probability vector of the system. This point is represented by an "$x$" in Figure 1.4.2.

Every monodesmic process will experience the contraction of the possible state probability vectors to a region consisting of only one point, the common row of $\Phi(\infty) = P^\infty = \Phi$. The processes differ in the particular points they approach, that is, their limiting state probability vectors, and in their rate of approach to these points. It is clear in the marketing example that the possible region is shrinking at the rate 1/2, the same rate we observed in the components of $\Phi(n)$. What determines this rate? We shall soon see.

### *The marketing example with trapping*

We can illustrate another important type of behavior with our marketing example. Suppose that when a consumer purchases brand $B$, he never again buys brand $A$, but that the process is otherwise unchanged. The transition probability matrix for the new process is

$$P = \begin{bmatrix} 0.8 & 0.2 \\ 0 & 1 \end{bmatrix}. \tag{1.4.2}$$

The transition diagram appears as Figure 1.4.3. We see immediately that the system will wind up in state 2 after many transitions regardless of where it is started.

To draw the state probability diagram, we compute the first few multistep transition probability matrices. We find

$$\Phi(0) = P^0 = \begin{bmatrix} 1 & 0 \\ 0 & 1 \end{bmatrix}, \qquad \Phi(1) = P = \begin{bmatrix} 0.8 & 0.2 \\ 0 & 1 \end{bmatrix},$$

$$\Phi(2) = P^2 = \begin{bmatrix} 0.64 & 0.36 \\ 0 & 1 \end{bmatrix}, \qquad \Phi(3) = P^3 = \begin{bmatrix} 0.512 & 0.488 \\ 0 & 1 \end{bmatrix}, \tag{1.4.3}$$

$$\Phi(4) = P^4 = \begin{bmatrix} 0.4096 & 0.5904 \\ 0 & 1 \end{bmatrix}.$$

The boundaries of the possible region for each $\pi(n)$ are given by the two rows of the matrix $\Phi(n)$; the state probability diagram is shown in Figure 1.4.4. Notice

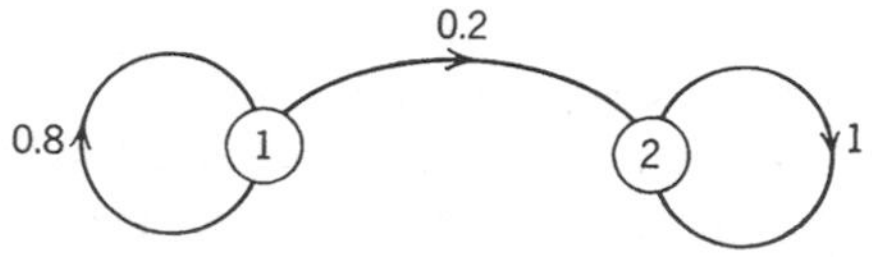


**Figure 1.4.3** Transition diagram for marketing example when brand *B* traps customer.

that the right end of the possible region is always at the point [0 1], and that the left end approaches this point at the rate 0.8. This result confirms our belief that state 2 is the only state that can be occupied after a large number of transitions.

***Transient and trapping states***

It is not unusual for a Markov process to exhibit such behavior. Some states in the system may have a zero probability of occupancy after a very large number of transitions; they are called "transient states." A transient state is a state for which all the entries in the corresponding column of the limiting multistep transition probability matrix $\Phi$ are zero. We see from the sequence of $\Phi(n)$ in Equation 1.4.3 or from the geometric diagram, Figure 1.4.4, that for the present problem

$$\Phi = \begin{bmatrix} 0 & 1 \\ 0 & 1 \end{bmatrix}; \tag{1.4.4}$$

therefore, state 1 is a transient state. We see also that in the special case of a monodesmic process, a transient state is one that has a zero entry in the vector $\pi$.

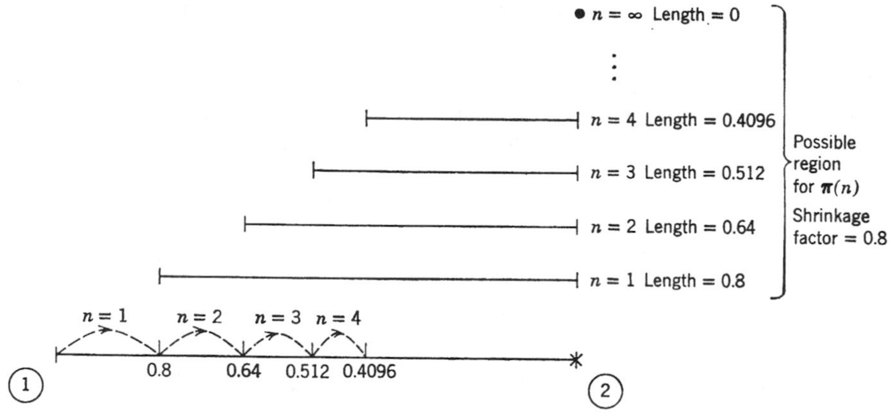


**Figure 1.4.4** State probability diagram for the marketing example when brand *B* traps customer.

The present example also reveals another important type of behavior. Some states in a process may have the property that once entered, they can never be left. These states are called "trapping states." A state $i$ is a trapping state if and only if $p_{ii} = 1$. Therefore, state 2 of our example is a trapping state.

The marketing process with trapping is monodesmic by our definition that the rows of $\Phi$ must be the same, even though a zero element (the 21 element) exists in all powers of $P$. The existence of transient states does not in itself imply that a system is not monodesmic. Equations 1.3.17 and 1.3.18 can still be used to find the limiting state probability vector, $\boldsymbol{\pi}$. To illustrate, from Equation 1.3.17 we obtain

$$\begin{aligned} \pi_1 &= 0.8\pi_1 \\ \pi_2 &= 0.2\pi_1 + \pi_2 \end{aligned} \tag{1.4.5}$$

Either of these equations implies that $\pi_1 = 0$; $\pi_2$ is left arbitrary. However, when we impose the condition 1.3.18

$$\pi_1 + \pi_2 = 1 \tag{1.4.6}$$

we find immediately

$$\pi_1 = 0, \qquad \pi_2 = 1 \tag{1.4.7}$$

or

$$\boldsymbol{\pi} = [0 \quad 1]. \tag{1.4.8}$$

This vector is just the common row of $\Phi$ in Equation 1.4.4.

### Geometric Interpretation of Many-State Processes

We can also use a state probability diagram to analyze Markov processes with more than two states. The three-state case will illustrate the idea. It is a property of the equilateral triangle that the sum of the perpendiculars to the three sides from any interior point is equal to the altitude of the triangle, a constant. If we consider an equilateral triangle with unit altitude, then any interior point can represent a state probability vector for a three-state Markov process. The diagram constructed on this basis is shown in Figure 1.4.5. A state probability vector $\boldsymbol{\pi}(n) = [0.3 \quad 0.2 \quad 0.5]$ is plotted to indicate the nature of the construction. We see that the vertices numbered 1, 2, and 3 correspond to starting the process in each of the three states.

Indeed we can imagine a state probability diagram for a process with any number of states $N$. When $N = 4$, for example, we can use a tetrahedron—the sum of the lengths of the perpendiculars from any interior point to each of the 4 faces is a constant. When $N$ is larger than 4, we find it difficult to visualize the resulting figure, but the general idea is still appropriate. The geometric figure for

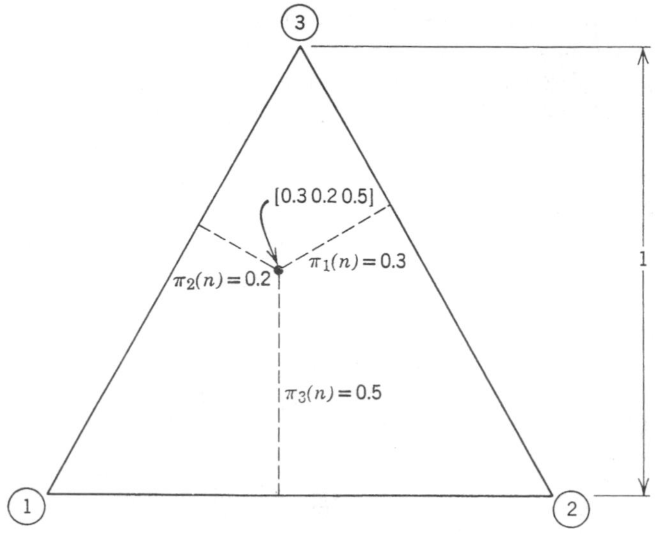


**Figure 1.4.5** A geometric representation of the state-probability vector $\boldsymbol{\pi}(n) = [0.3 \;\; 0.2 \;\; 0.5]$ in the state-probability diagram of a three-state process.

an $N$-state process occupies a physical space of $N - 1$ dimensions. The region of initial state probability vectors for a Markov process is called a simplex by mathematicians. The simplex achieved fame in the field of linear programming primarily for historical reasons.

### *A taxicab example*

Let us consider the three-state process with transition probability matrix

$$P = \begin{bmatrix} 0.3 & 0.2 & 0.5 \\ 0.1 & 0.8 & 0.1 \\ 0.4 & 0.4 & 0.2 \end{bmatrix} \tag{1.4.9}$$

and the transition diagram of Figure 1.4.6(a). The process could represent a three-brand marketing example, a three-pad lily pond, or one of many other physical systems. For convenience in our future references to this example, however, we shall interpret it as a model for a taxicab operation.

A taxicab driver conducts his business among three towns, 1, 2, and 3. We identify his being in each town with occupancy of corresponding states 1, 2, and 3.

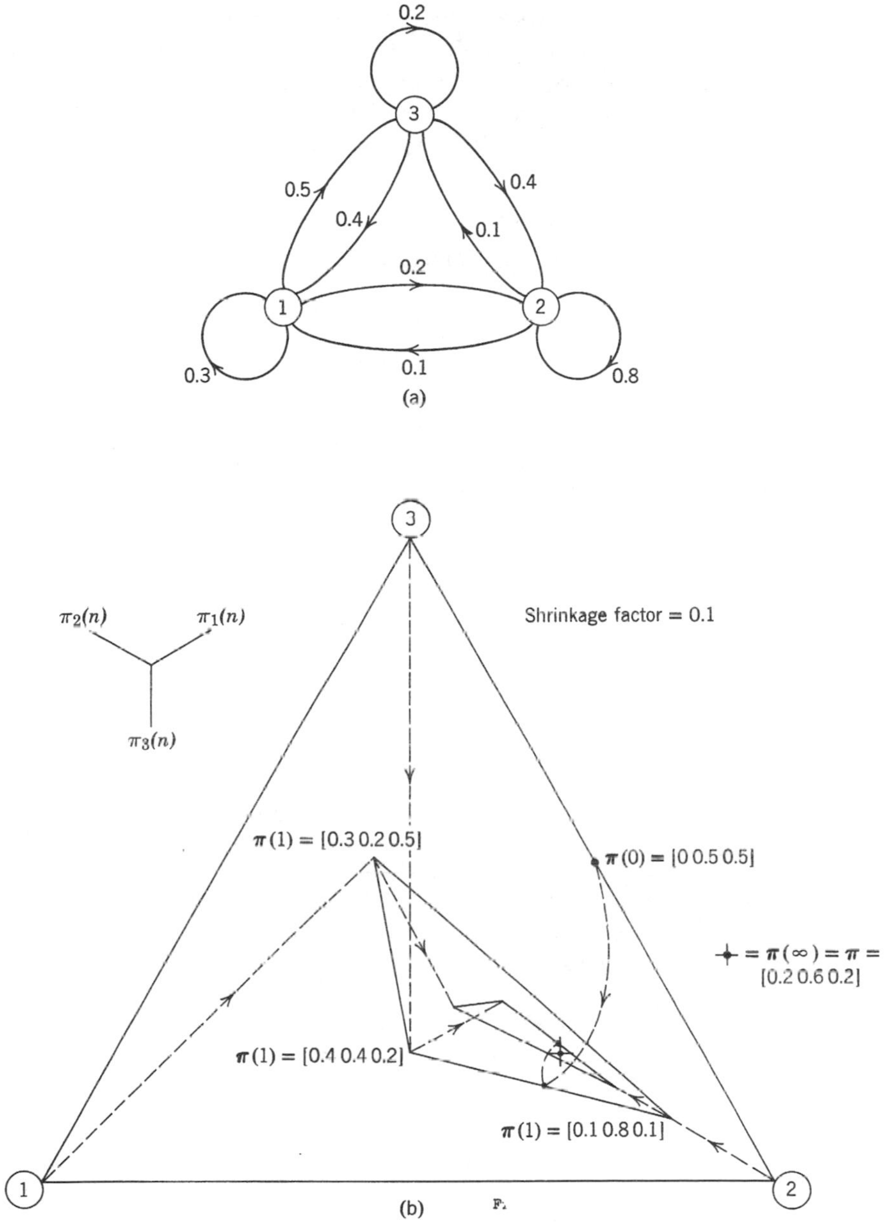


**Figure 1.4.6** (a) Transition diagram for taxicab example. (b) State probability diagram for taxicab example.

Thus a trip from town 2 to town 3 would correspond to the transition from state 2 to state 3; a trip within town 1 would correspond to the transition from state 1 to state 1. We see immediately that this three-state process is monodesmic since it is possible for the process to make a transition from any state to any other state. We are interested in how its state probabilities behave as more and more transitions are made.

**Multistep transition probabilities.** We begin by calculating the multistep transition probability matrices for the first few transitions. We find

$$\begin{aligned}
\Phi(0) &= P^0 = I = \begin{bmatrix} 1 & 0 & 0 \\ 0 & 1 & 0 \\ 0 & 0 & 1 \end{bmatrix} \\
\Phi(1) &= P^1 = P = \begin{bmatrix} 0.3 & 0.2 & 0.5 \\ 0.1 & 0.8 & 0.1 \\ 0.4 & 0.4 & 0.2 \end{bmatrix} \\
\Phi(2) &= P^2 = \begin{bmatrix} 0.31 & 0.42 & 0.27 \\ 0.15 & 0.70 & 0.15 \\ 0.24 & 0.48 & 0.28 \end{bmatrix} \\
\Phi(3) &= P^3 = \begin{bmatrix} 0.243 & 0.506 & 0.251 \\ 0.175 & 0.650 & 0.175 \\ 0.232 & 0.544 & 0.224 \end{bmatrix}.
\end{aligned} \tag{1.4.10}$$

Thus we see, for example, from the 12 element of the successive $\Phi(n)$ matrices that a taxicab driver who started in town 1 has probabilities 0.2, 0.42, and 0.506 of being in town 2 after 1, 2, and 3 trips. We further observe the expected tendency for the rows of $\Phi(n)$ to approach a constant vector as $n$ becomes large.

**Limiting state probabilities.** To explore this asymptotic behavior, we calculate the limiting state probability vector for this monodesmic process using Equations 1.3.17 and 1.3.18. We write

$$\begin{aligned}
\pi_1 &= 0.3\pi_1 + 0.1\pi_2 + 0.4\pi_3 \\
\pi_2 &= 0.2\pi_1 + 0.8\pi_2 + 0.4\pi_3 \\
\pi_3 &= 0.5\pi_1 + 0.1\pi_2 + 0.2\pi_3
\end{aligned} \tag{1.4.11}$$

and

$$\pi_1 + \pi_2 + \pi_3 = 1. \tag{1.4.12}$$

The simultaneous solution of Equations 1.4.11 and 1.4.12 produces the unique result

$$\pi_1 = 0.2, \qquad \pi_2 = 0.6, \qquad \pi_3 = 0.2 \tag{1.4.13}$$

or

$$\boldsymbol{\pi} = [0.2 \quad 0.6 \quad 0.2]. \tag{1.4.14}$$

We can now write the limiting multistep transition probability matrix, $\Phi$,

$$\Phi = \begin{bmatrix} 0.2 & 0.6 & 0.2 \\ 0.2 & 0.6 & 0.2 \\ 0.2 & 0.6 & 0.2 \end{bmatrix}. \tag{1.4.15}$$

The sequence $\Phi(n)$, $n = 0, 1, 2, 3, \ldots$ shown in Equation 1.4.10 indeed appears to be approaching the matrix $\Phi$. Thus after many, many trips—regardless of the town in which the taxi originally started—there is probability 0.2 that the taxi will be in each of towns 1 and 3, and probability 0.6 that the taxi will be in town 2. The reader might like to look at the transition diagram for the process to assure himself that this limiting behavior is reasonable.

**State probability diagram.** We now have the information necessary to construct the state probability diagram; it appears in Figure 1.4.6(b). The vertices of the large triangle transform into the vertices of the inner triangle; these vertices are given by the three rows of $\Phi(1)$ in Equation 1.4.10. Any point that lies within the original triangle will lie within the inner triangle, by an argument analogous to that for the 2-state case. Similarly, any point within the first inner triangle will be confined to the second inner triangle with vertices defined by the rows of $\Phi(2)$ after the second transition is made. As the process makes more and more transitions, the possible region for its state probability vector becomes a smaller and smaller triangle, each one lying within the preceding one. The possible region ultimately approaches the point [0.2 0.6 0.2], the limiting state probability vector of the system.

**The shrinkage phenomenon.** The following interpretation of this shrinkage is informative. Regardless of what probability vector we assign to the states initially, after 1, 2, 3, etc. transitions the only state probability vectors consistent with the information we now possess are those within the successive triangles. Of course, if we are able to observe the state of the process at some point, the state probability vector must again become the one corresponding to the appropriate vertex of the original triangle.

Notice that as the triangular regions shrink in size, they flip back and forth in orientation. The changing orientation is easy to see when we consider the trajectory of a point on the side of the original triangle like $\boldsymbol{\pi}(0) = [0 \quad 0.5 \quad 0.5]$, which corresponds to starting the process in states 2 and 3 with equal probability. This point, midway between 2 and 3 in the original triangle, always remains the midpoint of one side as shown in Figure 1.4.6(b). However, it appears on different sides of the perpendicular from 2 to the opposite side after every transition. What causes this change in orientation? This question will soon be answered.

The amount of shrinkage in the possible region that occurs with every transition is an important physical characteristic of a Markov process. Careful measurement of the areas of the triangles in Figure 1.4.6(b) reveals that each has an area one-tenth as large as its predecessor. The area of the triangle with vertices given by $\Phi(n)$ has an area $(1/10)^n$ of the area of the original triangle. After 10 transitions, for example, the possible region has an area $10^{-10}$ times its original value. For practical purposes, the limiting state probability vector of a Markov process is often achieved in a surprisingly small number of transitions. But why is the shrinkage factor for this process 1/10? This is yet another question to which we still seek an answer.

The shrinkage phenomenon in a Markov process has always reminded me of a baking powder can I saw as a child. The can had as part of its label a picture of the can, which, of course, displayed the can....

***An example with partial shrinkage***

An important type of special behavior occurs in the Markov process with the transition probability matrix

$$P = \begin{bmatrix} 1 & 0 & 0 \\ 0 & 1 & 0 \\ 0.3 & 0.2 & 0.5 \end{bmatrix} \tag{1.4.16}$$

and corresponding transition diagram of Figure 1.4.7(a). Inspection of the transition diagram or the matrix $P$ shows us that states 1 and 2 are trapping states, and that state 3 is a transient state. This insight is verified by calculating the first few multistep transition probability matrices,

$$\Phi(1) = P^1 = \begin{bmatrix} 1 & 0 & 0 \\ 0 & 1 & 0 \\ 0.3 & 0.2 & 0.5 \end{bmatrix} \qquad \Phi(2) = P^2 = \begin{bmatrix} 1 & 0 & 0 \\ 0 & 1 & 0 \\ 0.45 & 0.30 & 0.25 \end{bmatrix}$$

$$\Phi(3) = P^3 = \begin{bmatrix} 1 & 0 & 0 \\ 0 & 1 & 0 \\ 0.525 & 0.350 & 0.125 \end{bmatrix} \qquad \Phi(4) = P^4 = \begin{bmatrix} 1 & 0 & 0 \\ 0 & 1 & 0 \\ 0.5625 & 0.3750 & 0.0625 \end{bmatrix}$$

$$\Phi(5) = P^5 = \begin{bmatrix} 1 & 0 & 0 \\ 0 & 1 & 0 \\ 0.58125 & 0.38750 & 0.03125 \end{bmatrix}. \tag{1.4.17}$$

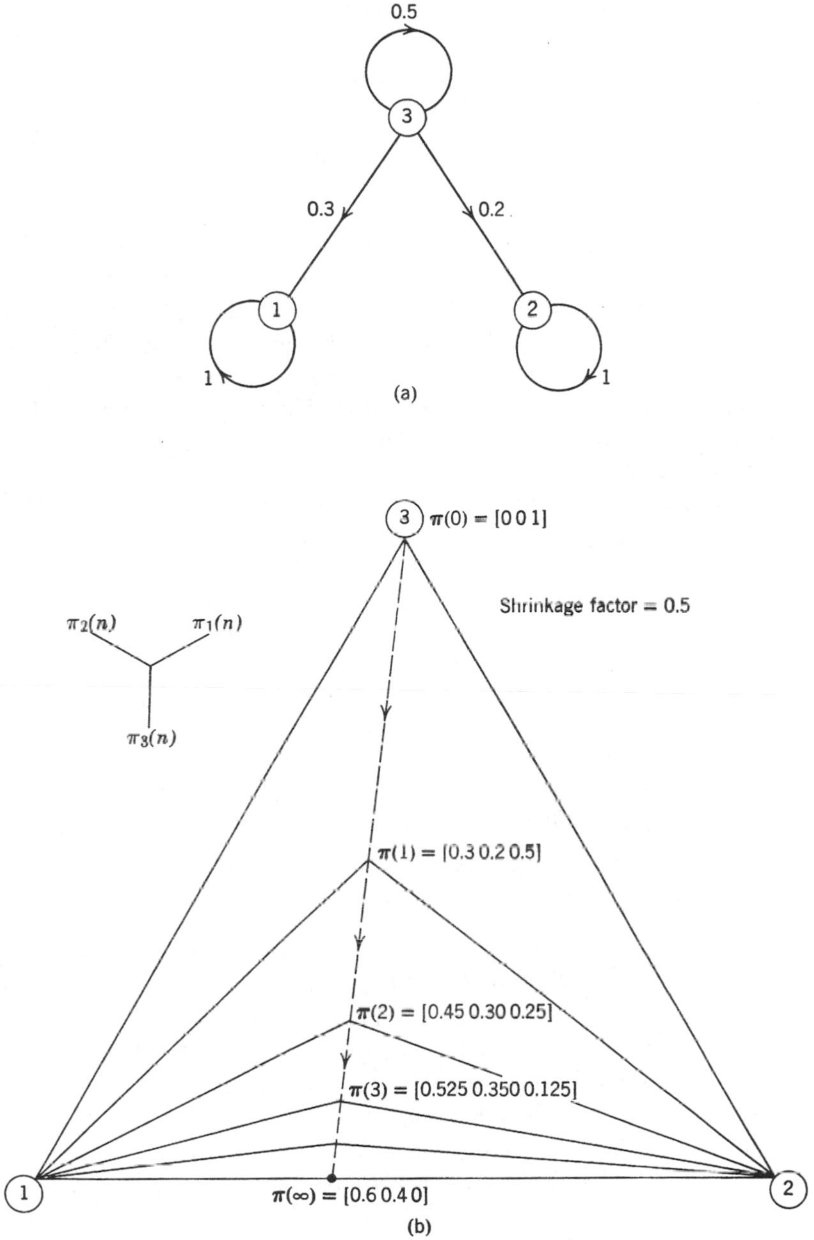


**Figure 1.4.7** (a) Transition diagram for a duodesmic process, (b) State probability diagram for a duodesmic process.

The $\Phi(n)$ sequence is approaching the matrix,

$$\Phi = \begin{bmatrix} 1 & 0 & 0 \\ 0 & 1 & 0 \\ 0.6 & 0.4 & 0 \end{bmatrix}, \tag{1.4.18}$$

when $n$ is very large.

Notice that the rows of $\Phi$ are not identical; this Markov process is not monodesmic. The limiting state probability vector of the system depends upon where it is started. As we can see from the transition diagram, if the system is started in either state 1 or 2, it will stay in those states indefinitely because they are trapping states. If the system is started in state 3, it will ultimately enter either state 1 or state 2, with the odds 3 to 2 in favor of state 1. The rows of the matrix $\Phi$ supply just this information. The zero entries in the third column of $\Phi$ confirm our finding that state 3 is a transient state.

The state probability diagram for this example appears in Figure 1.4.7(b). The vertices of the successive triangles are given by the rows of the $\Phi(n)$ matrices in Equation 1.4.17. Since the first two rows of these matrices are always the same, two vertices of the triangles are always at the points $[1 \quad 0 \quad 0]$ and $[0 \quad 1 \quad 0]$; only the third vertex changes. The possible region is therefore a nest of triangles with a common base, and whose altitude decreases at the rate 0.5.

The limiting form of these nested triangles is not a point, but a line—the line connecting points 1 and 2. Any point within the original triangle must lie on this line after a very large number of transitions; its position on the line will, however, depend on its starting point. The position is determined by finding the intersection of the base line with a line drawn parallel to the dashed line through the starting state probability vector.

### *Chain terminology*

To describe a process of this type, we need some new terminology. First we define a "chain." A chain is a kind of collective trapping state. It is a set of states with the property that when the process enters any of them, it can never leave the set. All Markov processes have at least one chain, most Markov processes in practice have exactly one, but some have more than one. The largest number of chains a process can have is its number of states.

As we have seen, the number of chains in a process affects the behavior we observe in the state probability diagram. If a process has $C$ chains, then $C - 1$ is the dimensionality of its limiting possible region. Thus if we consider three-state processes, whose state probability diagram is the unit altitude equilateral triangle, the limiting possible region will be an area, line, or point according to whether the number of chains in the process is three, two, or one. We have observed the last two cases and we will soon demonstrate the three-chain case. However, it is important to note that regardless of the number of states the limiting possible region is a point and the limiting state probability vector is unique only when the Markov process contains a single chain.

Because the number of chains in a process is an important descriptor of the process, we shall use a special adjective to designate this characteristic. The adjective is formed from the Greek word for chains, *desmos*, with a Greek prefix to designate the number of chains. Thus we refer to one-, two-, three-, etc. chain processes as monodesmic, duodesmic, tridesmic, etc. A process with more than one chain but with the exact number of chains unspecified will be called a polydesmic process.

We shall not think of transient states as being members of chains, but we shall associate them with chain. A transient state is associated with a chain if it is possible for the process to enter that chain from the transient state. Therefore, a transient state must be associated with at least one chain.

Since transient states have been excluded from chain membership, all states in a chain must have a non-zero probability of being occupied after many transitions provided that the system has entered that chain; they are called "recurrent" states. Furthermore, all states belonging to the same chain will have identical rows in the matrix $\Phi$. Together with any transient states associated only with that chain, each chain in a Markov process can itself be thought of as a monodesmic Markov process. The transient states associated with more than one chain will have a row in the matrix $\Phi$ that is the sum of the rows corresponding to each chain with which they are associated, weighted by the probabilities of entering each of these chains.

***A duodesmic process***

Now let us return to our example. Since a trapping state constitutes a chain by itself, states 1 and 2 are both chains consisting of one state; when either is entered, it can never be left. State 3 is a transient state that can enter either chain; therefore, it is associated with both chains. Our present example is therefore a two-chain or duodesmic process with one transient state. The first two rows in the matrix $\Phi$ are the limiting state probability vectors if the system is started in each chain. Because the transient state will ultimately enter state 1 with probability 0.6 and state 2 with probability 0.4, the third row of the matrix $\Phi$ is the sum of the first two rows taken with these weights.

**Limiting state probability equations.** What will happen if we attempt to find the limiting state probability vector of a duodesmic process like this by using Equations 1.3.17 and 1.3.18? The equations for this example are

$$\begin{aligned} \pi_1 &= \pi_1 + 0.3\pi_3 \\ \pi_2 &= \pi_2 + 0.2\pi_3 \\ \pi_3 &= 0.5\pi_3 \end{aligned} \tag{1.4.19}$$

and

$$\pi_1 + \pi_2 + \pi_3 = 1. \tag{1.4.20}$$

Equations 1.4.19 tell us only that $\pi_3 = 0$; Equation 1.4.20 adds that $\pi_1 + \pi_2 = 1$. Therefore, all we can say about the vector $\boldsymbol{\pi}$ is that its first two elements add to one, and that its third element is zero. All the rows of the $\Phi$ matrix in Equation 1.4.18 meet this condition.

Since we cannot obtain a unique solution for $\boldsymbol{\pi}$, we know that the process has more than one chain—it is polydesmic rather than monodesmic. Because only one element of the $\boldsymbol{\pi}$ vector can be selected arbitrarily, either $\pi_1$ or $\pi_2$, we also know that there are two chains in this process; it is duodesmic. In general, the number of chains is one greater than the number of arbitrary constants in the solution of the equations for $\boldsymbol{\pi}$.

***The identity process***

Only one transition probability matrix can produce three chains in a three-state Markov process; the matrix is

$$P = \begin{bmatrix} 1 & 0 & 0 \\ 0 & 1 & 0 \\ 0 & 0 & 1 \end{bmatrix} = I. \qquad (1.4.21)$$

The pertinent transition diagram appears as Figure 1.4.8.

The transition probability matrix is, of course, just the identity matrix. Each of the three states in the process is a trapping state and therefore constitutes a chain by itself. Any power of the identity matrix is again the identity matrix. Therefore, for this process

$$\Phi(n) = I \qquad n = 0, 1, 2, \ldots. \qquad (1.4.22)$$

The vertices of the possible region in the state probability diagram are always the vertices of the original triangle; there is no shrinkage—the limiting possible region

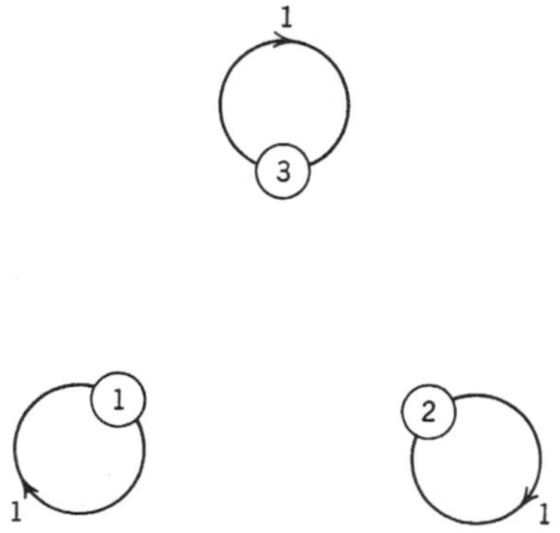


**Figure 1.4.8** Transition diagram for the identity process.

is the entire triangle. Any point within the triangle stays in the same place as successive transitions are made. If we attempt to find the limiting vector $\pi$ by applying Equations 1.3.17 and 1.3.18, we learn only that

$$\pi_1 + \pi_2 + \pi_3 = 1. \tag{1.4.23}$$

Since two of the three elements of the $\pi$ vector must be specified, the process is confirmed to have three chains; it is tridesmic.

***The periodic process***

We must now discuss one special type of behavior that we sometimes observe in a Markov process. We can illustrate it for the three-state process by the transition probability matrix

$$P = \begin{bmatrix} 0 & 1 & 0 \\ 0 & 0 & 1 \\ 1 & 0 & 0 \end{bmatrix}, \tag{1.4.24}$$

which implies the transition diagram of Figure 1.4.9(a). This process moves from 1 to 2, 2 to 3, then from 3 back to 1, over and over again; it is called a periodic process. The multistep transition probability matrices are:

$$\Phi(1) = P = \begin{bmatrix} 0 & 1 & 0 \\ 0 & 0 & 1 \\ 1 & 0 & 0 \end{bmatrix} \quad \Phi(2) = P^2 = \begin{bmatrix} 0 & 0 & 1 \\ 1 & 0 & 0 \\ 0 & 1 & 0 \end{bmatrix} \quad \Phi(3) = P^3 = \begin{bmatrix} 1 & 0 & 0 \\ 0 & 1 & 0 \\ 0 & 0 & 1 \end{bmatrix}$$

$$\Phi(4) = P^4 = P \qquad \Phi(5) = P^5 = P^2 \qquad \Phi(6) = P^6 = P^3 \tag{1.4.25}$$

The multistep transition probability matrices thus keep repeating with a period of three transitions.

The state probability diagram for this case appears in Figure 1.4.9(b). The triangles do not shrink as successive transitions are made—they rotate by 120°. After three transitions the triangle is right back where it started. In fact, if the process were observed only after every third transition, it would appear exactly like the identity process we just discussed.

**The immovable point.** Notice that as the triangle in Figure 1.4.9(b) is rotated, one and only one point does not move. That is the point in the geometric center of the triangle, the vector [1/3 1/3 1/3]. Every Markov process must have at least one such point that remains stationary when the $P$ transformation is applied.† It is the point or set of points satisfying the equation

$$\pi = \pi P \tag{1.4.26}$$

† This feature is a consequence of Brouwer's fixed point theorem.

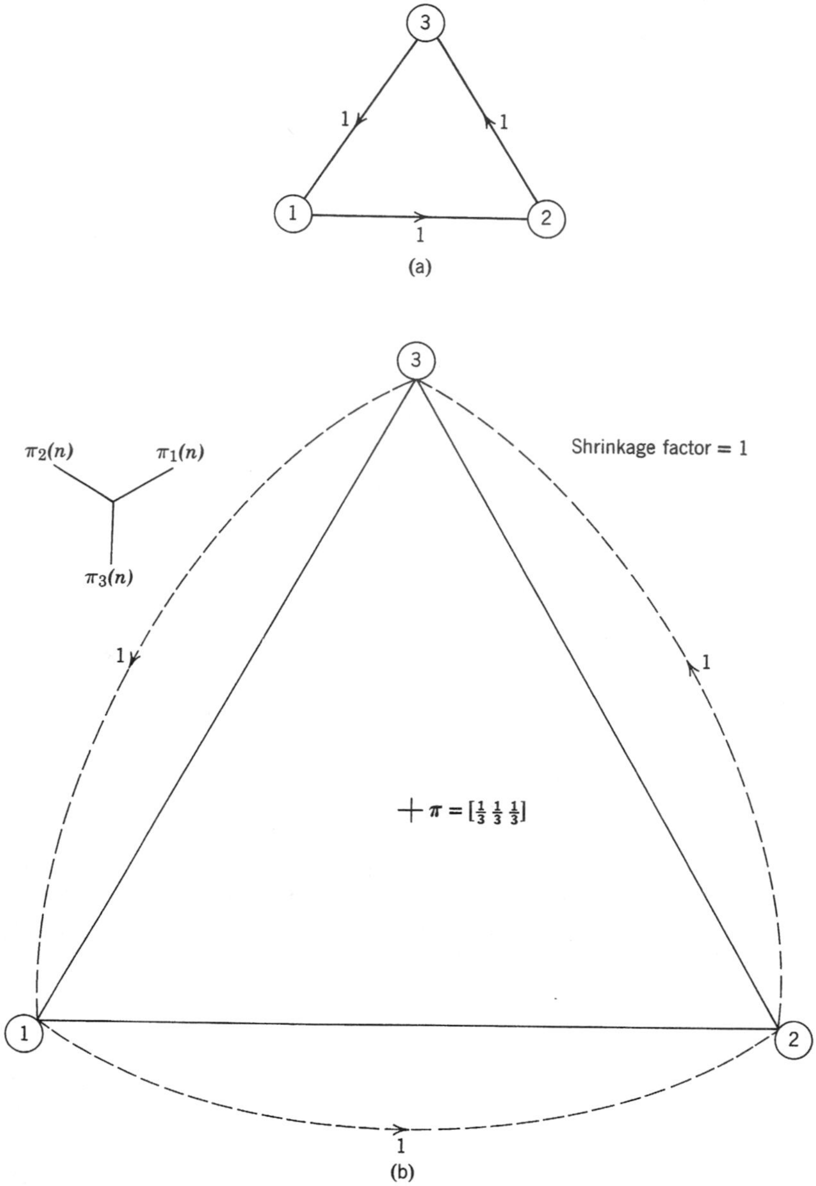


**Figure 1.4.9** (a) Transition diagram for the periodic process. (b) State probability diagram for the periodic process.

which, of course, defines the limiting state probability vector of the process. If for any process the initial state probability vector is a point lying in this set, the state probability vector will remain unchanged regardless of the number of transitions the process makes. For a monodesmic process the only point with this property is the limiting point of the successive possible regions; for a process with $P = I$, any point in the original region will reproduce itself.

**Interpretation of limiting state probabilities.** How shall we classify the periodic process? The sequence of matrices $\Phi(n)$ $n = 0, 1, 2, \ldots$ does not approach any limiting matrix $\Phi$ and yet if we solve Equations 1.3.17 and 1.3.18, we obtain the unique value of $\pi$,

$$\pi = [1/3 \quad 1/3 \quad 1/3]. \tag{1.4.27}$$

We shall find it convenient to consider a periodic process as a monodesmic process with a limiting state probability vector having equal elements. We can rationalize this decision in a number of ways. First, it is consistent with the stationary point concept. Second, the process will make approximately the same number of appearances in each state when the number of transitions is very large. Third, even though it is possible to know where the process is at each step if you know where it started, the equal limiting state probability interpretation is consistent with the situation where you have lost your place. We shall always be able to decide upon the interpretation to use in any application.

### *The multinomial process*

We can think of a multinomial process as the precursor of the Markov process. In a multinomial process, the probability of making a transition to each state is independent of the state already occupied; thus, all the rows of the transition probability matrix $P$ are identical. The transition probability matrix $P$ for a three-state multinomial process would have the transition probability matrix

$$P = \begin{bmatrix} p_1 & p_2 & p_3 \\ p_1 & p_2 & p_3 \\ p_1 & p_2 & p_3 \end{bmatrix} \tag{1.4.28}$$

where, of course, $p_1 + p_2 + p_3 = 1$. The transition diagram of this process appears as Figure 1.4.10(a). When we compute the multistep transition probability matrix, we find as a result of the special structure of $P$ that

$$\Phi(1) = P, \qquad \Phi(2) = P^2 = P, \qquad \Phi(3) = P^3 = P, \text{ etc.} \tag{1.4.29}$$

and in general

$$\Phi(n) = P \qquad 1 \leq n. \qquad \mathrm{n} = 1, 2, 3, \ldots. \tag{1.4.30}$$

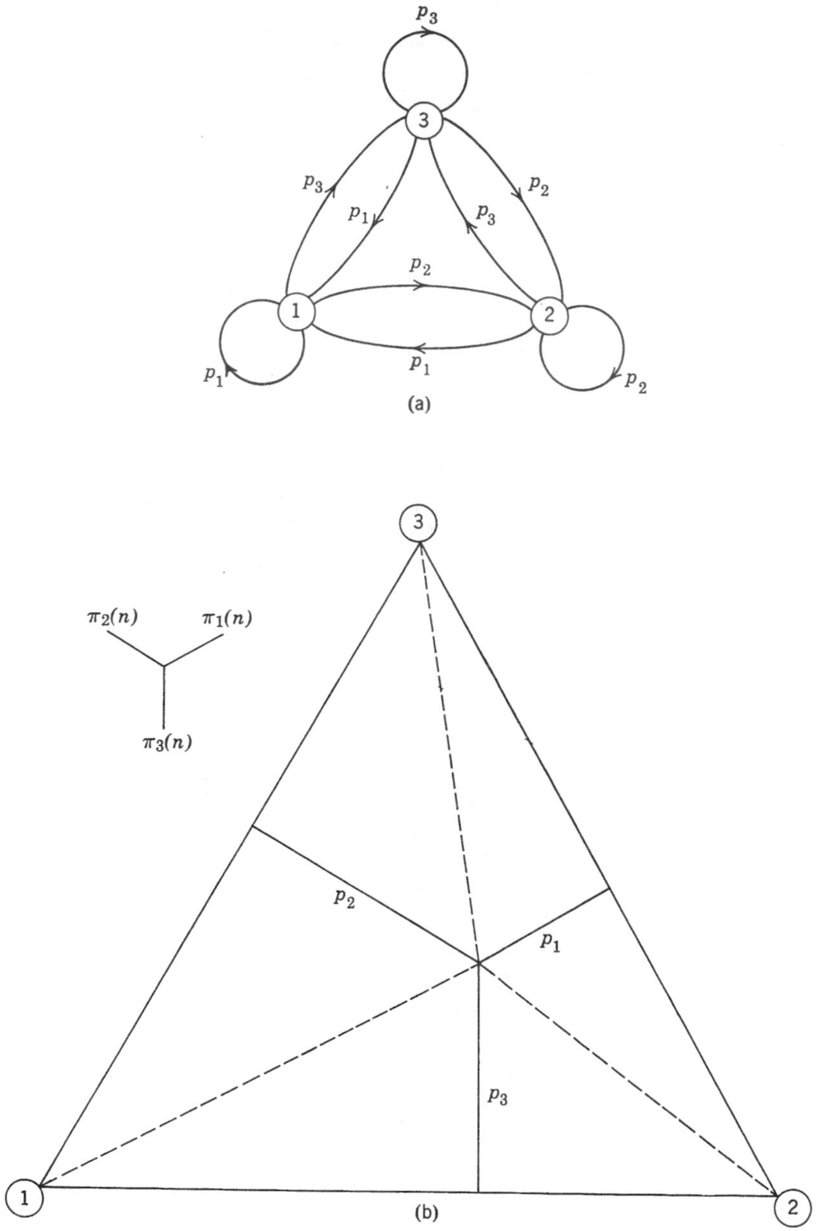


**Figure 1.4.10** (a) Transition diagram for a multinomial process. (b) State probability diagram for a multinomial process.

**Instant shrinkage.** This special behavior is illustrated clearly in the state probability diagram of Figure 1.4.10(b). The original triangle shrinks immediately to the point $[p_1 \quad p_2 \quad p_3]$ and stays there for all future transitions. The limiting state probability vector of the process must therefore be

$$\pi = [p_1 \quad p_2 \quad p_3], \tag{1.4.31}$$

a fact we confirm by noting that this vector satisfies the equations

$$\pi = \pi P, \qquad \pi_1 + \pi_2 + \pi_3 = 1. \tag{1.4.32}$$

**The multinomial projection.** We define the multinomial projection of a monodesmic Markov process as the multinomial process whose transition probability matrix has as its common row the limiting state probability vector of the Markov process. The multinomial projection of a Markov process is a useful baseline for determining the importance of any computation of the dependencies introduced by the Markov process.

***The doubly stochastic process***

A doubly stochastic process is a Markov process whose transition probability matrix has the property that both rows and columns sum to one. Thus, it is not only true as usual that

$$\sum_{j=1}^{N} p_{ij} = 1 \qquad i = 1, 2, \ldots, N \tag{1.4.33}$$

but also that

$$\sum_{i=1}^{N} p_{ij} = 1 \qquad j = 1, 2, \ldots, N. \tag{1.4.34}$$

We observe that the property that the elements in each column sum to one implies that a solution to the limiting state probability equations

$$\pi_j = \sum_{i=1}^{N} \pi_i p_{ij} \qquad j = 1, 2, \ldots, N \tag{1.4.35}$$

is

$$\pi_i = \pi_j \qquad \begin{matrix} i = 1, 2, \ldots, N \\ j = 1, 2, \ldots, N \end{matrix} \tag{1.4.36}$$

or after normalization, that

$$\pi_i = \frac{1}{N} \qquad i = 1, 2, \ldots, N. \tag{1.4.37}$$

Therefore the limiting state probabilities of a doubly stochastic monodesmic process are the same for all states.

For example, a three-state doubly stochastic process

$$P = \begin{bmatrix} 0.3 & 0.2 & 0.5 \\ 0.1 & 0.8 & 0.1 \\ 0.6 & 0 & 0.4 \end{bmatrix} \tag{1.4.38}$$

would have the limiting state probability vector

$$\pi = [1/3 \quad 1/3 \quad 1/3]. \tag{1.4.39}$$

We note in passing that identity and periodic processes of any number of states are all doubly stochastic.

The shrinkage properties of the monodesmic doubly stochastic process are notable only for the fact that the limiting possible region is the point at the center of the state probability diagram.

## Mergeable Processes

In some Markov processes we can divide all the states into groups that have the property that the transition probabilities from each state in a group $S_k$ to each of the states in another group $S_l$ when summed over all of the states in group $S_l$ are the same for each state of group $S_k$. Formally,

$$\sum_{j \in S_l} p_{ij} = \sum_{j \in S_l} p_{i'j} \qquad \text{for} \quad i, i' \in S_k. \tag{1.4.40}$$

When this relation holds for any pair of groups $k$, $l$, we say that the process is mergeable with respect to the grouping of states.

The idea behind mergeable processes is that a mergeable process behaves as a Markov process each of whose superstates $k$ is the group $S_k$ of the original process. When a Markov process is mergeable, we can often answer important questions about the process by performing our computations on the simpler merged process. The microstructure of each group is relevant only when we ask questions that concern states within a group. For example, we can find the multistep transition probability for the merged process and, hence, find the sum of the probabilities of being in the states of some group $S_k$ after the $n$th transition for a given starting condition. However, if we want to know the corresponding probability for a *particular* state of group $k$, we must return to the original process for the computation.

### *An example*

To illustrate the concept of merging, consider the transition probability matrix

$$P = \begin{array}{c} \\ 1 \\ 2 \\ 3 \\ 4 \\ 5 \\ 6 \\ 7 \\ 8 \end{array} \begin{array}{c} \begin{array}{cccccccc} 1 & 2 & 3 & 4 & 5 & 6 & 7 & 8 \end{array} \\ \begin{bmatrix} 0.1 & 0 & 0.1 & 0.2 & 0 & 0.1 & 0.3 & 0.2 \\ 0.1 & 0.2 & 0 & 0 & 0.3 & 0 & 0.1 & 0.3 \\ 0.1 & 0.1 & 0.1 & 0.3 & 0.1 & 0 & 0.1 & 0.2 \\ 0.3 & 0.1 & 0.2 & 0 & 0.1 & 0.1 & 0.2 & 0 \\ 0 & 0.6 & 0 & 0.1 & 0.1 & 0.1 & 0 & 0.1 \\ 0.2 & 0.1 & 0 & 0.2 & 0.2 & 0.1 & 0.1 & 0.1 \\ 0.3 & 0.2 & 0.1 & 0.1 & 0.1 & 0.1 & 0 & 0.1 \\ 0.1 & 0.3 & 0.1 & 0 & 0.3 & 0 & 0 & 0.2 \end{bmatrix} \end{array}. \qquad (1.4.41)$$

In this process we can form three groups of states: $S_1$, consisting of states 1 and 4; $S_2$, consisting of states 2, 5, and 8; and $S_3$, consisting of states 3, 6, and 7. We designate these three groups as superstates 1*, 2*, and 3*, respectively, in the merged process. The transition matrix $P^*$ for the merged process is then

$$P^* = \begin{array}{c} \\ 1^* \\ 2^* \\ 3^* \end{array} \begin{array}{c} \begin{array}{ccc} 1^* & 2^* & 3^* \end{array} \\ \begin{bmatrix} 0.3 & 0.2 & 0.5 \\ 0.1 & 0.8 & 0.1 \\ 0.4 & 0.4 & 0.2 \end{bmatrix} \end{array} \qquad (1.4.42)$$

with the transition diagram shown in Figure 1.4.11. We see that this merged process is exactly the three-state taxi process whose transition probability matrix appears in Equation 1.4.9.

Finding a grouping that will merge a transition matrix like that of Equation 1.4.41 so dramatically is not a simple manual task, although it can be performed efficiently by a computer. Regardless of how it is obtained, the merged process of Equation 1.4.42 will allow us to answer any question on the original process that is specified in terms of the groups of states rather than the original states.

For example, what is the probability that the process will occupy a state of group $S_1$ after many transitions? Since $\pi^*$, the limiting state probability $\pi^*$ of the merged process, appears explicitly in Equation 1.4.14,

$$\pi^* = [0.2 \quad 0.6 \quad 0.2], \qquad (1.4.43)$$

we see immediately that the probability of being in group $S_1$, that is, in either of states 1 and 4, equals 0.2. Similarly, the limiting probability of being in group $S_2$,

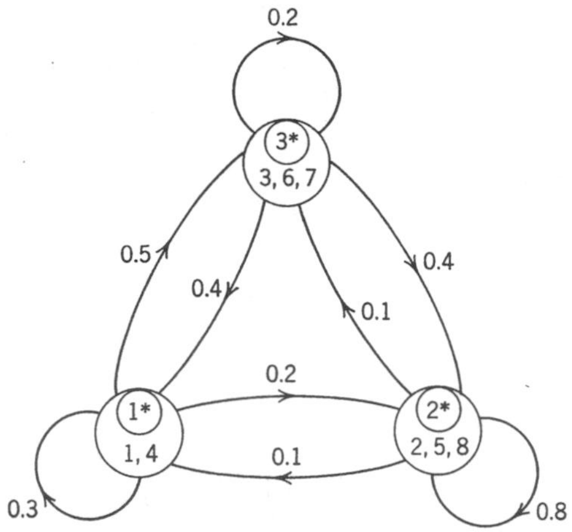


**Figure 1.4.11** A three-state merged version of an eight-state process.

in any of states 2, 5, and 8, equals 0.6, and the limiting probability of being in group $S_3$, in any of states 3, 6, and 7, equals 0.2.

***A smaller example***

Let us now turn to a more manageable example for the purpose of understanding merging. Suppose that there is a change in taxi traffic in the three-state example whose transition probability matrix appears in Equation 1.4.9. Transitions from state 2 to state 2 become less likely while those from state 2 to state 1 become more likely; the other transition probabilities are unchanged. We assume that the new pattern of traffic produces the transition probability matrix

$$P = \begin{bmatrix} 0.3 & 0.2 & 0.5 \\ 0.4 & 0.5 & 0.1 \\ 0.4 & 0.4 & 0.2 \end{bmatrix}. \tag{1.4.44}$$

We show the transition diagram for this process in Figure 1.4.12(a).

We observe that since both states 2 and 3 have the same probability 0.4 of a transition to state 1, we can merge them into a superstate 2*. State 1 then becomes a superstate 1*. The transition probability matrix $P^*$ for the merged process is

$$P^* = \begin{array}{c} \\ 1^* \\ 2^* \end{array} \begin{array}{c} \begin{array}{cc} 1^* & 2^* \end{array} \\ \begin{bmatrix} 0.3 & 0.7 \\ 0.4 & 0.6 \end{bmatrix} \end{array} \tag{1.4.45}$$

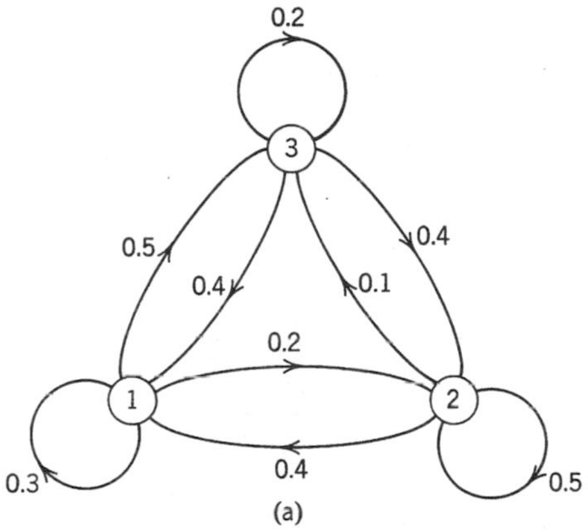


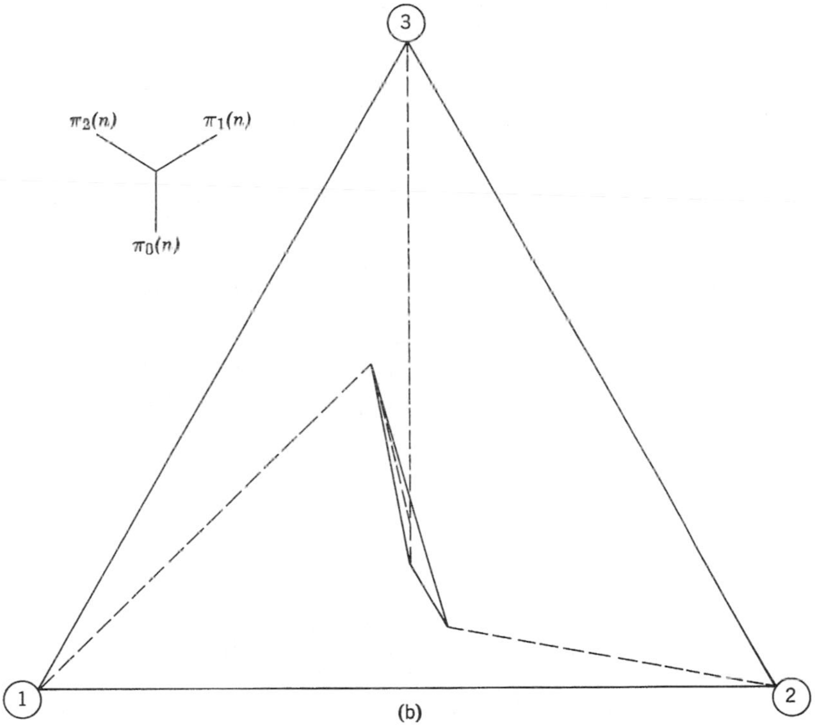


**Figure 1.4.12** (a) Transition diagram for a mergeable process. (b) State probability diagram for a mergeable process.

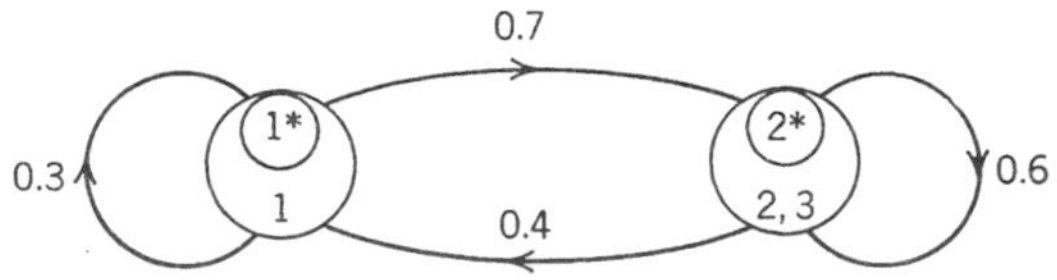


**Figure 1.4.13** Transition diagram for the merged process.

with the transition diagram of Figure 1.4.13. We readily find that the limiting state probability vector $\boldsymbol{\pi}^*$ for the merged process is

$$\boldsymbol{\pi}^* = [4/11 \quad 7/11]. \tag{1.4.46}$$

**Effect of mergeability on the state probability diagram.** Let us now explore the state probability diagram for the three-state mergeable process of Equation 1.4.44. We compute the multistep transition probabilities for the first few transitions:

$$\Phi(0) = I \qquad \Phi(1) = P$$

$$\Phi(2) = P^2 = \begin{bmatrix} 0.37 & 0.36 & 0.27 \\ 0.36 & 0.37 & 0.27 \\ 0.36 & 0.36 & 0.28 \end{bmatrix} \quad \Phi(3) = P^3 = \begin{bmatrix} 0.363 & 0.362 & 0.275 \\ 0.364 & 0.365 & 0.271 \\ 0.364 & 0.364 & 0.272 \end{bmatrix} \tag{1.4.47}$$

The state probability diagram corresponding to these multistep probability matrices appears in Figure 1.4.12(b). We observe a very rapid shrinkage; in fact, the shrinkage factor is 0.01. After only two transitions the possible region is only a ten-thousandth of the size of the original triangle. The possible region ultimately shrinks to the point

$$\boldsymbol{\pi} = [4/11 \quad 4/11 \quad 3/11] = [0.364 \quad 0.364 \quad 0.273], \tag{1.4.48}$$

the limiting state probability vector of the process obtained by solving the usual equations. By comparing Equations 1.4.46 and 1.4.48 we check that

$$\pi_1^* = \pi_1 = 4/11, \qquad \pi_2^* = \pi_2 + \pi_3 = 7/11. \tag{1.4.49}$$

What is interesting about the state probability diagram is not particularly the rapid shrinking, but rather the fact that the image of the line representing all possible combinations of starting states 2 and 3 on successive transitions is always perpendicular to the axis that measures $\pi_1(n)$. This shows that regardless of how the total probability for states 2 and 3 is divided between these states at any time, the probability of state 1 after the next transition depends only on the sum. Consequently, the entire state probability diagram can be projected onto the $\pi_1(n)$ axis to obtain a linear region that is adequate for any computation involving only the state probability of state 1. The transition probability matrix for the diagram after projection is, of course, the merged transition probability matrix of Equation 1.4.45.

***Some comments on merging***

Merging a Markov process can sometimes mean the difference between a computationally feasible problem and an intractable mess. However, the way to perceive merger possibilities is not initially to try all possible groupings until you find one that works. Usually the structure of the underlying physical process being modeled will suggest groupings without further work. If a grouping is observed after the process is formulated, it should be pursued for the insight into the process that it reveals.

Lastly, we should remember that the merging of a process helps us only with respect to relations between the merged groups and not with respect to the individual states. We are not getting something for nothing, but merely finding an answer to a broader set of questions.

## 1.5 TRANSFORM ANALYSIS

We have now gained considerable insight into how a Markov process can behave. However, we have left unanswered a number of questions about why it behaves the way it does. Most of these can be cleared up when we learn to write $\Phi(n)$, the $n$th power of the transition probability matrix, in a closed form involving the addition of at most $N$ matrices with appropriate coefficients, where $N$ is the number of states in the process. The key to obtaining such a closed form expression is transform analysis, the subject of this section.

### The Geometric Transform

Consider a discrete function like that shown in Figure 1.5.1. This is a function that can take on any real value, positive or negative, at any non-negative integer,

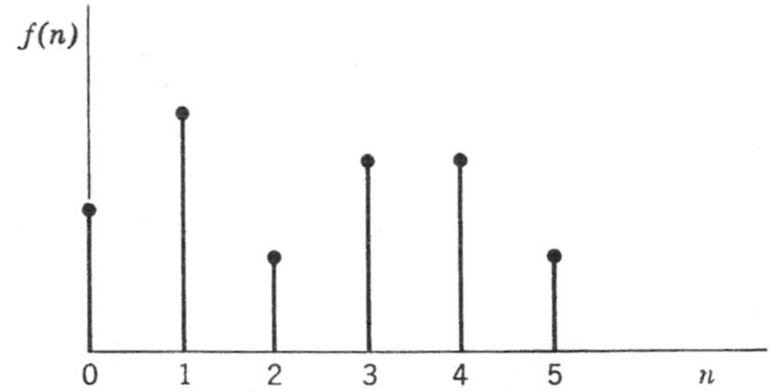


**Figure 1.5.1** A discrete function.

$n = 0, 1, 2, \ldots$. We shall find it convenient to define it to take on the value zero at all negative integers,

$$f(n) = 0 \qquad n < 0. \tag{1.5.1}$$

The geometric transform† of such a function is found by multiplying it term by term by a geometric sequence in the transform variable and summing. We shall use the transform variable $z$ and denote the transform of a function $f(n)$, $n = 0, 1, 2, \ldots$, by $f^g(z)$. The geometric transform is then defined by

$$f^g(z) = f(0) + f(1)z + f(2)z^2 + f(3)z^3 + \cdots = \sum_{n=0}^{\infty} f(n)z^n. \tag{1.5.2}$$

The geometric transform of a discrete function will exist as long as the function does not grow faster than geometrically, that is, as long as

$$\lim_{n \to \infty} \frac{|f(n)|}{b^n} = 0 \tag{1.5.3}$$

for some $b$.

### *Transform inversion*

We speak of the discrete function corresponding to a geometric transform as the inverse of the transform. The process of finding it we call transform inversion. In view of the defining Equation 1.5.2, it is clear that we could always, at least in theory, expand any geometric transform in a power series about $z = 0$ and then write the inverse discrete time function as the coefficients of successive powers of $z$. In some cases we can actually carry out this procedure by division when $f^g(z)$ is expressed as the ratio of two polynomials in $z$. In any case we know that $f(n)$, the discrete function at time $n$, is just the coefficient of $z^n$ in the series expansion of $f^g(z)$,

$$f(n) = \frac{1}{n!} \frac{d^n}{dz^n} f^g(z)\Big|_{z=0}. \tag{1.5.4}$$

Since the series expansion of $f^g(z)$ is unique, so is the relationship between the discrete function and its geometric transform.

### *A table of geometric transforms*

In practice, however, we find it convenient to have a table of discrete functions and their geometric transforms. Such a table, often referred to as a table of transform pairs, appears as Table 1.5.1. This table serves both in the process of geometric transformation of a discrete function and in the process of inversion of a

† Sometimes called the $z$-transform or the generating function.

**Table 1.5.1** A Table of Geometric Transforms

| Discrete Function | | Geometric Transform |
|---|---|---|
| 1 $f(n)$ | $n = 0, 1, 2, \ldots$ | $f^g(z) = f(0) + f(1)z + f(2)z^2 + \cdots = \sum_{n=0}^{\infty} f(n)z^n$ |
| 2 $\delta(n) = \begin{cases} 1 & n = 0 \\ 0 & n > 0 \end{cases}$ | Unit Impulse | $1$ |
| 3 $u(n) = 1$ | Unit Step | $\dfrac{1}{1 - z}$ |
| 4 $af(n)$ | $a$ is a constant | $af^g(z)$ |
| 5 $f_1(n) + f_2(n)$ | | $f_1{}^g(z) + f_2{}^g(z)$ |
| 6 $\sum_{m=0}^{n} f_1(m)f_2(n - m)$ | Convolution | $f_1{}^g(z)f_2{}^g(z)$ |
| 7 $f(n - k)$ | $k$ is a positive integer | $z^k f^g(z)$ |
| 8 $f(n + k)$ | $k$ is a positive integer | $z^{-k}[f^g(z) - f(0) - f(1)z - \cdots - f(k - 1)z^{k-1}]$ |
| 9 $nf(n)$ | | $z\dfrac{d}{dz}f^g(z)$ |
| 10 $a^n f(n)$ | | $f^g(az)$ |
| 11 $a^n$ | | $\dfrac{1}{1 - az}$ |
| 12 $na^n$ | | $\dfrac{az}{(1 - az)^2}$ |
| 13 $n$ | Unit Ramp | $\dfrac{z}{(1 - z)^2}$ |
| 14 $n^2 a^n$ | | $\dfrac{az(1 + az)}{(1 - az)^3}$ |
| 15 $n^2$ | Unit Parabolic Ramp | $\dfrac{z(1 + z)}{(1 - z)^3}$ |
| 16 $(n + 1)f(n + 1)$ | | $\dfrac{d}{dz}f^g(z)$ |
| 17 $(n + 1)a^n$ | | $\dfrac{1}{(1 - az)^2}$ |

| | Discrete Function | | Geometric Transform |
|---|---|---|---|
| 18 | $n + 1$ | | $\frac{1}{(1 - z)^2}$ |
| 19 | $\frac{1}{2}(n + 1)(n + 2)a^n$ | | $\frac{1}{(1 - az)^3}$ |
| 20 | $\frac{1}{2}(n + 1)(n + 2)$ | | $\frac{1}{(1 - z)^3}$ |
| 21 | $\frac{1}{k!}(n + 1)(n + 2)\cdots(n + k)a^n$ | | $\frac{1}{(1 - az)^{k+1}}$ |
| 22 | $\frac{1}{k!}(n + 1)(n + 2)\cdots(n + k)$ | | $\frac{1}{(1 - z)^{k+1}}$ |
| 23 | $f\left(\frac{n}{k}\right)$ | $n = 0, k, 2k, \ldots$ | $f^g(z^k)$ |
| 24 | $f(n) - f(n - 1)$ | | $(1 - z)f^g(z)$ |
| 25 | $\sum_{m=0}^{n} f(m)$ | | $\frac{1}{1 - z} f^g(z)$ |
| 26 | Summation Property | $\sum_{n=0}^{\infty} f(n) = f^g(1)$ | |
| 27 | Alternating Summation Property | $\sum_{n=0}^{\infty} (-1)^n f(n) = f^g(-1)$ | |
| 28 | Initial Value Property | $f(0) = f^g(0)$ | |
| 29 | Final Value Property | $f(\infty) = \lim_{z \to 1} (1 - z)f^g(z)$ If $f(\infty)$ exists | |
| 30 | $M(n)$—a matrix function | $n = 0, 1, 2, \ldots$ | $M^g(z) = M(0) + M(1)z + M(2)z^2 + \ldots = \sum_{n=0}^{\infty} M(n)z^n$ |
| 31 | $nM(n)$ | | $z \frac{d}{dz} M^g(z)$ |
| 32 | $A^n$ | | $[I - Az]^{-1}$ |
| 33 | $nA^n$ | | $z[I - Az]^{-1}A[I - Az]^{-1}$ |

geometric transform to produce the relevant discrete function. We should bear in mind throughout the discussion of the generation and interpretation of the table that the geometric transform is no more or less than the summation of the discrete function after multiplication by successive powers of the transform variable $z$.

**Unit impulse.** The first relation in the table is the defining property. It serves as the foundation for proving all other relationships. Relation 2 concerns a very special function, the unit impulse $\delta(n)$. This function is zero everywhere but at $n = 0$, and equals 1 at $n = 0$. The defining property shows immediately that the geometric transform of the unit impulse is 1.

**Unit step.** Another special function, the unit step, is the subject of relation 3. The unit step is equal to 1 for all non-negative arguments and, of course, equals 0 for negative arguments. A plot of the unit step appears in Figure 1.5.2. The geometric transform of the unit step is then

$$\sum_{n=0}^{\infty} u(n)z^n = \sum_{n=0}^{\infty} z^n = 1 + z + z^2 + \cdots = \frac{1}{1-z} \tag{1.5.5}$$

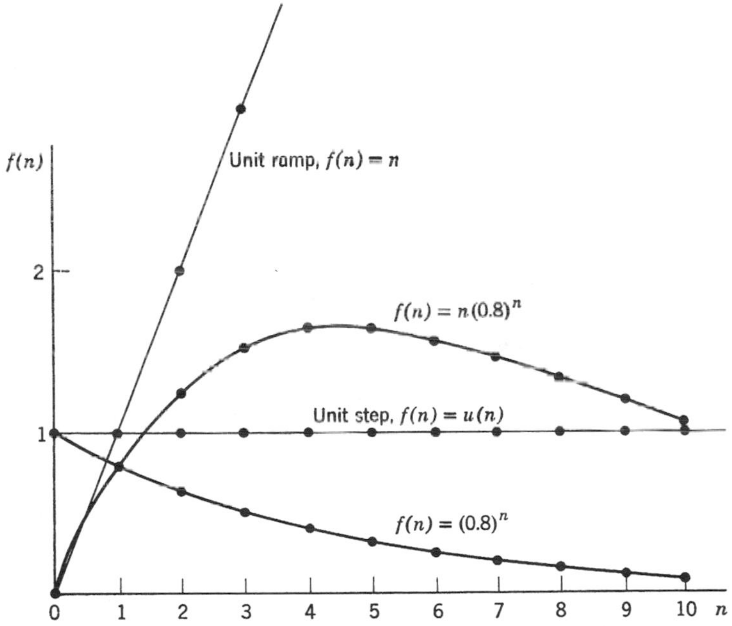


**Figure 1.5.2** Graphs of some discrete time functions appearing in Table 1.5.1.

where the summation is valid if we restrict $|z| < 1$; we shall comment on such restrictions later. This relationship also demonstrates that a transform written as a polynomial fraction can often be inverted by dividing the denominator into the numerator to obtain the series expansion.

Since all the discrete functions with which we shall deal are defined to be zero for negative arguments, we could emphasize this property by employing the unit step function to write each of them in the form $f(n)u(n)$. However, we shall use this type of notation only where necessary to preserve clarity.

**Linear combination.** Relations 4 and 5 follow directly from the transform definition. They demonstrate that multiplication of a discrete function by a constant and addition of discrete functions both carry over into the transform domain. In other words, any linear combination of discrete functions has a transform equal to the same linear combination of the individual function transforms.

**Convolution.** Relation 6 shows the discrete function operation that corresponds to transform multiplication. We call the operation convolution; we shall study it in detail in Chapter 2. To establish the relation, we write

$$\begin{aligned}
\sum_{n=0}^{\infty} z^n \sum_{m=0}^{\infty} f_1(m)f_2(n-m) &= \sum_{n=0}^{\infty}\sum_{m=0}^{n} f_1(m)z^m f_2(n-m)z^{n-m} \\
&= \sum_{m=0}^{\infty} f_1(m)z^m \sum_{n=m}^{\infty} f_2(n-m)z^{n-m} \\
&= \sum_{m=0}^{\infty} f_1(m)z^m \sum_{r=0}^{\infty} f_2(r)z^r \\
&= f_1^{g}(z)f_2^{g}(z). \qquad (1.5.6)
\end{aligned}$$

**Delay.** We can think of $f(n-k)$, where $k$ is a positive integer, as the discrete function $f(n)$ "delayed" by $k$ units. We readily establish the result of relation 7 that the geometric transform of such a function is the geometric transform of the original function multiplied by $z^k$,

$$\sum_{n=0}^{\infty} f(n-k)z^n = \sum_{r=-k}^{\infty} f(r)z^{r+k} = z^k \sum_{r=-k}^{\infty} f(r)z^r = z^k \sum_{r=0}^{\infty} f(r)z^r = z^k f^{g}(z). \qquad (1.5.7)$$

In establishing this relation, we have used the fact that our discrete functions are zero for negative arguments.

**Advance.** The function $f(n+k)$ is the function $f(n)$ advanced by $k$ units. Advancing the function may shift portions of it into the negative argument region; such portions will have to be eliminated from the transform. To see this formally and thus establish relation 8, we write

$$\begin{aligned}\sum_{n=0}^{\infty} f(n+k)z^n &= \sum_{r=k}^{\infty} f(r)z^{r-k} = z^{-k}\sum_{r=k}^{\infty} f(n)z^r \\ &= z^{-k}\left[\sum_{r=0}^{\infty} f(r)z^r - f(0) - f(1)z - \cdots - f(k-1)z^{k-1}\right] \\ &= z^{-k}[f^g(z) - f(0) - f(1)z - \cdots - f(k-1)z^{k-1}]. \end{aligned} \tag{1.5.8}$$

**Multiplication by argument.** Relation 9 shows the effect of multiplying a discrete function by its argument; namely, that it requires differentiation of the geometric transform and multiplication by the transform variable $z$. We establish the relation by differentiating the defining property with respect to $z$,

$$\frac{d}{dz} f^g(z) = \frac{d}{dz}\sum_{n=0}^{\infty} f(n)z^n = \sum_{n=0}^{\infty} nf(n)z^{n-1} = z^{-1}\sum_{n=0}^{\infty} nf(n)z^n \tag{1.5.9}$$

or

$$z\frac{d}{dz} f^g(z) = \sum_{n=0}^{\infty} nf(n)z^n. \tag{1.5.10}$$

**Multiplication by geometric sequence.** Multiplying a discrete function by a geometric sequence whose rate is $a$ requires replacing the transform variable $z$ by $az$. To establish this result as relation 10 we return to the transform definition,

$$\sum_{n=0}^{\infty} a^n f(n)z^n = \sum_{n=0}^{\infty} f(n)(az)^n. \tag{1.5.11}$$

It is clear that all our relations and this one in particular place requirements on allowable values of $z$. Again we postpone comment except to say that since these requirements will never cause any difficulty in our work, we can ignore them.

**Geometric sequence operations.** The groundwork we have laid will allow us to write several transform relationships directly. For example, as a consequence of relations 3 and 10 we find in relation 11 that the transform of the geometric sequence $a^n = a^n u(n)$ is $1/(1 - az)$. To find the transform of $na^n$ we apply relation 9 to this result and obtain $az/(1 - az)^2$ as shown in relation 12. When $a = 1$, $na^n$ becomes simply $n$, a function we call the unit ramp; its transform as reported in relation 13 is therefore $z/(1 - z)^2$. These time functions are illustrated in Figure 1.5.2.

Carrying this development one step further, we use relations 9 and 12 to establish in relation 14 that the transform of $n^2a^n$ is $az(1 + az)/(1 - az)^3$. When $a = 1$,

$n^2a^n$ becomes $n^2$, a function we call the unit parabolic ramp; its transform is therefore $z(1 + z)/(1 - z)^3$, as shown in relation 15.

From these basic results we can develop forms that will prove useful in special situations. For example, we use relation 8 with $k = 1$ and relation 9 to show that the transform of $(n + 1)f(n + 1)$ is given by relation 16, $z^{-1}\left[z\frac{d}{dz}f^g(z)\right] = \frac{d}{dz}f^g(z)$. Then, since $(n + 1)a^n = (1/a)(n + 1)a^{n+1}$ we can use this relation to write relation 17 for the transform of $(n + 1)a^n$ by applying it to relation 11,

$$\frac{1}{a}\frac{d}{dz}\frac{1}{1 - az} = \frac{1}{(1 - az)^2}. \tag{1.5.12}$$

Of course the same transform for $(n + 1)a^n$ results by adding the transforms in relations 11 and 12. With $a = 1$, this transform becomes the transform of the unit ramp advanced by one unit, as recorded in relation 18.

We can apply these results repetitively to develop still more complex transform pairs. For example the transform of $(n + 1)(n + 2)a^n = (1/a)(n + 1)[(n + 2)a^{n+1}]$ must as a consequence of relations 16 and 17 be given by

$$\frac{1}{a}\frac{d}{dz}\frac{1}{(1 - az)^2} = \frac{2}{(1 - az)^3}. \tag{1.5.13}$$

Or, after dividing by 2, the transform of $\frac{1}{2}(n + 1)(n + 2)a^n$ is therefore $1/(1 - az)^3$, as shown in relation 19. By continuing to use relation 16 in the bootstrap fashion we have employed in developing relations 17 and 19, we establish relation 21 showing that the discrete time function $(1/k!)(n + 1)(n + 2)\cdots(n + k)a^n$ has the transform $1/(1 - az)^{k+1}$. Our interest in this transform pair is that we often must invert geometric transforms that are powers of factors like $1/(1 - az)$. With $a = 1$, relations 19 and 21 assume the special forms of relations 20 and 22 that combine with relations 3 and 18 to demonstrate the inversion of geometric transforms that are powers of $1/(1 - z)$. We observe that the inverse discrete function is in each case a polynomial in $n$ of one degree lower than the power of $1/(1 - z)$.

**Spreading.** Relation 23 demonstrates the effect on the geometric transform of a discrete function of spreading the function by having it take on the same values at times $0, k, 2k, \ldots$ rather than $0, 1, 2, \ldots$. From the transform definition we have,

$$\sum_{n=0,k,2k}^{\infty} f\left(\frac{n}{k}\right)z^n = \sum_{m=0}^{\infty} f(m)z^{km} = \sum_{m=0}^{\infty} f(m)(z^k)^m = f^g(z^k) \tag{1.5.14}$$

and thus we establish that the sole effect is to replace the transform variable $z$ by $z^k$ in the original transform.

**Difference.** If we form a new discrete function $f(n) - f(n - 1)$ by taking the difference in successive values of $f(n)$, we see from relations 1, 5, and 7 with $k = 1$

that the geometric transform of the new discrete function will be $f^g(z) - zf^g(z) = (1 - z)f^g(z)$ as recorded in relation 24. The effect of successive differencing of a discrete function is therefore to multiply its geometric transform by $1 - z$.

**Sum.** If on the other hand, we form a new discrete function by summing the successive values of a discrete function, $\sum_{m=0}^{n} f(m)$, we find that the geometric transform of the new discrete function is

$$\begin{aligned}\sum_{n=0}^{\infty}\left(\sum_{m=0}^{n} f(m)\right)z^n &= \sum_{m=0}^{\infty} f(m)z^m \sum_{n=m}^{\infty} z^{n-m}\\ &= \sum_{m=0}^{\infty} f(m)z^m \sum_{r=0}^{\infty} z^r\\ &= f^g(z)\left(\frac{1}{1-z}\right).\end{aligned} \tag{1.5.15}$$

Thus summing a discrete function causes its geometric transform to be divided by $1 - z$, or multiplied by $1/(1 - z)$, as recorded in relation 25. This relation also follows from the convolution relation 6 by considering the discrete function $f_2(n)$ to be the unit step $u(n)$.

Since it is clear that forming a new discrete function by differencing an original discrete function and then summing the new discrete function would merely restore the original discrete function, relations 24 and 25 are consistent. We shall comment at much greater length on the implication of these observations in Chapter 2.

**Summation property.** The next relations in the table are properties included for completeness. Relation 26, the summation property, is established by taking $z = 1$ in the defining relation 1; we observe that the geometric transform evaluated at $z = 1$ is just the infinite sum of the discrete function, a sum that may or may not itself be finite. The summation property is at once trivial and important. If the discrete function is a probability mass function, for example, the property states that its geometric transform evaluated at $z = 1$ must equal one.

**Alternating summation property.** If we evaluate the defining relation 1 at $z = -1$, we obtain the alternating summation property of relation 27. This occasionally useful property is that the geometric transform of a discrete function evaluated at $z = -1$ is just the difference between the sum of the values of the discrete function at the even integers and the sum of the values at the odd integers.

**Initial value property.** The initial value property of relation 28 follows immediately from defining relation 1 by taking $z = 0$. The value of the discrete function at $n = 0$ is the value of the geometric transform at $z = 0$.

**Final value property.** The final value property of relation 29 is more difficult to see. It is intended to produce the limiting value $f(\infty) = \lim_{n\to\infty} f(n)$ of the discrete function if such a value exists. Perhaps the simplest approach is to consider the infinite sum of the difference discrete function of relation 24,

$$\sum_{m=0}^{\infty} [f(m) - f(m-1)]$$
$$= [f(0) - (f(-1) = 0)] + [f(1) - f(0)] + [f(2) - f(1)] + \cdots = f(\infty). \tag{1.5.16}$$

Thus we can consider the final value property to be a logical consequence of relation 24 and the summation property of relation 26. However, a more satisfactory interpretation of $f(\infty)$ is as the magnitude of the step component of $f(n)$ in cases where all other components vanish for large $n$. In this case the final value property is simply the procedure for finding the magnitude of the step component by partial fraction expansion. This comment should become meaningful after our discussion of partial fraction expansion in the next section.

**Matrix function.** Relation 30 is the defining relation of the geometric transform $M^g(z)$ of a discrete matrix function $M(n)$,

$$M^g(z) = \sum_{n=0}^{\infty} M(n)z^n. \tag{1.5.17}$$

The discrete matrix function $M(n)$ specifies the values of each element in the matrix at the non-negative integer $n$. We have therefore defined the geometric transform of a matrix to be the matrix of geometric transforms of the discrete time functions that are its elements,

$$m_{ij}{}^g(z) = \sum_{n=0}^{\infty} m_{ij}(n)z^n. \tag{1.5.18}$$

We can develop a series of matrix geometric transform pairs just as we developed the geometric transforms of discrete functions. In fact since most are direct analogs of the earlier relations, we shall concentrate only on those matrix transform pairs necessary in our further work.

**Matrix function multiplied by argument.** For example, consider the discrete matrix function $nM(n)$ obtained by multiplying the discrete matrix function $M(n)$ by the integer $n$. Since this discrete matrix function is analogous to the discrete function $nf(n)$ that appears in relation 9, we would expect it to have a geometric transform analogous to $z\frac{d}{dz}f^g(z)$. We proceed by differentiating the defining equation of the geometric transform for a discrete matrix function with respect to the transform variable $z$,

$$\frac{d}{dz} M^g(z) = \frac{d}{dz} \sum_{n=0}^{\infty} M(n)z^n = \sum_{n=0}^{\infty} nM(n)z^{n-1} \tag{1.5.19}$$

or

$$z\frac{d}{dz} M^g(z) = \sum_{n=0}^{\infty} nM(n)z^n, \tag{1.5.20}$$

and we have thus established that the geometric transform of the discrete matrix function $nM(n)$ is in fact $z$ times the derivative with respect to $z$ of the geometric transform of $M(n)$. Therefore relation 31 is a direct analog of relation 9.

**Power of a matrix.** To illustrate the computation of a matrix geometric transform, consider the case where the discrete matrix function $M(n)$ is a square matrix of constants $A$ raised to the $n$th power, $M(n) = A^n$. This discrete matrix function is the analog of the geometric sequence $a^n$ of relation 11; we already know that powers of square matrices play an important part in our study of Markov processes. If we apply the defining relation for the geometric transform we obtain

$$\sum_{n=0}^{\infty} M(n)z^n = \sum_{n=0}^{\infty} A^n z^n = I + Az + A^2z^2 + \cdots = [I - Az]^{-1} \tag{1.5.21}$$

where we have written the sum of the infinite matrix series as the inverse of the matrix $I - Az$, a matrix equal to the difference between the identity matrix and $A$ multiplied by $z$. Matrix theory shows that it is possible to write the infinite summation as this inverse only if $z$ is less in absolute value than the reciprocal of the characteristic value of $A$ that is largest in absolute value. In our work we shall never have to worry about such conditions any more than we did in the case of discrete functions. Consequently we write $[I - Az]^{-1}$ as the matrix geometric transform of $A^n$ in relation 32. Note that it reduces to relation 11 when $A$ is a simple constant $a$.

We can use relations 31 and 32 to produce the matrix geometric transform of the discrete matrix function $nA^n$, the result of multiplying $A^n$ by the integer $n$. They show that the transform must be given by $z\frac{d}{dz}[I - Az]^{-1}$. The only problem now is how to differentiate such an inverse matrix with respect to $z$.

**Differentiation of an inverse matrix.** Since the issue of differentiating inverse matrices will occur later in our studies, we shall develop the appropriate relationship now. Let $F^g(z)$ be a square matrix geometric transform and let $[F^g(z)]^{-1}$ be its inverse,

$$F^g(z)[F^g(z)]^{-1} = I. \tag{1.5.22}$$

If we now differentiate this expression with respect to $z$, we obtain

$$\frac{d}{dz}\{F^g(z)[F^g(z)]^{-1}\} = \left[\frac{d}{dz}F^g(z)\right][F^g(z)]^{-1} + [F^g(z)]\frac{d}{dz}[F^g(z)]^{-1}$$
$$= \frac{d}{dz}I = 0. \qquad (1.5.23)$$

Premultiplying by $[F^g(z)]^{-1}$ then shows that

$$\frac{d}{dz}[F^g(z)]^{-1} = -[F^g(z)]^{-1}\left[\frac{d}{dz}F^g(z)\right][F^g(z)]^{-1}. \qquad (1.5.24)$$

The derivative of the inverse of a matrix is the negative derivative of the matrix premultiplied and postmultiplied by the inverse of the matrix. Note in the case where $F^g(z)$ is simply an ordinary scalar function of $z$ that this relationship becomes the usual one for the derivative of the reciprocal of a function.

**Matrix geometric sequence operations.** When we apply Equation 1.5.24 to compute $z\frac{d}{dz}[I - Az]^{-1}$, the matrix geometric transform of $nA^n$, we obtain

$$\frac{d}{dz}[I - Az]^{-1} = -[I - Az]^{-1}[-A][I - Az]^{-1} = [I - Az]^{-1}A[I - Az]^{-1} \qquad (1.5.25)$$

and then

$$z\frac{d}{dz}[I - Az]^{-1} = z[I - Az]^{-1}A[I - Az]^{-1}. \qquad (1.5.26)$$

Consequently, as relation 33 shows, the matrix geometric transform of $nA^n$ is $z$ times the matrix $A$, premultiplied and postmultiplied by the matrix $[I - Az]^{-1}$. In the case where $A$ is a simple constant $a$, relation 33 reduces to relation 12.

We could easily develop a much more lengthy table of geometric transforms and geometric matrix transforms, but we shall find it more expeditious and informative to postpone such developments until they are required.

**Importance of restrictions on $z$.** Perhaps a few words here could allay some worries about the restrictions on $z$ that we are ignoring. As we shall see in Chapter 11, the geometric transform is a special case of the exponential transform. As such, $f^g(z)$ should be considered to be a function of a complex variable $z$. How have we avoided considering it in this way? Because the only time the complex nature of $z$ becomes important is when we desire to invert the transform $f^g(z)$ to find its corresponding discrete function. This inversion is formally accomplished by contour integration in the complex plane. The path taken in this contour integration depends vitally on the conditions on $z$ that we have been ignoring. We can thrive in our ignorance, however, as long as we can obtain the inverse of a geometric transform by using the relations of Table 1.5.1. Happily, this table will be more than adequate for our needs.

**Partial Fraction Expansion**

One algebraic technique will be especially helpful in getting the most mileage out of Table 1.5.1. This technique is partial fraction expansion. Let us refresh our memory on how it is accomplished. The most common problem is to write a fraction like

$$\frac{n(z)}{d(z)} = \frac{z}{(1 - az)(1 - bz)} \tag{1.5.27}$$

as the sum of two fractions, one with denominator $(1 - az)$ and one with denominator $(1 - bz)$. That is, we desire to find numbers $A$ and $B$ such that

$$\frac{n(z)}{d(z)} = \frac{z}{(1 - az)(1 - bz)} = \frac{A}{1 - az} + \frac{B}{1 - bz}. \tag{1.5.28}$$

A partial fraction expansion is possible only if the degree of the numerator polynomial $n(z)$ is at least one less than the degree of the denominator polynomial $d(z)$. For this case the numerator is of degree 1 in $z$, the denominator of degree 2, so we may proceed. If this condition were not met, then either we could, if possible, factor out enough powers of $z$ from the numerator to meet this condition and then perform a partial fraction expansion of the remaining factor, or we could divide the denominator into the numerator until the remainder was all of lower degree than the denominator and then perform a partial fraction expansion of the remainder. At least one of these methods will always work.

However, for the example we have no difficulty of this kind. Let us multiply both sides of Equation 1.5.28 by $(1 - az)$ alone and by $(1 - bz)$ alone. We obtain

$$\frac{z}{1 - bz} = A + \left(\frac{1 - az}{1 - bz}\right)B \tag{1.5.29}$$

and

$$\frac{z}{1 - az} = \left(\frac{1 - bz}{1 - az}\right)A + B. \tag{1.5.30}$$

Suppose $a$ is not equal to $b$. Then if we evaluate Equation 1.5.29 at $z = 1/a$ and Equation 1.5.30 at $z = 1/b$, we find

$$A = \left[\frac{z}{1 - bz}\right]_{z=1/a} = \frac{1}{a - b} \tag{1.5.31}$$

$$B = \left[\frac{z}{1 - az}\right]_{z=1/b} = \frac{1}{b - a} \tag{1.5.32}$$

and therefore

$$\frac{z}{(1-az)(1-bz)} = \frac{\frac{1}{a-b}}{1-az} + \frac{\frac{1}{b-a}}{1-bz}, \tag{1.5.33}$$

which the reader may verify. The formal procedure followed was:

$$A = \left\{(1-az)\left[\frac{z}{(1-az)(1-bz)}\right]\right\}_{z=1/a} \tag{1.5.34}$$

$$B = \left\{(1-bz)\left[\frac{z}{(1-az)(1-bz)}\right]\right\}_{z=1/b} \tag{1.5.35}$$

It works for any number of factors in the denominator of the fraction to be expanded as long as no two factors are identical. The method may be expressed in words as follows: To find the numerator of a factor in a partial fraction expansion, multiply the original fraction $n(z)/d(z)$ by that factor and evaluate the result at the value of the variable $z$ that makes the factor equal to zero.

***Repeated factors***

That is fine if all denominator factors are distinct—suppose they are not? Then factor out the extra powers and proceed iteratively. For example, consider this fraction where the factor $(1 - az)$ occurs two times,

$$\frac{z}{(1-az)^2(1-bz)}. \tag{1.5.36}$$

By factoring and making a regular partial fraction expansion, we can write it as

$$\begin{aligned}\frac{z}{(1-az)^2(1-bz)} &= \frac{1}{1-az}\left[\frac{z}{(1-az)(1-bz)}\right] \\ &= \frac{1}{1-az}\left[\frac{\frac{1}{a-b}}{1-az} + \frac{\frac{1}{b-a}}{1-bz}\right] \\ &= \frac{\frac{1}{a-b}}{(1-az)^2} + \frac{\frac{1}{b-a}}{(1-az)(1-bz)}.\end{aligned} \tag{1.5.37}$$

Now we make a regular partial fraction expansion of the second term,

$$\begin{aligned}\frac{z}{(1-az)^2(1-bz)} &= \frac{\frac{1}{a-b}}{(1-az)^2} + \frac{1}{b-a}\left[\frac{\frac{a}{a-b}}{1-az} + \frac{\frac{b}{b-a}}{1-bz}\right] \\ &= \frac{\frac{1}{a-b}}{(1-az)^2} + \frac{\frac{-a}{(a-b)^2}}{1-az} + \frac{\frac{b}{(a-b)^2}}{1-bz},\end{aligned} \tag{1.5.38}$$

and thus attain our goal. Factors of order higher than 2 and repetition of several factors are treated in the same way.

***Differentiation method for repeated factors***

In some situations it is convenient to have an alternate method of treating the case of repeated denominator factors. To develop this method, let us write the partial fraction expansion of the example of Equation 1.5.36 in a form with undetermined numerator coefficients,

$$\frac{n(z)}{d(z)} = \frac{z}{(1 - az)^2(1 - bz)} = \frac{A_0}{(1 - az)^2} + \frac{A_1}{1 - az} + \frac{B}{1 - bz}. \quad (1.5.39)$$

If we desire to find the constant $A_0$, we can multiply $n(z)/d(z)$ by $(1 - az)^2$ to produce

$$(1 - az)^2 \frac{n(z)}{d(z)} = \frac{z}{1 - bz} = A_0 + A_1(1 - az) + B\frac{(1 - az)^2}{1 - bz} \quad (1.5.40)$$

and then evaluate the result at $z = 1/a$ to obtain

$$A_0 = (1 - az)^2 \left.\frac{n(z)}{d(z)}\right|_{z=1/a} = \frac{1}{a - b}. \quad (1.5.41)$$

This is the coefficient of $\dfrac{1}{(1 - az)^2}$ developed in Equation 1.5.38 using repeated expansion.

Thus if we want to find the numerator of the highest power of a repeated factor, all we have to do is multiply $n(z)/d(z)$ by the highest power of that factor and evaluate the result at the value of $z$ that makes the factor equal zero.

Of course, this procedure reduces to our earlier technique whenever a factor is not repeated. To find $B$ in Equation 1.5.39, we write

$$B = (1 - bz)\left.\frac{n(z)}{d(z)}\right|_{z=1/b} = \left.\frac{z}{(1 - az)^2}\right|_{z=1/b} = \frac{b}{(a - b)^2} \quad (1.5.42)$$

and obtain the numerator of the $1 - bz$ factor in Equation 1.5.38.

We have now found $A_0$ and $B$ in Equation 1.5.39; it remains to determine $A_1$, the numerator of the factor $1 - az$ raised to the first power. Examination of Equation 1.5.40 that we used to develop $A_0$ shows that if we differentiate the equation with respect to $z$, we shall eliminate $A_0$, but not $A_1$; thus,

$$\frac{d}{dz}\left[(1 - az)^2 \frac{n(z)}{d(z)}\right] = \frac{d}{dz}\left[\frac{z}{1 - bz}\right] = 0 + A_1(-a) + (1 - az)g(z). \quad (1.5.43)$$

The term $(1 - az)g(z)$ represents all the others parts of the equation, each of which contains $1 - az$ to at least the first power. If we evaluate this equation at $z = 1/a$ we find

$$A_1 = \left(-\frac{1}{a}\right)\frac{d}{dz}\left[(1-az)^2\frac{n(z)}{d(z)}\right]\bigg|_{z=1/a}$$

$$= \left(-\frac{1}{a}\right)\frac{d}{dz}\left[\frac{z}{1-bz}\right]\bigg|_{z=1/a} = \frac{-a}{(a-b)^2}, \tag{1.5.44}$$

which is the numerator of the $1 - az$ term in the expansion of Equation 1.5.38.

***General form***

Now that we have seen the motivation behind the use of differentiation to evaluate the numerators of repeated factors, let us develop the general result. Suppose that a factor $(1 - az)$ is repeated $m \geq 1$ times in the denominator of a fraction suitable for partial fraction expansion. We express this situation as

$$\frac{n(z)}{d(z)} = \frac{n(z)}{(1-az)^m d_1(z)} = \frac{A_0}{(1-az)^m} + \frac{A_1}{(1-az)^{m-1}} + \frac{A_2}{(1-az)^{m-2}} + \cdots$$

$$+ \frac{A_{m-2}}{(1-az)^2} + \frac{A_{m-1}}{1-az} + f(z) \tag{1.5.45}$$

where $d_1(z)$ represents the other factors in the denominator $d(z)$, and $f(z)$ represents the remainder of the partial fraction expansion, again with no $(1 - az)$ denominator factors. If we now multiply Equation 1.5.45 by $(1 - az)^m$ we obtain

$$(1-az)^m\frac{n(z)}{d(z)} = \frac{n(z)}{d_1(z)} = A_0 + A_1(1-az) + A_2(1-az)^2 + \cdots$$

$$+ A_{m-2}(1-az)^{m-2} + A_{m-1}(1-az)^{m-1} + f(z)(1-az)^m. \tag{1.5.46}$$

Evaluation at $z = 1/a$ produces

$$A_0 = (1-az)^m\frac{n(z)}{d(z)}\bigg|_{z=1/a}. \tag{1.5.47}$$

If we differentiate Equation 1.5.46 with respect to $z$ and evaluate the result at $z = 1/a$, we find

$$A_1 = \left(-\frac{1}{a}\right)\frac{d}{dz}\left[(1-az)^m\frac{n(z)}{d(z)}\right]\bigg|_{z=1/a}. \tag{1.5.48}$$

Differentiating Equation 1.5.46 once more with respect to $z$ and evaluating the result at $z = 1/a$ shows that

$$A_2 = \frac{1}{2!}\left(-\frac{1}{a}\right)^2\frac{d^2}{dz^2}\left[(1-az)^m\frac{n(z)}{d(z)}\right]\bigg|_{z=1/a}. \tag{1.5.49}$$

We thus see that the numerator constant $A_k$ in Equation 1.5.45 is in general given by

$$A_k = \frac{1}{k!}\left(-\frac{1}{a}\right)^k \frac{d^k}{dz^k}\left[(1 - az)^m \frac{n(z)}{d(z)}\right]\Bigg|_{z=1/a} \qquad k = 0, 1, 2, \ldots, m - 1. \quad (1.5.50)$$

The coefficient $A_k$ is just the $k$th derivative with respect to $z$ of $(1 - az)^m$ times the original expression, evaluated at $z = 1/a$, multiplied by $(-1/a)^k$ and divided by $k!$.

The differentiation procedure for performing partial fraction expansion when faced with repeated denominator factors provides a useful alternative to the method described in the last section.

## 1.6 TRANSFORM ANALYSIS OF MARKOV PROCESSES

Now that we have the machinery of the geometric transform at our disposal, let us return to our major interest, the Markov process. The basic result of our previous study was that the multistep transition probability matrix $\Phi(n)$ is equal to the $n$th power of the transition probability matrix $P$. This result was expressed in Equation 1.2.14 and is repeated here:

$$\Phi(n) = P^n \qquad n = 0, 1, 2, \ldots. \quad (1.6.1)$$

Let us take the geometric transform of this equation. Relation 32 of Table 1.5.1 allows us to write immediately

$$\Phi^g(z) = [I - Pz]^{-1} = I + Pz + P^2z^2 + \cdots = \sum_{n=0}^{\infty} P^n z^n. \quad (1.6.2)$$

The transform of the multistep transition probability matrix can be obtained by multiplying each element of the transition probability matrix by $z$, substracting it from the identity matrix, and taking the inverse of the resulting matrix, or, of course, by direct summation of the matrix power series.

### An Alternate Derivation

Let us derive Equation 1.6.2 on another basis. Equation 1.2.8 states

$$\phi_{ij}(n + 1) = \sum_{k=1}^{N} \phi_{ik}(n)p_{kj} \qquad 1 \leq i, j \leq N, n = 0, 1, 2, \ldots. \quad (1.6.3)$$

The geometric transform of the left side of this equation is found from relation 8 in Table 1.5.1, of the right side from relations 4 and 5. We obtain

$$z^{-1}[\phi_{ij}^g(z) - \phi_{ij}(0)] = \sum_{k=1}^{N} \phi_{ik}^g(z)p_{kj} \qquad 1 \leq i, j \leq N. \quad (1.6.4)$$

The matrix form is

$$z^{-1}[\Phi^{g}(z) - \Phi(0)] = \Phi^{g}(z)P. \tag{1.6.5}$$

From Equation 1.2.12, $\Phi(0) = I$; therefore,

$$z^{-1}[\Phi^{g}(z) - I] = \Phi^{g}(z)P,$$
$$\Phi^{g}(z) - I = \Phi^{g}(z)Pz,$$
$$\Phi^{g}(z)[I - Pz] = I,$$

and finally

$$\Phi^{g}(z) = [I - Pz]^{-1}, \tag{1.6.6}$$

the same result as Equation 1.6.2.

**Matrix Inversion**

The problem of finding the geometric transforms of all multistep transition probabilities of a process is therefore reduced to the problem of inverting a matrix. Since we shall be inverting matrices in our examples, it might pay to review one method for finding the inverse of a matrix. The *ij*th element of the inverse of a matrix is equal to the *ij*th element of the adjoint matrix divided by the determinant of the matrix. The *ij*th element of the adjoint matrix is the cofactor of the *ji*th matrix element: $(-1)^{i+j}$ times the value of the determinant remaining when the *j*th row and *i*th column are deleted from the matrix. In the next chapter we shall find a more convenient method for such matrix inversion, but this will serve at present.

The common denominator of all elements of the inverse matrix is the determinant of the original matrix. Therefore, in finding the inverse of the matrix $[I - Pz]$, the determinant of this matrix, $|I - Pz|$, will play an important role. For an $N$-state system, this determinant will be an $N$th-order polynomial in $z$. If we set this polynomial equal to zero, we obtain the characteristic equation for the matrix, but the roots of this equation are the reciprocals of the characteristic values rather than the characteristic values themselves. Suppose the roots of $|I - Pz| = 0$ are $z_1, z_2, \ldots, z_n$. Then the characteristic values of the $P$ matrix, $\lambda_1, \lambda_2, \ldots, \lambda_N$ (the roots of the equation $|\lambda I - P| = 0$) will be given by $\lambda_1 = 1/z_1$, $\lambda_2 = 1/z_2, \ldots,$ $\lambda_N = 1/z_N$. The characteristic equation can then be written in factored form as

$$|I - Pz| = (1 - \lambda_1 z)(1 - \lambda_2 z)\cdots(1 - \lambda_N z) = 0. \tag{1.6.7}$$

Notice that the coefficients of the $z$'s in this equation *are* the characteristic values of the $P$ matrix. The pertinence of this observation to our studies will soon be apparent.

**The Marketing Example**

Let us apply the result of Equation 1.6.2 to the marketing problem we studied earlier. For this problem,

$$P = \begin{bmatrix} 0.8 & 0.2 \\ 0.3 & 0.7 \end{bmatrix} \tag{1.6.8}$$

and

$$I - Pz = \begin{bmatrix} 1 & 0 \\ 0 & 1 \end{bmatrix} - z\begin{bmatrix} 0.8 & 0.2 \\ 0.3 & 0.7 \end{bmatrix} = \begin{bmatrix} 1 - 0.8z & -0.2z \\ -0.3z & 1 - 0.7z \end{bmatrix}. \tag{1.6.9}$$

The determinant of $I - Pz$ is

$$\begin{aligned} |I - Pz| &= (1 - 0.8z)(1 - 0.7z) - (-0.2z)(-0.3z) \\ &= 1 - 1.5z + 0.5z^2 \\ &= (1 - z)[1 - (1/2)z]. \end{aligned} \tag{1.6.10}$$

Our discussion of characteristic values shows that the matrix $P$ has characteristic values of 1 and 1/2. Every stochastic matrix and therefore every transition probability matrix has at least one characteristic value equal to 1. Why? Because every Markov process has at least one starting vector $\pi(0)$ that satisfies the equation $\pi(0) = \pi(0)P$; the equation can only be satisfied if at least one characteristic value of $P$ is equal to one. This property of transition probability matrices assures us that when we write the characteristic equation $|I - Pz| = 0$, $z = 1$ will be a solution, and therefore $(1 - z)$ will be a factor of $|I - Pz|$.

The inverse of $I - Pz$ is the reciprocal of $|I - Pz|$ times the adjoint matrix. Applying the rule for the calculation of adjoint elements described above, we find

$$\Phi^{\mathscr{g}}(z) = [I - Pz]^{-1} = \frac{1}{(1 - z)[1 - (1/2)z]}\begin{bmatrix} 1 - 0.7z & 0.2z \\ 0.3z & 1 - 0.8z \end{bmatrix}. \tag{1.6.11}$$

Notice that the adjoint matrix for a 2 by 2 matrix is obtained by interchanging the elements on the main diagonal and changing the sign of those off the main diagonal. Next we perform a partial fraction expansion of each element of the matrix $[I - Pz]^{-1}$

$$[I - Pz]^{-1} = \frac{2}{1 - z}\begin{bmatrix} 0.3 & 0.2 \\ 0.3 & 0.2 \end{bmatrix} + \frac{-1}{1 - (1/2)z}\begin{bmatrix} -0.4 & 0.4 \\ 0.6 & -0.6 \end{bmatrix} \tag{1.6.12}$$

or

$$\Phi^{\mathscr{g}}(z) = [I - Pz]^{-1} = \frac{1}{1 - z}\begin{bmatrix} 0.6 & 0.4 \\ 0.6 & 0.4 \end{bmatrix} + \frac{1}{1 - (1/2)z}\begin{bmatrix} 0.4 & -0.4 \\ -0.6 & 0.6 \end{bmatrix}. \tag{1.6.13}$$

Equation 1.6.13 is the transform of the multistep transition probability matrix $\Phi(n)$ or, equivalently, $P^n$. We can easily find the matrix of inverse transforms by applying relations 3 and 11 in Table 1.5.1,

$$\Phi(n) = P^n = \begin{bmatrix} 0.6 & 0.4 \\ 0.6 & 0.4 \end{bmatrix} + (1/2)^n \begin{bmatrix} 0.4 & -0.4 \\ -0.6 & 0.6 \end{bmatrix} \qquad n = 0, 1, 2, \ldots \tag{1.6.14}$$

Thus we have managed to obtain a closed form expression for the $n$th power of the matrix $P$. We can easily check its validity for $n = 0$ and $n = 1$: for $n = 0$, $\Phi(0) = I$; for $n = 1$, $\Phi(1) = P$. Equation 1.6.14 in fact produces all the $\Phi(n)$ matrices computed in Equation 1.2.16.

The matrix $\Phi(n)$ is the sum of two terms, one that is constant as $n$ increases and one that decreases with $n$ at the rate 1/2. The constant term is just $\Phi$, the limiting multistep transition probability matrix for this process first expressed in Equation 1.2.19; it is the term arising from the characteristic value of 1. The other term is a matrix multiplied by a geometrically decreasing sequence. The rows of this matrix sum to zero; a matrix with this property is called a differential matrix.

### A Closed Form for $\Phi(n) = P^n$

We can summarize the behavior of $P^n$ for an $N$-state monodesmic process by the equation,

$$\Phi(n) = P^n = \Phi + T(n) \qquad n = 0, 1, 2, \ldots \tag{1.6.15}$$

The expansion of $P^n$ will consist of $N$ terms. One will be a constant term arising from the characteristic value 1; it will be the limiting multistep transition probability matrix for the process, $\Phi$. The other $N - 1$ terms are combined into a matrix $T(n)$ in Equation 1.6.15. They are transient terms whose effect will disappear when $n$ is large. Each of these terms is a differential matrix multiplied by a geometric sequence whose absolute value is continually decreasing. These sequences are generated by the remaining $N - 1$ characteristic values of the transition probability matrix. For a monodesmic process they must all be less than one in magnitude (the periodic case is again an exception). No transition probability matrix can ever have characteristic values larger than one because the component associated with such values would grow without bound as $n$ became large. This behavior would therefore violate the restriction that the sum of the elements in any row of $P^n$ must be equal to one.

Sometimes we shall want to use Equation 1.6.15 in its transformed form. We write

$$\Phi^{\mathscr{g}}(z) = \frac{1}{1 - z}\,\Phi + T^{\mathscr{g}}(z), \tag{1.6.16}$$

or in terms of components,

$$\phi_{ij}^{\mathscr{G}}(z) = \frac{1}{1-z}\phi_{ij} + t_{ij}^{\mathscr{G}}(z). \tag{1.6.17}$$

We shall find that the matrix $T^{\mathscr{G}}(z)$ evaluated at $z = 1$ is a very important quantity in finding Markov process statistics.

We can learn more about the structure of the closed form expression for $P^n$ if we review all our earlier examples from the transform point of view. Recall that the shrinkage factor for the marketing example as shown in Figure 1.4.2 was 1/2. One characteristic value of the $P$ matrix was 1/2. We are getting closer to the solution of the shrinkage factor mystery.

### The General Two-State Process

Let us solve the general two-state Markov process. We shall write the transition probability matrix in the form

$$P = \begin{bmatrix} 1-a & a \\ b & 1-b \end{bmatrix}. \tag{1.6.18}$$

Figure 1.6.1 shows the corresponding transition diagram. The two numbers $0 \le a \le 1$, $0 \le b \le 1$, specify the process completely; the only restriction on these numbers necessary to assure a monodesmic process is that both $a$ and $b$ not be zero, i.e., $a + b \neq 0$. This requirement avoids the 2-chain situation represented by the identity matrix. To find a closed form for $\Phi(n) = P^n$, we follow the 5-step process—1) form the matrix $[I - Pz]$, 2) find its determinant, 3) invert $[I - Pz]$ to obtain $[I - Pz]^{-1}$, 4) perform a partial fraction expansion of $[I - Pz]^{-1}$, 5) invert this set of geometric transforms to produce $\Phi(n) = P^n$.

*Computation of multistep transition probability matrix*

The matrix $[I - Pz]$ for the general two-state problem is

$$[I - Pz] = \begin{bmatrix} 1-(1-a)z & -az \\ -bz & 1-(1-b)z \end{bmatrix}. \tag{1.6.19}$$

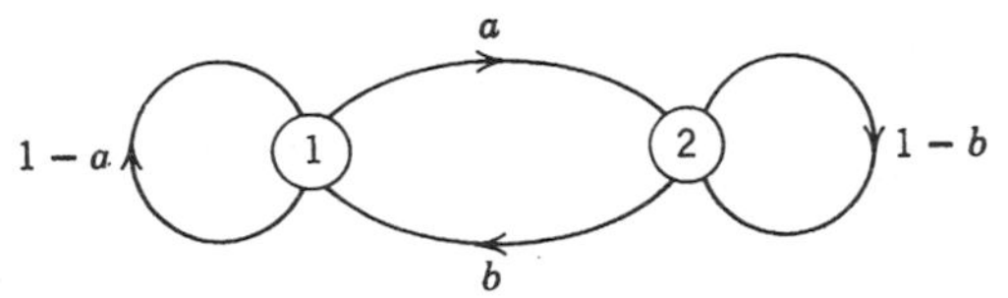


**Figure 1.6.1** Transition diagram for general two-state process.

Its determinant is

$$
\begin{aligned}
|I - Pz| &= [1 - (1 - a)z][1 - (1 - b)z] - (-az)(-bz) \\
&= 1 - (2 - a - b)z + (1 - a - b)z^2 \\
&= (1 - z)[1 - (1 - a - b)z]. \qquad (1.6.20)
\end{aligned}
$$

We see that the two characteristic values of the matrix $P$ are 1 and $1 - a - b$. The inverse matrix is then

$$
\begin{aligned}
\Phi^{\mathscr{g}}(z) &= [I - Pz]^{-1} \\
&= \frac{1}{(1 - z)[1 - (1 - a - b)z]} \begin{bmatrix} 1 - (1 - b)z & az \\ bz & 1 - (1 - a)z \end{bmatrix}. \qquad (1.6.21)
\end{aligned}
$$

Partial fraction expansion of $[I - Pz]^{-1}$ is feasible using the simple rule because the restriction that $a + b \neq 0$ eliminates the possibility of having a denominator factor $(1 - z)^2$. The expansion yields

$$
\Phi^{\mathscr{g}}(z) = \frac{\dfrac{1}{a + b}}{1 - z} \begin{bmatrix} b & a \\ b & a \end{bmatrix} + \frac{-\left(\dfrac{1 - a - b}{a + b}\right)}{1 - (1 - a - b)z} \begin{bmatrix} \dfrac{-a}{1 - a - b} & \dfrac{a}{1 - a - b} \\ \dfrac{b}{1 - a - b} & \dfrac{-b}{1 - a - b} \end{bmatrix}
$$

$$
= \frac{1}{1 - z} \begin{bmatrix} \dfrac{b}{a + b} & \dfrac{a}{a + b} \\ \dfrac{b}{a + b} & \dfrac{a}{a + b} \end{bmatrix} + \frac{1}{1 - (1 - a - b)z} \begin{bmatrix} \dfrac{a}{a + b} & \dfrac{-a}{a + b} \\ \dfrac{-b}{a + b} & \dfrac{b}{a + b} \end{bmatrix}. \qquad (1.6.22)
$$

We now use Table 1.5.1 to perform the inverse geometric transformation. We find

$$
\Phi(n) = P^n = \begin{bmatrix} \dfrac{b}{a + b} & \dfrac{a}{a + b} \\ \dfrac{b}{a + b} & \dfrac{a}{a + b} \end{bmatrix} + (1 - a - b)^n \begin{bmatrix} \dfrac{a}{a + b} & \dfrac{-a}{a + b} \\ \dfrac{-b}{a + b} & \dfrac{b}{a + b} \end{bmatrix}
$$

$$
n = 0, 1, 2, \ldots. \qquad (1.6.23)
$$

This equation illustrates again the general form for $P^n$: A constant stochastic matrix $\Phi$ plus $N - 1$ differential matrices multiplied by geometric coefficients. We see that the expansion satisfies $\Phi(0) = I$, $\Phi(1) = P$.

***Limiting state probabilities***

Since

$$
\Phi = \begin{bmatrix} \dfrac{b}{a + b} & \dfrac{a}{a + b} \\ \dfrac{b}{a + b} & \dfrac{a}{a + b} \end{bmatrix}, \qquad (1.6.24)
$$

the limiting state probability vector of the process is

$$\pi = \begin{bmatrix} \dfrac{b}{a+b} & \dfrac{a}{a+b} \end{bmatrix}. \tag{1.6.25}$$

We have therefore found that the limiting state probabilities for a two-state monodesmic process are given by

$$\pi_1 = \frac{p_{21}}{p_{12} + p_{21}}, \qquad \pi_2 = \frac{p_{12}}{p_{21} + p_{21}}, \tag{1.6.26}$$

which is consistent with the equilibrium of probabilistic flows argument.

***Possible behavior***

A two-state process can have only one rate of geometric decay; the engineer would say that such a process can have only one time constant. This geometric decay occurs at the rate $1 - a - b$.

If $a + b = 1$, the rows of the matrix $P$ are identical and we have a multinomial or, in this case, binomial process. Consequently,

$$P^n = \Phi = P \qquad n = 0, 1, 2, \ldots \tag{1.6.27}$$

Each row of the matrix $P$ is the unique limiting state probability vector of the process. As we know, the possible region of state probability vectors for this process shrinks to its limiting point at the first transition; there is no transient behavior.

If $0 < a + b < 1$, then the transient is a geometrically decreasing sequence whose influence is more prolonged the closer $a + b$ is to zero. If $1 < a + b < 2$, the sequence is the same except that its components are negative when $n$ is odd; the sequence oscillates.

***Periodic case***

The most extreme form of oscillation occurs when $a + b = 2$. Then the transient term coefficient is $(-1)^n$; it oscillates but does not diminish as $n$ increases. However, for this case, we have

$$P = \begin{bmatrix} 0 & 1 \\ 1 & 0 \end{bmatrix}, \tag{1.6.28}$$

the transition probability matrix for the periodic two-state process. Equation 1.6.23 then becomes

$$P^n = \begin{bmatrix} 1/2 & 1/2 \\ 1/2 & 1/2 \end{bmatrix} + (-1)^n \begin{bmatrix} 1/2 & -1/2 \\ -1/2 & 1/2 \end{bmatrix} \qquad n = 0, 1, 2, \ldots \tag{1.6.29}$$

which is equivalent to:

$$P^n = \begin{cases} \begin{bmatrix} 1 & 0 \\ 0 & 1 \end{bmatrix} & n = 0, 2, 4, 6, \ldots \\ \begin{bmatrix} 0 & 1 \\ 1 & 0 \end{bmatrix} & n = 1, 3, 5, 7, \ldots \end{cases} \tag{1.6.30}$$

Continually multiplying $P$ by itself confirms these results.

The form of $P^n$ demonstrates that the geometric interpretation of a two-state periodic process in terms of the line of unit length of Figure 1.4.1 corresponds to a possible region that reverses itself end-to-end at each transition, but does not shrink. The only point that remains stationary is the point $[1/2 \quad 1/2]$, that is, the common row of $\Phi$ from Equation 1.6.29. Thus we find that the geometric transform analysis produces for the two-state periodic process a $\Phi(n)$ that is correct and whose interpretation is consistent with that of the three-state periodic process.

***The marketing example***

We can apply Equation 1.6.23 to the two-state processes we have already solved numerically. For the marketing problem where

$$P = \begin{bmatrix} 0.8 & 0.2 \\ 0.3 & 0.7 \end{bmatrix} \tag{1.6.31}$$

we take $a = 0.2$, $b = 0.3$, and obtain result 1.6.14 directly.

***Variants of the marketing example***

Suppose that the probabilities of staying in a state and leaving it were interchanged for this marketing example. The transition probability matrix would then be

$$P = \begin{bmatrix} 0.2 & 0.8 \\ 0.7 & 0.3 \end{bmatrix}. \tag{1.6.32}$$

Now $a = 0.8$, $b = 0.7$; Equation 1.6.23 becomes

$$\Phi(n) = \begin{bmatrix} 7/15 & 8/15 \\ 7/15 & 8/15 \end{bmatrix} + (-1/2)^n \begin{bmatrix} 8/15 & -8/15 \\ -7/15 & 7/15 \end{bmatrix}. \tag{1.6.33}$$

The transient part of the multistep matrix decays at the rate $-1/2$; every term is opposite in sign to the one that precedes it.

The marketing example where customers remained with brand $B$ after buying it for the first time had transition probability matrix

$$P = \begin{bmatrix} 0.8 & 0.2 \\ 0 & 1 \end{bmatrix}. \tag{1.6.34}$$

With $a = 0.2$, $b = 0$, Equation 1.6.23 becomes

$$\Phi(n) = \begin{bmatrix} 0 & 1 \\ 0 & 1 \end{bmatrix} + (0.8)^n \begin{bmatrix} 1 & -1 \\ 0 & 0 \end{bmatrix} \qquad n = 0, 1, 2, \ldots \qquad (1.6.35)$$

This equation reproduces exactly the calculations of Equation 1.4.3. The presence of transient states is no problem.

***Shrinkage***

The shrinkage factor for the original marketing problem was 0.5, as shown in Figure 1.4.2. For the marketing problem we have just discussed where a customer gets trapped by brand $B$, the shrinkage factor is 0.8, as we found in Figure 1.4.4. The shrinkage factor for the two-state periodic case where there is no shrinking is,

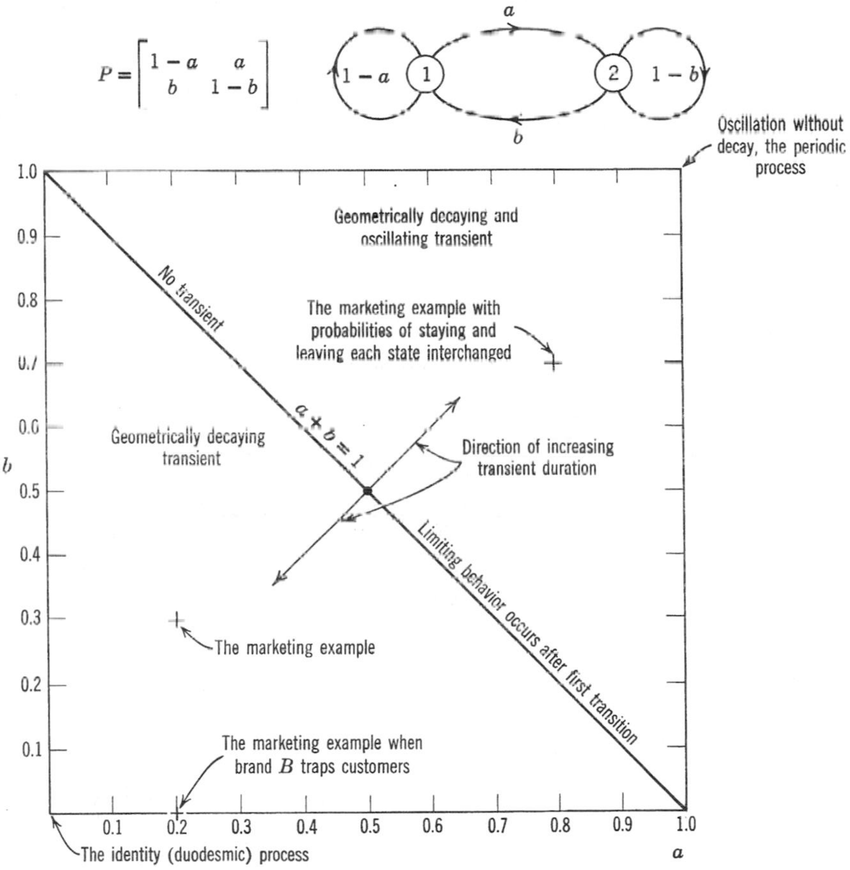


**Figure 1.6.2** Possible behavior of the general two-state process.

of course, 1. For all these cases the shrinkage factor is the absolute value of $1 - a - b$, the coefficient of the differential matrix. The mystery is diminishing and will soon be solved.

### *Summary of findings*

Now that we have constructed the general solution for the two-state case, we can write down the multistep transition probabilities for any two-state problem by inspection. All we have to remember is the form of the limiting state probability vector given by Equation 1.6.26 and that the decay factor is $1 - a - b$ or $1 - p_{12} - p_{21}$. With this information we can write the two rows of $\Phi$, and obtain the entries in the differential matrix from the requirement that $P^0 = I$. The relation $P^1 = P$ serves as a check. The reader might like to try his hand at a few two-state processes to fix this idea.

Our findings for the two-state process are summarized in Figure 1.6.2. This figure shows the type of behavior to expect for any pair of values $(a, b)$. Figure 1.6.3 is a plot of the elements $\phi_{11}(n)$ and $\phi_{21}(n)$, $n = 0, 1, 2, \ldots$, for each of the

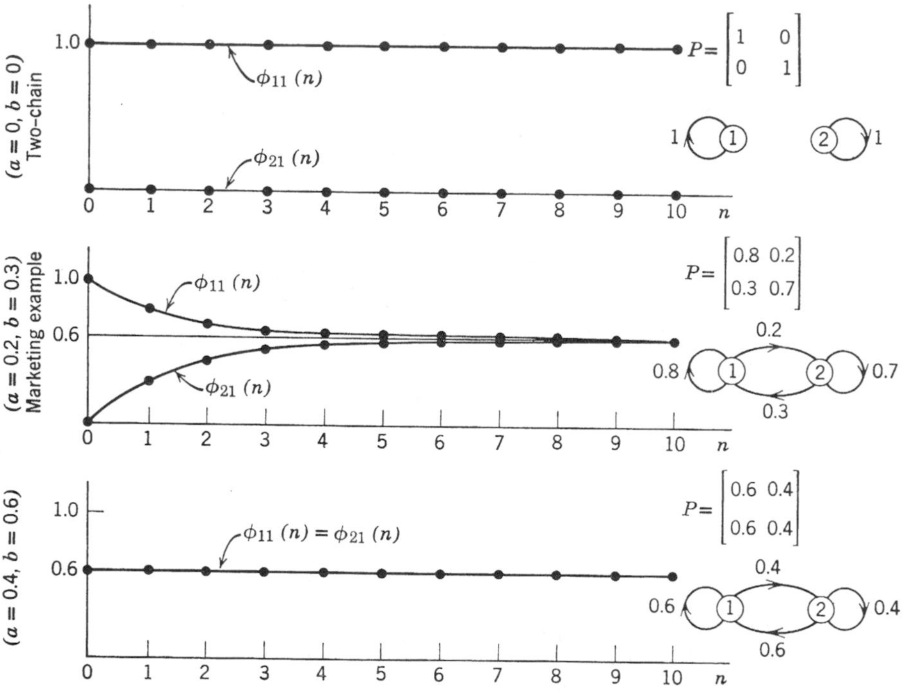


**Figure 1.6.3** Elements of $\Phi(n)$ as discrete time functions for certain two-state processes. (Continued on page 69.)

examples we have discussed. It is sufficient to plot only these entries of the $\Phi(n)$ matrix because the other entry in the first row is just $\phi_{12}(n) = 1 - \phi_{11}(n)$, and in the second row, $\phi_{22}(n) = 1 - \phi_{21}(n)$. These graphs of the elements of $\Phi(n)$ as discrete time functions further reinforce our understanding of typical behavior.

### The Taxicab Example

We now advance to three-state processes. The transition probability matrix for our three-state taxicab example was

$$P = \begin{bmatrix} 0.3 & 0.2 & 0.5 \\ 0.1 & 0.8 & 0.1 \\ 0.4 & 0.4 & 0.2 \end{bmatrix}. \tag{1.6.36}$$

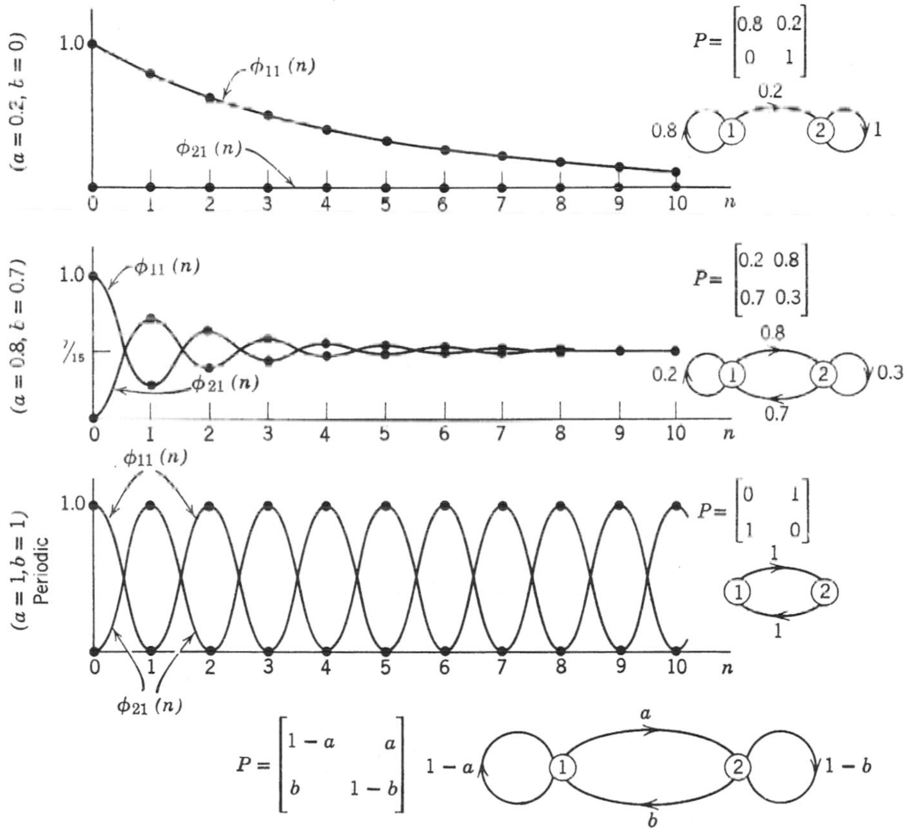

Proceeding in our usual way, we write

$$[I - Pz] = \begin{bmatrix} 1 - 0.3z & -0.2z & -0.5z \\ -0.1z & 1 - 0.8z & -0.1z \\ -0.4z & -0.4z & 1 - 0.2z \end{bmatrix}. \tag{1.6.37}$$

Its determinant is

$$\begin{aligned} |I - Pz| &= 1 - 1.3z + 0.2z^2 + 0.1z^3 \\ &= (1 - z)[1 - (1/2)z][1 + (1/5)z]. \end{aligned} \tag{1.6.38}$$

The characteristic values of $P$ are therefore 1, 1/2, and $-1/5$. The inverse of $[I - Pz]$ is

$$\Phi^{g}(z) = [I - Pz]^{-1} = \frac{1}{(1 - z)[1 - (1/2)z][1 + (1/5)z]}$$

$$\times \begin{bmatrix} 1 - z + 0.12z^2 & 0.2z + 0.16z^2 & 0.5z - 0.38z^2 \\ 0.1z + 0.02z^2 & 1 - 0.5z - 0.14z^2 & 0.1z + 0.02z^2 \\ 0.4z - 0.28z^2 & 0.4z - 0.04z^2 & 1 - 1.1z + 0.22z^2 \end{bmatrix}. \tag{1.6.39}$$

We shall find it more convenient to perform the partial fraction expansion if we first write $\Phi^{g}(z)$ in the form

$$\Phi^{g}(z) = [I - Pz]^{-1} = \frac{1}{(1 - z)[1 - (1/2)z][1 + (1/5)z]}$$

$$\times \left\{ \begin{bmatrix} 1 & 0 & 0 \\ 0 & 1 & 0 \\ 0 & 0 & 1 \end{bmatrix} + z \begin{bmatrix} -1 & 0.2 & 0.5 \\ 0.1 & -0.5 & 0.1 \\ 0.4 & 0.4 & -1.1 \end{bmatrix} + z^2 \begin{bmatrix} 0.12 & 0.16 & -0.38 \\ 0.02 & -0.14 & 0.02 \\ -0.28 & -0.04 & 0.22 \end{bmatrix} \right\}. \tag{1.6.40}$$

The partial fraction expansion yields

$$\Phi^{g}(z) = [I - Pz]^{-1}$$

$$= \frac{5/3}{1 - z} \begin{bmatrix} 0.12 & 0.36 & 0.12 \\ 0.12 & 0.36 & 0.12 \\ 0.12 & 0.36 & 0.12 \end{bmatrix} + \frac{-5/7}{1 - (1/2)z} \begin{bmatrix} -0.52 & 1.04 & -0.52 \\ 0.28 & -0.56 & 0.28 \\ -0.32 & 0.64 & -0.32 \end{bmatrix}$$

$$+ \frac{1/21}{1 + (1/5)z} \begin{bmatrix} 9 & 3 & -12 \\ 0 & 0 & 0 \\ -9 & -3 & 12 \end{bmatrix}, \tag{1.6.41}$$

or, in the form of Equation 1.6.16,

$$\Phi = \begin{bmatrix} 0.2 & 0.6 & 0.2 \\ 0.2 & 0.6 & 0.2 \\ 0.2 & 0.6 & 0.2 \end{bmatrix} \tag{1.6.42}$$

$$T^{g}(z) = \frac{-5/7}{1 - (1/2)z}\begin{bmatrix} -0.52 & 1.04 & -0.52 \\ 0.28 & -0.56 & 0.28 \\ -0.32 & 0.64 & -0.32 \end{bmatrix} + \frac{1/21}{1 + (1/5)z}\begin{bmatrix} 9 & 3 & -12 \\ 0 & 0 & 0 \\ -9 & -3 & 12 \end{bmatrix}. \tag{1.6.43}$$

Finally we invert the geometric transforms to obtain

$$\Phi(n) = P^n$$

$$= \begin{bmatrix} 1/5 & 3/5 & 1/5 \\ 1/5 & 3/5 & 1/5 \\ 1/5 & 3/5 & 1/5 \end{bmatrix} + (1/2)^n \begin{bmatrix} 13/35 & -26/35 & 13/35 \\ -1/5 & 2/5 & -1/5 \\ 8/35 & -16/35 & 8/35 \end{bmatrix}$$

$$+ (-1/5)^n \begin{bmatrix} 3/7 & 1/7 & -4/7 \\ 0 & 0 & 0 \\ -3/7 & -1/7 & 4/7 \end{bmatrix} \qquad n = 0, 1, 2, \ldots \tag{1.6.44}$$

This expression for $\Phi(n)$ checks for $n = 0$, and $n = 1$: $\Phi(0) = I$, $\Phi(1) = P$. Further, it produces the values of $\Phi(n)$ computed in Equation 1.4.10.

There are three matrices in the closed form for $P^n$ because this is a three-state monodesmic process. The first matrix is the stochastic matrix $\Phi$ we first found for this system in Equation 1.4.15. It arose from the characteristic value of the $P$ matrix that was equal to one. The other two matrices are differential matrices multiplied by geometric sequences with rates equal to the other two characteristic values of the $P$ matrix. The contributions of these differential matrices $\Phi(n)$ vanish when $n$ is large.

### *A new geometric interpretation*

We shall find it informative to give a new geometric interpretation to Equation 1.6.44. We shall represent any point in the original state probability triangle diagram as the sum of two vectors constructed on a base point. The base point will be the limiting state probability vector of the system, $\boldsymbol{\pi} = [1/5 \quad 3/5 \quad 1/5]$. One vector will be $\mathbf{a}(n) = (1/2)^n[1/5 \quad -2/5 \quad 1/5]$ in our nonorthogonal coordinate system; the other will be $\mathbf{b}(n) = (-1/5)^n[3/7 \quad 1/7 \quad -4/7]$. Equation 1.6.44 can now be placed in the form,

$$\Phi(n) = P^n = \begin{bmatrix} \boldsymbol{\pi} \\ \boldsymbol{\pi} \\ \boldsymbol{\pi} \end{bmatrix} + \begin{bmatrix} (13/7)\mathbf{a}(n) \\ -\mathbf{a}(n) \\ (8/7)\mathbf{a}(n) \end{bmatrix} + \begin{bmatrix} \mathbf{b}(n) \\ 0 \\ -\mathbf{b}(n) \end{bmatrix} \qquad n = 0, 1, 2, \ldots. \tag{1.6.45}$$

In particular, for $n = 0$,

$$\Phi(0) = P^0 = I = \begin{bmatrix} \boldsymbol{\pi} \\ \boldsymbol{\pi} \\ \boldsymbol{\pi} \end{bmatrix} + \begin{bmatrix} (13/7)\mathbf{a}(0) \\ -\mathbf{a}(0) \\ (8/7)\mathbf{a}(0) \end{bmatrix} + \begin{bmatrix} \mathbf{b}(0) \\ 0 \\ -\mathbf{b}(0) \end{bmatrix}. \tag{1.6.46}$$

Equation 1.6.46 is interpreted graphically in Figure 1.6.4. The three vertices of the triangle, which correspond to the rows of $\Phi(0) = I$, are represented as appropriate linear combinations of vectors $\mathbf{a}(0)$ and $\mathbf{b}(0)$ constructed upon the point $\boldsymbol{\pi}$.

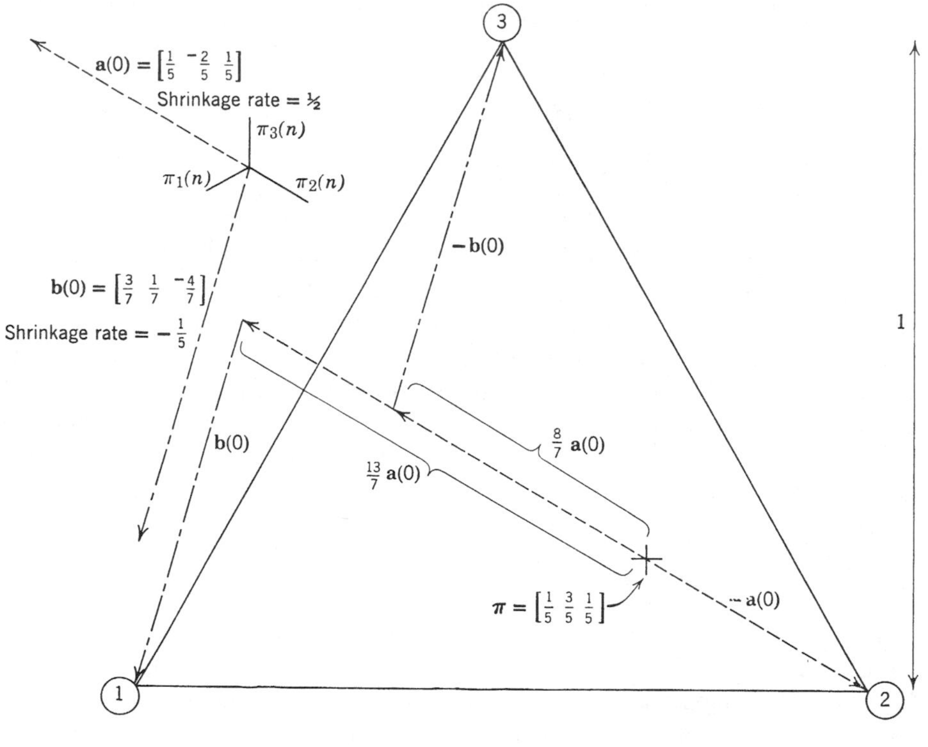


**Figure 1.6.4** A new geometric interpretation of the taixcab problem.

Suppose one transition has been made. Then

$$\Phi(1) = P = \begin{bmatrix} \boldsymbol{\pi} \\ \boldsymbol{\pi} \\ \boldsymbol{\pi} \end{bmatrix} + \begin{bmatrix} (13/7)\mathbf{a}(1) \\ -\mathbf{a}(1) \\ (8/7)\mathbf{a}(1) \end{bmatrix} + \begin{bmatrix} \mathbf{b}(1) \\ 0 \\ -\mathbf{b}(1) \end{bmatrix}. \tag{1.6.47}$$

The vector $\mathbf{a}(1)$ is half as long as $\mathbf{a}(0)$; the vector $\mathbf{b}(1)$ is one-fifth as long as $\mathbf{b}(0)$ and in the opposite direction. If we trace the effect of these changes in Figure 1.6.4 we find that the vertices of the triangle have been moved within it. Their new locations are just the vertices of the first inner triangle in Figure 1.4.6(b). If another transition is made, the shrinking vectors form the vertices of the second inner triangle, etc. As the number of transitions grows larger and larger, the vectors $\mathbf{a}(n)$ and $\mathbf{b}(n)$ approach zero length; the possible region shrinks to the point $\boldsymbol{\pi}$.

We now have a much better feel for the shrinkage process. We know that the reversal in orientation of the possible region that we observed in Figure 1.4.6(b) is a consequence of the transition probability matrix's having a negative characteristic value. The area of the possible region shrinks by 1/10 with each transition because the triangle is shrinking at the rate 1/2 in one direction, and at the rate 1/5 in the other. The fact that these shrinkages occur in nonorthogonal directions does not affect this result because any projection of the vectors shrinks at the same rate.

### Shrinkage Factor Evaluation

We thus find that the shrinkage factor for this and our earlier problems is the absolute value of the product of the characteristic values of the transition probability matrix $P$. Is this result true in general? Yes. We shall now prove it. By an elementary theorem of analytic geometry, the area $A(n)$ of the possible region triangle with vertices defined by $\Phi(n)$ for a three-state process is given by†

$$A(n) = K \left\| \begin{matrix} \phi_{11}(n) & \phi_{12}(n) & 1 \\ \phi_{21}(n) & \phi_{21}(n) & 1 \\ \phi_{31}(n) & \phi_{32}(n) & 1 \end{matrix} \right\|. \tag{1.6.48}$$

We find this equation for a process with an arbitrary number of states by writing $\Phi(n)$ with its last column replaced by a column of ones as the matrix whose determinant is to be found. The constant $K$ depends upon the dimensionality of the matrix and the orientation of the system of coordinates in which the area is plotted, which need not concern us here.

† The double vertical line refers to the absolute value of a determinant.

An important property of determinants is that their value is unchanged if we multiply any row or column by a constant and add it to another row or column. If we multiply the first and second columns of the determinant in Equation 1.6.48 by $-1$ and add the result to the third column, we find that

$$A(n) = K \begin{Vmatrix} \phi_{11}(n) & \phi_{12}(n) & \phi_{13}(n) \\ \phi_{21}(n) & \phi_{22}(n) & \phi_{23}(n) \\ \phi_{31}(n) & \phi_{32}(n) & \phi_{33}(n) \end{Vmatrix} = K \|\Phi(n)\| \qquad n = 0, 1, 2, \ldots \tag{1.6.49}$$

The area of the possible region after the $n$th transition is therefore proportional to the absolute value of the determinant of the $n$-step transition probability matrix. Consequently,

$$A(n) = K\|P^n\| \qquad n = 0, 1, 2, \ldots. \tag{1.6.50}$$

The area is proportional to the absolute value of the determinant of the $n$th power of the transition probability matrix $P$. We need only two more properties of determinants. The first is that the determinant of the product of two square matrices is the product of their determinants. Therefore,

$$A(n) = K\|P\|^n \qquad n = 0, 1, 2, \ldots. \tag{1.6.51}$$

The shrinkage factor, sf, is defined by

$$\text{sf} = \frac{A(n+1)}{A(n)} \qquad n = 0, 1, 2, 3, \ldots. \tag{1.6.52}$$

Therefore, from Equation 1.6.51,

$$\text{sf} = \|P\|. \tag{1.6.53}$$

We have now found that the shrinkage factor is equal to the absolute value of the determinant of the transition probability matrix $P$.

Now we need the second result on determinants. It is that the determinant of a square matrix is equal to the product of its characteristic values. Let the $N$ characteristic values of an $N$ by $N$ matrix $P$ be given by $\lambda_1, \lambda_2, \ldots, \lambda_N$. These values satisfy the equation,

$$|P - \lambda I| = (\lambda_1 - \lambda)(\lambda_2 - \lambda)\cdots(\lambda_N - \lambda) = 0. \tag{1.6.54}$$

If we take $\lambda = 0$ in this equation, we obtain immediately

$$|P| = \prod_{i=1}^{N} \lambda_i. \tag{1.6.55}$$

The determinant of the matrix $P$ is the product of its characteristic values. Therefore, we find as the final relation for the shrinkage factor

$$\text{sf} = \|P\| = \left|\prod_{i=1}^{N} \lambda_i\right| = \prod_{i=1}^{N} |\lambda_i|. \tag{1.6.56}$$

The shrinkage factor for any Markov process is the absolute value of the product of the characteristic values of its transition probability matrix.† All our examples have confirmed this result.

### The Duodesmic Example

Now let us apply transform analysis to the duodesmic example. The transition probability matrix is

$$P = \begin{bmatrix} 1 & 0 & 0 \\ 0 & 1 & 0 \\ 0.3 & 0.2 & 0.5 \end{bmatrix}. \tag{1.6.57}$$

We calculate $[I - Pz]$,

$$[I - Pz] = \begin{bmatrix} 1 - z & 0 & 0 \\ 0 & 1 - z & 0 \\ -0.3z & -0.2z & 1 - 0.5z \end{bmatrix} \tag{1.6.58}$$

and its determinant

$$|I - Pz| = (1 - z)^2(1 - 0.5z). \tag{1.6.59}$$

We see that the matrix $P$ has two characteristic values equal to 1, and one equal to 0.5. The shrinkage factor for the process is therefore 0.5, as we saw in Figure 1.4.7(b).

The matrix $\Phi^{g}(z)$ is given by:

$$\Phi^{g}(z) = [I - Pz]^{-1}$$

$$= \frac{1}{(1 - z)^2(1 - 0.5z)} \begin{bmatrix} (1 - z)(1 - 0.5z) & 0 & 0 \\ 0 & (1 - z)(1 - 0.5z) & 0 \\ 0.3z(1 - z) & 0.2z(1 - z) & (1 - z)^2 \end{bmatrix}$$

$$= \begin{bmatrix} \dfrac{1}{1 - z} & 0 & 0 \\ 0 & \dfrac{1}{1 - z} & 0 \\ \dfrac{0.3z}{(1 - z)(1 - 0.5z)} & \dfrac{0.2z}{(1 - z)(1 - 0.5z)} & \dfrac{1}{1 - 0.5z} \end{bmatrix}. \tag{1.6.60}$$

† The shrinkage factor is also the Jacobian of the transformation on the state probability vectors implied by the matrix $P$.

A partial fraction expansion yields

$$\Phi^{\mathcal{G}}(z) = \frac{1}{1-z}\begin{bmatrix}1 & 0 & 0\\ 0 & 1 & 0\\ 0.6 & 0.4 & 0\end{bmatrix} + \frac{1}{1-0.5z}\begin{bmatrix}0 & 0 & 0\\ 0 & 0 & 0\\ -0.6 & -0.4 & 1\end{bmatrix}. \quad (1.6.61)$$

Finally we invert the geometric transform to obtain

$$\Phi(n) = P^n = \begin{bmatrix}1 & 0 & 0\\ 0 & 1 & 0\\ 0.6 & 0.4 & 0\end{bmatrix} + (0.5)^n\begin{bmatrix}0 & 0 & 0\\ 0 & 0 & 0\\ -0.6 & -0.4 & 1\end{bmatrix}. \quad (1.6.62)$$

Equation 1.6.62 is consistent with the values of $\Phi(n)$ we found for this process in Equation 1.4.17. The limiting multistep transition probability matrix $\Phi$ of Equation 1.4.18 is the first matrix in Equation 1.6.62.

***Effect of multiple chains***

The effect of multiple chains is to decrease the number of matrices in the closed form expression for $P^n$; they cause no difficulty in the transform analysis. In an $N$-state, $C$-chain process, there will be $N - C + 1$ matrices in this expression. One of these will be the limiting multistep matrix $\Phi$, the other $N - C$ matrices will be differential matrices with geometrically decreasing coefficients, except in the periodic case. In the present example there is only one transient matrix; it decays at the rate 0.5.

The number of chains in a Markov process, $C$, is just the number of characteristic values of the transition probability matrix $P$ that are equal to 1. We can see the truth of this statement from the shrinkage factor argument. Each characteristic value equal to 1 in addition to the one that is always present increases the dimensionality of the limiting possible region by one because it prevents shrinking in one direction. Therefore, we have another way of describing a monodesmic process: a monodesmic process has a transition probability matrix that has exactly one characteristic value equal to one.

## The Identity Process

The three-state, three-chain or tridesmic process is the one with a transition probability matrix that is the identity matrix. We can use our transform analysis even in this unusual case. We write

$$P = \begin{bmatrix}1 & 0 & 0\\ 0 & 1 & 0\\ 0 & 0 & 1\end{bmatrix}, \quad (1.6.63)$$

$$[I - Pz] = \begin{bmatrix} 1 - z & 0 & 0 \\ 0 & 1 - z & 0 \\ 0 & 0 & 1 - z \end{bmatrix}, \tag{1.6.64}$$

$$|I - Pz| = (1 - z)^3, \tag{1.6.65}$$

$$[I - Pz]^{-1} = \frac{1}{(1 - z)^3} \begin{bmatrix} (1 - z)^2 & 0 & 0 \\ 0 & (1 - z)^2 & 0 \\ 0 & 0 & (1 - z)^2 \end{bmatrix}, \tag{1.6.66}$$

and finally,

$$\Phi^{g}(z) = [I - Pz]^{-1} = \frac{1}{1 - z} \begin{bmatrix} 1 & 0 & 0 \\ 0 & 1 & 0 \\ 0 & 0 & 1 \end{bmatrix}. \tag{1.6.67}$$

The inverse geometric transform is

$$\Phi(n) = P^n = \begin{bmatrix} 1 & 0 & 0 \\ 0 & 1 & 0 \\ 0 & 0 & 1 \end{bmatrix}, \tag{1.6.68}$$

the result we first found in Equation 1.4.22. Because there are three chains in the process, there is only one matrix in the expression for $P^n$. Since all three characteristic values are equal to one, there is no shrinkage; the limiting possible region is the whole triangle.

### The Periodic Process

The next three-state process we should analyze is the periodic process. We proceed as usual

$$P = \begin{bmatrix} 0 & 1 & 0 \\ 0 & 0 & 1 \\ 1 & 0 & 0 \end{bmatrix}, \tag{1.6.69}$$

$$[I - Pz] = \begin{bmatrix} 1 & -z & 0 \\ 0 & 1 & -z \\ -z & 0 & 1 \end{bmatrix}, \tag{1.6.70}$$

$$|I - Pz| = 1 - z^3, \tag{1.6.71}$$

and

$$\Phi^{\mathscr{G}}(z) = [I - Pz]^{-1} = \frac{1}{1 - z^3}\begin{bmatrix} 1 & z & z^2 \\ z^2 & 1 & z \\ z & z^2 & 1 \end{bmatrix}. \tag{1.6.72}$$

We can invert this equation in its present form. The transform $1/(1 - z^3)$ can be written

$$\frac{1}{1 - z^3} = 1 + z^3 + z^6 + z^9 + \cdots. \tag{1.6.73}$$

This is the transform of a discrete function that takes on the value 1 when $n = 0, 3, 6, 9, \ldots$ and is zero elsewhere. The diagonal elements of $\Phi(n)$ are exactly this function. By relation 7 of Table 1.5.1 the elements $\phi_{12}(n)$, $\phi_{23}(n)$, and $\phi_{31}(n)$ will each then be this discrete function delayed by one unit so that it is equal to one at $n = 1, 4, 7, 10, \ldots$. By the same relation, the elements $\phi_{13}(n)$, $\phi_{21}(n)$, and $\phi_{32}(n)$ are each the function delayed by two units; it is one when $n = 2, 5, 8, 11, \ldots$. Therefore $\Phi(n)$ has exactly the values shown by Equation 1.4.25.

Equation 1.6.72 could have been inverted by finding the three roots of $1 - z^3 = 0$, two of which are imaginary. Then $1 - z^3$ could have been written as the product of three factors, a partial fraction expansion could have been performed, and the resulting transform inverted by using Table 1.5.1. This procedure is just an algebraically more complicated way of arriving at the same result.

### The Multinomial and Doubly Stochastic Processes

We simplify the transform analysis of the three-state multinomial process whose transition matrix is

$$P = \begin{bmatrix} p_1 & p_2 & p_3 \\ p_1 & p_2 & p_3 \\ p_1 & p_2 & p_3 \end{bmatrix} \tag{1.6.74}$$

by recalling that for this process $P^n = P$, $n = 1, 2, 3, \ldots$. Then

$$\Phi^{\mathscr{G}}(z) = [I - Pz]^{-1} = \sum_{n=0}^{\infty} P^n z^n = I + P\sum_{n=1}^{\infty} z^n$$

$$= I + P\frac{z}{1 - z}$$

$$= \frac{1}{1 - z}P + (I - P) \tag{1.6.75}$$

or

$$\Phi = P \qquad T^{\mathscr{G}}(z) = I - P. \tag{1.6.76}$$

The entire transient behavior occurs on the first transition, and $\Phi(n)$ assumes the form

$$\Phi(n) = P^n = \Phi + T(n) = P + (I - P)\,\delta(n). \tag{1.6.77}$$

This result agrees, of course, with Equation 1.4.30 and the requirement that $\Phi(0) = I$.

The transform analysis of the monodesmic doubly stochastic process has no special features except that the limiting multistep transition probability matrix $\Phi$, the coefficient of $1/(1 - z)$ in the matrix expansion of $\Phi^{\mathscr{g}}(z)$, has all elements equal.

### A Mergeable Process

We finally perform the transform analysis of the three-state mergeable process whose transition matrix appears in Equation 1.4.44 as

$$P = \begin{bmatrix} 0.3 & 0.2 & 0.5 \\ 0.4 & 0.5 & 0.1 \\ 0.4 & 0.4 & 0.2 \end{bmatrix}. \tag{1.6.78}$$

We write

$$[I - Pz] = \begin{bmatrix} 1 - 0.3z & -0.2z & -0.5z \\ -0.4z & 1 - 0.5z & -0.1z \\ -0.4z & -0.4z & 1 - 0.2z \end{bmatrix}, \tag{1.6.79}$$

whose determinant is

$$\begin{aligned} |I - Pz| &= 1 - z - 0.01z^2 + 0.01z^3 \\ &= (1 - z)(1 - 0.1z)(1 + 0.1z). \end{aligned} \tag{1.6.80}$$

The characteristic values of $P$ are therefore 1, 0.1, and $-0.1$. The shrinkage factor of the process is consequently $|(1)(0.1)(-0.1)| = 0.01$. The inverse of $[I - Pz]$ is computed to be:

$$\Phi^{\mathscr{g}}(z) = [I - Pz]^{-1} = \frac{1}{(1 - z)(1 - 0.1z)(1 + 0.1z)}$$

$$\times \begin{bmatrix} 1 - 0.7z + 0.06z^2 & 0.2z + 0.16z^2 & 0.5z - 0.23z^2 \\ 0.4z - 0.04z^2 & 1 - 0.5z - 0.14z^2 & 0.1z + 0.17z^2 \\ 0.4z - 0.04z^2 & 0.4z - 0.04z^2 & 1 - 0.8z + 0.07z^2 \end{bmatrix}$$

$$= \frac{1}{(1 - z)(1 - 0.1z)(1 + 0.1z)}$$

$$\times \left\{ \begin{bmatrix} 1 & 0 & 0 \\ 0 & 1 & 0 \\ 0 & 0 & 1 \end{bmatrix} + z \begin{bmatrix} -0.7 & 0.2 & 0.5 \\ 0.4 & -0.5 & 0.1 \\ 0.4 & 0.4 & -0.8 \end{bmatrix} + z^2 \begin{bmatrix} 0.06 & 0.16 & -0.23 \\ -0.04 & -0.14 & 0.17 \\ -0.04 & -0.04 & 0.07 \end{bmatrix} \right\} \tag{1.6.81}$$

Partial fraction expansion produces

$$\Phi^{g}(z) = [I - Pz]^{-1} = \frac{100/99}{1 - z}\begin{bmatrix} 0.36 & 0.36 & 0.27 \\ 0.36 & 0.36 & 0.27 \\ 0.36 & 0.36 & 0.27 \end{bmatrix} + \frac{-1/18}{1 - 0.1z}\begin{bmatrix} 0 & 18 & -18 \\ 0 & -18 & 18 \\ 0 & 0 & 0 \end{bmatrix}$$

$$+ \frac{1/22}{1 + 0.1z}\begin{bmatrix} 14 & 14 & -28 \\ -8 & -8 & 16 \\ -8 & -8 & 16 \end{bmatrix}. \qquad (1.6.82)$$

Lastly, we invert the transforms to obtain

$$\Phi(n) = P^{n} = \begin{bmatrix} 4/11 & 4/11 & 3/11 \\ 4/11 & 4/11 & 3/11 \\ 4/11 & 4/11 & 3/11 \end{bmatrix} + (0.1)^{n}\begin{bmatrix} 0 & -1 & 1 \\ 0 & 1 & -1 \\ 0 & 0 & 0 \end{bmatrix}$$

$$+ (-0.1)^{n}\begin{bmatrix} 7/11 & 7/11 & -14/11 \\ -4/11 & -4/11 & 8/11 \\ -4/11 & -4/11 & 8/11 \end{bmatrix} \qquad n = 0, 1, 2, \ldots. \qquad (1.6.83)$$

This expression passes the checks that $\Phi(0) = I$, $\Phi(1) = P$, and $\Phi(\infty) = \Phi$. We thus have the analytic expression that underlies the computations of Equation 1.4.47 and the state probability diagram of Figure 1.4.12(b).

***Effect of arbitrary starting probability vector***

To demonstrate the merging character of the process, we compute the state probability vector at time $n$, $\boldsymbol{\pi}(n)$, for an arbitrary starting probability vector $\boldsymbol{\pi}(0) = [\pi_1(0) \quad \pi_2(0) \quad \pi_3(0)]$,

$$\begin{aligned} \boldsymbol{\pi}(n) = \boldsymbol{\pi}(0)\Phi(n) = {} & [4/11 \quad 4/11 \quad 3/11] + (0.1)^{n}[0 \quad -\pi_1(0)+\pi_2(0) \quad \pi_1(0)-\pi_2(0)] \\ & + (-0.1)^{n}[(7/11)\pi_1(0) - (4/11)(\pi_2(0)+\pi_3(0)) \ (7/11)\pi_1(0) - (4/11)(\pi_2(0)+\pi_3(0)) \\ & -(14/11)\pi_1(0) + (8/11)(\pi_2(0)+\pi_3(0))]. \qquad n = 0, 1, 2, \ldots. \qquad (1.6.84) \end{aligned}$$

We obtain for $\pi_1(n)$,

$$\pi_1(n) = 4/11 + (-0.1)^{n}[(7/11)\pi_1(0) - (4/11)(\pi_2(0)+\pi_3(0))] \qquad n = 0, 1, 2, \ldots \qquad (1.6.85)$$

or since $\pi_1(0) + \pi_2(0) + \pi_3(0) = 1$,

$$\pi_1(n) = 4/11 + (-0.1)^{n}[\pi_1(0) - 4/11] \qquad n = 0, 1, 2, \ldots. \qquad (1.6.86)$$

Thus the probability of being in state 1 after the $n$th transition depends only on the probability of starting in state 1 and not in any way on the manner in which the remaining starting probability is spread between states 2 and 3. Consequently, merging states 2 and 3 into a single state can have no effect on the probability we assign to being in state 1, an observation we made originally in Section 1.4.

## 1.7 CONCLUSION

We have now worked enough examples using transform analysis to get a good idea of its power. Like people who have never had to rub two sticks together to light a fire, we find it difficult to imagine the other complicated methods that have been proposed for finding a closed form expression for $P^n$. Most of these have relied on the theory of matrix functions. The matrix function approach becomes quite complex when it has to deal with repeated characteristic values of the matrix $P$, a problem that is bound to arise in polydesmic processes.

The transform analysis presented in this chapter avoids such difficulties—it always works, at least from a theoretical point of view. We may find that the algebra gets a little tedious when the number of states gets up around 5 or 6, but we are solving these problems to gain understanding, not to vie with computing machines. However, there is a hope for reducing the amount of calculation necessary even for our small problems. We find a marked advantage in realizing that the probability transformation represented by any Markov process is a linear system and that the powerful graphical techniques so useful in system analysis can be applied to Markov processes.

First, however, we shall have to learn these techniques for analyzing linear systems; they are presented in the next chapter. As you first read this material you may consider it a digression from our main topic. It is. I assure you that you will find it rewarding.

## PROBLEMS

**1.** a) $\begin{bmatrix} 1/3 & 2/3 \\ 1/8 & 7/8 \end{bmatrix}$

b) $\begin{bmatrix} 1/6 & 1/3 & 1/2 \\ 1/6 & 1/3 & 1/2 \\ 1/6 & 1/3 & 1/2 \end{bmatrix}$

c) $\begin{bmatrix} 1/4 & 3/8 & 3/8 \\ 0 & 0 & 1 \\ 0 & 1 & 0 \end{bmatrix}$

d) $\begin{bmatrix} 1/2 & 0 & 1/2 \\ 0 & 1 & 0 \\ 2/5 & 0 & 3/5 \end{bmatrix}$

e) $\begin{bmatrix} 1/4 & 1/4 & 1/2 \\ 1/3 & 1/3 & 1/3 \\ 1/2 & 0 & 1/2 \end{bmatrix}$

f) $\begin{bmatrix} 2/3 & 0 & 0 & 1/3 \\ 0 & 1/5 & 4/5 & 0 \\ 0 & 1/4 & 3/4 & 0 \\ 1/8 & 0 & 0 & 7/8 \end{bmatrix}$

Each of the above matrices represents the transition probabilities for a Markov process. For each process draw the transition diagram; determine the chain structure, and identify transient and trapping states. Find and interpret the limiting state probabilities for all states in each process.

For *only* the matrix a) find $\phi_{12}(n)$ by a suitable method.

For *only* the matrix d) perform the following:

i) Find $\Phi(1)$ and $\Phi(2)$.

ii) Sketch (reasonably large scale) the state probability diagram and plot the two triangles corresponding to $\Phi(1)$ and $\Phi(2)$. Indicate on the diagram the limiting possible region.

iii) What is the ratio of the areas of two consecutive triangles; what is the shrinkage factor?

**2.** You are given a 3-state Markov process with the following properties:

i) the $P$ matrix has characteristic values of 1, 0.25, and $-0.5$

ii) $\phi_{13}(\infty) = 1/2$

iii) $\phi_{13}(2) = 9/32$

a) Using your knowledge of the *form* of $\phi_{13}(n)$ and the above conditions, find $\phi_{13}(n)$ in closed form.

*Remember:* $\Phi(0) = I$

b) If $\phi_{12}(n) = 1/8 + (5/3)(1/4)^n - (43/24)(-1/2)^n$, find a closed form expression for $\phi_{11}(n)$, using the result of part a).

**3.** A certain communications system is transmitting a string of binary digits. The string consists of sequences of the four codes: 0, 10, 110, and 111 selected with probabilities, $p_1$, $p_2$, $p_3$, and $p_4$ respectively. At the completion of each code, a new code is selected according to these probabilities independently of the previous codes that were transmitted.

a) Can the transmission of these binary digits be modeled with a Markov process?

b) In a very long string, what fraction of the digits will be 0's?

c) In a very long string, what fraction of successive digits will be 01?

d) In a very long string, what is the probability that a randomly observed sequence of four digits will be 0111?

**4.** Consider a Markov process with the following transition probability matrix:

$$P = \begin{bmatrix} 1/8 & 1/8 & 3/4 \\ 1/2 & 1/4 & 1/4 \\ 3/4 & 0 & 1/4 \end{bmatrix}$$

a) If the process starts in state 1 and a large number of transitions occur, what fraction of these transitions are from state 2 to state 3?

b) Repeat part a) if the process starts in state 2.

c) What should be the initial state probabilities, $\pi(0) = [\pi_1(0), \pi_2(0), \pi_3(0)]$, in order for the process to be in the steady state after *one* transition?

**5.** A repair shop can do either job $A$ or job $B$ but not both simultaneously; job $A$ takes two days to finish, job $B$ takes one day. The possible states of the shop are thus: 1 = no job; 2 = job $A$, first day; 3 = job $A$, second day; 4 = job $B$. The probability of a new demand for job $A$ at the beginning of each day is $a$; for job $B$ it is $b$. There is no waiting line; if the shop is halfway through a job $A$, any new demand arriving that day is sent elsewhere.

The only ambiguity arises when the shop finishes a job at the end of a day and has both a job $A$ and a job $B$ waiting at the beginning of the next day. The two possible policies are:

1) Always take job $A$ in preference to job $B$;
2) Always take job $B$ in preference to job $A$.

a) For policy 1 show that the transition probability matrix is

$$P = \begin{bmatrix} (1-a)(1-b) & a & 0 & b(1-a) \\ 0 & 0 & 1 & 0 \\ (1-a)(1-b) & a & 0 & b(1-a) \\ (1-a)(1-b) & a & 0 & b(1-a) \end{bmatrix}$$

Does this process have any transient states? If so, which ones?

b) Find the limiting state probabilities for this process.

c) Find the transition probability matrix and limiting state probabilities for policy 2.

d) What is the limiting ratio of the expected number of idle days for policy 1 to that for policy 2?

e) If the profits from job $A$ and job $B$ are $r_A$ and $r_B$, respectively, what is the condition on $r_A$ and $r_B$ for policy 2 to be more profitable than policy 1?

f) What fraction of job $A$ and what fraction of job $B$ are turned away under each of the two policies?

**6.** Consider an $N$ state, stationary, monodesmic Markov process that is in the steady state.

a) Find the probability that a randomly observed transition does not involve the $i$th state (i.e., is neither to nor from the $i$th state). Express your answer in terms of $\pi_i$ and $p_{ii}$.

b) A stationary, monodesmic Markov model for a process has been proposed for which the limiting state probability of the $i$th state is 0.6 and the probability of a transition from state $i$ to state $i$ is 0.2. Is this a reasonable model? Why?

c) Find a nontrivial inequality relationship between $\pi_i$ and $p_{ii}$ for a stationary, monodesmic Markov process.

**7.** Suppose a Markov process has been operating without observation for so long that our knowledge of it is in the steady state. What is the probability that if we observe the next transition of the process we shall see it make that transition from state $i$ to state $j$? Find this probability for each possible transition in the taxicab example of Section 1.4. Which trips are the most and the least likely to be observed?

**8.** The Run 'em Down Taxi Company has ten drivers working between towns $A$ and $B$. The company has found that if a driver is in town $A$, the probability that his next trip will be to town $B$ is 0.4; if he is in town $B$, the probability that his next trip will be to town $A$ is 0.1.

a) All the drivers have started in town $A$ and have made $n$ trips. If they operate independently of one another, what is the mean number of drivers in town $A$?

b) If a driver has made a large number of trips, what fraction of his trips will be from town $A$ to town $B$?

c) If a driver starts the day in town $A$, what is the probability that his $n$th trip will be from town $A$ to town $B$?

**9.** The following Markov process starts in state 1.

$$P = \begin{bmatrix} 0 & 0.5 & 0.5 \\ 0.4 & 0 & 0.6 \\ 0 & 0.2 & 0.8 \end{bmatrix}$$

Draw the transition diagram for the process and then find by inspection the probability that:

a) The process is in state 3 after three transitions.

b) The process arrives in state 3 for the first time on the $n$th transition.

c) The process has not entered state 2 by the $n$th transition.

d) After the third transition from state 3 to state 2 the next two transitions are either $(2 \to 1 \to 3)$ or $(2 \to 3 \to 3)$.

e) The process enters state 2 exactly once in the first three transitions.

f) The process makes the $1 \to 2$ transition exactly once in the first three transitions.

g) The expected number of times the process will enter state 2 in the first three ransitions.

**10.** Consider the following genetic model:

There are two types of genes: $G$ and $B$. Every person is born with two genes, one taken from each parent to give combinations $GG$, $GB$, or $BB$. A person has green eyes, if and only if, his gene pair is $GG$; otherwise his eyes are brown. The selection of a gene from each parent's pair is random, i.e., any parental gene is passed on with probability 1/2. Genes have no effect on selection of a mate.
For a gene selected at random

$$\mathscr{P}(G) = p \qquad \mathscr{P}(B) = 1 - p.$$

a) Find the transition probability matrix $P$ for the genetic transmission of the three possible states of *one* parent to the three states of the child. Give an interpretation, in terms of genes of ancestors, of the $GG \to GB$ element of $P^n$. (*Don't bother to evaluate the element.*)

b) If $A$'s maternal grandfather, Mr. Smith, had the gene pair $BB$, what is the probability that $A$ has green eyes?

c) $Z$ has a distant relative $X$, who is in the same generation as $Z$. They have only one ancestor in common, a grandmother. If $Z$ has green eyes, what is the probability that $X$ also has green eyes?

**11.** In his monthly journeys from planet to planet, Astronaut Jones has found that his trips among four planets can be described by a Markov process having the following transition probability matrix:

$$\begin{array}{ll} & \begin{array}{cccc} 1 & 2 & 3 & 4 \end{array} \leftarrow \text{to planet} \\ \text{from planet} \rightarrow & \begin{array}{c} 1 \\ 2 \\ 3 \\ 4 \end{array} \begin{bmatrix} 0.4 & 0.1 & 0.5 & 0 \\ 0 & 0.5 & 0.2 & 0.3 \\ 0.4 & 0.2 & 0 & 0.4 \\ 0.5 & 0.1 & 0.3 & 0.1 \end{bmatrix} \end{array}$$

a) If astronaut Jones starts at planet 1 and five months later is at planet 4, what is the probability that his next trip will be to planet 3?

b) For economic reasons, planets 1 and 2 have combined to form common market $\alpha$ while planets 3 and 4 have combined to form common market $\beta$. If Jones starts at planet 1 and five months later is in common market $\beta$, what is the probability that his next trip will be to common market $\alpha$?

c) Can Jones' trips from one common market to another be described by a 2-state Markov process, and why? If so, what is the transition probability matrix for this process?

**12.** Consider the Markov process with the following transition probability matrix:

$$P = \begin{bmatrix} 1/3 & 2/3 \\ 3/4 & 1/4 \end{bmatrix}$$

To start the process we toss a coin having probability $p$ of falling heads. If it falls heads, we start in state 1 (i.e., $s(0) = 1$); if not, we commence in state 2.

a) Determine the probability of being in state 2 at time $n$.

b) If transition $k$ occurs between times $k - 1$ and $k$, what is the probability that transition $n + 1$ is from state 2 to state 1?

c) What value(s) of $p$ will make the limiting state probability of being in state 1 equal to that for state 2?

**13.** For a certain Markov process the time between transitions is exactly one second. We are provided with a $P^*$ matrix that represents the transition probabilities for the situation where the system is only observed every two seconds, i.e.,

$$p_{ij}^* = \mathscr{P}\{s(n + 2) = j | s(n) = i\}.$$

a) Using this $P^*$ matrix could we obtain the matrix $Q$ whose elements we defined by

$$q_{ij} = \mathscr{P}\{s(n + 4) = j | s(n) = i\}?$$

b) Using the $P^*$ matrix could we find a $P$ matrix of the underlying Markov process where

$$p_{ij} = \mathscr{P}\{s(n + 1) = j | s(n) = i\}?$$

What are the conditions for such a process to exist?

c) Illustrate your answers to parts a) and b) by finding $Q$ and $P$ for

$$P^* = \begin{bmatrix} 1/2 & 1/2 \\ 1/4 & 3/4 \end{bmatrix}.$$

**14.** a) A certain traveling salesman operates in $N$ cities. Every week he decides which city to visit by drawing a slip of paper out of a hat. There are many slips of paper in the hat and the fraction of slips with the $i$th city's name on them is $p_i$. Thus,

$$\sum_{i=1}^{N} p_i = 1.$$

What is the transition probability matrix for the Markov process that describes the salesman's visits from city to city? Draw a transition diagram for $N = 4$. What is $\Phi(n)$ for this process? What are the limiting state probabilities for this process? Why is this such a simple Markov process?

b) The salesman is told by the head office that he is spending too much consecutive time in the cities. Therefore, he decides that before drawing a slip of paper from the hat he will temporarily remove those slips of paper with the name of the city he is currently visiting. After each drawing all slips are replaced in the hat. What are the transition probability matrix and the limiting state probability vector for this new process?

**15.** a) What are the necessary and sufficient conditions on the transition probabilities of a two-state monodesmic Markov process for its state probability vector to be its limiting state probability vector after only one transition?

b) Consider a three-state monodesmic Markov process whose state probability vector becomes its limiting state probability vector after only two transitions (but not necessarily after only one transition). Assume that $p_{ij} \neq 0$ for $1 \leq i, j \leq 3$. What is the shrinkage factor for the areas of the possible region? What will the possible region look like after only one transition? What are the conditions on the transition probabilities for this to occur? What are the characteristic values of the $P$ matrix for this situation? Why?

**16.** An interesting property of any matrix is that the sum of the diagonal elements (called the "trace") is equal to the sum of the characteristic values. Now consider the following three-state Markov process:

$$P = \begin{bmatrix} 0.5 & 0.3 & 0.2 \\ 0.1 & 0.6 & 0.3 \\ 0.1 & 0.7 & 0.2 \end{bmatrix}.$$

a) How many characteristic values are equal to one for this process?

b) What will be the product of the characteristic values for the process?

c) What are the characteristic values for the process?

d) How many transitions must we wait to ensure that the transient terms in $\Phi(n)$ are less than 1% of their original value?

**17.** For a certain three-state monodesmic Markov process the determinant of $I - Pz$ is given by

$$|I - Pz| = (1 - z)(1 - 0.5z)^2.$$

We know that

$$[I - Pz]^{-1} = \frac{1}{|I - Pz|}\begin{bmatrix}\text{adjoint}\\ \text{matrix}\end{bmatrix}.$$

a) $\Phi(n)$ will be a sum of how many matrices?

b) What will be the coefficients of these matrices?

c) Using $a, b, c, d, \ldots$ etc. to denote the elements of these matrices, write a closed form expression for $\Phi(n)$.

d) Using your knowledge of the properties of these matrices, rewrite c) using only *eight* arbitrary constants for elements of the matrices.

**18.** If $1/(1 - az)(1 - bz)^2 = A/(1 - az) + B/(1 - bz) + C/(1 - bz)^2$, then

a) Find $A$ by letting $z = 1/a$.

b) Find $C$ by letting $z = 1/b$.

c) Find $B$ directly by letting $z = 0$!

**19.** Prove that a fraction of the form $N(z)/(1 - az)^n$ can be expanded in the form

$$\sum_{i=1}^{n} \frac{A_i M_i(z)}{(1 - az)^i}$$

where $N(z)$ is a polynomial of order less than $n$ and $M_i(z)$ is an arbitrary nonzero polynomial of order less than $i$ for which $M_i(1/a) \neq 0$.

**20.** Show that the four-state Markov process with transition probability matrix

$$P = \begin{bmatrix} 0.4 & 0.1 & 0.4 & 0.1 \\ 0.1 & 0.2 & 0.2 & 0.5 \\ 0.6 & 0 & 0.2 & 0.2 \\ 0.2 & 0.4 & 0.1 & 0.3 \end{bmatrix}$$

is mergeable when we consider a superstate $S_1$ consisting of states 1 and 3 and a superstate $S_2$ consisting of states 2 and 4. Demonstrate the consistency of the limiting state probabilities of the original and merged processes. Compare the merged process with the marketing example of Section 1.2. What can you say about the shrinkage behavior of the four-state process?

# 2 SYSTEMS ANALYSIS OF LINEAR PROCESSES

In this chapter we develop an approach to the study of dynamic probabilistic systems based upon the techniques of linear system analysis that have proven so useful in the control of physical systems. An important feature of this methodology is its visualization power, so useful in human thought. Through graphical techniques the investigator can gain an insight into the interaction of system components that would be difficult to attain by any other approach.

## 2.1 PROPERTIES AND MODELS OF SIMPLE SYSTEMS

Development of a clear notation will be one of our primary goals in outlining linear system theory. Since we are embarking on a study of systems analysis, we shall begin with a definition of "system." A system is an operator that acts on one signal called the input to produce another signal called the output. The nature of the system is specified when we know the kind of input signals the system will accept, the type of output it will generate, and how these two signals are related.

### Discrete Time Operation

Figure 2.1.1 represents a system graphically. We shall discuss only those systems whose input and output signals are discrete functions. As we recall from Chapter 1, a discrete function is a function that is defined on the integers. We are especially concerned with discrete functions that can take on any real value, positive or negative, at the points 0, 1, 2, . . . ; we assume that any discrete function is equal to zero if it has a negative argument.

We shall use $f(n)$ and $g(n)$ to denote the discrete functions at time $n$ at the input and output of a system, respectively. Figure 2.1.2 shows a graphical representation

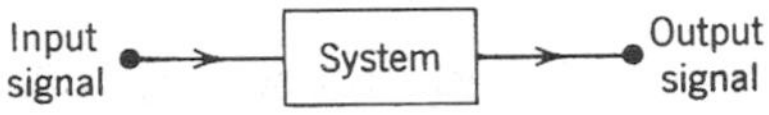


**Figure 2.1.1** A system.

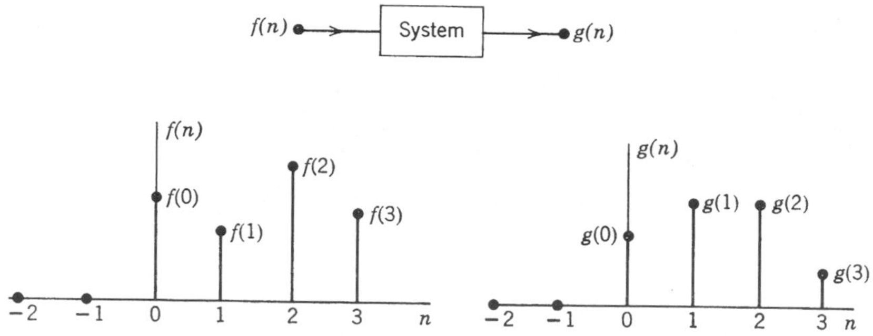


**Figure 2.1.2** A discrete system: discrete input and output.

of a system, with typical input and output functions indicated. Because the input and output of this system are discrete functions, we shall refer to it as a discrete system.

Examples of discrete functions and discrete systems are easy to find in everyday life. A radar system may receive returns from the target at regular intervals measured in millionths of a second. It may use these returns to make predictions of target position that appear at the output of the system as a discrete function. In the world of business, we often find that the sales of an enterprise are reported regularly, say on a monthly basis. The production control system may operate on these monthly sales figures to determine monthly production, also a discrete function. Such discrete systems play a role of special importance as a result of the widespread use of digital computers.

## Operator Interpretation

It will be convenient to think of the input and output signals of a system as vectors. Thus, we can call the input signal a vector $f$ with a countably infinite number of components $f(0), f(1), f(2), \ldots$. Similarly, the output signal can be represented by a vector $g$. The vector $f$ is operated upon by the system to produce the vector $g$; the system operator will be given the symbol $S$. Then we can write

$$g = S(f). \tag{2.1.1}$$

A system is thus a transformation that operates on one vector to produce another. The symbolic relationship of Equation 2.1.1 is interpreted graphically in Figure 2.1.3.

It is important to note that the system operator is directional. In general, the effect on the input of placing a given signal at the output of a system cannot be predicted. In terms of our earlier example, causing the production system to

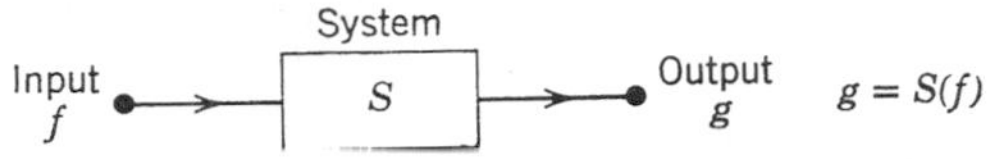


**Figure 2.1.3** The system as a vector transformation.

produce will not cause the sales pattern that might have required this production any more than shouting into a radio loudspeaker will cause a disturbance at the transmitting station.

## Linearity

Not all discrete systems are of interest to us—only those that make certain kinds of transformations on their input signal vectors. The primary requirement we shall place on the system is that the transformation it applies to the input vector must be linear; such a system will be called a linear system.

We can illustrate the linearity requirement very easily. Suppose that some input vector $f_1$ has been applied to the system, and that an output vector $g_1$ resulted. Then a different input vector $f_2$ was applied and the system produced the output vector $g_2$. Finally, a new input vector is formed by multiplying $f_1$ by the constant $a$, $f_2$ by the constant $b$, adding the vectors, and then applying the resultant composite vector to the system. If for any $a$ and $b$ we choose, and for any pair of inputs $f_1$ and $f_2$, the output obtained from the system is $ag_1 + bg_2$, then we say that the system is a linear system. The sequence of events is shown graphically in Figure 2.1.4. This defining feature of linear systems is called the superposition property, because superposition of inputs causes superposition of corresponding outputs.

In real life, very few systems are strictly linear. Automobiles do not travel twice as fast if the depression of the accelerator is doubled. It is unlikely that a production system could maintain linear behavior over a large range of sales patterns.

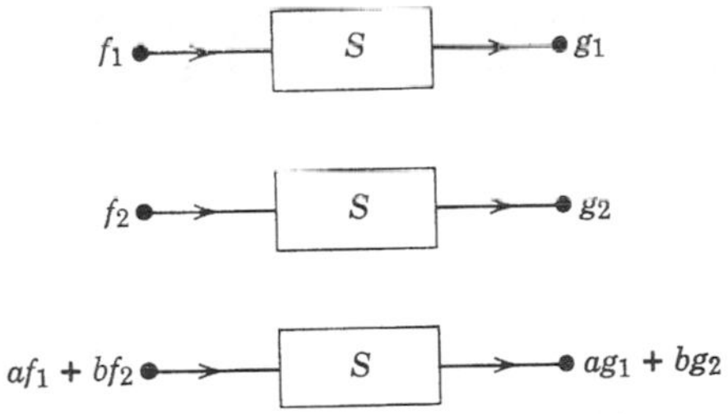


If $g_1 = S(f_1)$ and $g_2 = S(f_2)$, then $ag_1 + bg_2 = S(af_1 + bf_2)$

**Figure 2.1.4** A linear system.

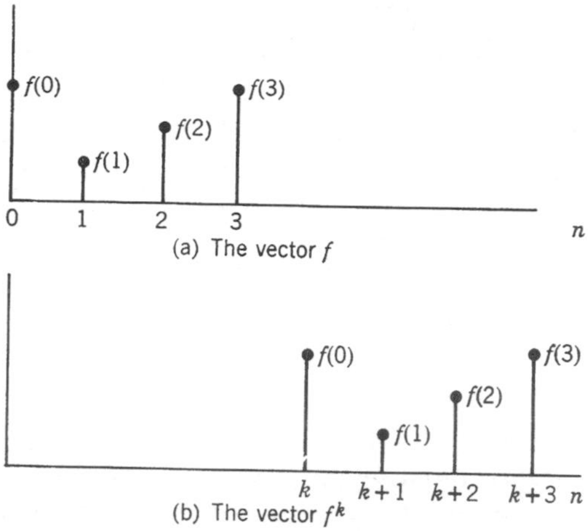


**Figure 2.1.5** Illustration of delay by $k$ units.

In both cases, one of the main causes of nonlinearity is the presence of saturating elements in the respective system. The existence of capacity limitations in every system will prevent truly linear operation. However, almost all systems behave very much like linear systems for some range of operating conditions. If this range includes the cases we must analyze, then what follows is relevant. We are fortunate that in many cases of practical interest, the linear assumption is justified.

## Time Invariance

Even the class of linear discrete systems is too large for the present analysis. We must restrict it further by requiring that the systems we consider be time-invariant. Let us define the vector $f^k$ to be the vector $f$ with all components delayed by $k$ units of time; that is, $f^k = e$ where $e(n) = f(n - k)$. Thus, when a vector is given a superscript $k$ all its components are shifted $k$ units into the future and, of course, zeros must be introduced into $f^k(0)$ through $f^k(k - 1)$ because $f(n)$ is

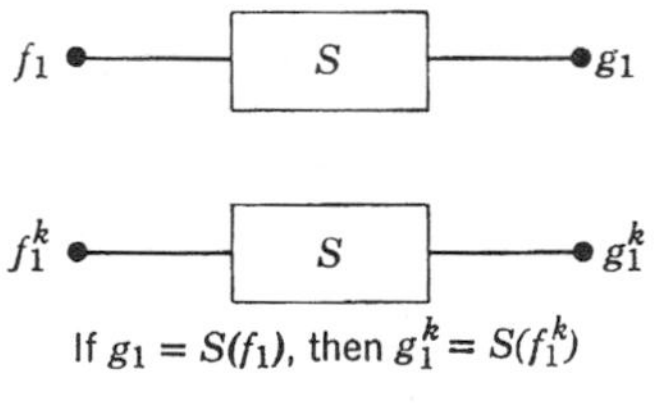


**Figure 2.1.6** A time-invariant system.

defined to be zero for $n < 0$. The nature of the superscript notation is illustrated in Figure 2.1.5. Time-invariance can now be expressed very simply: Suppose $g_1$ is the output when some signal $f_1$ is applied; if when $f_1$ is delayed by $k$ units, $g_1$ experiences the same delay for any positive $k$ and any $f_1$, then the system is said to be time-invariant. Figure 2.1.6 provides a graphical illustration of time-invariance.

## System Characterization

Our universe of discourse has now been limited considerably. We are going to deal only with linear, time-invariant discrete systems. We still must face the problem of characterizing the behavior of such systems. How can the transformation $S$ necessary to specify the output vector for a given input vector be described? The fact that this question has such a simple answer is one of the major reasons why this class of systems was chosen for study.

To develop the answer, we first recall a special discrete function, the unit impulse, defined in Chapter 1. The unit impulse function, $\delta(n)$, takes on the value one when $n = 0$ and the value zero everywhere else. Its vector representation is $\delta$ with components as graphed in Figure 2.1.7. If the unit impulse is applied to a system, then some vector output will be generated—we shall call this output the impulse response vector and give it the symbol $h$. Thus $h = g$ when $f = \delta$ or, more simply,

$$h = S(\delta). \tag{2.1.2}$$

We shall see that a basic property of linear time-invariant systems is that they are completely characterized by their response to the impulse function. Thus, the output for an arbitrary input may be easily calculated when the impulse response is known.

Since a system to which $\delta$ is applied receives no excitation until $n = 0$, it would be a strange system indeed that exhibited any output response before $n = 0$. Such a system would have to "laugh before it was tickled." We shall consider only physically realizable systems, those for which $h(n) = 0$ for $n < 0$.

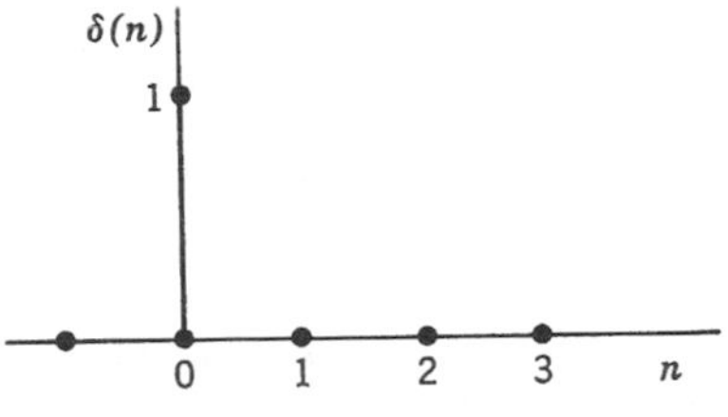


**Figure 2.1.7** The unit impulse vector $\delta$.

## Convolution

Let us now see how the impulse response may be used to find the response to an arbitrary input vector. First we note that any input vector may be represented in terms of the vector $\delta$ and its translations. Thus $f$ may be written in the form:

$$f = \sum_{k=0}^{\infty} f(k)\,\delta^k. \tag{2.1.3}$$

The sequence of steps that reveals how the response to an arbitrary input may be found is shown in Figure 2.1.8. If $\delta$ is applied to the system, then by definition $h$ will be the output. If $\delta^k$ is applied to the system, then $h^k$ will be the output because of the time-invariance property. If now the input $\delta^k$ is multiplied by $f(k)$, the $k$th component of the input vector, the output $h^k$ will be multiplied by the same factor because of the superposition property. When such inputs are summed over all the values of $k$, superposition requires that the output to this summed input be $\sum_{k=0}^{\infty} f(k)h^k$. However, the summed input $\sum_{k=0}^{\infty} f(k)\,\delta^k$ is just the alternate representation of $f$ given by Equation 2.1.3. Thus the output $g$ to an arbitrary input $f$ is given by:

$$g = \sum_{k=0}^{\infty} f(k)h^k \tag{2.1.4}$$

where $h$ is the impulse response of the system. Equation 2.1.4 states that the output to an arbitrary input may be found by delaying the impulse response by 0, 1, 2, 3, . . . units, weighting each shifted impulse response with the corresponding

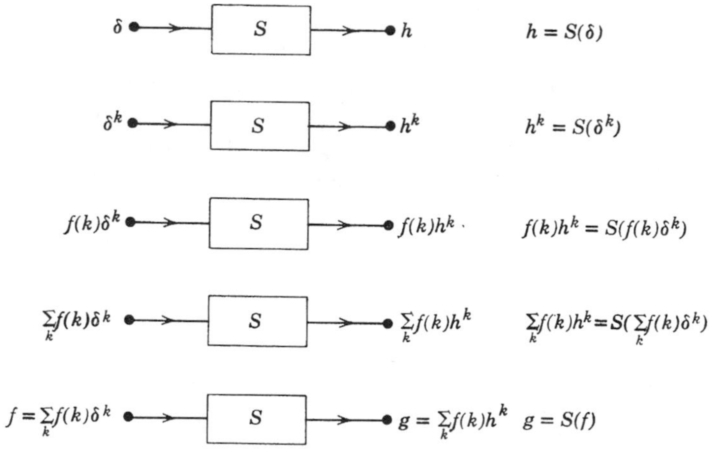


**Figure 2.1.8** The evolution of convolution.

component of $f$, and then adding all of these contributions to obtain the final result.

Equation 2.1.4 relates the vector $g$ to the vector $h$ and the components of $f$. This equation may be written in terms of the components of all vectors in the form:

$$g(n) = \sum_{k=0}^{n} f(k)h(n-k) \qquad n = 0, 1, 2, \ldots. \tag{2.1.5}$$

Note that the upper limit of the summation in this equation has been changed to $n$ to incorporate the fact that $h(n) = 0$ for $n < 0$ as the result of our physical realizability requirement. The often-encountered operation of Equation 2.1.5 is known as the convolution summation. It is important in the theory of probability because if $f(\cdot)$ and $h(\cdot)$ were the probability mass functions of two independent non-negative discrete random variables, then we would calculate the probability mass function of their sum $g(\cdot)$ by using Equation 2.1.5.

The convolution transformation operates on two vectors to produce a third. The output vector $g$ is the convolution of the input vector $f$ and the impulse response vector $h$. It is convenient to define a special symbol, a star, for the convolution operator. We can then write Equation 2.1.4 in the form

$$g = f * h. \tag{2.1.6}$$

***Properties***

It is easy to show that the convolution operator is commutative. Thus

$$\begin{aligned} g(n) &= \sum_{k=0}^{n} f(k)h(n-k) = \sum_{j=0}^{n} f(n-j)h(j) \\ &= \sum_{j=0}^{n} h(j)f(n-j) \end{aligned} \tag{2.1.7}$$

and

$$g = f * h = h * f. \tag{2.1.8}$$

Similarly it may be shown that the convolution operator is associative and distributive, and that we can write

$$(f * h_1) * h_2 = f * (h_1 * h_2) \qquad f * (h_1 + h_2) = f * h_1 + f * h_2. \tag{2.1.9}$$

These properties will have special significance later when we consider systems of increasing complexity.

**Graphical Representation**

It is advisable now to incorporate the impulse response in the graphical representation of a linear system. Such a representation is shown in Figure 2.1.9; we

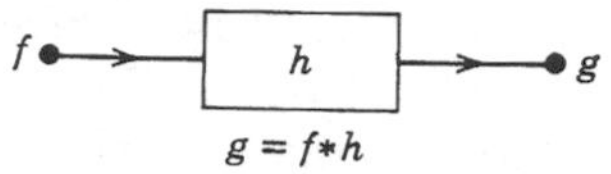


**Figure 2.1.9** Block diagram of linear system.

call it a system block diagram. Here the input and output points are represented by two small solid circles that we shall call nodes. It is understood that the output vector is to be calculated by convolving the input vector with the impulse response vector according to Equation 2.1.5.

## An Example: The Decadent Publisher

Before proceeding further into system theory, let us pause for an example of the type of system we have been discussing. A magazine publisher finds that of all who begin subscribing to his magazine in any year, only 1/3 subscribe the following year, $(1/3)^2 = 1/9$ the next year, and so on. Furthermore, while new subscriptions numbered one million for the zeroth year of operation, they were only 1/2 million for the year 1; $(1/2)^2 = 1/4$ million for year 2; etc. Our problem is to find how many million subscriptions the unfortunate publisher could claim in year $n$, $n=0,1,2,\ldots$.

We model this problem by treating it as a linear time-invariant discrete system whose basic time period is measured in years. We let the impulse response of the system be the individual subscription renewal pattern, $h(n) = (1/3)^n$, $n = 0, 1, 2, \ldots$. The input to the system is then the new subscription function, $f(n) = (1/2)^n$, $n = 0, 1, 2, \ldots$, where we have measured in units of millions of subscriptions. The problem is to find the output of the system, $g(n)$, the number of subscriptions (in millions) serviced by the publisher in each year.

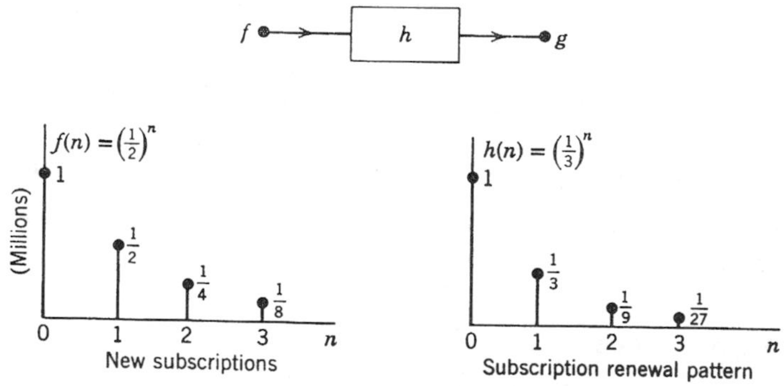


**Figure 2.1.10** The decadent publisher.

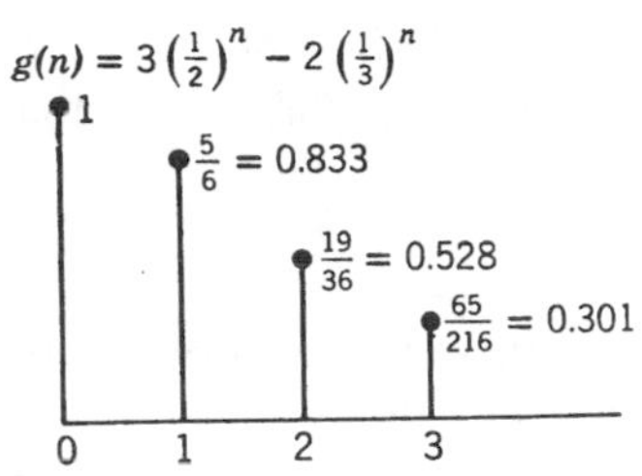


**Figure 2.1.11** Total subscriptions.

The problem is summarized in Figure 2.1.10. We find the components of the output vector by applying Equation 2.1.5,

$$g(n) = \sum_{k=0}^{n} f(k)h(n-k) = \sum_{k=0}^{n} (1/2)^k(1/3)^{n-k}$$

$$= (1/3)^n \sum_{k=0}^{n} (3/2)^k = (1/3)^n\left(\frac{1-(3/2)^{n+1}}{1-3/2}\right)$$

$$g(n) = 3(1/2)^n - 2(1/3)^n \qquad n = 0, 1, 2, \ldots \tag{2.1.10}$$

The output vector specified by Equation 2.1.10 is shown in Figure 2.1.11. The output signal begins at 1 and falls gradually to an ultimate value of zero. We observe that the publisher will handle 1 million subscriptions during year zero, 0.833 million during year 1, and 0.528 million during year 2. The difference between total subscriptions $g(n)$ and new subscriptions $f(n)$ in year $n$ is, of course, the number of subscriptions renewed by former subscribers, $r(n)$,

$$\begin{aligned} r(n) &= g(n) - f(n) \\ &= 3(1/2)^n - 2(1/3)^n - (1/2)^n \\ &= 2(1/2)^n - 2(1/3)^n \\ &= 2[(1/2)^n - (1/3)^n] \qquad n = 0, 1, 2, \ldots \end{aligned} \tag{2.1.11}$$

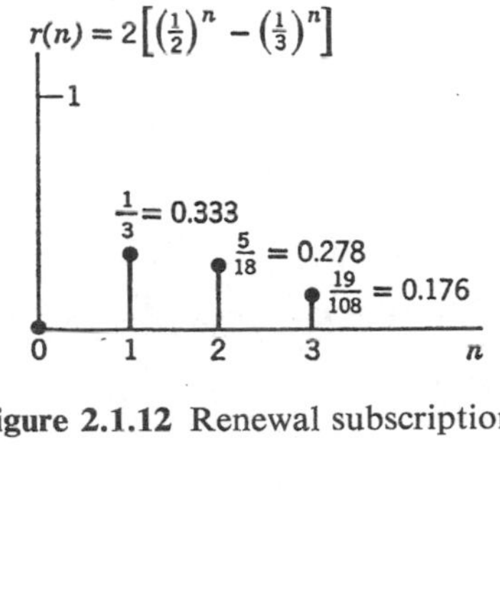


**Figure 2.1.12** Renewal subscriptions.

This function appears in Figure 2.1.12. Note that renewal subscriptions increase to 1/3 of a million in one year and then gradually decrease.

### Graphical Convolution

It is instructive to derive the result of Equation 2.1.10 by actually performing the convolution operation on the input and impulse response vectors. This calculation can be carried out in accordance with the result of Equation 2.1.7,

$$g(n) = \sum_{j=0}^{n} h(j)f(n-j), \tag{2.1.12}$$

which can be interpreted as follows. To compute the $n$th component of the output vector, the input vector is first reversed so that it runs from $-\infty$ to 0 and represents the function $f(-j)$. It is then shifted to the right $n$ time units so that its 0th component is aligned with the $n$th component of the impulse response vector; it now represents $f(n-j)$. As indicated by Equation 2.1.12, the two vectors are multiplied, component by component, and the terms are summed. The result of this calculation is the $n$th component of the output vector. The nature of this calculation for the $n = 2$ component of the output vector is shown in Figure 2.1.13. It is easy to

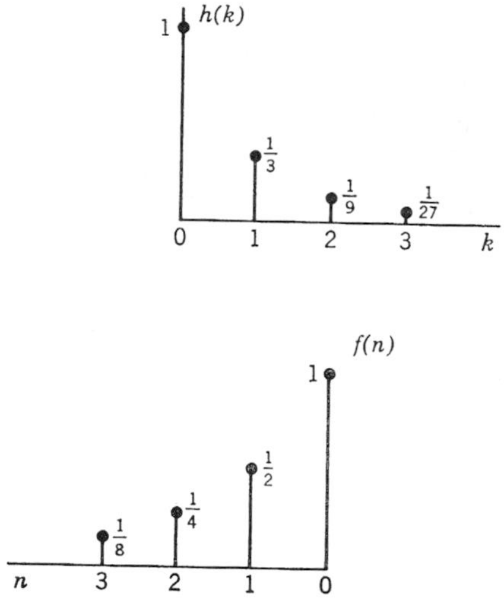


$$g(2) = \sum_{k=0}^{2} h(k)f(2-k) = 1 \times \tfrac{1}{4} + \tfrac{1}{3} \times \tfrac{1}{2} + \tfrac{1}{9} \times 1 = \tfrac{19}{36} = 0.528$$

**Figure 2.1.13** Solution by graphical convolution.

understand how the convolution operation received its name when we study the graphical procedure of Figure 2.1.13.

### Special Systems

An important class of systems will be those whose impulse response vectors are of the form $\delta^m$. Such a system has an impulse response time function that is a unit impulse function delayed by $m$ periods: $h(n) = \delta(n - m)$. Equation 2.1.12 quickly shows that for this system $g = f^m$; the output is equal to the input delayed by $m$ time periods. Two cases of special interest are where $m = 1$, called the simple delay system, and where $m = 0$, called the identity system because it leaves the input unchanged. The block diagrams of these systems would have impulse response vectors $\delta^1$ and $\delta^0 = \delta$, respectively.

## 2.2 COMPLEX SYSTEMS

Now that we have explored the phenomenon of convolution, we can begin to study how simple systems are interconnected to form complex systems. It will be necessary to expand our graphical notation to cover this case. The only additional feature we must define is: If two systems have their output at the same node, the signal at that node is the sum of the signals emerging from the two systems. This definition is illustrated in Figure 2.2.1. A system $h_1$ (we may as well call systems by their impulse response vector because they are completely characterized by that vector) receives an input $f_1$ while another system $h_2$ is subjected to an input $f_2$. The outputs from both systems are combined at a single node. The combination is additive, by definition.

### Parallel Systems

We can now examine the kinds of interconnection found in large systems. One of the most common is parallel connection, where two systems receive the same

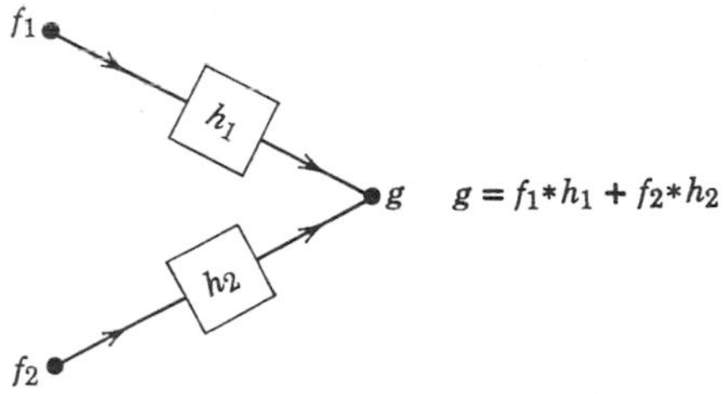


**Figure 2.2.1** Additivity of node signals.

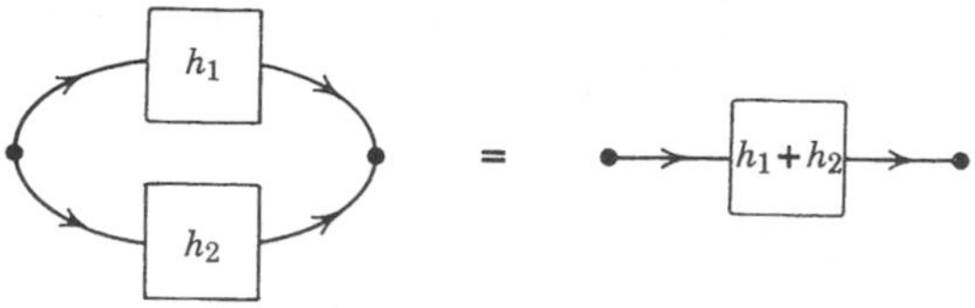


**Figure 2.2.2** Two parallel systems.

input and have their outputs added. Two production systems producing the same product in response to the same sales pattern form an example of this type of system. Figure 2.2.2 shows two such systems, $h_1$ and $h_2$. In the process of simplifying large systems of which such a structure may be a part, it is of interest to determine whether this type of connection of branches can be replaced by a single system that has an impulse response simply related to those of the original systems. It is easy to show that parallel systems may be replaced by a single system whose impulse response is the sum of the impulse responses of the component systems, as indicated in Figure 2.2.2. If a unit impulse is applied to the input node of the parallel connection, then the $h_1$ system will produce an output vector $h_1$; the $h_2$ system, an output vector $h_2$. Since the total system output is the sum of these two quantities, the total response to an impulse is $h_1 + h_2$. Consequently, the impulse response of two systems in parallel is the sum of their individual impulse responses. Since the two parallel systems constitute a linear and time-invariant system, the impulse response is a complete characterization of composite behavior.

### Series Systems

It is also possible to connect two systems in series. Two assembly operations that followed each other in a manufacturing process could fall in this category. The output of the first system is the input to the second. Figure 2.2.3 shows two systems $h_1$ and $h_2$ connected in series. Once again, it is possible to replace such a structure by a single system. In this case, the operation required to form the equivalent impulse response of the simplified system is not addition, but convolution. If a unit impulse is presented to the first system of the sequence, $h_1$, its output vector will be $h_1$. This output vector becomes the input vector of the second system $h_2$. We know that the output of the second system must then be $h_1 * h_2$;

**Figure 2.2.3** Two series systems.

consequently, the impulse response of the series structure is obtained by convolving the impulse responses of the component systems.

### Feedback Systems

If all complex systems were composed of systems in series and parallel, our analysis problem would be simple indeed. We could reduce all such systems to a single equivalent system simply by successively adding and convolving impulse responses. However, a third type of structure often appears in systems, a structure that cannot be treated by series-parallel reduction. This structure is the feedback loop. A typical feedback system is shown in Figure 2.2.4. The distinguishing feature of this system is that, in addition to the input and output nodes, it contains a third node that serves as both input and output node for a system $h$. Although this system looks like a dog chasing its own tail, it is still a completely acceptable system with respect to our previous definitions. We may assure ourselves of this fact by carrying out a step-by-step analysis of the system.

The input node is connected by an identity system to an intermediate node whose signal is designated by $e$. This node is both the input and output node for the system $h$; indeed, the arrow on the feedback loop can be drawn in either direction. The signal $e$ is then transferred by another identity system to the output node $g$. The signal $e$ is consequently composed of two additive parts; the input $f$ received through the identity system and the output of the system $h$. The output of $h$ is equal to its input $e$ convolved with its impulse response $h$. Thus

$$\begin{aligned} e &= f * \delta + e * h \\ &= f + e * h \end{aligned} \tag{2.2.1}$$

where we have used the fact that an identity system leaves its input unchanged. Since $e$ and $g$ are also related by an identity system, it follows that $g = e$ and, therefore, that we can write this equation in the form

$$g = f + g * h \tag{2.2.2}$$

or in the more interesting form

$$g * (\delta - h) = f. \tag{2.2.3}$$

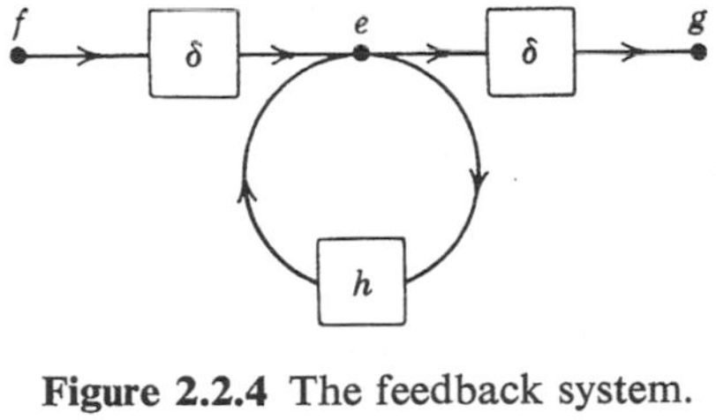


**Figure 2.2.4** The feedback system.

Equation 2.2.3 reveals that the input and output of a feedback system are not related to each other in a simple way. If this equation showed that $g$ were equal to $f$ convolved with some expression, then it would be clear that the equivalent system we seek would have an impulse response given by that expression. If we let $h_{eq}$ be the impulse response vector of this desired equivalent system, then we would be able to write

$$g = f * h_{eq}. \tag{2.2.4}$$

However, Equation 2.2.3 states just the reverse of this situation. It says that the input is equal to the output convolved with $\delta - h$. But since, from Equation 2.2.4 $g = h_{eq}$ when $f = \delta$, it follows from Equation 2.2.3 that $h_{eq}$ is related to $h$ by the equation

$$h_{eq} * (\delta - h) = \delta. \tag{2.2.5}$$

It will be necessary to find an inverse of the convolution operation if a closed form expression for $h_{eq}$ is to be obtained. It is generally difficult to solve for $h_{eq}$ in Equation 2.2.5 unless $h$ has a very simple form.

***A simple feedback system***

For an example of where such a solution is possible, consider the case where $h$ is a simple unit delay system, $h = \delta^1$. Equation 2.2.5 becomes

$$h_{eq} * (\delta - \delta^1) = \delta. \tag{2.2.6}$$

If we rewrite this equation in terms of the components of the vectors, we obtain

$$\sum_{k=0}^{n} h_{eq}(k)[\delta(n-k) - \delta(n-k-1)] = \delta(n) \qquad n = 0, 1, 2, \ldots. \tag{2.2.7}$$

Since $\delta(j) = 0$ for $j \neq 0$ and $\delta(0) = 1$, we note that for $n = 0$ Equation 2.2.7 implies

$$h_{eq}(0) = 1 \tag{2.2.8}$$

while for $n > 0$ we obtain

$$h_{eq}(n) - h_{eq}(n-1) = 0. \tag{2.2.9}$$

Equations 2.2.8 and 2.2.9 together show that

$$h_{eq}(n) = 1 \qquad n = 0, 1, 2, \ldots. \tag{2.2.10}$$

Thus if the loop in the feedback system of Figure 2.2.4 represents a system producing a delay of one time period, the over-all system will have an impulse response that is equal to one at every time point $n \geq 0$; in other words, a step function. This result is shown graphically in Figure 2.2.5. We shall have more to say about this particular system.

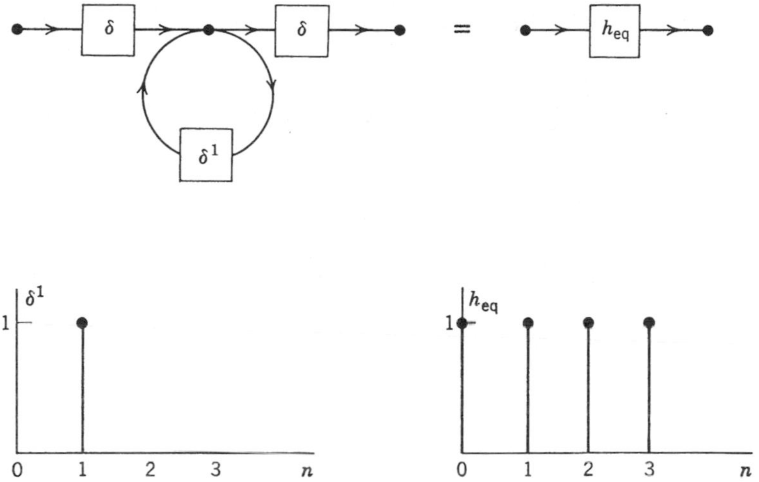


**Figure 2.2.5** A feedback example.

Although we were fortunate in the example just discussed, the effort required to solve Equation 2.2.5 using algebraic techniques in most cases is prohibitive. Consequently, a more promising solution method for the feedback situation is needed. The basis of our difficulty is clear. Since convolution is a kind of vector multiplication, solution of Equation 2.2.5 would require vector division. If Equation 2.2.5 could be changed to a scalar equation, then ordinary division would suffice for its solution. The device that will perform this collapsing of a vector into a scalar is the geometric transform, defined in the last chapter.

## 2.3 TRANSFORM ANALYSIS OF SYSTEMS

As you recall, geometric transformation is accomplished simply by multiplying the $k$th component of the vector to be transformed by the $k$th power of $z$, the transform variable, and then summing all components. We shall presently use several of the function-transform relationships developed in Table 1.5.1.

### The Geometric Transform in Convolution

Let us now examine the implication of the geometric transform for the systems we have discussed. For the basic system of Figure 2.1.9, the three vectors $f$, $g$, and $h$ are related by $g = f * h$. Suppose that each of these is transformed. What is the relation among the transformed signals? To answer this question we shall take the geometric transform of the component form of the convolution operation, Equation 2.1.5. Thus

$$\sum_{n=0}^{\infty} g(n)z^n = \sum_{n=0}^{\infty} z^n \sum_{k=0}^{n} f(k)h(n-k)$$

$$= \sum_{k=0}^{\infty} f(k)z^k \sum_{n=k}^{\infty} h(n-k)z^{n-k}$$

$$= \sum_{k=0}^{\infty} f(k)z^k \sum_{j=0}^{\infty} h(j)z^j$$

or

$$g^g(z) = f^g(z)h^g(z). \tag{2.3.1}$$

The effect of the geometric transform is therefore to change the relationship between input and impulse response from one of convolution to one of multiplication, a result we have previously observed as relation 6 of Table 1.5.1. This change of convolution to multiplication is exactly that experienced when, to avoid convolution, we multiply the so-called "generating functions" of discrete independent random variables to obtain the generating function of their sum. "Generating function" is an alternate name for the geometric transform.

***The transfer function***

The geometric transform of the output of a discrete time-invariant linear system is thus equal to the product of the transforms of its input and impulse response. It is convenient to assign a name to the transform of the impulse response, $h^g(z)$; we shall use the name "transfer function." Equivalently, as we can see from Equation 2.3.1, the transfer function is the ratio of output to input transforms,

$$g^g(z) = \frac{g^g(z)}{f^g(z)}. \tag{2.3.2}$$

## Transform Domain Analysis

We are now able to analyze systems entirely in the transform domain. That is, we can represent all signals in the system by their transforms, provided that we operate upon these transformed signals appropriately. We have already seen that the basic system can be characterized by its transfer function and that multiplication then becomes the operation necessary to find the transformed output of the system given the transformed input.

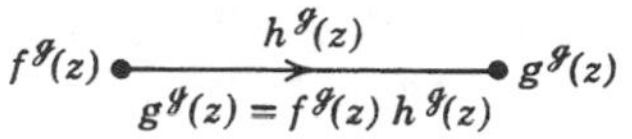
$h^{\mathscr{G}}(z)$
$f^{\mathscr{G}}(z)$
$g^{\mathscr{G}}(z)$
$g^{\mathscr{G}}(z) = f^{\mathscr{G}}(z)\, h^{\mathscr{G}}(z)$

**Figure 2.3.1** The basic flow graph.

### *The flow graph*

We can incorporate this simple relationship in the graphical representation of the system. The representation appears in Figure 2.3.1, where we abandon the idea of using a box to denote the system. Joining the input and output nodes is a line segment, called a branch, directed from input to output. We label the branch with the system transfer function. The transformed output signal is the transformed input signal multiplied by the transfer function. We call this system representation a flow graph. If two system flow graphs have the same output node, we define the transformed signal at that node to be the sum of the transformed outputs of the system, as shown in Figure 2.3.2. The flow graph serves as a compact, convenient graphical representation of the transform relationship, Equation 2.3.1, which governs linear time-invariant discrete systems.

We shall sometimes call $h^{\mathscr{G}}(z)$ the transmission of the branch rather than the transfer function. When we have many systems connected together to form a more complex system, we shall speak interchangeably of the transfer function or transmission from input to output of the complex system.

### *The decadent publisher*

Even if we did not need the geometric transform to help us with the feedback situation, we might still find it useful just to avoid the convolution algebra. Let us solve the publishing example of Figure 2.1.10 by the transform method. The transform of the output $g^{\mathscr{G}}(z)$ will be equal to the product of the transform of the

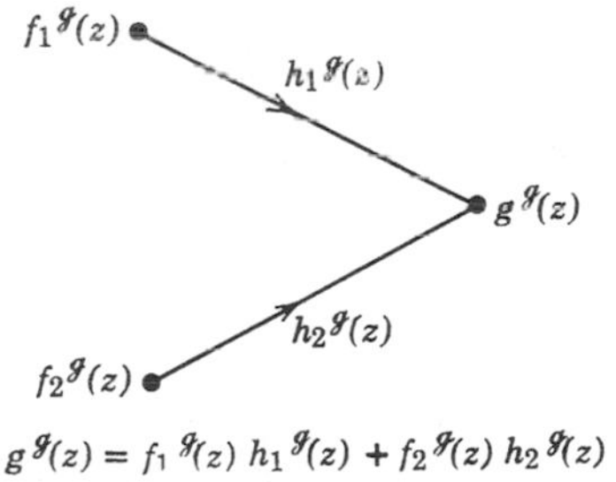
$f_1^{\mathscr{G}}(z)$
$h_1^{\mathscr{G}}(z)$
$g^{\mathscr{G}}(z)$
$h_2^{\mathscr{G}}(z)$
$f_2^{\mathscr{G}}(z)$
$g^{\mathscr{G}}(z) = f_1^{\mathscr{G}}(z)\, h_1^{\mathscr{G}}(z) + f_2^{\mathscr{G}}(z)\, h_2^{\mathscr{G}}(z)$

**Figure 2.3.2** Additivity in flow graphs.

input $f^g(z)$ and the transfer function $h^g(z)$. The transforms of input and impulse response are easily obtained from Table 1.5.1:

$$f^g(z) = \frac{1}{1 - (1/2)z} \qquad h^g(z) = \frac{1}{1 - (1/3)z} \tag{2.3.3}$$

and

$$g^g(z) = f^g(z)h^g(z) = \frac{1}{(1 - (1/2)z)(1 - (1/3)z)}. \tag{2.3.4}$$

Partial fraction expansion of this expression yields:

$$g^g(z) = \frac{3}{1 - (1/2)z} + \frac{-2}{1 - (1/3)z}. \tag{2.3.5}$$

Finally, by inverse transformation, we obtain

$$g(n) = 3(1/2)^n - 2(1/3)^n \qquad n = 0, 1, 2, \ldots \tag{2.3.6}$$

in accordance with the result of Equation 2.1.10.

***Parallel systems***

We shall now return to the complex system reduction problem to explore further the impact of the transform approach. Systems with parallel or series branches are easily disposed of. We know that two systems in parallel have an equivalent single branch system whose impulse response is the sum of the individual impulse responses. From Table 1.5.1, we see that the transform of the sum of two functions is the sum of their transforms. Consequently, two systems in parallel must be equivalent to a single branch system whose transfer function is the sum of the transfer functions of the individual branches. Thus, addition is the relevant mode of combination of parallel systems—whether we work in the time or transform domains.

***Series systems***

Because two systems in series have an equivalent impulse response equal to the convolution of their individual impulse responses, and because convolution becomes multiplication in the transform domain, it is clear that two series systems have an equivalent transfer function equal to the product of the transfer functions of each system. Figure 2.3.3 summarizes our relations for parallel and series systems in the transform domain.

***Feedback systems***

Armed with the geometric transform we can resume our discussion of the feedback situation of Figure 2.2.4. The feedback system in the transform domain is

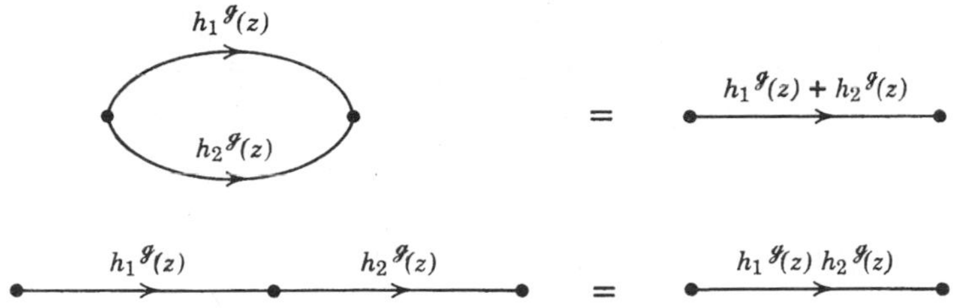


**Figure 2.3.3** Flow graph equivalents for parallel and series systems.

shown in Figure 2.3.4. Note that the identity systems now have a transfer function of one, an appropriate result. The output of each branch is equal to its input multiplied by its transfer function. Systems that share an output node still add their contribution at that node. The signal $e^{\mathscr{G}}(z)$ has two components, the input $f^{\mathscr{G}}(z)$ and the output of the feedback loop. The output of the loop is equal to the product of its input $e^{\mathscr{G}}(z)$ and its transfer function $h^{\mathscr{G}}(z)$. Thus we obtain

$$e^{\mathscr{G}}(z) = f^{\mathscr{G}}(z) + e^{\mathscr{G}}(z)h^{\mathscr{G}}(z). \tag{2.3.7}$$

Since $g^{\mathscr{G}}(z) = e^{\mathscr{G}}(z)$, we can write:

$$g^{\mathscr{G}}(z) = f^{\mathscr{G}}(z) + g^{\mathscr{G}}(z)h^{\mathscr{G}}(z),$$
$$g^{\mathscr{G}}(z)(1 - h^{\mathscr{G}}(z)) = f^{\mathscr{G}}(z)$$

and finally,

$$g^{\mathscr{G}}(z) = f^{\mathscr{G}}(z)\frac{1}{1 - h^{\mathscr{G}}(z)}. \tag{2.3.8}$$

The crucial operation necessary to express the output of a feedback system in terms of its input has been performed; in the transform domain, it turns out to be a simple division. A feedback loop has a transfer function equal to one over one minus the loop transfer function. Our victory over the feedback system is illustrated in Figure 2.3.5.

Figure 2.3.6 summarizes the correspondence between time and transform domain representations.

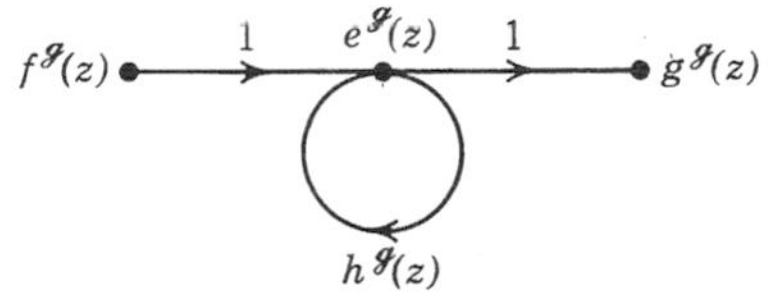


**Figure 2.3.4** Flow graph for the feedback system.

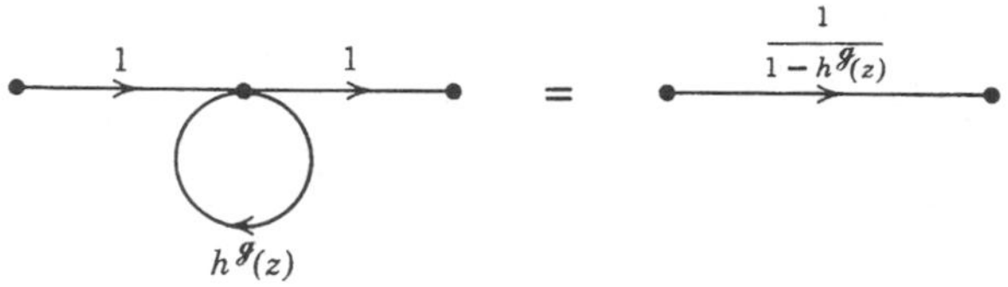


**Figure 2.3.5** Flow graph equivalent of the feedback system.

***A physical interpretation of graph transmissions***

We may abstract a general property of system transmissions and verify it in our three simple systems. We define a "route" in a complex system as a sequence of directed branch transmissions that lead from input to output, and the route transmission as the product of the branch transmissions that constitute the route. Our general property is this: the transmission of any graph is the sum of the route transmissions for all possible routes.

Applying this principle to the parallel and series systems of Figure 2.3.3, we find readily that the transmission of the parallel system is just the sum of the two route transmissions $h_1{}^g(z)$ and $h_2{}^g(z)$, and that the transmission of the series system is the transmission $h_1{}^g(z)h_2{}^g(z)$ of the single route that traverses both systems. When we come to the feedback system of Figure 2.3.4, however, we may have some difficulty in seeing what routes exist. The answer, according to the definition of route, is that every sequence of branches that leads from input to output is a route. Nothing is said to prevent a route's traversing the same node more than once

Figure 2.3.7 illustrates that there are, in fact, an infinite number of routes in the feedback system. First there is the route that does not traverse the feedback system at all. Then there is a route that traverses it exactly once; another that traverses it exactly twice, and so on. The transmission of the graph is, as we have said, the sum of these route transmissions, or

$$1 + h^g(z) + [h^g(z)]^2 + [h^g(z)]^3 + \cdots = \frac{1}{1 - h^g(z)} \tag{2.3.9}$$

where we have summed the infinite series to produce the result of Equation 2.3.8.

Although the interpretation of the graph transmission as the sum of the route transmissions is always applicable, it is usually of more use in theoretical discussions than in practical graph reduction. Fortunately, we have another reduction technique based on the route transmission interpretation that is eminently practical. We shall discuss it in Section 2.5.

***A simple feedback system revisited***

The example of Figure 2.2.5 can now be easily treated by using the result of Equation 2.3.8. Because the feedback system was a simple delay of one time

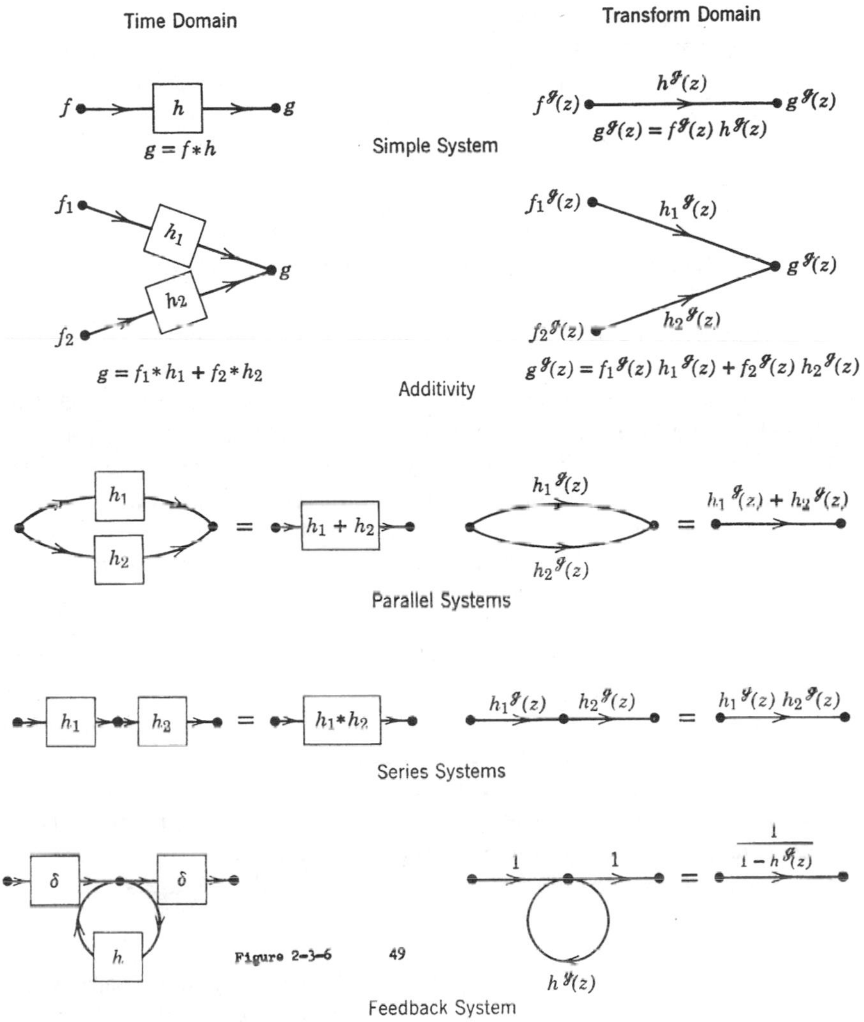


**Figure 2.3.6** Time and transform domain system representations.

period, the transfer function of the feedback loop is $z$. Therefore, the transfer function of the feedback system must be $1/(1 - z)$; its flow graph appears as Figure 2.3.8. Table 1.5.1 shows that the time function corresponding to this transform is the unit step function that is given in Figure 2.2.5. It will be interesting to investigate the properties of the system of Figure 2.3.8.

Suppose that this system is itself driven by a unit step input. What will the output be? Figure 2.3.9 shows the computation of the result. Since both the input function

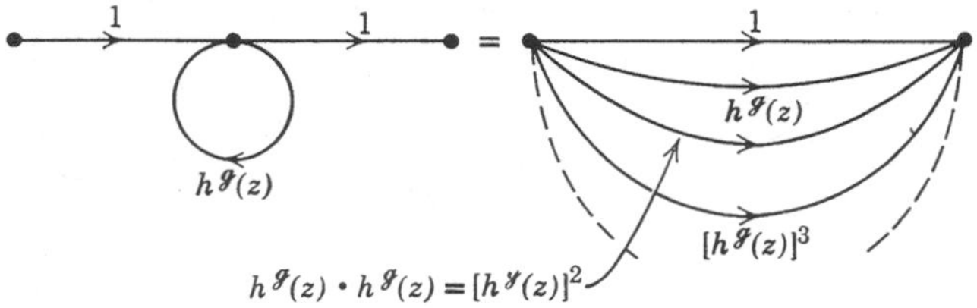


**Figure 2.3.7** Route structure of the feedback system.

and the impulse response of the system are unit step functions, both have transforms $1/(1 - z)$, and the output has the transform $1/(1 - z)^2$. Table 1.5.1 shows that this transform corresponds to a unit ramp time function that has been advanced by one time period. Consequently, the output is $g(n) = n + 1$ for $n \geq 0$, $g(n) = 0$ for $n < 0$. When we compare the output of this system with its input, we see that the action of the system can be described very simply. The output at any time is just the sum of all inputs up to that time, a property that is revealed to hold for an arbitrary input by relation 25 of Table 1.5.1. As a result, this particular system is called a "summer." Remember that it was created simply by using a delay of one unit in the feedback loop. (The reader may like to verify that the system with the transfer function $1 - z$ is properly called a "differencer.")

If two summers were placed in series, their joint transfer function would be $[1/(1 - z)]^2$. Therefore, their joint impulse response would be the output time function of Figure 2.3.9. A unit impulse applied to such a system would produce a linearly increasing output. Further cascading of summers would create a system that reacted even more energetically to a unit impulse input. These observations regarding summers are the discrete analogs of those that may be made about continuous variable control systems containing cascaded integrators.

***On raising rabbits***

An interesting and illuminating example of a feedback system occurs when we consider a certain rabbit population. Each pair of rabbits of this particular type reproduces exactly twice: one pair of offspring at the end of one time period and another pair at the end of the following time period. We can therefore draw the

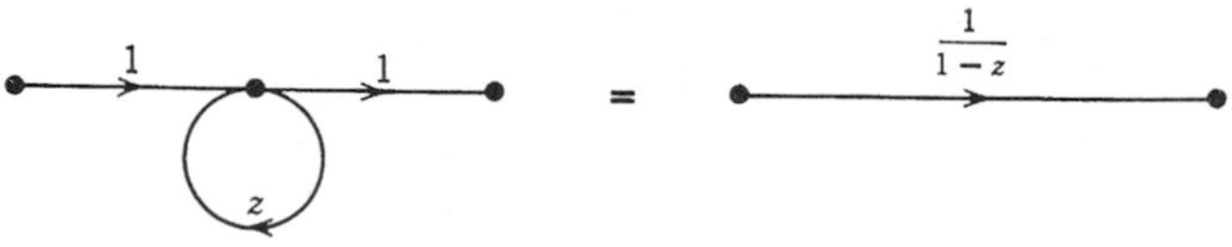


**Figure 2.3.8** The summer.

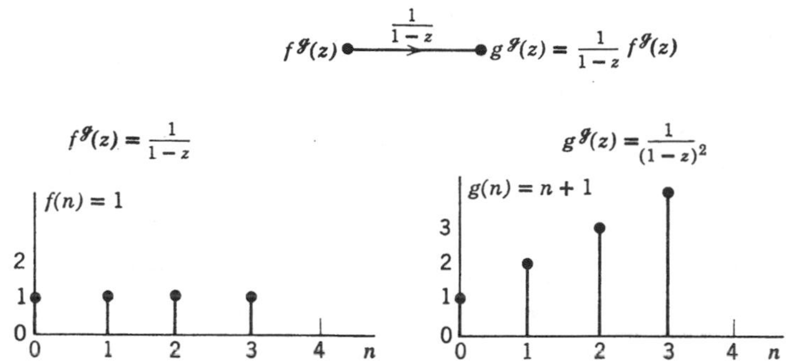


**Figure 2.3.9** Response of summer to step input.

impulse response of one rabbit pair as shown in Figure 2.3.10—the impulse response would have a geometric transform $z + z^2$. The problem we pose is this. Suppose that we start with one pair of rabbits of this type and that the rabbit pairs that descend from this pair in all future generations reproduce in the same way. How many rabbit pairs will we have after $n$ time periods?

To answer this problem we note that any rabbit pairs produced by a rabbit pair become future unit inputs to the system that occur after one and two time periods. Therefore, the transfer function $h^{\mathscr{g}}(z)$ of the entire rabbit system must be just the transmission of a simple feedback system whose feedback transmission is the transfer function $z + z^2$ appropriate to one rabbit pair. The flow graph for this system appears as Figure 2.3.11. We observe immediately that the transfer function for the rabbit system is

$$h^{\mathscr{g}}(z) = \frac{1}{1 - z - z^2}. \tag{2.3.10}$$

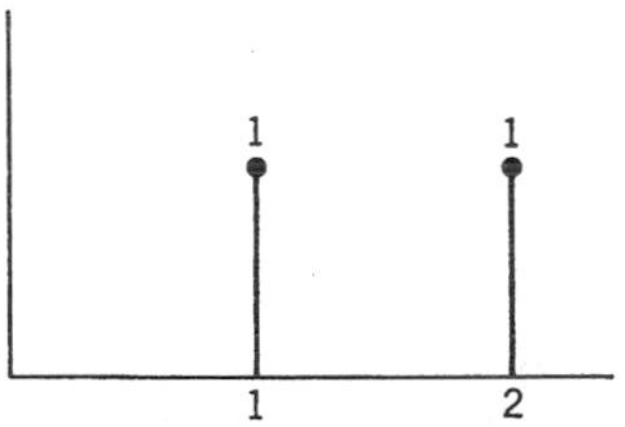

**Figure 2.3.10** Impulse response of one rabbit pair.

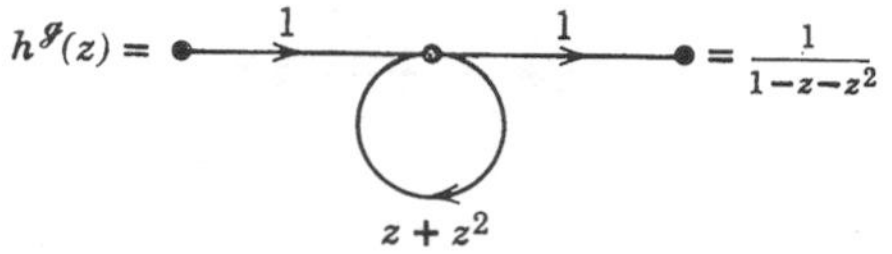


**Figure 2.3.11** Transfer function of rabbit system.

We have many ways for finding $h(n)$, the number of rabbit pairs after $n$ time periods that result from the original rabbit pair at time zero and its descendants. One way is to write Equation 2.3.10 in the form

$$h^g(z) = zh^g(z) + z^2h^g(z) + 1. \tag{2.3.11}$$

Then by inverse geometric transformation,

$$h(n) = h(n-1) + h(n-2) + \delta(n), \qquad n = 0, 1, 2, \ldots. \tag{2.3.12}$$

This equation implies

$$\begin{aligned} h(0) &= 1 \\ h(1) &= h(0) = 1 \\ h(n) &= h(n-1) + h(n-2), \qquad n = 2, 3, 4, \ldots. \end{aligned} \tag{2.3.13}$$

Thus $h(n)$ represents a sequence whose first two terms are each equal to 1 and whose succeeding terms are each the sum of the previous two terms. Thus the terms of the sequence are

$$1, 1, 2, 3, 5, 8, 13, 21, 34, 55, 89, 144, 233, \ldots. \tag{2.3.14}$$

We can expect to be overrun by rabbits.

Another way to obtain this result is to expand Equation 2.3.10 in a power series by dividing the denominator into the numerator,

$$\begin{array}{r|l} & 1 + z + 2z^2 + 3z^3 + \cdots \\ \hline 1 - z - z^2 & 1 \\ & \underline{1 - z - z^2} \\ & z + z^2 \\ & \underline{z - z^2 - z^3} \\ & 2z^2 + z^3 \\ & \underline{2z^2 - 2z^3 - 2z^4} \\ & 3z^3 + 2z^4 \end{array}$$

and then recognize that $h(n)$ is the coefficient of $z^n$ in the expansion.

**Fibonacci numbers.** The sequence $h(n)$ defined by Equation 2.3.12 and illustrated by 2.3.14 is a famous mathematical sequence called the Fibonacci numbers, for it was Fibonacci (*circa* 1175–1250) who first considered this problem in rabbit reproduction. It has been found useful in several situations whose importance transcends the raising of rabbits.

Let us now turn to the problem of finding a closed form expression for computing $h(n)$, the $n$th term of the sequence, directly. We can accomplish this by finding the inverse transform of Equation 2.3.10 in closed form. We first write the denominator of the transfer function as the product of simple factors,

$$h^{g}(z) = \frac{1}{1 - z - z^2} = \frac{1}{\left[1 - \left(\frac{1 + \sqrt{5}}{2}\right)z\right]\left[1 - \left(\frac{1 - \sqrt{5}}{2}\right)z\right]}, \qquad (2.3.15)$$

and then perform a partial fraction expansion,

$$h^{g}(z) = \frac{\frac{1}{2\sqrt{5}}(1 + \sqrt{5})}{1 - \left(\frac{1 + \sqrt{5}}{2}\right)z} + \frac{-\frac{1}{2\sqrt{5}}(1 - \sqrt{5})}{1 - \left(\frac{1 - \sqrt{5}}{2}\right)z}. \qquad (2.3.16)$$

Finally, we write the inverse geometric transform of each term to produce

$$\begin{aligned} h(n) &= \frac{1}{2\sqrt{5}}(1 + \sqrt{5})\left(\frac{1 + \sqrt{5}}{2}\right)^n - \frac{1}{2\sqrt{5}}(1 - \sqrt{5})\left(\frac{1 - \sqrt{5}}{2}\right)^n \\ &= \frac{1}{\sqrt{5}}\left(\frac{1 + \sqrt{5}}{2}\right)^{n+1} - \frac{1}{\sqrt{5}}\left(\frac{1 - \sqrt{5}}{2}\right)^{n+1} \qquad n = 0, 1, 2, \ldots. \qquad (2.3.17) \end{aligned}$$

We can readily verify by computation that this expression is just the closed form necessary to generate the sequence 2.3.14. Since $\frac{1}{2}(1 + \sqrt{5}) = 1.618$ and $\frac{1}{2}(1 - \sqrt{5}) = -0.618$, it is clear that when $n$ is large the second term in Equation 2.3.17 will tend to zero and the terms in the sequence will be given by

$$h(n) = \frac{1}{\sqrt{5}}\left(\frac{1 + \sqrt{5}}{2}\right)^{n+1}, \quad n \text{ large}. \qquad (2.3.18)$$

**Fibonacci fractions.** The Fibonacci fractions are a sequence defined as the ratio of successive Fibonacci numbers, $h(n)/h(n + 1)$, $n = 0, 1, 2, \ldots$. From sequence 2.3.14 the first few Fibonacci fractions are

$$\frac{1}{1}, \frac{1}{2}, \frac{2}{3}, \frac{3}{5}, \frac{5}{8}, \frac{8}{13}, \frac{13}{21}, \frac{21}{34}, \frac{34}{55}, \frac{55}{89}, \frac{89}{144}, \frac{144}{233}, \ldots. \qquad (2.3.19)$$

The Fibonacci fractions appear often in nature, as, for example, in the distribution of leaves around the circumference of a plant. Inspection of the sequence 2.3.19 reveals that the fractions appear to approach a limit as $n$ grows large. We can find the limit from the result of Equation 2.3.18 as

$$\lim_{n \to \infty} \frac{h(n)}{h(n + 1)} = \frac{2}{1 + \sqrt{5}} = \frac{\sqrt{5} - 1}{2} = 0.618. \qquad (2.3.20)$$

This limit is called the "golden mean," and was used by the ancients as the ratio of length to breadth of the "golden section," the rectangle that was supposedly most pleasing to the eye. We find the golden section in modern times in Mondrian's painting "Black, White, and Red," so perhaps the ancients were right.

## 2.4 FLOW GRAPH REDUCTION

Now that we thoroughly understand how to deal with parallel, series, and feedback connections in systems, can we reduce any complex multibranch system? The answer is a qualified "yes." The reason for the qualification is apparent in the example of Figure 2.4.1. In this example we see a quite respectable system that is not, at first glance, reducible by our three standard reductions. However, as we shall soon show, even this case may be reduced by using a little circumspection.

We note in passing that it is important in reducing the more complex systems that we clearly specify the input and output nodes. Usually this will be done by using an identity branch at both input and output as shown in Figure 2.4.1. However, in some cases, we shall omit the unity transfer function on these branches for simplicity.

### Evolutionary Reduction

To reduce the example of Figure 2.4.1, let us first consider the general form of the example, shown in Figure 2.4.2(a). Here a system with transfer function $h_1{}^{\mathscr{g}}(z)$ connects the input to the output; however, a system with transfer function $h_2{}^{\mathscr{g}}(z)$ connects the output to the input. The problem is to find the equivalent single branch system. We begin by duplicating the branch $h_1{}^{\mathscr{g}}(z)$, and taking the input for this new branch at the same point from which the original branch $h_1{}^{\mathscr{g}}(z)$ drew its input (Figure 2.4.2(b)). This tampering does not change the relationship between input and output, because none of the signals at the nodes of the original graph has been changed. Then we take the input for the branch $h_2{}^{\mathscr{g}}(z)$ from the output of the new $h_1{}^{\mathscr{g}}(z)$ branch (Figure 2.4.2(c)). Since the outputs produced by both $h_1{}^{\mathscr{g}}(z)$ systems are identical, no harm has been done. The two systems in the feedback loop are in series so their transfer functions can be multiplied to yield a single feedback transfer function $h_1{}^{\mathscr{g}}(z)h_2{}^{\mathscr{g}}(z)$ (Figure 2.4.2(d)). Finally, the results of our

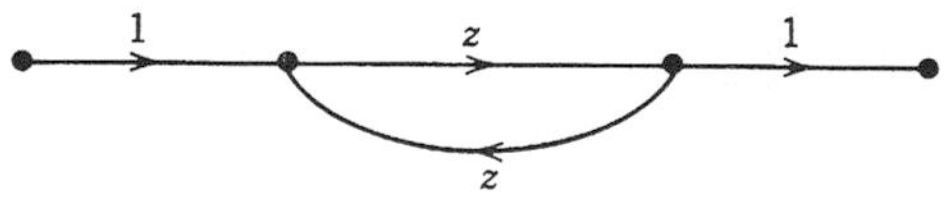


**Figure 2.4.1** An example.

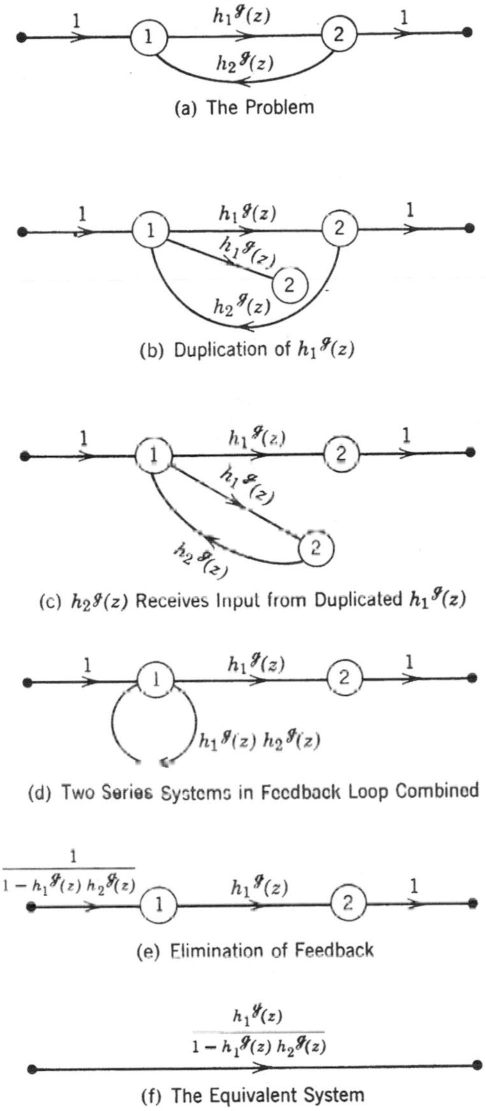
1
$h_1\mathcal{G}(z)$
1
$h_2\mathcal{G}(z)$
(a) The Problem
1
$h_1\mathcal{G}(z)$
1
$h_1\mathcal{G}(z)$
$h_2\mathcal{G}(z)$
(b) Duplication of $h_1\mathcal{G}(z)$
1
$h_1\mathcal{G}(z)$
1
$h_1\mathcal{G}(z)$
$h_2\mathcal{G}(z)$
(c) $h_2\mathcal{G}(z)$ Receives Input from Duplicated $h_1\mathcal{G}(z)$
1
$h_1\mathcal{G}(z)$
1
$h_1\mathcal{G}(z)\, h_2\mathcal{G}(z)$
(d) Two Series Systems in Feedback Loop Combined
$\frac{1}{1 - h_1\mathcal{G}(z)\, h_2\mathcal{G}(z)}$
$h_1\mathcal{G}(z)$
1
(e) Elimination of Feedback
$\frac{h_1\mathcal{G}(z)}{1 - h_1\mathcal{G}(z)\, h_2\mathcal{G}(z)}$
(f) The Equivalent System

**Figure 2.4.2** A reduction example.

standard feedback and series reductions are used (Figure 2.4.2(e), (f)) to obtain the equivalent single branch transfer function,

$$\frac{h_1^{\mathcal{G}}(z)}{1 - h_1^{\mathcal{G}}(z)h_2^{\mathcal{G}}(z)}.$$

Note that this expression represents the forward transfer function $h_1^{\mathcal{G}}(z)$ divided by one minus the loop transfer function.

*Node splitting*

Fortunately, we have at our disposal several ways to reduce a flow graph. For instance, the graph of Figure 2.4.2(a) could be reduced by "splitting" node 1. This procedure appears in Figure 2.4.3. The essential observation is that the signal

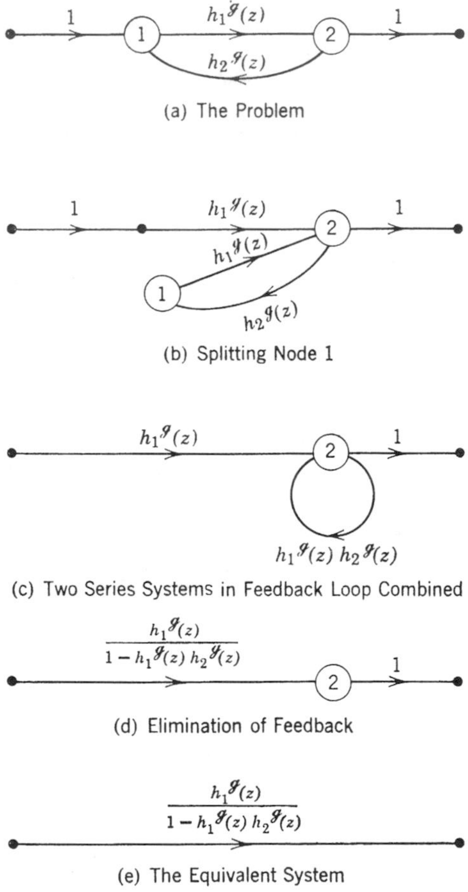


**Figure 2.4.3** An alternate reduction.

that passes through the branch labeled $h_1{}^g(z)$ in Figure 2.4.3(a) is the sum of the input signal operated upon by $h_1{}^g(z)$ and the signal through $h_2{}^g(z)$ operated upon by $h_1{}^g(z)$. Consequently, if we split node (1) in the way shown in Figure 2.4.3(b), we shall not have disturbed the relationship between the input and output of the over-all system. Of course, we must realize that the signal on neither of the two nodes generated by the split is the original signal at node 1.

We now proceed much as we did in Figure 2.4.2. The two branches in series with transfer functions $h_1{}^g(z)$ and $h_2{}^g(z)$ are combined in a single feedback loop (Figure 2.4.3(c)). Finally, this feedback loop is replaced by its equivalent (Figure 2.4.3(d)); the resulting system in Figure 2.4.3(e) is the same as that in Figure 2.4.2(f).

Although both forms of reduction have produced the same final result, it would be incorrect to assume that they are equivalent at the intermediate stages of reduction. For example, comparison of the graphs in Figures 2.4.2(d) and 2.4.3(c) shows that in the first reduction, we have maintained the integrity of node 1 and represented the system as a feedback loop around node 1 followed by a series system with transfer function $h_1{}^g(z)$. However, in the second reduction we have preserved the character of node 2 by constructing a feedback transfer function $h_1{}^g(z)h_2{}^g(z)$ around it and preceding the node by a series system with transfer function $h_1{}^g(z)$. Thus, the signal on the intermediate node is different in Figures 2.4.2(d) and 2.4.3(c). In Figure 2.4.2(d) it is the signal at node 1 in the original graph, but in Figure 2.4.3(c) it is the signal at node 2.

Usually when we perform a reduction of a system flow graph, we are not interested in the signal at internal nodes, and hence the possible loss of a node's identity at various stages of the reduction causes no concern. However, when we do have an interest in some particular internal node, we can always manage to carry out the reduction so that its nature is preserved until the final step of the reduction process.

***Route transmission sum method***

We might consider the route transmission sum method as a final way of reducing the system of Figure 2.4.2(a). By summing the transmission of all routes we find that the transmission of the graph is

$$\begin{aligned} h_1{}^g(z) + h_1{}^g(z)h_2{}^g(z)h_1{}^g(z) + h_1{}^g(z)[h_2{}^g(z)h_1{}^g(z)]^2 + \cdots \\ = h_1{}^g(z)\{1 + h_2{}^g(z)h_1{}^g(z) + [h_2{}^g(z)h_1{}^g(z)]^2 + \cdots\} \\ = \frac{h_1{}^g(z)}{1 - h_1{}^g(z)h_2{}^g(z)}. \end{aligned} \tag{2.4.1}$$

Thus the route transmission sum method is perhaps the most direct route to the answer. However, the confusion that may arise in applying it to more complicated systems makes more methodical procedures desirable.

## The Encysted System Concept

This is a timely place to make a further point about reduction in general. The fundamental idea becomes clear if we consider any portion of a graph (including the entire graph) obtained by enclosing any part of the graph within a boundary surface that passes through a set of nodes as shown in Figure 2.4.4. The boundary nodes are those at the boundary; the nodes outside the boundary are the external nodes. We require that the surface be drawn in such a way that any nodes contained within it be connected by branches only to each other or to boundary nodes; such nodes are called internal nodes. If we find after drawing the surface that a node which appeared internal is, in fact, connected to an external node, then that falsely-named internal node must be designated a boundary node. Of course, we always have the option of drawing the boundary surface through some other set of nodes.

We call this procedure the creation of an encysted system. The point of the procedure is this: We can make any reduction within an encysted system without changing the relationships between and among boundary and external nodes *provided that the transmission of the encysted system between every pair of boundary nodes remains unchanged.* In other words, if we meet this proviso, we can consider the encysted system separately, reduce it, perhaps by eliminating all internal nodes, and then reconnect the encysted system into the original graph without changing any relationships involving boundary and external nodes.

The encysting principle is a fundamental aid to the reduction of any graph. By applying it over and over again, we can reduce a graph of any size given enough time.

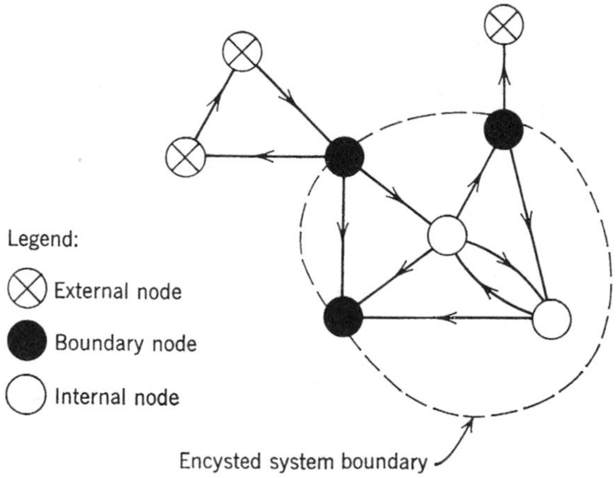


**Figure 2.4.4** Creation of an encysted system.

### *Isolated encysted systems*

We can use the encysting principle not only to reduce a particular graph, but also to gain a conceptual understanding of graph structures in general. For example, consider the system of Figure 2.4.5. The encysted system at the right of the figure receives inputs from the external system, but provides no outputs to it. As a consequence of the route transmission sum interpretation, we see that, regardless of the internal structure of the encysted system, the encysted system cannot affect, in any way, the relationships among the external nodes. We may, therefore, eliminate the encysted system from the system graph and, in addition, remove all the branches that lead to its boundary nodes without affecting in the least the external relationships. We shall call an encysted system of this type an isolated encysted system.

Another type of isolated encysted system that may arise is the encysted system that provides inputs to the external nodes, but receives no outputs from them. If we change the direction of the arrows on the branches that lead from the external nodes to the boundary nodes in Figure 2.4.5, we have constructed an isolated encysted system of this type. Here again, we can eliminate the isolated encysted system and the branches emanating from its boundary nodes to external nodes without changing the relationships in the external system in any way.

Thus, a useful step in flow graph reduction is to identify and remove the isolated encysted systems. Only encysted systems that are non-isolated need be retained during reduction. A non-isolated system is, of course, one that has at least one input from and at least one output to the external system.

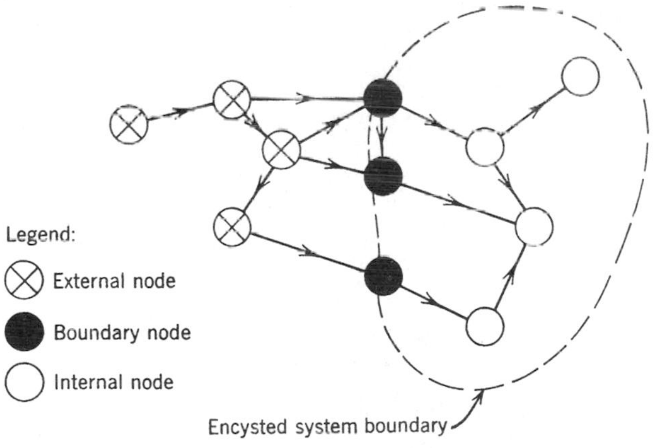


**Figure 2.4.5** An isolated encysted system.

## Pushing Transfer Functions

Two other special consequences of the encysting principle appear in Figure 2.4.6. Part (a) of this figure shows an encysted system that has been reduced by elimination of the internal node 2. We sometimes speak of this reduction as "pushing the transfer function $h_0^g(z)$ forward through node 2." Part (b) demonstrates a related reduction that we call "pushing the transfer function $h_0^g(z)$ backward through node 2." Of course, the external system relationships are unaffected in both cases.

Figure 2.4.7 demonstrates a more complicated case of pushing transfer functions through a node. The original encysted system consists of an internal node that receives inputs from two boundary nodes 1 and 2 through transfer functions $h_1^g(z)$ and $h_2^g(z)$, and that serves as the input node for a system with transfer function $h_3^g(z)$ whose output appears at another boundary node 3.

The first reduction shown is that of pushing the transfer function $h_3^g(z)$ back through the intermediate node. The resulting system has the same transfer function as the original between each pair of nodes 1, 2, and 3.

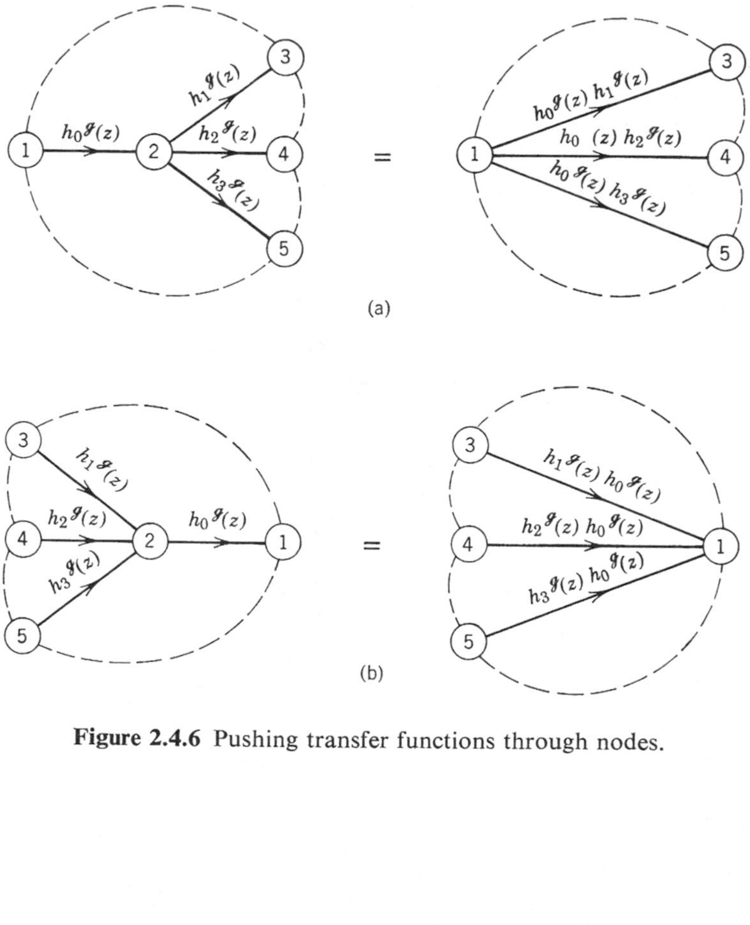


**Figure 2.4.6** Pushing transfer functions through nodes.

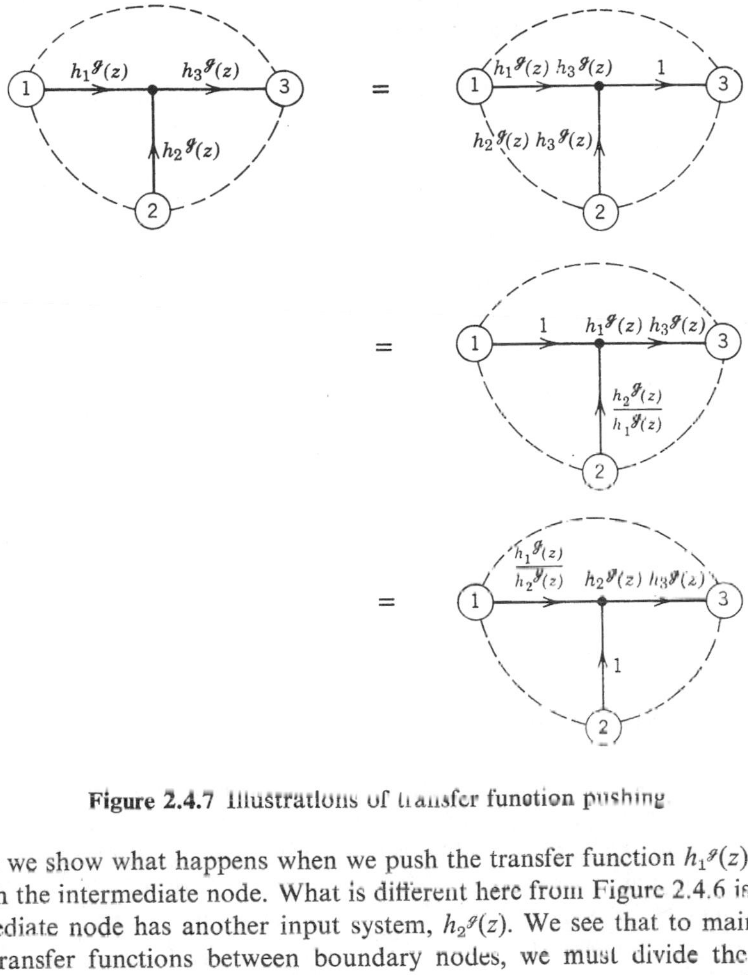


**Figure 2.4.7** Illustrations of transfer function pushing

Next we show what happens when we push the transfer function $h_1^{g}(z)$ forward through the intermediate node. What is different here from Figure 2.4.6 is that the intermediate node has another input system, $h_2^{g}(z)$. We see that to maintain the same transfer functions between boundary nodes, we must divide the transfer function $h_2^{g}(z)$ by $h_1^{g}(z)$ and multiply the transfer function $h_3^{g}(z)$ by $h_1^{g}(z)$.

A similar operation is necessary if we desire to push the transfer function $h_2^{g}(z)$ forward through the intermediate node as shown in the last part of the figure. The transfer function $h_1^{g}(z)$ of the other input to the intermediate node must be divided by $h_2^{g}(z)$, and the transfer function of the output branch must be multiplied by $h_2^{g}(z)$ to maintain the correct transfer functions between boundary nodes.

As we can see from this example, and from the general route transmission sum interpretation, the rule governing pushing transfer functions through nodes is as follows: If a transfer function $h^{g}$ is pushed forward through a node, all output branches of that node must have their transfer functions multiplied by $h^{g}$ while all input branches, (including the $h^g$ branch) must have their transfer functions divided by $h^g$; conversely, if a transfer function is pushed backward through a node, all output branches of that node (including the $h^g$ branch) must have their transfer functions divided by $h^g$ while all input branches must have their transfer functions multiplied by $h^g$. Of course, in both of these operations, the signal on the node is changed.

Now that we understand the principle of pushing transfer functions through nodes, we can reduce the system of Figure 2.4.2(a) quite easily. All we do is push the transfer function $h_1^g(z)$ forward through node 2 to obtain Figure 2.4.2(d) directly. Alternately, we can push $h_1^g(z)$ backward through node 1 to produce Figure 2.4.3(b).

**Example**

In the course of this discussion, we have solved in principle the problem posed in Figure 2.4.1. By taking $h_1^g(z) = h_2^g(z) = z$, the reduction of Figure 2.4.2 shows us that the transfer function for the system in Figure 2.4.1 is $z/(1 - z^2)$. This transform can be expanded in a power series, as follows:

$$\frac{z}{1 - z^2} = z + z^3 + z^5 + z^7 + \cdots. \tag{2.4.2}$$

An examination of the coefficients of this expansion indicates that the time function corresponding to this transform takes on the value one at times 1, 3, 5, 7, . . . and zero everywhere else. This impulse reponse is shown graphically in Figure 2.4.8.

The impulse response could also have been obtained by inspection of the flow graph. If a unit impulse were applied to the input, the output would not respond immediately, but the output would become 1 after a delay of one time interval. The input would no longer be excited, so after another delay of one period the output would be zero. However, the delay system leading from output back to

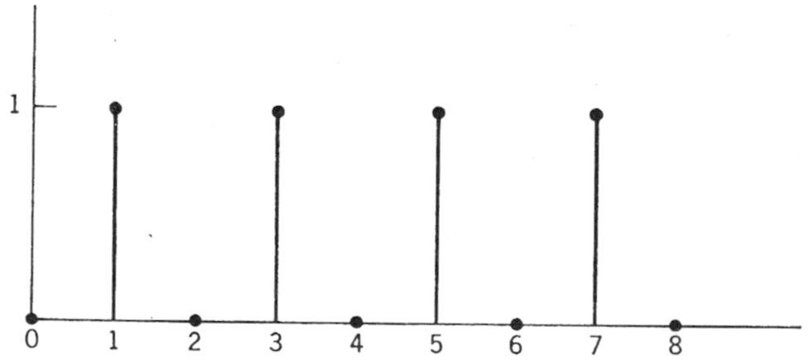


**Figure 2.4.8** Impulse response of system in Figure 2.4.1.

input would have produced a 1 at the input at the end of the second period. This 1 would then appear at the output after three periods had passed. As a result, an output of 1 is produced on every odd integer.

## Reduction of Large Graphs

As we might expect, the reduction of very large graphs is a major undertaking. Reducing a graph of even moderate size can be tedious at its best. A typical problem is the one shown in Figure 2.4.9. For convenience, we use lower case letters to represent the transmissions of the branches, and the symbol $\vec{t}$ for the transmission of the graph. The evolutionary reduction technique in the figure follows our previous work and requires no comment.

However, let us solve the same problem by the route transmission sum method. We begin by noting that if the branch with transmission $b$ did not exist, then the

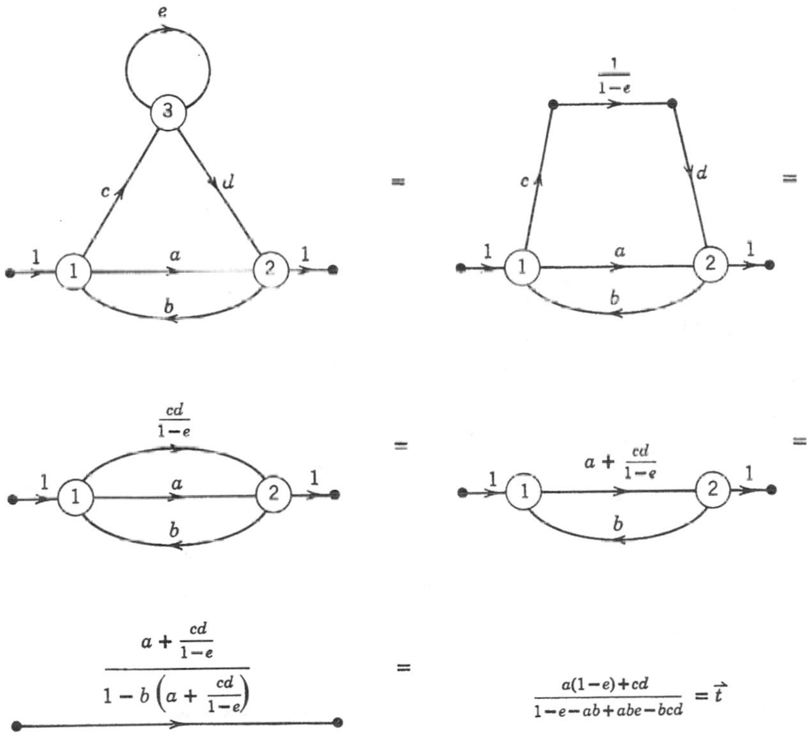


**Figure 2.4.9** A reduction example.

route transmission sum method would give as the transmission $\vec{t}_{b'}$ of the resulting graph

$$\begin{aligned}\vec{t}_{b'} &= a + cd + ced + ce^2d + \cdots \\ &= a + cd(1 + e + e^2 + \cdots) \\ &= a + \frac{cd}{1 - e}. \end{aligned} \tag{2.4.3}$$

The routes of the original graph are composed of the routes of this graph plus the routes generated by passing through the branch with transmission $b$ 1, 2, 3, . . . times. Thus the transmission of the original graph $\vec{t}$ is

$$\begin{aligned}\vec{t} &= \vec{t}_{b'}[1 + b\vec{t}_{b'} + (b\vec{t}_{b'})^2 + \cdots] \\ &= \frac{\vec{t}_{b'}}{1 - b\vec{t}_{b'}}. \end{aligned} \tag{2.4.4}$$

When we write this expression using Equation 2.4.3 for $\vec{t}_{b'}$, we find

$$\begin{aligned}\vec{t} &= \frac{a + \dfrac{cd}{1 - e}}{1 - b\left(a + \dfrac{cd}{1 - e}\right)} \\ &= \frac{a(1 - e) + cd}{1 - e - ab + abe - bcd} \end{aligned} \tag{2.4.5}$$

which is in complete agreement with the results of the evolutionary reduction.

After viewing these approaches to large graph reduction, we see that it would be very desirable to have a reduction technique that avoided the graphical manipulation necessary with our present methods. Fortunately, such a technique is available; it was developed to simplify the analysis of physical control systems. This approach allows us to reduce a graph like the one in Figure 2.4.9 by inspection. We shall now present this method without proof.

## 2.5 AN INSPECTION METHOD FOR FLOW GRAPH REDUCTION

We shall illustrate our procedure with the example just worked, as shown in Figure 2.5.1. The first step in the inspection procedure is to identify the "simple loops" of the original graph. A simple loop is defined to be a closed string of branches head-to-tail that touches no node more than once. The graph under consideration has only three simple loops: $(a, b)$, $(e)$, and $(b, c, d)$. The "loop product" for these simple loops is defined to be the *negative* product of the transmissions of all the branches included in the loop. For our example, these loop products are $-ab$, $-e$, and $-bcd$, as shown in Figure 2.5.1.

$$\Delta = 1 + \text{sum of loop products} = 1 - e - ab - bcd + abe$$

| Path index $k$ | Graphical representation of path $k$ | Path transmission for path $k$ $\bar{t}_k$ | 1 + sum of loop products of loops not touched by path $k$ $\Delta^k$ |
|---|---|---|---|
| 1. | ①——ᵃ——② | $u$ | $1 - e$ |
| 2. | ③ ↙c ↘d ① ② | $cd$ | $1$ |

$$\text{Graph transmission } \bar{t} = \frac{\sum\limits_{k \in P} \bar{t}_k \Delta^k}{\Delta} = \frac{a(1-e) + cd}{1 - e - ab - bcd + abe}$$

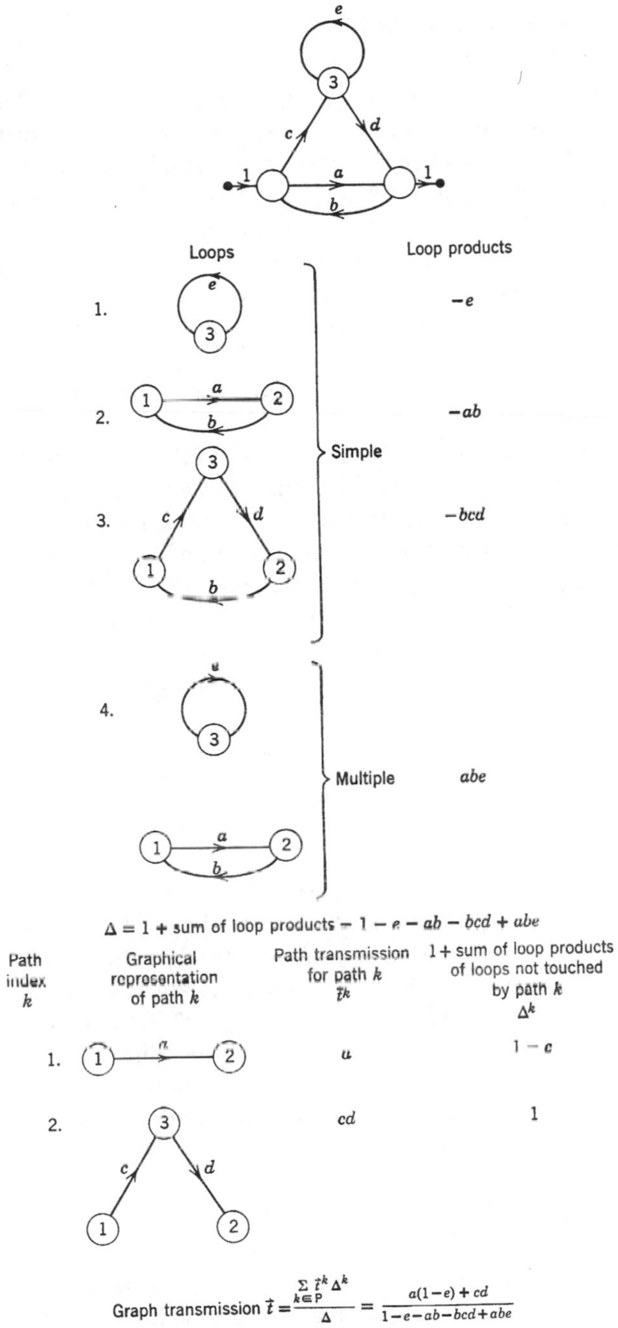

**Figure 2.5.1** Inspection method for flow graph reduction.

Next we must consider multiple loops. These are sets of simple loops that have no nodes in common. This type of loop may occur in pairs, triplets, etc., up to $N$-tuples, where $N$ is the number of nodes in the graph. In our example, there is only one multiple loop, the loop $(a, b; e)$. A loop product is computed for each multiple loop by multiplying together the loop products of the two or more simple loops of which the multiple loop is composed. The multiple loop $(a, b; e)$ has therefore the loop product $abe$. We then calculate a quantity $\Delta$ equal to one plus the sum of the loop products of all loops in the graph, simple or multiple. For the example, $\Delta = 1 - e - ab - bcd + abe$.

We now define a new graph component called a "path." A path is a string of branches that leads from the input to the output in the direction of the branch arrows, but touches no node more than once. Paths are therefore a subset of routes. Our flow graph has two paths, $a$ and $cd$. For each path $k$ we compute a path transmission $\vec{t}^k$ equal to the product of the transmissions of the branches in the path. We define $\mathsf{P}$ as the set of paths associated with the transmission. Finally, a quantity $\Delta^k$ equal to one plus the sum of the loop products of all loops that share no node with path $k$ is determined. For our example, path number 1 touches all loops but loop 2 with loop product $-e$; therefore, $\Delta^1 = 1 - e$. Since path 2 traverses every node in the graph, it touches all loops and $\Delta^2 = 1$.

When all this preliminary work has been completed, the actual writing of the transmission is almost trivial. The transmission of the graph is given by

$$\vec{t} = \frac{1}{\Delta} \sum_{k \in \mathsf{P}} \vec{t}^k \, \Delta^k. \tag{2.5.1}$$

Comparison of Figures 2.4.9 and 2.5.1 reveals that the two computations do, in fact, agree. Of course, one may ask at this point about whether we have not done more work with the so-called inspection method than we did with our more primitive techniques. The answer would be yes, if we actually drew all the loops, paths, and made tables of path transmissions and loop products. In practice, the final transmission expression for the whole graph is written directly by inspection of its form; no intermediate steps are required. It is usually convenient to write the denominator of the transmission expression first and then to use parts of this calculation in writing the numerator.

### The Graph Transmission Matrix

If we refer to the flow graph reduced in Figure 2.5.1 and imagine it without the input and output branches indicating that the transmission between node 1 and node 2 is required, we see that there is the possibility of selecting any node as the input node, any node as the output node, and then finding the transmission of the resulting graph. In our following work we shall often want to consider several different transmissions of the same flow graph, so it will be to our benefit to have a more appropriate general notation.

We shall define $\vec{t}_{ij}$ as the transmission of a flow graph from node $i$ to node $j$, that is, when $i$ is considered as the input node, $j$ as the output node. We extend our path notation by defining $\vec{t}_{ij}^{\,k}$ as the path transmission of the $k$th path from node $i$ to node $j$ and $\mathsf{P}_{ij}$ as the set of paths leading from node $i$ to node $j$. When $i = j$, $\mathsf{P}_{ii}$ consists of only one path and for this path $\vec{t}_{ii}^{\,1} = 1$.

The notation extends directly to the loop products. The quantity $\Delta_{ij}^{k}$ is equal to one plus the sum of the loop products of all loops that share no node with the $k$th path from node $i$ to node $j$. The quantity $\Delta$, which is equal to one plus the sum of the loop products of all loops in the graph, is the same regardless of which transmission $\vec{t}_{ij}$ we are seeking. Finally, we can write the $ij$th transmission as

$$\vec{t}_{ij} = \frac{1}{\Delta} \sum_{k \in \mathsf{P}_{ij}} \vec{t}_{ij}^{\,k} \Delta_{ij}^{k}. \tag{2.5.2}$$

If the flow graph has $N$ nodes, then there are $N^2$ possible transmissions. We sometimes choose to display these in a square graph transmission matrix $\vec{T}$ whose $ij$th element is the transmission $\vec{t}_{ij}$. The graph transmission matrix has important uses in our future work.

The question naturally arises as to why the inspection method works. Put most simply, it is an orderly procedure for summing all the route transmissions in the route transmission sum interpretation of a graph transmission. The proof that the procedure works is not constructive in pursuing our main aim in this work, and so we shall omit it.†

In summary, the inspection method of reduction considerably enhances the flow graph as an investigational tool. We can easily determine the effects of a change in graph structure on the transmission expression. The reader might like to verify that the inspection method is consistent with the results we have obtained in all examples up to this point.

## 2.6 AN INVENTORY SYSTEM EXAMPLE

To make the theory we have developed more tangible, let us examine an extremely simple inventory system. In this system sales and production are recorded at monthly intervals. Let $s(n)$ and $p(n)$ be, respectively, the sales and production during the $n$th month. The inventory at the end of the $n$th month, $i(n)$, will be

† For a proof, see the book by Mason and Zimmermann listed in the references.

equal to the cumulative difference between sales and production from month 0 through month $n$. Thus,

$$i(n) = \sum_{j=0}^{n} [p(j) - s(j)]. \tag{2.6.1}$$

We shall assume that production, sales, and inventory are measured *relative to some reference value* to obviate any physical requirement that these quantities be non-negative. Naturally, it will be necessary to limit the conclusions we draw from this model to the case where the fluctuations in these variables are small in comparison with their customary values.

The inventory control rule that we shall use for this system is that $p(n)$, the production in month $n$, is to be proportional to the negative sum of the inventories at the end of the two preceding months, $n - 1$ and $n - 2$. If we take the proportionality constant to be $\alpha$, then we can write

$$p(n) = -\alpha[i(n-1) + i(n-2)]. \tag{2.6.2}$$

This system is representative of a large class of commonly encountered control systems in which prediction of production requirements is based on average inventory levels. If $\alpha$ is taken to be 1/2 in Equation 2.6.2, then production equals the amount by which average inventory over the past two months is less than the desired value of this quantity.

## Block Diagram Representation

We can represent this system by a block diagram relating $s$, $p$, and $i$, if we first rewrite Equation 2.6.1 in the form

$$i(n) = i(n-1) + p(n) - s(n). \tag{2.6.3}$$

Figure 2.6.1 shows a block diagram implied by Equations 2.6.2 and 2.6.3. The $s$ vector has been shown entering at the left because it is the input to the system; $p$ is shown at the right as the system output. The difference between $s$ and $p$ at any time undergoes a change in sign and is then presented as one contribution to the inventory signal $i$. The other contribution is the signal at $i$ the time before, as indicated by the simple unit delay system $\delta^1$. The inventory signal is passed through two parallel systems providing, respectively, one and two units of delay. It is then multiplied by $-\alpha$ and becomes the production signal. Every implication of the defining equations is revealed by this graph; it is, in every sense, an alternate representation of Equations 2.6.2 and 2.6.3. In fact, it could have been drawn directly from the statement of the problem without writing any equations at all.

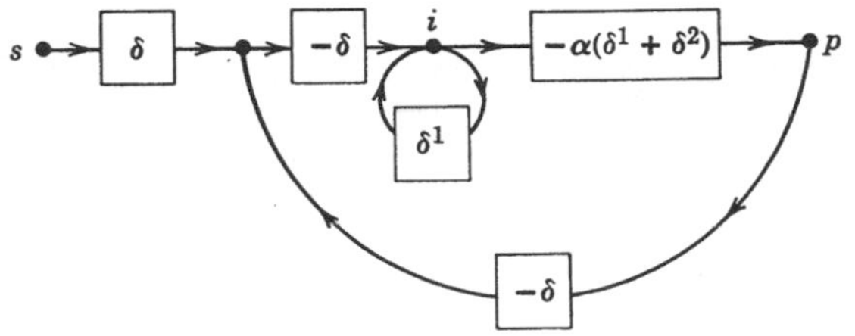


**Figure 2.6.1** Block diagram of inventory system.

The ability to depict system relationships by inspection is one of the major advantages of the graphical approach.

## Transform Analysis

As we might expect, it will be informative to analyze this system by using transformed vectors rather than the original signals. In our usual notation $p^g(z)$, $s^g(z)$, and $i^g(z)$ will be the transformed production, sales, and inventory. A sequence of flow graphs based on Figure 2.6.1 is shown in Figure 2.6.2. As a first step in simplification, the feedback delay around node $i$ is replaced by a system with transfer function $1/(1 - z)$ preceding node $i$. This shows that the inventory signal is simply the result of a summer operating on the difference between production and sales. In fact, the whole flow graph could also have been developed directly from the defining equations and knowledge of the operational characteristics of systems with transfer functions like $1/(1 - z)$, $z^2$, etc.

To continue the reduction, the flow graph is rearranged to place it in the form of a feedback loop around node $i$ plus series systems. Finally, this feedback is replaced by its equivalent forward transmission to produce a form of the flow graph without feedback that contains only the three nodes corresponding to sales, inventory, and production. The transfer functions between sales and inventory and between sales and production are then:

$$\frac{i^g(z)}{s^g(z)} = \frac{-1}{1 - (1 - \alpha)z + \alpha z^2}, \tag{2.6.4}$$

$$\frac{p^g(z)}{s^g(z)} = \frac{\alpha z(1 + z)}{1 - (1 - \alpha)z + \alpha z^2}. \tag{2.6.5}$$

These transfer functions allow us to calculate the inventory and production response to an arbitrary sales signal. If we find the impulse responses associated with

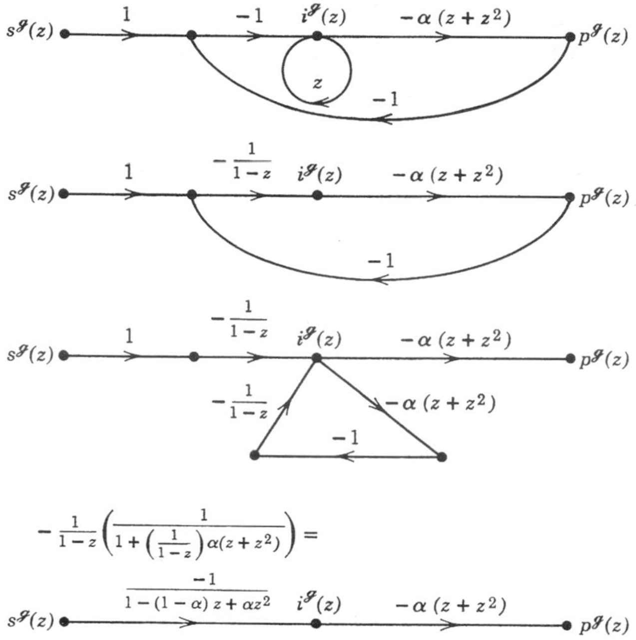


**Figure 2.6.2** Equivalent flow graphs for inventory system.

these transforms, then we can find the responses to any other input sequence by using the superposition techniques described earlier.

## Recursive Evaluation

Recursive evaluation is the easiest computational method for determining impulse responses. We can perform recursive evaluation even without developing the flow graph by returning to the basic Equations 2.6.2 and 2.6.3 that govern the system. The first equation,

$$p(n) = -\alpha[i(n-1) + i(n-2)], \tag{2.6.6}$$

allows us to compute the production in any time period from the inventory during the previous two time periods. The second equation,

$$i(n) = i(n-1) + p(n) - s(n), \tag{2.6.7}$$

shows that the inventory in any time period is just the previous inventory plus the difference between production and sales.

In Table 2.6.1 we apply these results to find the impulse response of the example for $\alpha = 1/2$ using recursive evaluation. We begin by setting the sales equal to 1 in time period 0 and zero at all other times, and by setting both inventory and production equal to zero for negative times. Then according to Equation 2.6.6, production at time zero is equal to $-1/2$ times the sum of the inventories at time $-1$ and $-2$; namely, zero. Inventory at time zero is the sum of previous inventory plus the difference between production and sales at time zero, or $-1$.

As we move on to time period 1, we note that sales are zero from this point onward, so that we need no longer consider the term representing sales in our equations. Production in time period 1 is minus 1/2 the sum of inventory in periods 0 and $-1 = -(1/2)(-1 + 0) = 1/2$. Inventory at time 1 is the sum of inventory at time 0 and production in time period $1 = -1 + 1/2 = -1/2$.

At time 2 we find

$$p(2) = (1/2)[i(1) + i(0)] = -(1/2)[-1/2 - 1] = 3/4$$

$$i(2) = i(1) + p(2) - s(2) = -1/2 + 3/4 - 0 = 1/4.$$

The remainder of the entries in the table are analogously computed. We can therefore extend computation of the production and inventory impulse responses as far in time as we like.

Figure 2.6.3 shows the system impulse responses graphically. Both the inventory and production impulse responses appear to have decaying, oscillatory envelopes. Such oscillation is not unusual in systems containing delays.

### Recursive Equations from Transforms

Note that Equations 2.6.6 and 2.6.7 relate production to inventory, and then inventory to previous inventory and current production and sales. In most circumstances, we find it more convenient to develop impulse responses recursively

**Table 2.6.1** Recursive Evaluation of Impulse Responses ($\alpha = 1/2$)

| Time Period $n$ | Sales $s(n)$ | Production $p(n)$ | Inventory $i(n)$ |
|---|---|---|---|
| <0 | 0 | 0 | 0 |
| 0 | 1 | 0 | $-1$ |
| 1 | 0 | $1/2 = 0.500$ | $-1/2 = -0.500$ |
| 2 | 0 | $3/4 = 0.750$ | $1/4 = 0.250$ |
| 3 | 0 | $1/8 = 0.125$ | $3/8 = 0.375$ |
| 4 | 0 | $-5/16 = -0.313$ | $1/16 = 0.063$ |
| 5 | 0 | $-7/32 = -0.219$ | $-5/32 = -0.156$ |
| 6 | 0 | $3/64 = 0.047$ | $-7/64 = -0.109$ |

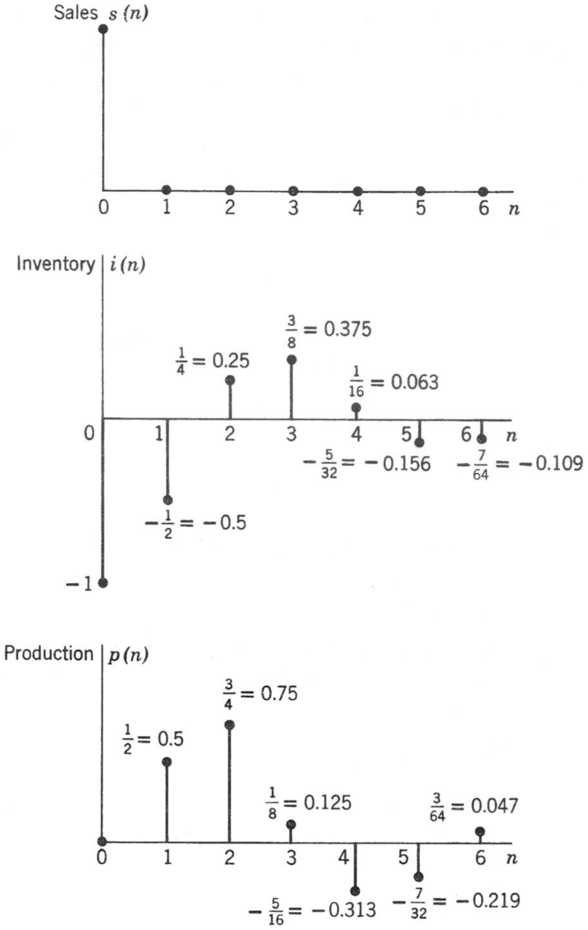


**Figure 2.6.3** The inventory and production responses to an impulse in sales ($\alpha = 1/2$).

by using an equation that relates the output directly to the input; in this case, production directly to sales, or inventory directly to sales. Fortunately, we have a means of achieving this end with little effort.

### *Effect of sales on production*

All we do to determine the effect of sales on production is write the transfer function of Equation 2.6.5 that relates production to sales, and then manipulate it until the quantity $p^{g}(z)$ is expressed in terms of all other quantities, including itself multiplied by powers of $z$. Thus we obtain

$$\frac{p^{g}(z)}{s^{g}(z)} = \frac{\alpha z(1 + z)}{1 - (1 - \alpha)z + \alpha z^2}$$

$$p^{g}(z)[1 - (1 - \alpha)z + \alpha z^2] = \alpha z(1 + z)s^{g}(z)$$

$$p^{g}(z) - (1 - \alpha)zp^{g}(z) + \alpha z^2 p^{g}(z) = \alpha z s^{g}(z) + \alpha z^2 s^{g}(z)$$

$$p^{g}(z) = (1 - \alpha)zp^{g}(z) - \alpha z^2 p^{g}(z) + \alpha z s^{g}(z) + \alpha z^2 s^{g}(z). \qquad (2.6.8)$$

Now we can invert this equation immediately by recalling that any transform multiplied by $z^m$ represents the time function delayed by $m$ periods. We find

$$p(n) = (1 - \alpha)p(n - 1) - \alpha p(n - 2) + \alpha s(n - 1) + \alpha s(n - 2). \qquad (2.6.9)$$

The production in the $n$th period is related by this equation only to production and sales in the two preceding periods. For $\alpha = 1/2$, the equation becomes

$$p(n) = (1/2)[p(n - 1) - p(n - 2)] + 1/2[s(n - 1) + s(n - 2)]. \qquad (2.6.10)$$

Production in the $n$th time period is equal to one-half the difference of production in the last two periods plus one-half the sum of sales in those periods. Reference to Table 2.6.1 reveals that this equation reproduces our earlier results, this time without any reference to inventory.

***Effect of sales on inventory***

If we are interested in the effect of a sales pattern on inventory, we can apply the same method to finding the inventory impulse response. We start with the inventory-sales transfer function of Equation 2.6.4 and rearrange it,

$$\frac{i^{g}(z)}{s^{g}(z)} = \frac{-1}{1 - (1 - \alpha)z + \alpha z^2}$$

$$i^{g}(z)[1 - (1 - \alpha)z + \alpha z^2] = -s^{g}(z)$$

$$i^{g}(z) - (1 - \alpha)zi^{g}(z) + \alpha z^2 i^{g}(z) = -s^{g}(z)$$

$$i^{g}(z) = (1 - \alpha)zi^{g}(z) - \alpha z^2 i^{g}(z) - s^{g}(z). \qquad (2.6.11)$$

Inverse transformation produces

$$i(n) = (1 - \alpha)i(n - 1) - \alpha i(n - 2) - s(n). \qquad (2.6.12)$$

The inventory in period $n$ depends on the inventory in the two preceding periods and on sales in period $n$. With $\alpha = 1/2$, we have

$$i(n) = (1/2)[i(n - 1) - i(n - 2)] - s(n). \qquad (2.6.13)$$

The inventory in period $n$ is equal to one-half the difference of the inventory in the last two periods less the sales in period $n$. Table 2.6.1 verifies this relationship.

Using transfer functions is thus an efficient method of developing the difference equations used in recursive evaluation of impulse responses. Recursive evaluation is, of course, naturally suited to computer analysis.

### Step Response

Once we have found system impulse responses, then responses to a different input signal, such as a step, may also be easily computed. According to the superposition relation, the step response at any time is just equal to the sum of the impulse responses up to that time. When we apply this result to the time functions of Figure 2.6.3, we obtain those of Figure 2.6.4. The step responses are, of course, also oscillatory. Notice that the inventory signal is decaying to a value of $-1$. This means that when such a system is subjected to the sudden change in sales rate represented by a step, the initial inventory discrepancy will never be eliminated. On the other hand, the production signal does approach its proper value of 1 as time passes. If maintenance of desired inventory levels under step changes in sales rate were an important goal for the managers of this system, they would be very much dissatisfied with its performance.

### Analytic Results

Although step-by-step observation of flow through the graph will produce impulse responses, it is often desirable for purposes of analysis to obtain analytic expressions for the impulse response functions. Such expressions can be derived by taking the inverse geometric transform of the system transfer function. Let us find a closed form expression for $p(n)$ by this method. A useful transform pair derivable from Table 1.5.1 by using de Moivre's theorem is

$$\begin{aligned} f(n) &= r^n(A\cos n\theta + B\sin n\theta) \qquad n = 0, 1, 2, \ldots \\ f^g(z) &= \frac{A - r(A\cos\theta - B\sin\theta)z}{1 - 2rz\cos\theta + r^2z^2}. \end{aligned} \tag{2.6.14}$$

By using the fact that a delay of one time period in the time function corresponds to multiplication by $z$ in its transform, we can also write the pair:

$$\begin{aligned} f(n) &= r^{n-1}(A\cos(n-1)\theta + B\sin(n-1)\theta) \qquad n = 1, 2, 3, \ldots \\ f^g(z) &= \frac{Az - r(A\cos\theta - B\sin\theta)z^2}{1 - 2rz\cos\theta + r^2z^2} \end{aligned} \tag{2.6.15}$$

This geometric transform is of the same form as the transfer function between sales and production shown in Equation 2.6.5. If $\alpha = 1/2$, we take $r = 1/\sqrt{2}$,

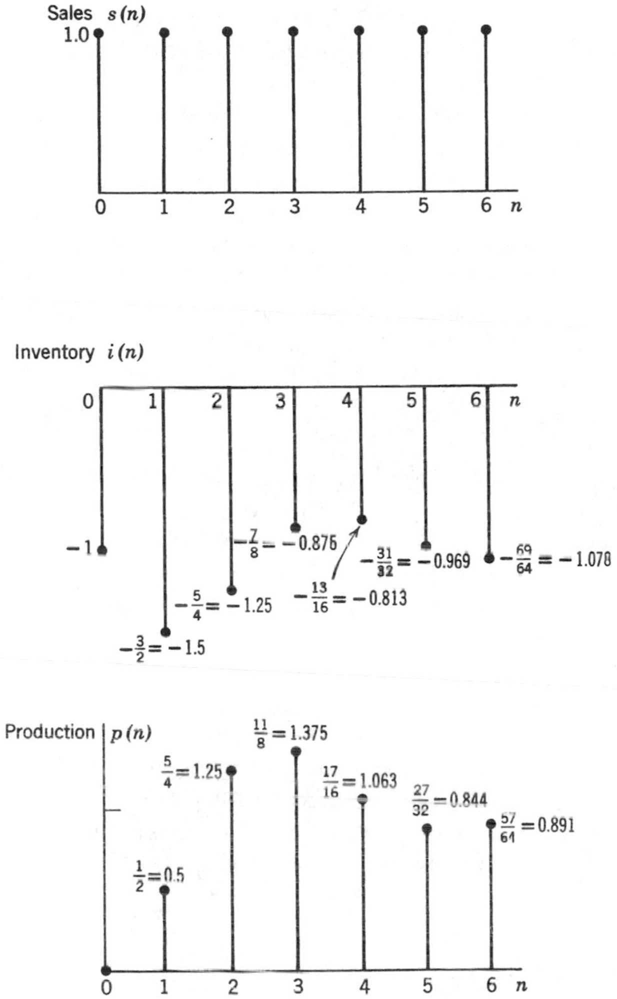


**Figure 2.6.4** The inventory and production responses to a step in sales ($\alpha = 1/2$).

$\theta = \cos^{-1} 1/2\sqrt{2}$, $A = 1/2$, $B = 5/2\sqrt{7}$, and write the production impulse response as

$$p(n) = \left(\frac{1}{\sqrt{2}}\right)^{n-1}\left[\frac{1}{2}\cos(n-1)\theta + \frac{5}{2\sqrt{7}}\sin(n-1)\theta\right]. \qquad (2.6.16)$$

Equation 2.6.16 brings the decaying sinusoidal nature of $p(n)$ into full view. The period of the sinusoid is $2\pi/\theta = 5.2$. The sine wave envelope of $p(n)$ thus has a zero crossing every 5.2 time intervals, and illustrates the fundamental frequency of this inventory system. This result is consistent with Figure 2.6.3; the envelope decays in amplitude by $1/\sqrt{2} = 0.707$ per time interval. The frequency and decay rate are the same for the inventory and production impulse responses in Figure 2.6.3 and for the inventory and production step responses of Figure 2.6.4.

We could continue the analysis of this example in more detail, but the main points have been made. It is easy to derive the flow graph of an inventory system of the class we are considering and it is possible to derive from it all relevant transfer functions. Since these transfer functions characterize the system, they constitute a complete solution to the analysis problem.

## 2.7 MATRIX SYSTEMS

Up to this point we have discussed systems that have only one input and one output. It is not uncommon for a system to have multiple inputs, multiple outputs, or both. We shall need to extend our previous discussion to include this situation; we shall call systems of this class "matrix systems."

We can consider any system as a matrix system. For instance, in the inventory example we have just treated, the input was sales and one output was production. This interpretation would correspond to the scalar, or one input-one output matrix system. Another scalar matrix system would have the single input sales and the single output inventory. However, if we consider sales as the input and both inventory and production as outputs, then we have a one input-two output matrix system. Another interesting matrix system in this example would consist of sales and production as inputs, inventory as the output, a two input-one output system. We shall explore these possibilities in detail after developing a terminology for matrix systems.

### The Matrix System Model

Let us consider as a general model the system with $N$ inputs and $M$ outputs. We shall use an index $i$ ($i = 1, 2, \ldots, N$) for the inputs and an index $j$ ($j = 1, 2, \ldots, M$) for the outputs. Thus, $f_i(n)$ is the value of the $i$th input at time $n$, while $g_j(n)$ is the value of the $j$th output at that time. Matrix systems have impulse responses just like simple systems; in fact, they have a whole matrix of impulse responses. We shall use $h_{ij}(n)$ to indicate the response at output $j$ at time $n$ as a result of a unit impulse applied to the $i$th input at time 0 with the understanding that all inputs with the exception of $i$ remain unexcited, that is,

$$h_{ij}(n) = g_j(n) \quad \text{when } f_i(n) = \delta(n) \text{ and } f_k(n) = 0 \text{ for } k \neq i \quad n = 0, 1, 2, \ldots \qquad (2.7.1)$$

From our earlier work we know that if $i$ were the only node excited, then the signal at output $j$ due to this excitation would be given by the convolution of the signal at node $i$ with the impulse response $h_{ij}(\cdot)$. Thus we obtain

$$g_j(n) = \sum_{m=0}^{n} f_i(m)h_{ij}(n - m) \qquad \text{if } f_k(n) = 0 \text{ for } k \neq i \qquad n = 0, 1, 2, \ldots. \tag{2.7.2}$$

It will not be unusual for several of the input nodes to be excited at the same time. In this case, by superposition, the total signal at output node $j$ will be the sum of the outputs caused by each input acting separately. Thus, for an arbitrary set of $N$ input signals, $f_i(n)$, we find that the total output at node $j$ would be given by

$$g_j(n) = \sum_{i=1}^{N} \sum_{m=0}^{n} f_i(m)h_{ij}(n - m) \qquad j = 1, 2, \ldots, M, \; n = 0, 1, 2, \ldots. \tag{2.7.3}$$

***Matrix system notation***

It is convenient in matrix systems to use matrix notation. We shall let $\mathbf{f}(n)$ and $\mathbf{g}(n)$ be row vectors whose components represent the value of the input and output time functions at the corresponding input and output at time $n$. The vector $\mathbf{f}(n)$ will have $N$ components; $\mathbf{g}(n)$ will have $M$ components. Since each input and output signal must be specified for an infinite number of time periods, each component of $\mathbf{f}$ and $\mathbf{g}$ is itself a vector, a vector time function in the sense of our earlier discussion.

The impulse response of the system can likewise be placed in matrix form. Let $H(n)$ be an $N$ by $M$ matrix whose $ij$th element is the quantity $h_{ij}(n)$. With this definition $H$ is a matrix, each of whose elements is a vector time function, the response at output $j$ to a unit impulse at input $i$ with all other inputs unexcited. We can now write in matrix notation the output vector $\mathbf{g}$ that will result from a given input vector $\mathbf{f}$ in its time component form as

$$\mathbf{g}(n) = \sum_{k=0}^{n} \mathbf{f}(n)H(n - k). \tag{2.7.4}$$

Equation 2.7.4 is the form of the convolution operation for matrix systems. The only change is that the multiplicative part of the convolution process is now being carried out using the operations of matrix algebra. Equation 2.7.4 can also be written using the convolution operator as

$$\mathbf{g} = \mathbf{f} * H. \tag{2.7.5}$$

This equation then becomes the defining relation for the basic matrix block diagram shown in Figure 2.7.1. Under each node is written in parentheses the

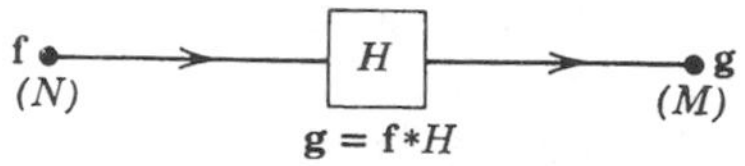


**Figure 2.7.1** The basic matrix block diagram.

number of signals appearing at that node. These nodes consequently define the dimensionality of the impulse response matrix.

It is important to remember that it is no longer true that the convolution operation is commutative. The output signal is generally no longer equal to $H * \mathbf{f}$, the impulse response convolved with the input signal. In most cases examination of dimensionality alone will make clear the futility of attempting convolution in reverse order. We shall find that the noncommutability of matrix graphical operations is the main reason for studying them separately from their simpler brethren.

## Transform Analysis of Matrix Systems

Transform analysis is just as useful for matrix systems as for the systems discussed earlier. If we take the geometric transform of Equation 2.7.3 we obtain

$$g_j^{\mathcal{g}}(z) = \sum_{i=1}^{N} f_i^{\mathcal{g}}(z) h_{ij}^{\mathcal{g}}(z) \qquad j = 1, 2, \ldots, M \tag{2.7.6}$$

because, as we already know, convolution becomes multiplication in the transform domain. In Equation 2.7.6 $g_j^{\mathcal{g}}(z)$ is the transform of the signal at the $j$th output node, $f_i^{\mathcal{g}}(z)$ is the transform of the signal at the $i$th input node, and $h_{ij}^{\mathcal{g}}(z)$ is the transfer function of the system between input $i$ and output $j$. In matrix form Equation 2.7.6 can be written

$$\mathbf{g}^{\mathcal{g}}(z) = \mathbf{f}^{\mathcal{g}}(z) H^{\mathcal{g}}(z). \tag{2.7.7}$$

Here $\mathbf{g}^{\mathcal{g}}(z)$ is the $M$-dimensional vector of output signal transforms, $\mathbf{f}^{\mathcal{g}}(z)$ is the $N$-dimensional vector of input signal transforms, and $H^{\mathcal{g}}(z)$ is the $N$ by $M$ matrix of transfer functions that completely describes the system. Equation 2.7.7 is the basic defining relation for matrix systems in the transform domain. Figure 2.7.2 illustrates the fundamental relation for the simplest matrix system. Note that the order of multiplication is very definitely important.

### *Interpretation of the transfer function matrix*

The nature of a matrix system is clarified if we consider how its impulse response could be measured. If a unit impulse were applied to the first input of the system,

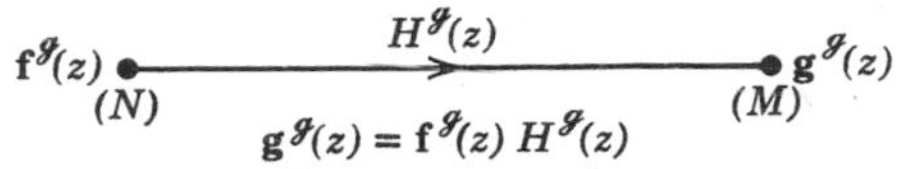


**Figure 2.7.2** The basic matrix flow graph.

but all other inputs were unexcited, then the transform of the signals at each output node would constitute the first row of the transfer function matrix $H^g(z)$. That is, if the transformed $N$-dimensional input vector were $[1000\cdots0]$, then the transformed output would be the first row of $H^g(z)$. Similarly, if the transformed input vector were $[0100\cdots0]$, the transformed output signals would be the second row of $H^g(z)$, and so on. In other words, the $i$th row of the transfer function matrix $H^g(z)$ is the transformed output vector when the $i$th input is a unit impulse and all other inputs are zero. We can therefore think of $H^g(z)$ as the matrix of transformed output vectors when the matrix of transformed input vectors is the $N$ by $N$ identity matrix $I$.

***The identity matrix system***

With this interpretation in mind, it is easy to characterize certain often encountered systems. For example, the $N$ input-$N$ output system that makes no change in any of its inputs will have an impulse response matrix that has $\delta(n)$ on the main diagonal and zero everywhere else, and a transfer function matrix that is the $N$ by $N$ identity matrix $I$; we shall call this system the identity matrix system. The system whose $N$ by $N$ transfer function matrix is zero except for $z$'s on the main diagonal would delay each of the input signals by one time unit before presenting them at the output. These matrix systems are analogous to the identity and delay systems we encountered earlier.

## The Inventory Example

We are now in a position to specify certain of the matrix systems pertinent to the inventory example. The matrix system with the single input, sales, and two outputs, inventory and production, has a 1 by 2 matrix transfer function that we shall denote by ${}^{s/i,p}H^g(z)$. Equations 2.6.4 and 2.6.5 show that this transfer function is given by

$$ {}^{s/i,p}H^g(z) = \left[\frac{-1}{1-(1-\alpha)z+\alpha z^2} \quad \frac{\alpha z(1+z)}{1-(1-\alpha)z+\alpha z^2}\right]. \tag{2.7.8} $$

We can also represent the 1 by 2 system with the single input, sales, and two outputs, sales and production; we use ${}^{s/s,p}H^g(z)$ for its transfer function. By referring again to the transfer function between sales and production, we write this matrix transfer function as

$$ {}^{s|s,p}H^{\mathscr{g}}(z) = \begin{bmatrix} 1 & \dfrac{\alpha z(1+z)}{1-(1-\alpha)z+\alpha z^2} \end{bmatrix}. \tag{2.7.9} $$

Note that the first element in this matrix merely transfers the sales signal to the first element of the output vector.

If we consider sales and production to be inputs and inventory to be the output, then we have a 2 by 1 system with matrix transfer function designated by ${}^{s,p|i}H^{\mathscr{g}}(z)$. By transforming Equations 2.6.1 or 2.6.3, or by inspecting the flow graph of Figure 2.6.2, we find that

$$ i^{\mathscr{g}}(z) = \frac{1}{1-z}[-s^{\mathscr{g}}(z) + p^{\mathscr{g}}(z)]. \tag{2.7.10} $$

Consequently, the matrix transfer function for this system is,

$$ {}^{s,p|i}H^{\mathscr{g}}(z) = \begin{bmatrix} \dfrac{-1}{1-z} \\[2ex] \dfrac{1}{1-z} \end{bmatrix}. \tag{2.7.11} $$

If for some reason we wanted to consider both inventory and production as outputs, then the resulting 2 by 2 system would have the matrix transfer function ${}^{s,p|i,p}H^{\mathscr{g}}(z)$ given by

$$ {}^{s,p|i,p}H^{\mathscr{g}}(z) = \begin{bmatrix} \dfrac{-1}{1-z} & 0 \\[2ex] \dfrac{1}{1-z} & 1 \end{bmatrix} \tag{2.7.12} $$

where the second column serves only to transfer the production signal to the output vector.

We observe that the matrix system notation provides useful flexibility in system representation.

## 2.8 COMPLEX MATRIX SYSTEMS

We may now proceed to consider interconnections of matrix systems. Once more the signal at a node upon which two matrix system outputs impinge is defined

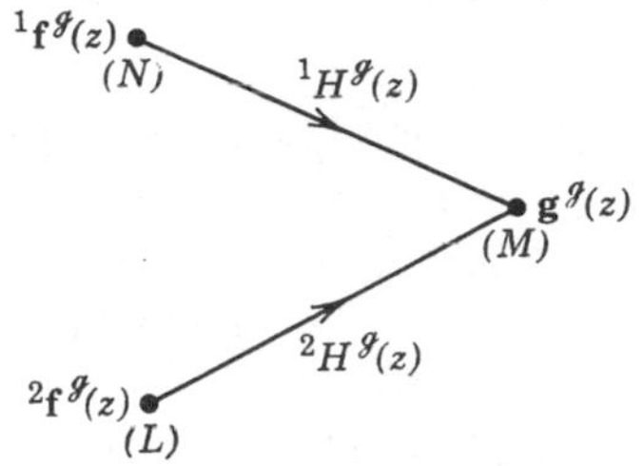


$$\mathbf{g}^{\mathscr{g}}(z) = {}^1\mathbf{f}^{\mathscr{g}}(z)\ {}^1H^{\mathscr{g}}(z) + {}^2\mathbf{f}^{\mathscr{g}}(z)\ {}^2H^{\mathscr{g}}(z)$$

**Figure 2.8.1** Additivity in matrix flow graphs

as the sum of the two outputs. However, since we are dealing with matrix systems, we must require that each system impinging on the node have the same number of outputs. This requirement is shown pictorially in Figure 2.8.1. Here we have used superscripts to indicate the inputs and transfer functions of the two different matrix systems that have different numbers of inputs. However, both systems in this figure have $M$ outputs and so it is permissible to write the total output of the system as the sum of the individual outputs

## Parallel Matrix Systems

A parallel matrix system configuration is shown in Figure 2.8.2. The two systems, ${}^1H$ and ${}^2H$, each have $N$ inputs and $M$ outputs. If each is excited by an $N$ by $N$ matrix of transformed input vectors equal to the identity matrix, then the transformed outputs will be ${}^1H^{\mathscr{g}}(z)$ and ${}^2H^{\mathscr{g}}(z)$ for each system separately. These two outputs are additive at the output node of the matrix system graph and so the parallel combination of the systems may be replaced by a single branch matrix system whose transfer function matrix is the sum of the two individual system transfer function matrices. This result is in complete correspondence with our findings in the original flow graph work. Here we must remember, however, that two systems can be placed in parallel only if both have the same number of inputs and outputs.

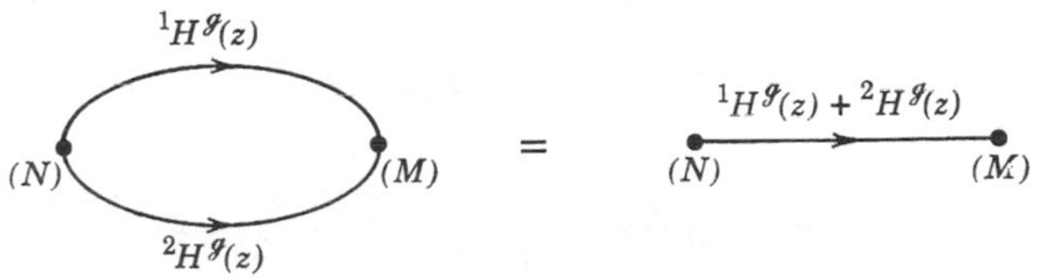


**Figure 2.8.2** Parallel matrix systems.

### Series Matrix Systems

Matrix systems in series are also found in practice. Two such systems are shown in Figure 2.8.3. The first system in the sequence, ${}^{1}H$, has $N$ inputs and $L$ outputs; the second, ${}^{2}H$, has $L$ inputs and $M$ outputs. It is a requirement of systems in series that the number of outputs of each system agree with the number of inputs of the following system. If an arbitrary transformed input vector $\mathbf{f}^{\mathscr{g}}(z)$ is applied to the first system, the transformed output at the intermediate node will be

$$\mathbf{e}^{\mathscr{g}}(z) = \mathbf{f}^{\mathscr{g}}(z)\, {}^{1}H^{\mathscr{g}}(z) \tag{2.8.1}$$

in accordance with the basic result of Equation 2.7.7. The transformed output $\mathbf{g}^{\mathscr{g}}(z)$ is related to $\mathbf{e}^{\mathscr{g}}(z)$ by

$$\mathbf{g}^{\mathscr{g}}(z) = \mathbf{e}^{\mathscr{g}}(z)\, {}^{2}H^{\mathscr{g}}(z) \tag{2.8.2}$$

in correspondence with the same equation. Thus $\mathbf{g}^{\mathscr{g}}(z)$ is obtained from $\mathbf{f}^{\mathscr{g}}(z)$ by the operation

$$\mathbf{g}^{\mathscr{g}}(z) = \mathbf{f}^{\mathscr{g}}(z)\, {}^{1}H^{\mathscr{g}}(z)\, {}^{2}H^{\mathscr{g}}(z). \tag{2.8.3}$$

Since Equation 2.8.3 must hold for an arbitrary input, it follows that the single branch matrix system that is equivalent to two systems in series has a matrix transfer function equal to the product of the two individual matrix transfer functions taken in order. This result is shown in Figure 2.8.3. We note that once more two systems in series have transfer functions that multiply, but, in distinction to the scalar systems considered earlier, the commutability of the multiplication is not assured.

### The Inventory Problem

We can now illustrate the relevance of these results to the inventory problem in matrix form. Equation 2.7.9 shows the matrix transfer function ${}^{s/s,p}H^{\mathscr{g}}(z)$ of the matrix system that transforms the single input sales to the vector output of sales and production. Equation 2.7.12 specifies the matrix transfer function ${}^{s,p/i,p}H^{\mathscr{g}}(z)$ of the matrix system that converts the vector input of sales and production into the vector output of inventory and production. If these two matrix systems are connected in series as shown in Figure 2.8.4, what is the matrix transfer function ${}^{s/i,p}H^{\mathscr{g}}(z)$ of the equivalent matrix system that converts the single input sales into the vector output of inventory and production?

In accordance with our results for series matrix systems, all we need do is multiply together the matrix transfer functions of the individual systems in the proper

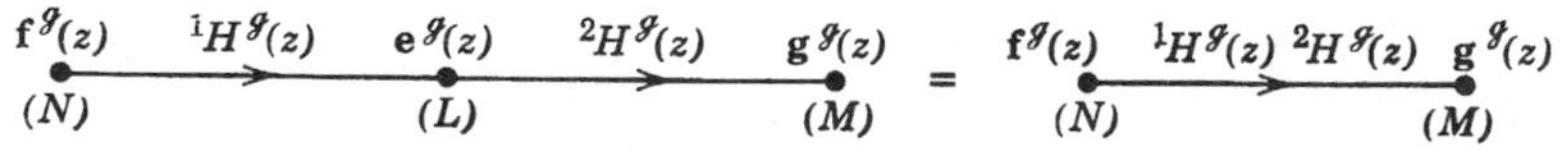


**Figure 2.8.3** Series matrix systems.

$[s^{\mathscr{G}}(z)]$ 　　　　　　　$[s^{\mathscr{G}}(z), p^{\mathscr{G}}(z)]$ 　　　　　　　$[i^{\mathscr{G}}(z), p^{\mathscr{G}}(z)]$

(1) 　　　　　　　　　　　(2) 　　　　　　　　　　　(2)

$${}^{s/s,\,p}H^{\mathscr{G}}(z) = \begin{bmatrix} 1 & \dfrac{\alpha z(1+z)}{1-(1-\alpha)z+\alpha z^2} \end{bmatrix} \qquad {}^{s,\,p/i,\,p}H^{\mathscr{G}}(z) = \begin{bmatrix} \dfrac{-1}{1-z} & 0 \\ \dfrac{1}{1-z} & 1 \end{bmatrix}$$

$[s^{\mathscr{G}}(z)]$ 　　　　　　　　　　　　　　　　　　　　　$[i^{\mathscr{G}}(z), p^{\mathscr{G}}(z)]$

$=$ 　(1) 　　　　　　　　　　　　　　　　　　　　　　　　(2)

$${}^{s/i,\,p}H^{\mathscr{G}}(z) = {}^{s/s,\,p}H^{\mathscr{G}}(z)\,{}^{s,\,p/i,\,p}H^{\mathscr{G}}(z)$$

$$= \begin{bmatrix} 1 & \dfrac{\alpha z(1+z)}{1-(1-\alpha)z+\alpha z^2} \end{bmatrix} \begin{bmatrix} \dfrac{-1}{1-z} & 0 \\ \dfrac{1}{1-z} & 1 \end{bmatrix}$$

$$= \begin{bmatrix} \dfrac{-1}{1-(1-\alpha)z+\alpha z^2} & \dfrac{\alpha z(1+z)}{1-(1-\alpha)z+\alpha z^2} \end{bmatrix}$$

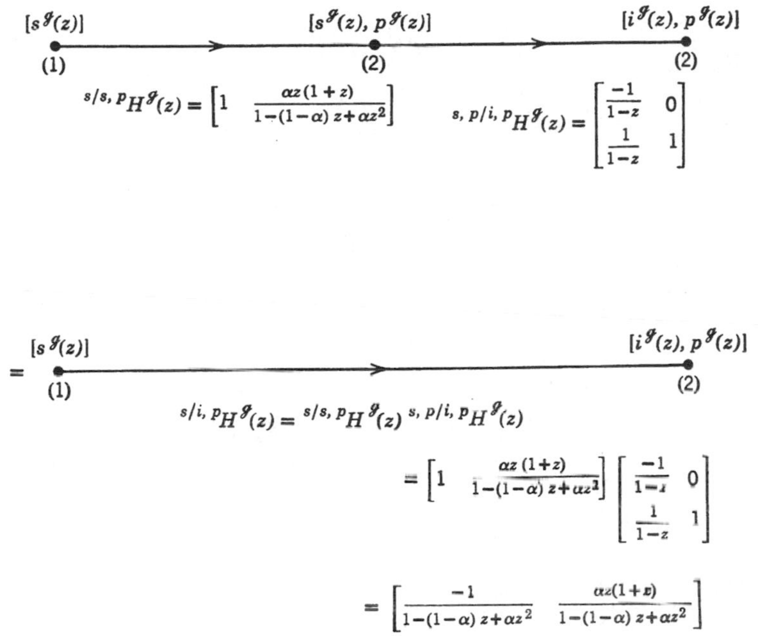

**Figure 2.8.4** A matrix system interpretation of the inventory problem.

order; this operation is performed in Figure 2.8.4. We find that the matrix transfer function ${}^{s/i,\,p}H^{\mathscr{G}}(z)$ of the equivalent matrix system is exactly that given by Equation 2.7.8. Note that multiplying the individual transfer functions in reverse order would be impossible because they are not commutable.

### Feedback Matrix Systems

Our third basic configuration, the feedback condition, can also be found in matrix systems. Figure 2.8.5 shows a typical system of this sort. It is essential that the matrix system providing the feedback, $H^{\mathscr{G}}(z)$, have the same number of inputs and outputs. The systems with matrix transfer functions $I$ are the identity systems whose outputs bear the same signal as the corresponding inputs. The fundamental equations for this system are

$$e^{\mathscr{G}}(z) = f^{\mathscr{G}}(z) + e^{\mathscr{G}}(z)H^{\mathscr{G}}(z) \tag{2.8.4}$$

and

$$g^{\mathscr{G}}(z) = e^{\mathscr{G}}(z). \tag{2.8.5}$$

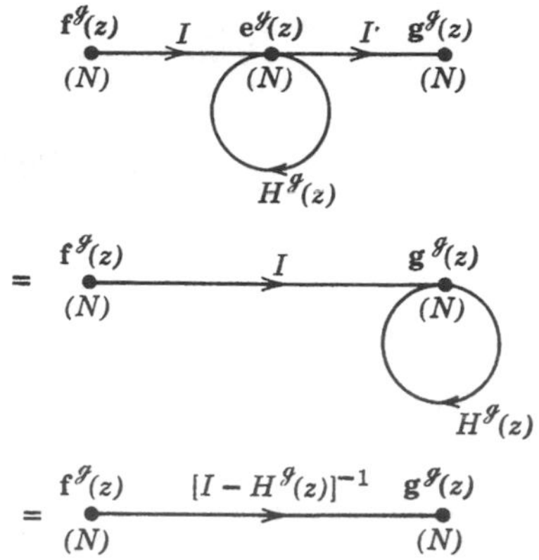


**Figure 2.8.5** The feedback matrix system.

Therefore, we have

$$\mathbf{g}^{\mathscr{G}}(z) = \mathbf{f}^{\mathscr{G}}(z) + \mathbf{g}^{\mathscr{G}}(z)H^{\mathscr{G}}(z) \tag{2.8.6}$$

and finally

$$\mathbf{g}^{\mathscr{G}}(z) = \mathbf{f}^{\mathscr{G}}(z)[I - H^{\mathscr{G}}(z)]^{-1}. \tag{2.8.7}$$

This equation states that a feedback matrix system has a matrix transfer function equal to the inverse of the matrix $I - H^{\mathscr{G}}(z)$, where $H^{\mathscr{G}}(z)$ is the matrix transfer function of the feedback loop. We note that if we were dealing with simple systems rather than matrix systems, the result of Equation 2.8.7 would reduce to that obtained earlier.

***Generality of the feedback matrix system***

However, the feedback matrix system flow graph is more general than it looks—it is, in fact, the most general of flow graphs. To see this, consider a general $N$-node flow graph. The transformed signal at any node $j$, $g_j^{\mathscr{G}}(z)$, must be equal to the sum of the transformed signals that it receives. From node $i$ it receives the signal $g_i^{\mathscr{G}}(z)$ on node $i$ multiplied by the transfer function from node $i$ to node $j$, $h_{ij}^{\mathscr{G}}(z)$; this contribution must be summed over all nodes $i$ of the graph. The final contribution to be added is any external input $f_j^{\mathscr{G}}(z)$ to node $j$. Thus we have

$$g_j^{\mathscr{G}}(z) = f_j^{\mathscr{G}}(z) + \sum_{i=1}^{N} g_i^{\mathscr{G}}(z)h_{ij}^{\mathscr{G}}(z) \qquad j = 1, 2, \ldots, N \tag{2.8.8}$$

or, in matrix form, Equation 2.8.6.

We see, therefore, that any flow graph can be represented by the feedback matrix system flow graph of Figure 2.8.5. The transmission matrix for the system

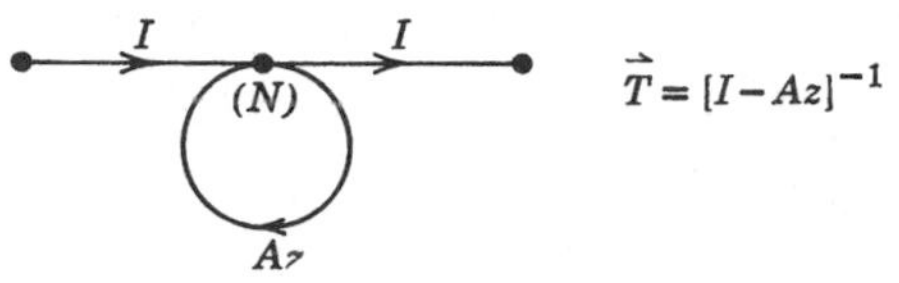


**Figure 2.8.6** An important feedback matrix system.

is always the inverse $[I - H^{\mathscr{g}}(z)]^{-1}$, where $h_{ij}^{\mathscr{g}}(z)$ is the branch transmission from node $i$ to node $j$ of the graph. Furthermore, in computing the $ij$th inverse element $\vec{t}_{ij}$ by the inspection method according to Equation 2.5.2, one plus the sum of the loop products, $\Delta$, is just the determinant of $I - H^{\mathscr{g}}(z)$,

$$\Delta = |I - H^{\mathscr{g}}(z)|, \tag{2.8.9}$$

while

$$\sum_{k \in P_{ij}} \vec{t}_{ij}^{\,k} \, \Delta_{ij}^{\,k}$$

is the $ij$th element in the adjoint of $[I - H^{\mathscr{g}}(z)]$. The interpretation of any flow graph as a feedback matrix system can be a powerful conceptual and computational aid.

*A feedback matrix system example*

A very interesting feedback matrix system is

$$H^{\mathscr{g}}(z) = Az \tag{2.8.10}$$

where $A$ is a square matrix of constants; it is shown as Figure 2.8.6. The outputs of this system will be linear combinations of the inputs delayed by one time interval. The equivalent transfer function of such a system will be

$$H_{\text{eq}}^{\mathscr{g}}(z) = [I - Az]^{-1}. \tag{2.8.11}$$

In view of relation 32 of Table 1.5.1, we find immediately that

$$H_{\text{eq}}(n) = A^n \qquad n = 0, 1, 2, \ldots. \tag{2.8.12}$$

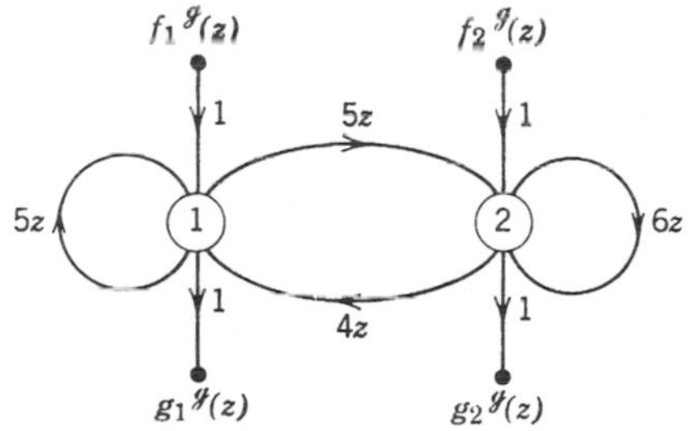


**Figure 2.8.7** Flow graph for the difference equations.

This type of feedback system is particularly important in the solution of systems of simultaneous linear difference equations. Consider the following set of equations:

$$\begin{aligned} g_1(n) &= 5g_1(n-1) + 4g_2(n-1) + f_1(n), \\ g_2(n) &= 5g_1(n-1) + 6g_2(n-1) + f_2(n). \end{aligned} \tag{2.8.13}$$

In the transform domain these equations could be represented either by the simple flow graph of Figure 2.8.7 or by the matrix flow graph of Figure 2.8.8. The equivalent transfer function $[I - Az]^{-1}$ of the matrix flow graph is computed as follows:

$$[I - Az] = \begin{bmatrix} 1 - 5z & -5z \\ -4z & 1 - 6z \end{bmatrix}$$

$$[I - Az]^{-1} = \frac{1}{(1 - z)(1 - 10z)} \begin{bmatrix} 1 - 6z & 5z \\ 4z & 1 - 5z \end{bmatrix}$$

$$= \frac{1}{1 - z}\begin{bmatrix} 5/9 & -5/9 \\ -4/9 & 4/9 \end{bmatrix} + \frac{1}{1 - 10z}\begin{bmatrix} 4/9 & 5/9 \\ 4/9 & 5/9 \end{bmatrix} \tag{2.8.14}$$

The reader should verify that the $ij$th element of the $[I - Az]^{-1}$ matrix is the transmission from $i$ to $j$ of the flow graph of Figure 2.8.7. If we take the inverse transform of Equation 2.8.14, we obtain the impulse response of the matrix system as

$$H_{\text{eq}}(n) = \begin{bmatrix} 5/9 & -5/9 \\ -4/9 & 4/9 \end{bmatrix} + (10)^n \begin{bmatrix} 4/9 & 5/9 \\ 4/9 & 5/9 \end{bmatrix}. \tag{2.8.15}$$

This expression can be used to compute the output functions $g_1$ and $g_2$ that correspond to any set of input functions $f_1$ and $f_2$. For example, the outputs for the trivial input time functions $f_1(n) = \delta(n)$, $f_2(n) = 0$ are given directly by the first row of the $H_{\text{eq}}(n)$ matrix as

$$g_1(n) = 5/9 + 4/9(10)^n \qquad g_2(n) = -5/9 + 5/9(10)^n \qquad n = 0, 1, 2, \ldots. \tag{2.8.16}$$

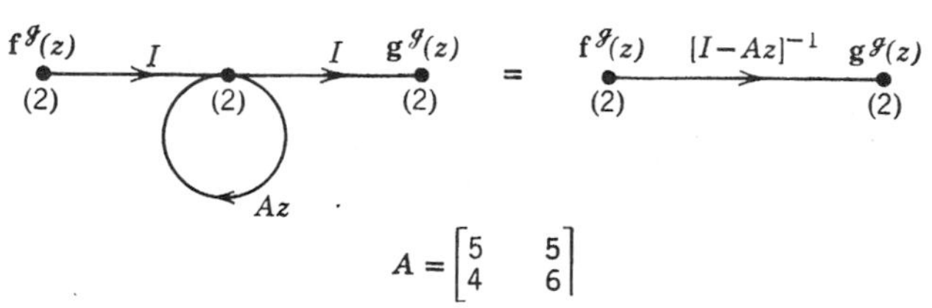


**Figure 2.8.8** Flow graph for the difference equations.

The outputs for more complicated inputs may be found by convolution or by transform methods.

***Simultaneous linear equations***

If a feedback matrix system has no delay in its feedback loop but merely a square matrix of constants $A$, then it represents a set of $N$ linear simultaneous equations. Consequently, sets of linear equations that have a solution can be solved by the flow graph methods we have discussed. Although this type of system is trivial in the sense that time dependence is not important, the fact that a solution can be obtained using the flow graph methods is very convenient in practice.

***Matrix inversion***

As we might expect as an extension of this result, we can use the flow graph formalism to find the inverse of a matrix. By taking $z = 1$ in Equations 2.8.10 and 2.8.11, we observe that if we construct a flow graph corresponding to a square matrix $A$ by labeling the branch from $i$ to $j$ with the $ij$th element $a_{ij}$ of matrix, then the transmission of the entire flow graph from node $i$ to node $j$ will be the $ij$th element of $[I - A]^{-1}$. In compact form, we can write that the transmission matrix is just the inverse of $I - A$,

$$\vec{T} = [I - A]^{-1}. \tag{2.8.17}$$

Therefore, if we want to find the inverse of a nonsingular square matrix $M$, our first step is to let

$$M = I - A \tag{2.8.18}$$

or

$$A = I - M. \tag{2.8.19}$$

That is, we construct a flow graph for which the transmission of the branch from node $i$ to node $j$ is just $\delta_{ij} - m_{ij}$ for every $i, j$ pair. The transmission matrix of this flow graph will be the required inverse, $\vec{T} = M^{-1}$.

To illustrate the procedure, consider the nonsingular matrix

$$M = \begin{bmatrix} 1 & 3 \\ 2 & 7 \end{bmatrix}. \tag{2.8.20}$$

We form

$$A = I - M = \begin{bmatrix} 0 & -3 \\ -2 & -6 \end{bmatrix} \tag{2.8.21}$$

and construct the corresponding flow graph of Figure 2.8.9. By inspection, we readily establish that the transmission matrix of this flow graph is

$$\vec{T} = \begin{bmatrix} 7 & -3 \\ -2 & 1 \end{bmatrix} = [I - A]^{-1} = M^{-1}. \tag{2.8.22}$$

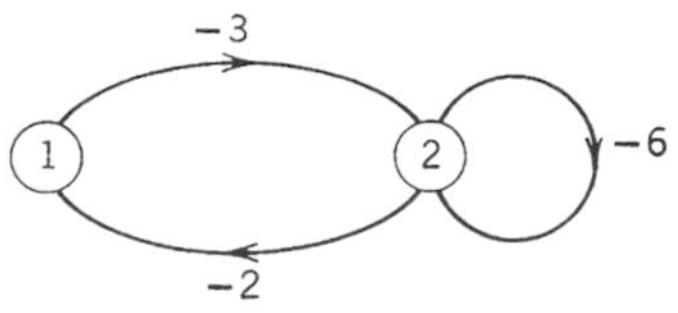


**Figure 2.8.9** Matrix inversion by flow graph reduction.

Direct calculation verifies that this result is, in fact, the inverse of $M$. We note that in the course of this calculation we have developed the determinant of $M$ as $\Delta$, one plus the sum of loop products:

$$\Delta = |I - A| = |M| = 1. \tag{2.8.23}$$

**Reduction of Interconnected Matrix Systems**

A very interesting situation arises when we attempt to combine several matrix systems with more complicated connections than those we have already discussed. Consider the configuration of Figure 2.8.10. Here we have a system with an $N$ by $M$ forward transfer function ${}^1H^g$ and an $M$ by $N$ feedback transfer function ${}^2H^g$. If we choose to reduce this system by applying the same argument used in the simple case, then we duplicate the system ${}^1H^g$, use the output of the duplicated system as the input to ${}^2H^g$, and thus obtain the reduction shown in Figure 2.8.10 as form (I). The system transfer function using this reduction is given by $[I - {}^1H^g\,{}^2H^g]^{-1}\cdot{}^1H^g$.

On the other hand, there is an alternate reduction. If we say that the total input to the $M$-signal node is given by the sum of ${}^1H^g$ acting on the input and ${}^1H^g$ acting on the output of ${}^2H^g$, then we can draw the equivalent system in form (II). The transfer function of the system in this form is ${}^1H^g[I - {}^2H^g\,{}^1H^g]^{-1}$. Thus we obtain two apparently different $N$ by $M$ matrix transfer functions for the same system. (We must remember, of course, that the $I$ matrix is an $N$ by $N$ matrix in form (I), and an $M$ by $M$ matrix in form (II).) In fact, these two transmissions are identical.

To show this, let us write the transform equations implied by the original system. We have

$$\mathbf{g}^g = {}^1H^g\mathbf{e}^g \tag{2.8.24}$$

and

$$\mathbf{e}^g = \mathbf{f}^g + {}^2H^g\mathbf{g}^g. \tag{2.8.25}$$

If we substitute $\mathbf{e}^g$ from this equation into Equation 2.8.24, we obtain

$$\mathbf{g}^g = {}^1H^g(\mathbf{f}^g + {}^2H^g\mathbf{g}^g). \tag{2.8.26}$$

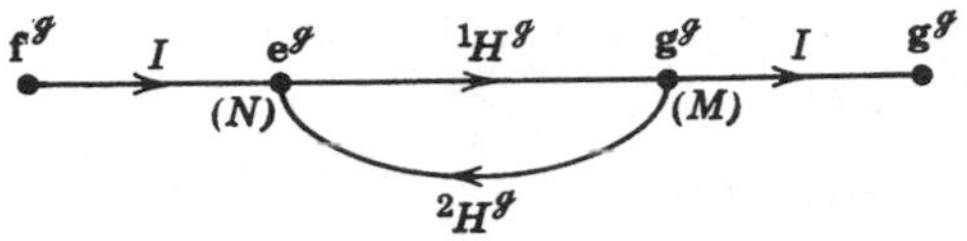


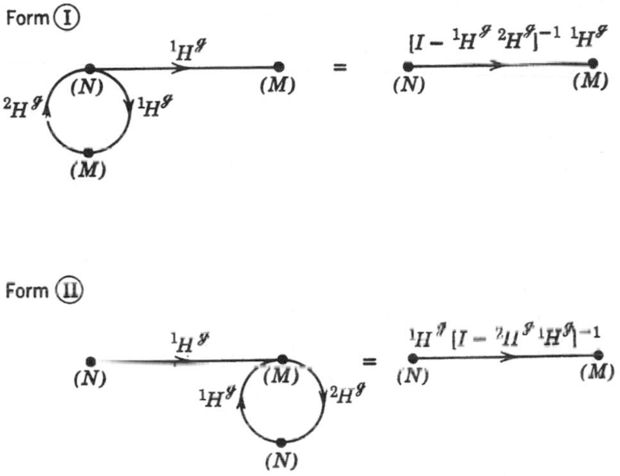


**Figure 2.8.10** An ambiguity.

Solving for $\mathbf{g}^{\mathscr{G}}$, we find

$$\mathbf{g}^{\mathscr{G}} = [I - {}^1H^{\mathscr{G}}\,{}^2H^{\mathscr{G}}]^{-1}\,{}^1H^{\mathscr{G}}\mathbf{f}^{\mathscr{G}} \tag{2.8.27}$$

in accordance with form (I) of the reduction. If, on the other hand, we substitute $\mathbf{g}^{\mathscr{G}}$ from Equation 2.8.24 into Equation 2.8.25, we can write

$$\mathbf{e}^{\mathscr{G}} = \mathbf{f}^{\mathscr{G}} + {}^2H^{\mathscr{G}}\,{}^1H^{\mathscr{G}}\mathbf{e}^{\mathscr{G}} \tag{2.8.28}$$

and

$$\mathbf{e}^{\mathscr{G}} = [I - {}^2H^{\mathscr{G}}\,{}^1H^{\mathscr{G}}]^{-1}\mathbf{f}^{\mathscr{G}}. \tag{2.8.29}$$

When this result is substituted into Equation 2.8.24, we obtain

$$\mathbf{g}^{\mathscr{G}} = {}^1H^{\mathscr{G}}[I - {}^2H^{\mathscr{G}}\,{}^1H^{\mathscr{G}}]^{-1}\mathbf{f}^{\mathscr{G}}, \tag{2.8.30}$$

an equation that corresponds to form (II) of the reduction. Thus, both reductions are correct and in any problem we can choose the more convenient one. For example, the first reduction requires inversion of an $N$ by $N$ matrix, the second of an $M$ by $M$ matrix. Generally, we would choose the reduction that involves inversion of the smaller matrix.

***Route transmission sum reduction***

Since the route transmission sum interpretation preserves the order in which branches are traversed, it is equally applicable to matrix flow graph reduction. We can trace the ambiguity of form we have just observed to the route transmission sum solution of the problem. By inspection of the original graph in Figure 2.8.10, we write the transmission matrix in the route transmission sum form

$$\vec{T} = {}^1H^g + {}^1H^g\,{}^2H^g\,{}^1H^g + {}^1H^g\,{}^2H^g\,{}^1H^g\,{}^2H^g\,{}^1H^g + \cdots. \tag{2.8.31}$$

If we regard this sum as a matrix expression containing ${}^1H^g$ as a postmultiplier, we obtain

$$\begin{aligned}\vec{T} &= [I + {}^1H^g\,{}^2H^g + {}^1H^g\,{}^2H^g\,{}^1H^g\,{}^2H^g + \cdots]{}^1H^g \\ &= [I + ({}^1H^g\,{}^2H^g) + ({}^1H^g\,{}^2H^g)^2 + \cdots]{}^1H^g \\ &= [I - {}^1H^g\,{}^2H^g]^{-1}\,{}^1H^g \end{aligned} \tag{2.8.32}$$

which is form (I) in Figure 2.8.10. However, if we regard the sum as a matrix expression containing ${}^1H^g$ as a premultiplier, then we have

$$\begin{aligned}\vec{T} &= {}^1H^g[I + {}^2H^g\,{}^1H^g + {}^2H^g\,{}^1H^g\,{}^2H^g\,{}^1H^g + \cdots] \\ &= {}^1H^g[I + ({}^2H^g\,{}^1H^g) + ({}^2H^g\,{}^1H^g)^2 + \cdots] \\ &= {}^1H^g[I - {}^2H^g\,{}^1H^g]^{-1} \end{aligned} \tag{2.8.33}$$

which is form (II) of Figure 2.8.10. Thus we see that such ambiguities can merely be considered as the options exercised in factoring an infinite matrix sum.

***A complex matrix system example***

As a somewhat larger example, let us reduce the matrix form of the three-node system considered earlier. It is shown in Figure 2.8.11 with capital letters to indicate transfer functions. The first steps in this reduction are simple applications of our series and feedback equivalent relationships for matrix systems, with due regard to the commutability of the matrix transfer functions. Finally, we produce a graph like that of the system of Figure 2.8.10, and then we have a choice concerning the form of the equivalent transmission. Both forms are shown in Figure 2.8.11. If $M$ were smaller than $N$, then form (II) of the reduction would probably be preferred.

The fact that the commutative property of matrix systems must be preserved makes our inspection reduction method inapplicable. Although both transmission expressions in Figure 2.8.11 reduce to that of the earlier example in Figure 2.4.9 if the capital letters are taken to be scalar rather than matrix quantities, it is not clear how matrix transfer functions can be obtained from corresponding scalar

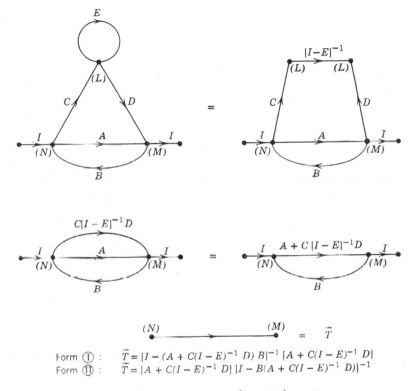

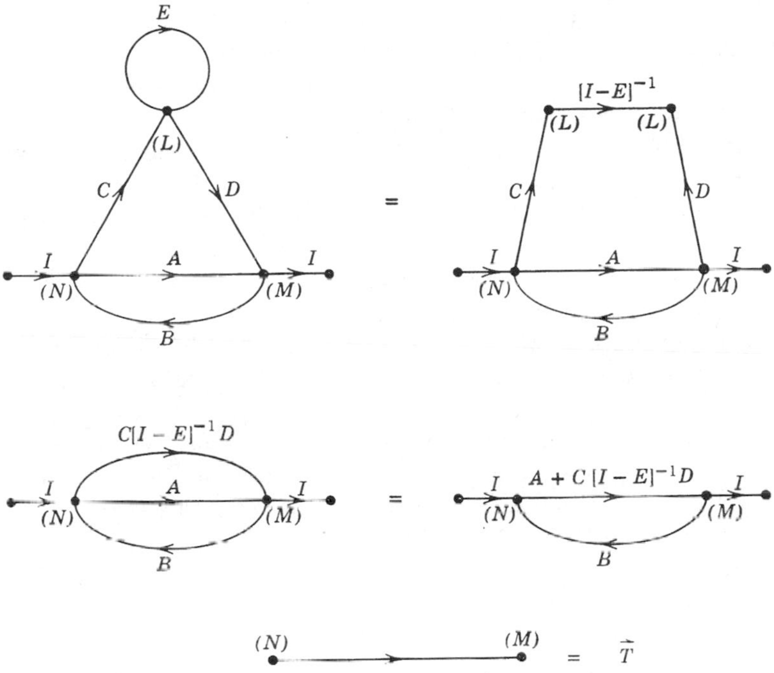


Form (I) : $\vec{T} = [I - (A + C(I - E)^{-1} D) B]^{-1} [A + C(I - E)^{-1} D]$

Form (II) : $\vec{T} = [A + C(I - E)^{-1} D] [I - B(A + C(I - E)^{-1} D)]^{-1}$

**Figure 2.8.11** A more complex matrix system.

equations. Some inspection methods for matrix systems do exist, but actually they consist of orderly procedures for applying our simple reductions rather than being a true inspection method such as the one we used for scalar systems. The best plan for the reduction of matrix systems would seem to be to follow the residual node method.

***Residual node reduction***

The residual node method is a sequential reduction procedure based on the encysted system concept. If we desire to reduce an $S$-node graph by this method, then $S - 1$ of the nodes in the graph, including the input and output nodes, are chosen as boundary nodes and designated "residual nodes." The residual nodes should be chosen in such a way that most of the feedback loops in the graph pass through them. When this has been done, the only internal node of the encysted system is the node that has not been designated as residual. The feedback around this node is changed into a series transmission on all its input or output branches.

Then, the transmissions between each pair of residual nodes and around each residual node itself are computed and drawn on the graph. However, in this computation any connection between or around residual nodes that requires traversal of another residual node is not included. When this step is completed, $S - 2$ nodes are designated as residual nodes and the process is repeated until only the input and output nodes remain. At this point the system can be placed in the form shown in Figure 2.8.10 and either of the two forms of the transfer function may be written for it.

The residual node method will, of course, work for the scalar systems considered earlier. In this case, the method corresponds to solving the underlying set of simultaneous equations by successive substitution. The variable corresponding to the node eliminated at each step is being expressed in terms of the remaining variables.

We can illustrate the residual node method easily by using the example of Figure 2.8.11. In this three-node problem, we are forced to select the input and output nodes 1 and 2 as residual nodes for the first step of reduction. Then we convert the path that passes through the eliminated node 3 into an equivalent simple branch transmission. Finally, we represent the system in terms of transmissions between and around the residual nodes 1 and 2. The resulting system has the structure of Figure 2.8.10; its transmission may be written in either of the alternate forms.

The significant reduction in dimensionality caused by originally characterizing a system as an interconnection of matrix systems ensures that we can expect to encounter in practice matrix systems with only a few nodes. Such systems are easily handled by the simple reductions or by the residual node method.

## 2.9 CONCLUSION

Many possible extensions of the analytic techniques we have discussed have been developed. For example, it is clear that the restriction to discrete time processes is not essential. All the time functions appearing in our work could be continuous in time without introducing any real change in our results. We may regard such time functions as vectors with a nondenumerable number of components.

In Chapter 11 (Volume II) we shall discuss this extension in detail; however, it is easy to anticipate our future development. The output vector from a linear system will be obtained by convolving the input vector with the impulse response vector, as before. However, the convolution operation will be defined by a convolution integral rather than the convolution summation. The relevant transformation will be an exponential (or Laplace) transform instead of the geometric transform. We shall find it possible to consider the geometric transform as a special case of the exponential transform. Especially important is the observation that all of the flow graph relationships we have developed are identical for the continuous-time formulation.

Historically, the use of flow graph techniques in continuous-time systems preceded their use for discrete-time systems. It is interesting to see how a technique developed originally for the description of servomechanisms became an operational tool for the solution of systems governed by linear constant coefficient differential equations, and then an equally important tool for working with systems described by linear constant coefficient difference equations of the type we have discussed earlier. A central theme of our present work is application of this approach to the Markov process and its generalizations.

Our treatment of linear system analysis has been all too brief. Hopefully, however, the present discussion has indicated the value of using the graphical power of the human mind to make easier and more meaningful the analysis of an important class of complex systems.

## PROBLEMS

**1.** Find the output $g(n)$ of a discrete system with impulse response

$$h(n) = \begin{cases} (-1/2)^n & n \geq 0 \\ 0 & n < 0 \end{cases} \quad \text{to an input of the form} \quad f(n) = \begin{cases} n & n \geq 0 \\ 0 & n < 0 \end{cases}.$$

**2.** Find the transmission of the following flow graph:

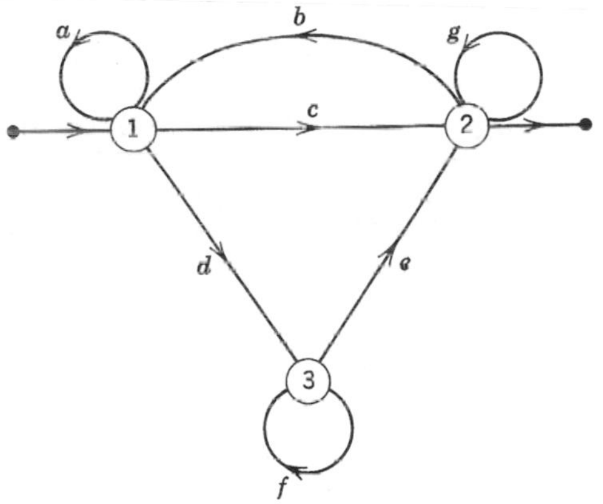


**3.** a) Find the output $g(n)$ of the linear system characterized by the impulse response

$$h(n) = \begin{cases} (1/3)^n & n \geq 0 \\ 0 & n < 0 \end{cases}$$

to the input

$$f(n) = \begin{cases} c^n & n \geq 0 \\ 0 & n < 0 \end{cases}.$$

For what values of $c$ does $g(n)$ remain finite as $n$ goes to infinity? What is the form of the response if $c = 1/3$?

b) Find the output $g(n)$ to a step input for a system with a transfer function equal to

$$h^g(z) = \frac{1}{1 - (6/5)z + (1/5)z^2}.$$

c) Find the input $f(n)$ that produces the output

$$g(n) = \begin{cases} (1/2)^n - (2/3)^n & n \geq 0 \\ 0 & n < 0 \end{cases}$$

in the linear system:



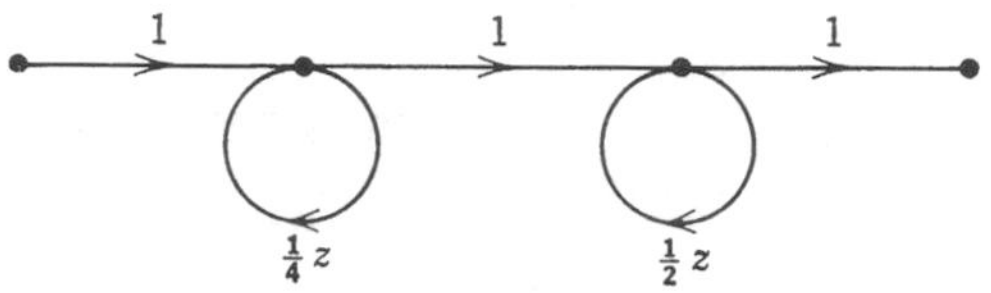

**4.** Consider the following system:



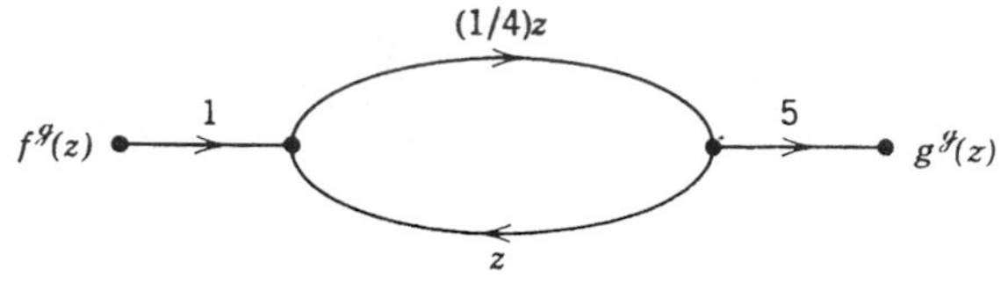

If we are told that

$$g(n) = -3(1/3)^n + 2.5(1/2)^n + 0.5(-1/2)^n \qquad n \geq 0,$$

then find $f(n)$.

**5.** Find the impulse response characterizing the system for which an input

$$f(n) = \begin{cases} 1 + (0.6)^n & n \geq 0 \\ 0 & n < 0 \end{cases}$$

produces the output

$$g(n) = \begin{cases} -3(0.4)^n & n \geq 0 \\ 0 & n < 0 \end{cases}.$$

**6.** A discrete system with impulse response

$$h(n) = \begin{cases} 1 & n \geq 0 \\ 0 & n < 0 \end{cases}$$

has an output

$$g(n) = \begin{cases} (1/3)^n & n \geq 0. \\ 0 & n < 0 \end{cases}$$

What was the input, $f(n)$?

**7.** An input **f** to a discrete system results in the output **g**. If

$$f(n) = \begin{cases} 2n & n \geq 0, \\ 0 & n < 0 \end{cases}$$

and

$$g(n) = \begin{cases} 4n/3 + (4/9)[1 - (-1/2)^n] & n \geq 0, \\ 0 & n > 0 \end{cases}$$

find $h(n)$, the impulse response of the system.

**8.** a) Consider the following discrete system

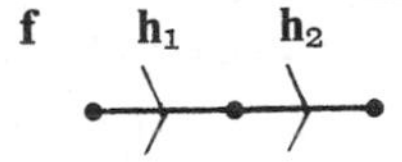


Find the output $g(n)$ if

$$f(n) = \begin{cases} (1/4)^n & n \geq 0 \\ 0 & n < 0 \end{cases}$$

$$h_1(n) = \begin{cases} (1/3)^n & n \geq 0 \\ 0 & n < 0 \end{cases}$$

$$h_2(n) = \begin{cases} (1/8)^n & n \geq 0 \\ 0 & n < 0 \end{cases}.$$

**9.** Reduce the following flow graph:

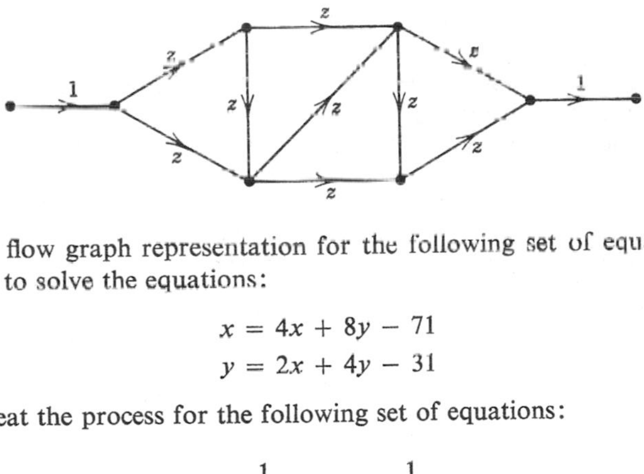


**10.** a) Draw a flow graph representation for the following set of equations, and use the flow graph to solve the equations:

$$x = 4x + 8y - 71$$
$$y = 2x + 4y - 31$$

b) Now repeat the process for the following set of equations:

$$x = \frac{1}{1-4}\cdot 8y - \frac{1}{1-4}\cdot 71$$

$$y = \frac{1}{1-4}\cdot 2x - \frac{1}{1-4}\cdot 31$$

How are these equations and their flow graph related to those in a)?

c) Repeat for the set of equations,

$$x = 4ax + 8ay - 71$$
$$y = 2ax + 4ay - 31$$

where $a$ is a constant.

d) Repeat once more for:

$$x = a \cdot p_{11}x + a \cdot p_{21}y + 1$$
$$y = a \cdot p_{12}x + a \cdot p_{22}y$$

where the $p$'s are known numbers and $a$ is a constant. (Notice the similarity of this result to the transition diagram for a two-state Markov process).

**11.** Consider the following system:

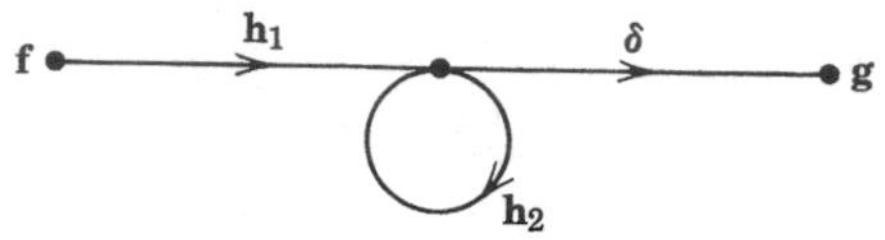


Let

$$\mathbf{f} = 4\delta,$$

i.e., an impulse of magnitude 4 at time zero.

$$h_1(n) = (1/5)^n \qquad n \geq 0$$
$$h_2(n) = 3\left(\frac{a}{2}\right)^n \qquad n \geq 0$$

a) Obtain a closed form expression for $g(n)$, the output at time $n$.

b) For what values of $a$ is $g(\infty)$ infinite?

c) Evaluate $g(n)$ for the case $a = -4/5$.

**12.** a) A system whose flow graph is

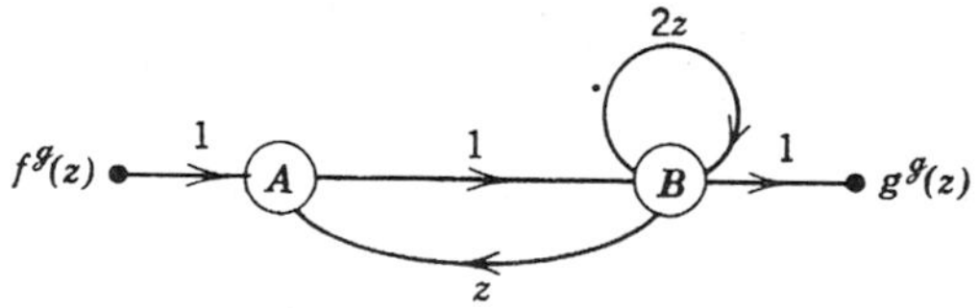


has input

$$f(n) = \begin{cases} 1 & n \geq 0 \\ 0 & n < 0 \end{cases}.$$

Find the output $g(n)$.

b) Using $a(n)$ and $b(n)$ to represent the signals at the intermediate nodes $A$ and $B$, write a set of difference equations relating $f(n)$, $a(n)$, $b(n)$, and $g(n)$. Eliminate $a(n)$

and $b(n)$ from these equations to obtain a single difference equation relating $f(n)$ and $g(n)$. Verify the solution for $g(n)$ obtained above for the specified $f(n)$. Explain the initial condition situation.

**13.** Is $\bar{t}_{12}$ the same for the following two flow graphs? Why?

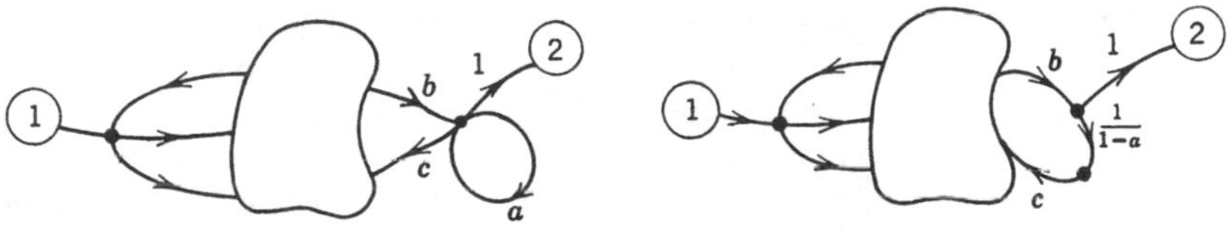


**14.** Consider the following flow graph.

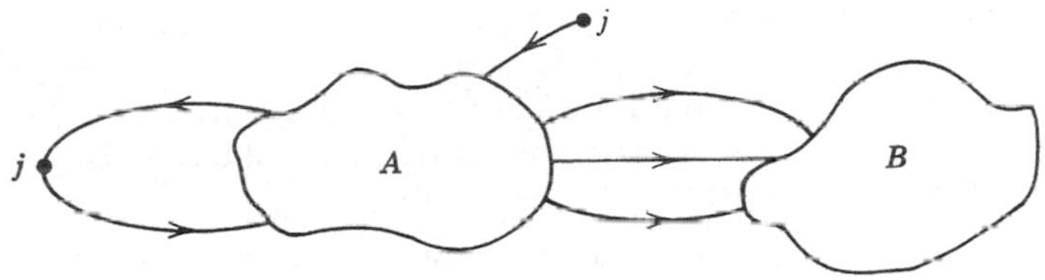


Let: $t_{ij}^k$ equal the transmission of the $k$th path from node $i$ to node $j$.

${}_A\Delta_{ij}^k$ equal one plus the sum of the loop products of all loops in component $A$ of the graph that share no node with the $k$th path from node $i$ to node $j$.

${}_A\Delta$ equal one plus the sum of the loop products of component $A$ of the graph (and let ${}_B\Delta$ be the same for component $B$).

Show that the transmission from node $i$ to node $j$ may be written as

$$\bar{t}_{ij} = \frac{\sum_k t_{ij}^k \, {}_A\Delta_{ij}^k \, {}_B\Delta}{{}_A\Delta \, {}_B\Delta}$$

and thus that component $B$ may be neglected in the determination of this transmission.

**15.** a) Solve the following difference equation by flow graph techniques

$$f(n) = g(n) - \frac{3}{2}g(n-1) + \frac{1}{2}g(n-2)$$

where

$$f(n) = \begin{cases} (1/3)^n & n \geq 0 \\ 0 & n < 0 \end{cases}$$

and

$$f(-1) = g(-2) = 0.$$

Do this by constructing nodes corresponding to the transforms of $g(n)$, $g(n-1)$, and $g(n-2)$ and connecting the nodes with appropriate transmissions. Consider $f(n)$ to be the input to this system and take the output from node $g(n)$.

b) How can the graph be modified to account for initial conditions for $g(n)$ such as

$$g(-1) = 1$$
$$g(-2) = 0$$

and what is the modified response?

**16.** a) A certain four year high school is interested in the total number of students to be in attendance during the next school year. Let $f(n)$ = number of new students entering the school at the beginning of school year $n$ and $s(n)$ = total number of students in attendance during school year $n$. Assuming that all students pass on to the next grade (or graduate) at the end of each school year, draw a flow graph relating the geometric transforms of $f(n)$ and $s(n)$.

b) The school committee has decided that the total number of school desks available at the beginning of each school year will be equal to the total number of students in attendance during the previous year. Draw a modification of the flow graph of part a) that may be used to obtain the transform of $d(n)$, the desk deficiency during year $n$.

c) If there is a unit increase in the number of entering students, i.e., $f(n) = 1$ for $n \geq 0$, what will be the response, $d(n)$?

**17.** Consider the following inventory control problem. A certain industry has sales of $s(n)$ during the $n$th month and produces $p(n)$ items during the $n$th month. Its inventory shortage, $i(n)$ is defined as:

$$i(n) = \sum_{k=0}^{n} [s(k) - p(k)]$$

Each of these quantities, $s(n)$, $p(n)$, and $i(n)$ are incremental, i.e., they are deviations from some arbitrary norm, so that it is possible for each of them to be positive or negative. The problem then is to find the production policy that will produce an appropriate response to the expected sales. We will consider the following two policies:

a) $p(n) = a\,i(n - 1)$, production proportional to last month's inventory shortage. ($a$ is a constant.)

(b) $p(n) = i(n - 1) + \sum_{k=0}^{n-1} i(k)$, production proportional to the sum of last month's inventory shortage and the cumulative inventory shortage. For each of these two policies draw a relevant flow graph and find the incremental inventory shortage and production for $s(n)$ = unit impulse at $n = 0$, and for $s(n) = n$. Would you consider these good production policies? If time permits, investigate the effect of an extra month's delay in changing the production.

**18.** In Section 2.6 we looked at a linear, discrete-time, time-invariant causal model of a production-inventory system. The notation used was:

$i(n)$ = inventory at the end of the $n$th month
$s(n)$ = sales during the $n$th month
$p(n)$ = production during the $n$th month

where these are incremental quantities and all can be both positive and negative. By the definition of the inventory we have

$$i(n) = i(n - 1) + p(n) - s(n).$$

The production foreman has decided to use a production policy of the form:

$$p(n + 1) = \alpha i(n) + \beta s(n)$$

a) Find $p^{\mathcal{g}}(z)/s^{\mathcal{g}}(z)$ for zero initial conditions.
b) Find $i^{\mathcal{g}}(z)/s^{\mathcal{g}}(z)$ for zero initial conditions.
c) What are the conditions on $\alpha$ and $\beta$ for $p(n) = s(n - 1)$?
d) If the system is subjected to a unit step increase in sales [i.e., $s(n) = u(n)$], for what values of $\alpha$ and $\beta$ will $i(n) \to 0$ as $n \to \infty$?

**19.** A certain committee gives an oral examination to one student at the end of every month if there is anyone on the waiting list. The probability of $\ell$ students becoming eligible to take the exam during any month is

$$p(\ell) = \frac{2}{3}\left(\frac{1}{3}\right)^{\ell} \qquad \text{for } \ell \geq 0.$$

It has been determined that a student taking the exam has a probability 2/3 of passing; if he fails, he goes to the end of the waiting list. It is assumed that there are no arrivals during the exam and that the end of the exam occurs simultaneously with the end of the month. There is no one on the waiting list at the beginning of the first month. If $p(k, n)$ is the probability that there are $k$ people on the waiting list at the end of the $n$th month (after the exam),

a) Write the difference equations for $p(k, n)$ in terms of $p(\cdot, n - 1)$. Do this for $k = 0$, $k = 1$, and $k > 1$.
b) By letting $n \to \infty$ in the difference equations, find a set of equations for $\pi(k) = \lim_{n\to\infty} p(k, n)$. Find the transform $\pi^{\mathcal{g}}(z)$ and invert it to obtain a closed form expression for $\pi(k)$.
c) What will be the average number of students on the waiting list in the steady state?

# 3 SYSTEMS ANALYSIS OF MARKOV PROCESSES

The probability transformation represented by a Markov process is a linear system. The type of Markov process we discussed in Chapter 1 is in fact a time-invariant, physically realizable linear system. We can therefore apply the systems analysis techniques of Chapter 2, including flow graph theory, to the study of Markov processes.

## 3.1 FLOW GRAPH ANALYSIS OF MARKOV PROCESSES

We know from Equation 1.6.6 that the geometric transform of the multistep transition probability matrix is equal to an inverse matrix constructed by subtracting $z$ times the transition probability matrix from the identity matrix,

$$\Phi^{g}(z) = [I - Pz]^{-1}. \tag{3.1.1}$$

Figure 2.8.6 shows that the inverse matrix is equal to the transmission matrix of a feedback matrix system whose feedback transmission is $Pz$, a system that appears as Figure 3.1.1. The flow graph corresponding to the matrix system is illustrated in detail in Figure 3.1.2; we call it the flow graph of the Markov process. The flow graph is the Markov process transition diagram of Figure 1.1.3 with each branch labeled $p_{ij}z$ instead of $p_{ij}$.

We have thus found that the geometric transform of the multistep transition probability from state $i$ to state $j$, $\phi_{ij}^{g}(z)$, is equal to the transmission $\vec{t}_{ij}$ from node

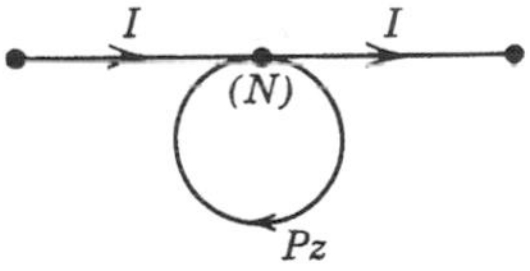


$$\vec{T} = [I - Pz]^{-1} = \Phi^{g}(z)$$

**Figure 3.1.1** Matrix flow graph for an $N$-state Markov process.

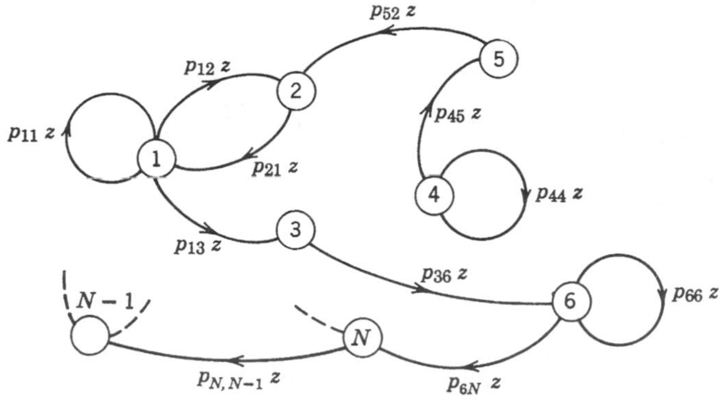


**Figure 3.1.2** Flow graph for an $N$-state Markov process.

$i$ to node $j$ of the flow graph formed from the process transition diagram by multiplying each transition probability by $z$.

## The Marketing Example

The marketing example of Chapter 1 will provide an opportunity to try the new method. The transition diagram for this 2-state process appeared in Figure 1.2.1. This transition diagram is readily converted into the flow graph for the process by multiplying the numbers on all its branches by $z$; the result is shown in Figure 3.1.3.

Suppose we wish to calculate $\phi_{11}(n)$, the probability that the process will be in state 1 after $n$ transitions, given that it started in state 1. The transform of this quantity, $\phi_{11}{}^{\mathscr{g}}(z)$, is $\vec{t}_{11}$, the transmission of the flow graph from node 1 to node 1. This transmission can be found either by using the basic reduction theorems or by the inspection method. We shall employ both techniques.

### *Solution for $\phi_{11}{}^{\mathscr{g}}(z)$ using basic reductions*

Figure 3.1.4 shows the sequence of steps in finding $\vec{t}_{11}$ by using basic reductions. Figure 3.1.4(a) illustrates that the required transmission is through node 1. The

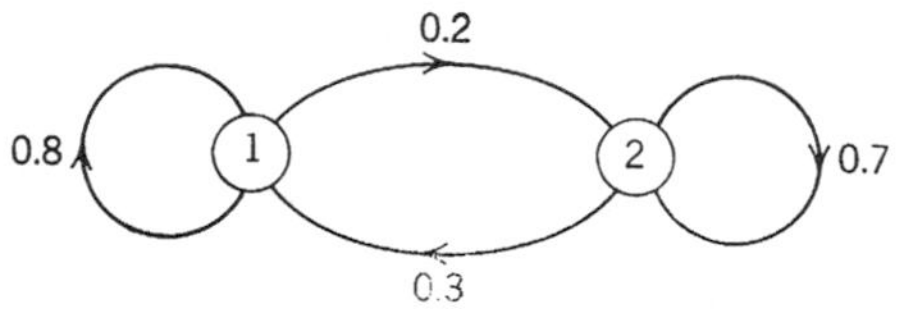


**Figure 3.1.3** Flow graph for marketing example.

$$\tilde{t}_{11} = \frac{1-0.7\,z}{(1-z)(1-\tfrac{1}{2}z)} = \phi_{11}\ (z)$$

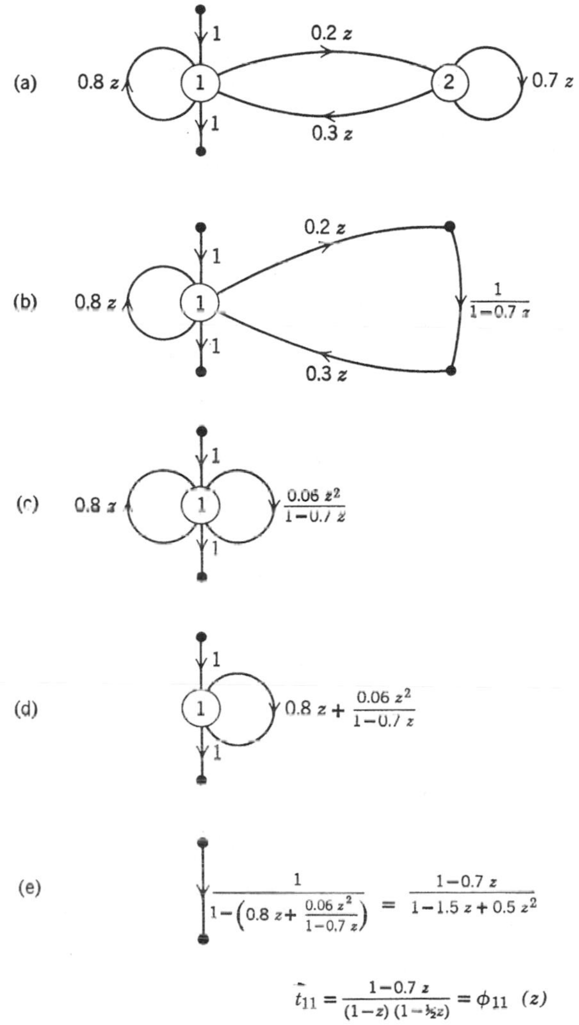

**Figure 3.1.4** Calculation of $\tilde{t}_{11} = \phi_{11}{}''(z)$ for marketing example using basic reductions.

feedback loop around node 2 is changed into a simple transmission in Figure 3.1.4(b) by using the basic feedback reduction. This transmission then becomes the central of three systems in series that are replaced in Figure 3.1.4(c) by a single system with transmission equal to the product of the three branch transmissions. Now there are two feedback loops in parallel around node 1; these are replaced

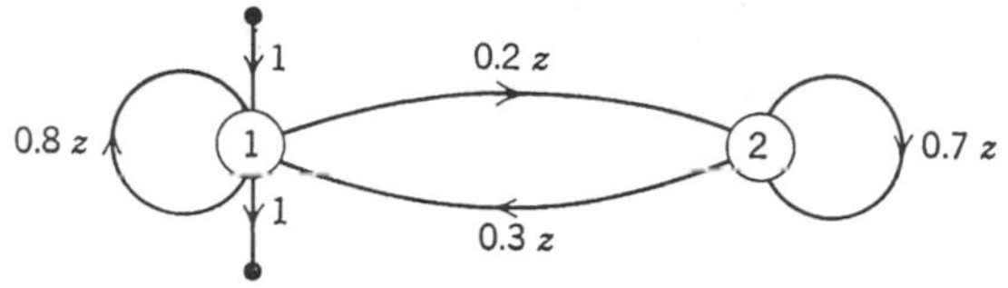


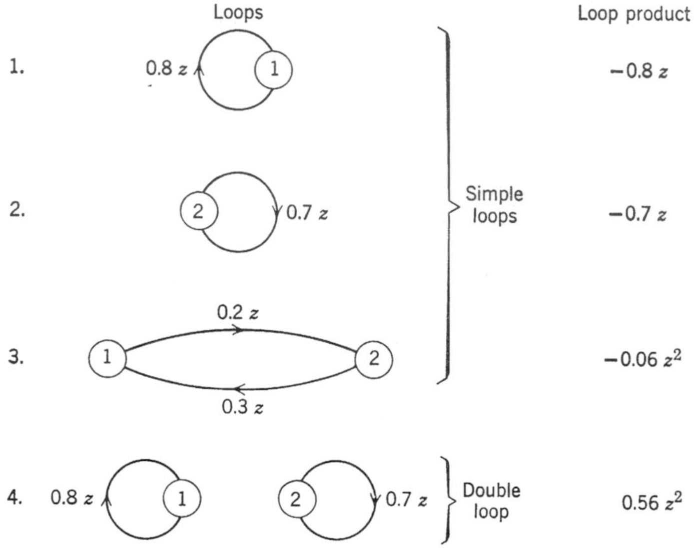


$$\Delta = 1 + \text{sum of loop products} = 1 - 0.8\,z - 0.7\,z - 0.06z^2 + 0.56\,z^2 = 1 - 1.5\,z + 0.5\,z^2$$
$$= (1 - z)\,(1 - \tfrac{1}{2}\,z)$$

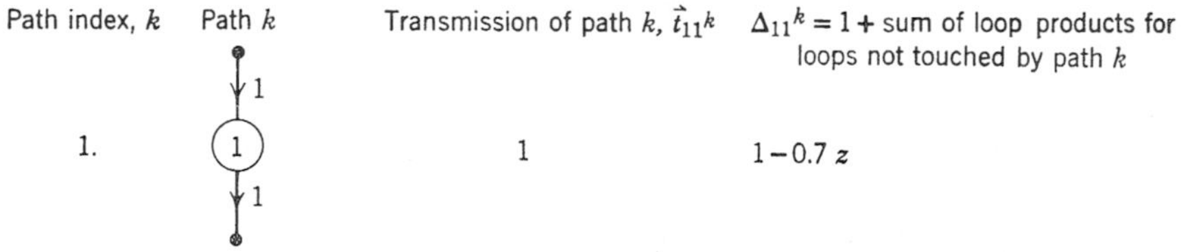


$$\vec{t}_{11} = \frac{\sum_k \vec{t}_{11}{}^k \Delta_{11}{}^k}{\Delta} = \frac{1 - 0.7\,z}{(1 - z)(1 - \frac{1}{2}\,z)} = \phi_{11}{}^{\mathscr{G}}(z)$$

**Figure 3.1.5** Calculation of $\vec{t}_{11} = \phi_{11}{}^{\mathscr{G}}(z)$ for marketing example using inspection method.

in Figure 3.1.4(d) by a single feedback loop with a transmission equal to their sum. Finally, Figure 3.1.4(e) shows this feedback loop replaced by a simple system whose transfer function is $\phi_{11}{}^{g}(z)$, the 11 element of the matrix $\Phi^{g}(z) = [I - Pz]^{-1}$ first found in Equation 1.6.11.

***Solution for $\phi_{11}{}^{g}(z)$ by inspection method***

We calculate the same element of $\Phi^{g}(z)$ in Figure 3.1.5 using the inspection method. The figure shows every step in the calculation. There are three simple loops in the graph; they have a loop product equal to the negative product of the transmissions around them. The one multiple loop has a loop product equal to the products of the loop products of the simple loops from which it is composed. There is only one path; it has a path transmission equal to one; it touches all loops except the feedback loop around node 2. The required transmission is then one plus the loop product for this loop divided by one plus the sum of loop products for the whole graph. We obtain the same result for $\phi_{11}{}^{g}(z)$ found earlier.

Since all the loop visualization and loop product calculation can be done mentally, the transmission of the graph can be written directly, at least in the form with unfactored denominator. The inspection method therefore provides a significant saving in the time required for the calculation of the transform of multistep transition probabilities in simple problems.

***Solution for $\phi_{21}{}^{g}(z)$ using basic reductions***

We need calculate only one element in the second row of the matrix $\Phi^{g}(z)$ for this problem to have the whole matrix because the sum of the rows of $\Phi^{g}(z)$ is $1/(1 - z)$. Let us calculate $\phi_{21}{}^{g}(z)$. This transform will be equal to $\hat{t}_{21}$, the transmission of the flow graph for the Markov process from node 2 to node 1. The calculation is carried out using simple reductions in Figure 3.1.6 to illustrate how to treat the situation where feedback occurs around a node with more than one input or output.

As we know from our discussion of pushing transfer functions through nodes, the simple transmission equivalent to the feedback loop may be placed in series with either all inputs or all outputs of a node. In this example node 1 has 1 input and 2 outputs exclusive of the feedback loop around it. Therefore the equivalent system will appear in fewer places if it is placed in series with the single input. Node 2, on the other hand, has 2 inputs and 1 output exclusive of its feedback loop. The equivalent system will consequently have fewer appearances if it is placed in series with the single output.

The effect of these transformations is the flow graph of Figure 3.1.6(b). According to our discussion in Chapter 2, the signal at each original node is now the signal at the output of the feedback equivalent system. Because the identification of

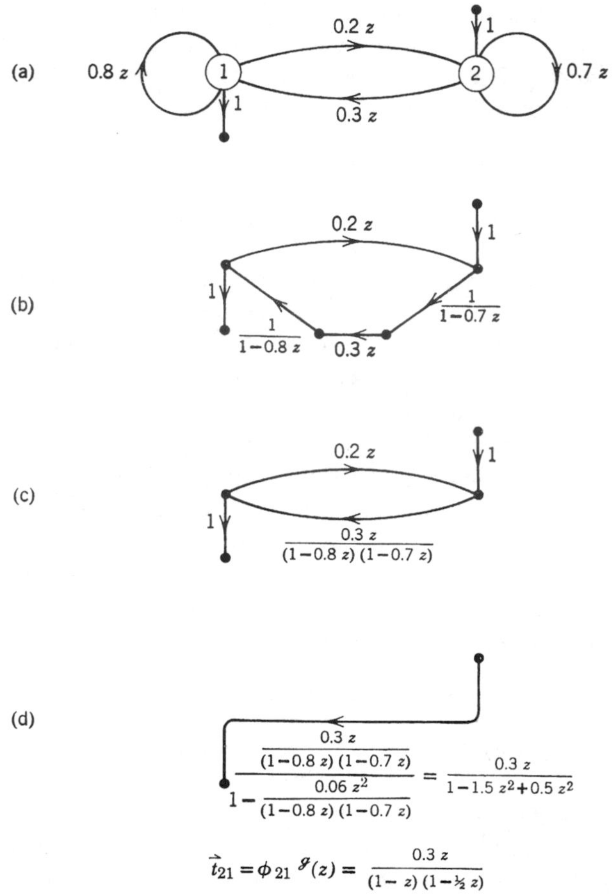


**Figure 3.1.6** Calculation of $\vec{t}_{21} = \phi_{21}{}^{g}(z)$ for marketing example using basic reductions.

these signals will not aid in finding the transmission we require, the node numbers are not shown in this or the remaining parts of Figure 3.1.6.

The three systems in series that provide the forward transmission from node 2 to node 1 are replaced in Figure 3.1.6(c) with a simple system whose transmission is the product of their transmissions. Finally, we apply the result found in Figure 2.4.2 for a system of this form. The result is the system of Figure 3.1.6(d), a simple system whose transfer function is indeed the $\phi_{21}{}^{g}(z)$ we observed in Equation 1.6.11.

***Solution for $\phi_{21}{}^{g}(z)$ by inspection method***

The calculation of $\phi_{21}{}^{g}(z)$ by the inspection method is trivial. The denominator of the transmission expression, $\Delta = (1 - z)(1 - \frac{1}{2}z)$, is unchanged because the loop structure of the graph is the same. Once one transmission has been calculated for any flow graph, we have the denominator for them all. This result follows from a basic property of flow graphs discussed in Chapter 2: The quantity $\Delta$ is just the determinant of the set of linear equations satisfied by the flow graph. Consequently, for Markov processes,

$$\Delta = |I - Pz|. \tag{3.1.2}$$

The numerator is now very simple: there is only one path, with transmission $0.3z$, running from node 2 to node 1; since it touches all the nodes in the graph, it touches all the loops in the graph and is thus multiplied only by one. We obtain immediately the $\phi_{21}{}^{g}(z)$ of Figure 3.1.6. The entries $\phi_{12}{}^{g}(z)$ and $\phi_{22}{}^{g}(z)$ of $\Phi^{g}(z)$ can be written from the requirement that each row of $\Phi^{g}(z)$ sum to $1/(1 - z)$ to make the correspondence with Equation 1.6.11 complete.

The general 2-state Markov process flow graph is no more difficult to solve than this example. The reader can easily verify that the transform of the multistep transition probability matrix formed for a 2-state process in Equation 1.6.21 is just the set of transmissions of the flow graph based on the transition diagram of Figure 1.6.1.

### A Physical Interpretation

But why does a flow graph work for analyzing Markov processes—from a physical rather than mathematical point of view? Well, recall that $z$ is just the transform of a unit delay. Multiplying all the transition probabilities by $z$ corresponds to the delay associated with each transition. In fact, as far as the calculation of multistep probabilities is concerned, we could make a completely deterministic physical equivalent of the Markov process. The only requirement would be a collection of delay lines of the type used in primitive electronic computers; let us suppose that they all provide the same amount of delay. These lines would be connected together like the branches of the flow graph. At each node we would add the outputs of the incoming delay lines electronically. Then we would divide the sum among the lines emanating from each node in proportion to the transition probabilities leaving the node, using, for example, some type of voltage divider arrangement. We would need to provide some small amount of amplification at the nodes to make up for losses in the electronic system.

But what a model we would have! We could apply a unit impulse to, say, node $i$ and then observe, using an oscilloscope, the impulses at each node of the system at each multiple of the delay time. These quantities would just be $\phi_{ij}(n)$, the elements of the $i$th row of the matrix $\Phi(n)$. We could then short out the signals in the system and apply the impulses to another node. In this way we could obtain every row of the matrix $\Phi(n)$.

It is not likely that anyone will ever be willing to pay for such a Markov process model, but it is a good "thought" model for understanding the physical characteristics of the process. It shows, for example, that the Markov process can be used to model situations where a unit amount of some commodity is being divided in different ways at different places in a network, even if this commodity is not probability as it is in most of our studies.

## The Taxicab Example

Let us gain more experience in analyzing Markov processes using flow graphs.

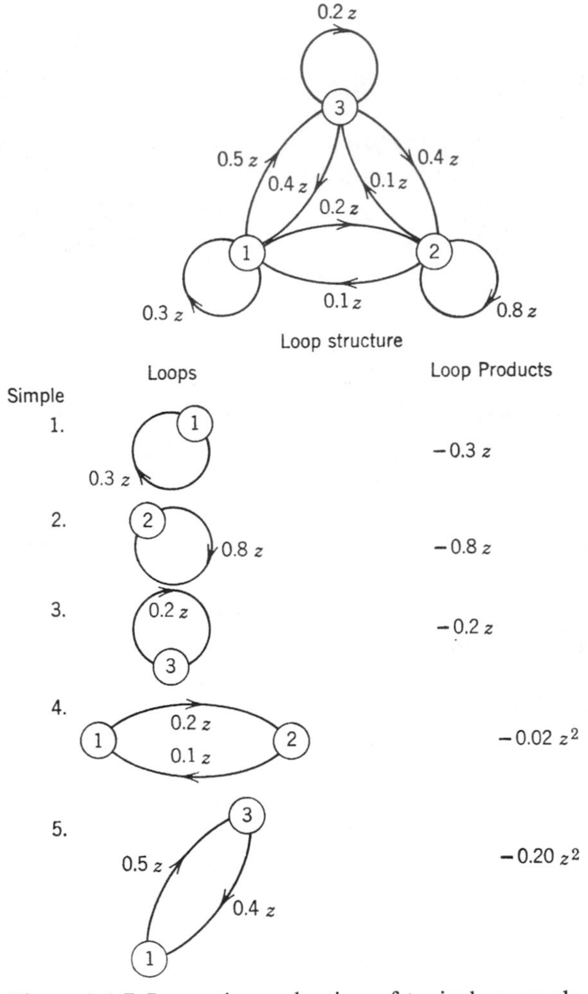


**Figure 3.1.7** Inspection reduction of taxicab example flow graph.

We shall investigate the three-state taxicab example we worked in Chapter 1 whose transition probability matrix was

$$P = \begin{bmatrix} 0.3 & 0.2 & 0.5 \\ 0.1 & 0.8 & 0.1 \\ 0.4 & 0.4 & 0.2 \end{bmatrix}; \tag{3.1.3}$$

the transition diagram appears as Figure 1.4.6(a). All calculations necessary to find the complete multistep transition probability matrix $\Phi^{\mathscr{I}}(z)$ from the flow graph appear in Figure 3.1.7. We see that the loop structure of this process is very complicated in comparison with that of our two-state examples. In fact, this loop structure is as complicated as the loop structure for a three-state process can get. There are 8

Simple

6.

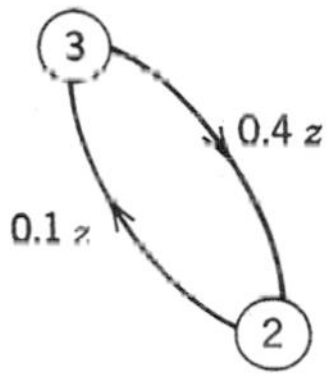


$-0.04\,z^2$

7.

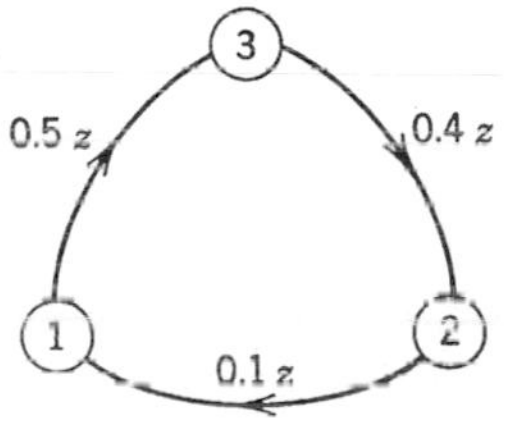


$-0.020\,z^3$

8.

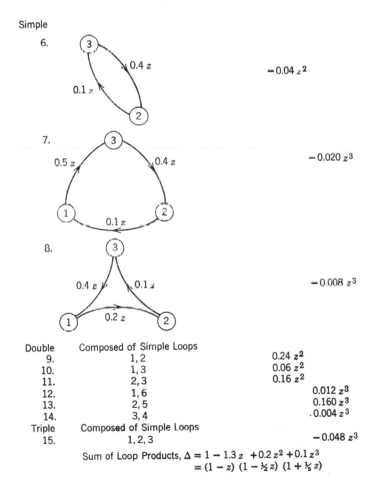

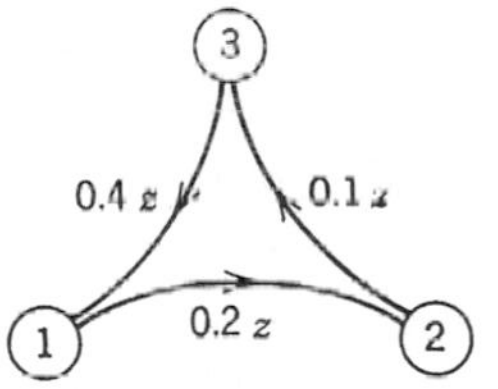


$-0.008\,z^3$

| | Composed of Simple Loops | | |
|---|---|---|---|
| Double | | | |
| 9. | 1, 2 | $0.24\,z^2$ | |
| 10. | 1, 3 | $0.06\,z^2$ | |
| 11. | 2, 3 | $0.16\,z^2$ | |
| 12. | 1, 6 | | $0.012\,z^3$ |
| 13. | 2, 5 | | $0.160\,z^3$ |
| 14. | 3, 4 | | $0.004\,z^3$ |
| Triple | Composed of Simple Loops | | |
| 15. | 1, 2, 3 | | $-0.048\,z^3$ |

Sum of Loop Products, $\Delta = 1 - 1.3z + 0.2z^2 + 0.1z^3$
$= (1 - z)(1 - \tfrac{1}{2}z)(1 + \tfrac{1}{5}z)$

| Input $i$ | Output $j$ | Path Structure: Path set $P_{ij}$ | Path index $k$ | Path transmission $\vec{t}_{ij}{}^k$ | Loops not touched by path $k$ | 1 + Sum of loop products of loops not touched by path $k$ $\Delta_{ij}{}^k$ | $\sum_{k \in P_{ij}} \vec{t}_{ij}{}^k \Delta_{ij}{}^k$ |
|---|---|---|---|---|---|---|---|
| 1 | 1 | 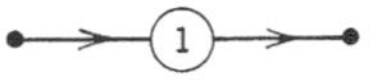  | 1 | 1 | 2, 3, 6, 11 | $1-z+0.12z^2$ | $1-z+0.12z^2$ |
| 1 | 2 | 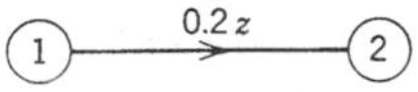  | 1 | $0.2z$ | 3 | $1-0.2z$ | |
| | | 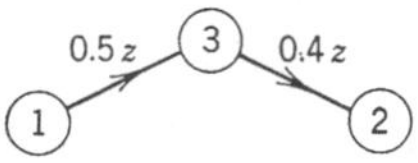  | 2 | $0.20z^2$ | – | 1 | $0.2z+0.16z^2$ |
| 1 | 3 | 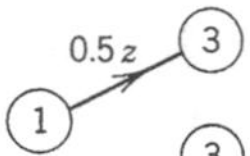  | 1 | $0.5z$ | 2 | $1-0.8z$ | |
| | | 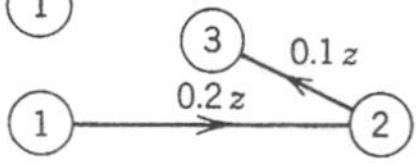  | 2 | $0.02z^2$ | – | 1 | $0.5z-0.38z^2$ |
| 2 | 1 | 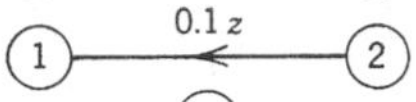  | 1 | $0.1z$ | 3 | $1-0.2z$ | |
| | | 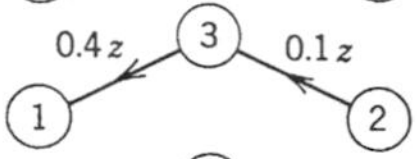  | 2 | $0.04z^2$ | – | 1 | $0.1z+0.02z^2$ |
| 2 | 2 | 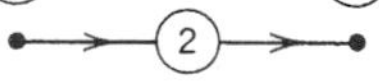  | 1 | 1 | 1, 3, 5, 10 | $1-0.5z-0.14z^2$ | $1-0.5z-0.14z^2$ |
| 2 | 3 | 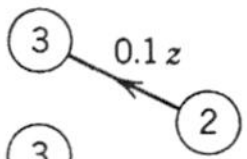  | 1 | $0.1z$ | 1 | $1-0.3z$ | |
| | | 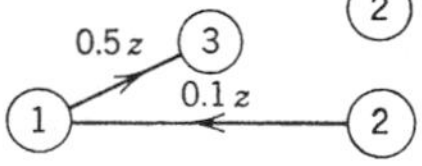  | 2 | $0.05z^2$ | – | 1 | $0.1z+0.02z^2$ |
| 3 | 1 | 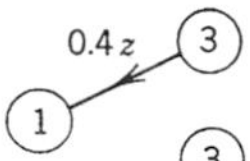  | 1 | $0.4z$ | 2 | $1-0.8z$ | |
| | | 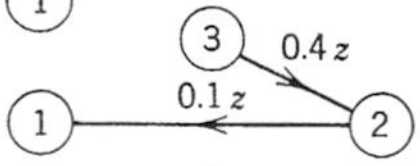  | 2 | $0.04z^2$ | – | 1 | $0.4z-0.28z^2$ |
| 3 | 2 | 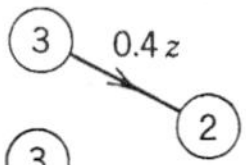  | 1 | $0.4z$ | 1 | $1-0.3z$ | |
| | | 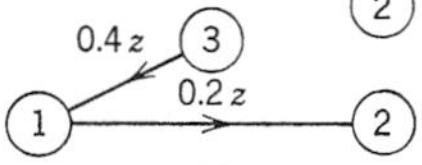  | 2 | $0.08z^2$ | – | 1 | $0.4z-0.04z^2$ |
| 3 | 3 |   | 1 | 1 | 1, 2, 4, 9 | $1-1.1z+0.22z^2$ | $1-1.1z+0.22z^2$ |

$$\vec{t}_{ij} = \frac{1}{\Delta} \sum_{k \in P_{ij}} \vec{t}_{ij}{}^k \Delta_{ij}{}^k = \phi_{ij}{}^g(z)$$

$$\vec{T} = \Phi^g(z) = \frac{1}{(1-z)(1-\frac{1}{2}z)(1+\frac{1}{5}z)} \begin{bmatrix} 1-z+0.12z^2 & 1-0.2z+0.16z^2 & 0.5z-0.38z^2 \\ 0.1z+0.02z^2 & 1-0.5z-0.14z^2 & 0.1z+0.02z^2 \\ 0.4z-0.28z^2 & 0.4z-0.04z^2 & 1-1.1z+0.22z^2 \end{bmatrix}$$

simple loops, 6 double loops, and 1 triple loop. Each of the 8 simple loops is drawn; each loop product is the negative product of transmissions around the loop. The double and triple loops are indicated by the numbers of the simple loops they contain; each of their loop products is the product of the loop products of these simple loops. By summing all the loop products for the graph and then adding one, we obtain the denominator of all transmission expressions; it is, of course, just the determinant $|I - Pz|$.

Every transmission for this problem involves either one or two paths; each path is drawn in Figure 3.1.7. The path transmissions and one plus the sum of loop products not touched by each path are calculated, multiplied, and summed over the paths pertinent to each transmission. When the resulting quantity is divided by $\Delta$, we have the corresponding transmission of the graph. This is found in every case to check with the $\Phi^{\mathscr{g}}(z)$ entries in Equation 1.6.39.

The procedure we have followed is not quite so complicated as it seems because many of the steps made explicit here are really mental. However, it is complicated and there seems to be little hope of carrying it out when the number of states is around 5 or 6. Fortunately, most of our problems have some regularity that allows us to find all transmissions with less difficulty than the size of the problem would indicate. We shall see many examples of such problems in our later work. In fact the other three-state processes we have already considered are examples.

### The Duodesmic Process

Recall the duodesmic three-state process we studied in Chapter 1; its transition matrix was

$$P = \begin{bmatrix} 1 & 0 & 0 \\ 0 & 1 & 0 \\ 0.3 & 0.2 & 0.5 \end{bmatrix}; \tag{3.1.4}$$

its transition diagram appears in Figure 1.4.7(a). The flow graph for this process is shown in Figure 3.1.8. We could apply the inspection method directly to this graph, but we would obtain many cancellations because the graph is so disconnected. The analysis of Figure 3.1.8 takes advantage of this special structure by using the results for isolated encysted systems. Each transmission in the graph is shown as the transmission of a much smaller graph equivalent to the large one for this particular transmission. The transmissions from states 1 or 2 and the transmission $\vec{t}_{33}$ are trivial calculations after this decomposition is made. The transmissions $\vec{t}_{31}$ and $\vec{t}_{32}$ are the transmissions of three systems in series; the end systems are feedback loops readily replaced by their simple equivalents. Thus the whole

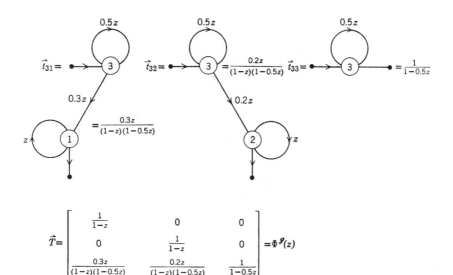

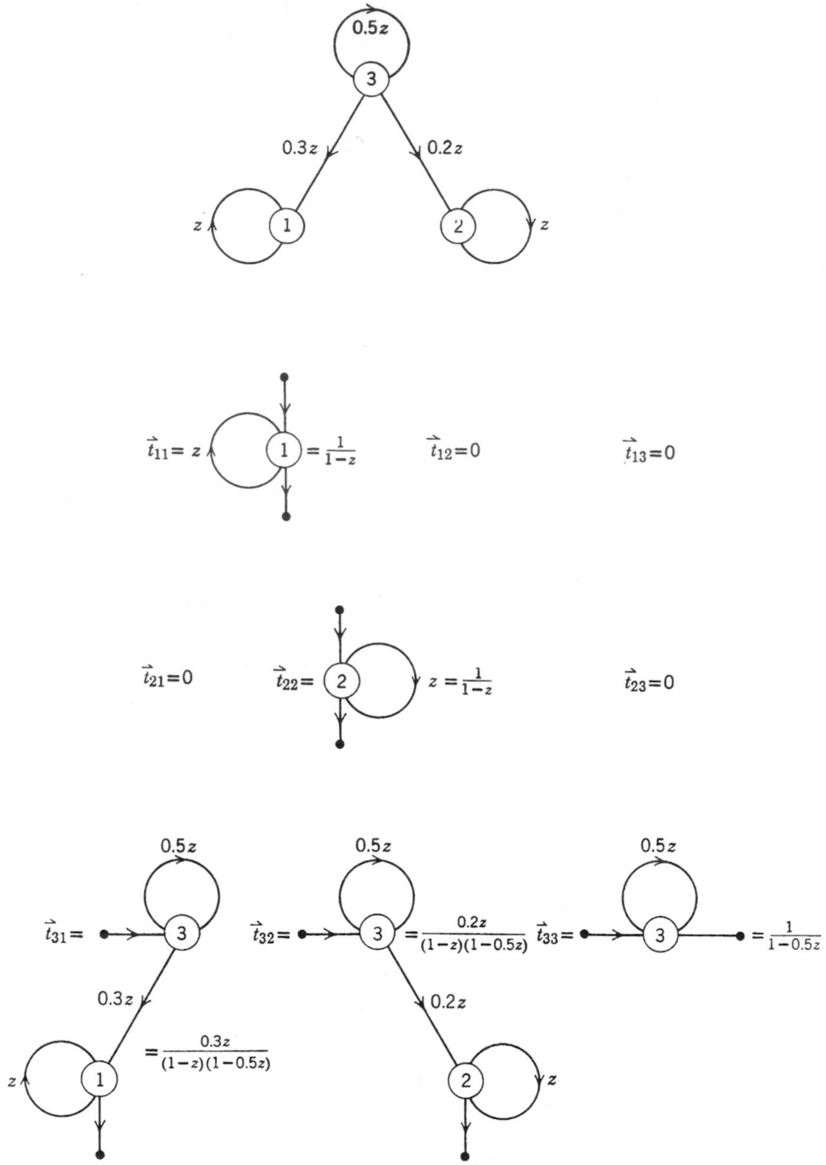


$$\vec{T} = \begin{bmatrix} \frac{1}{1-z} & 0 & 0 \\ 0 & \frac{1}{1-z} & 0 \\ \frac{0.3z}{(1-z)(1-0.5z)} & \frac{0.2z}{(1-z)(1-0.5z)} & \frac{1}{1-0.5z} \end{bmatrix} = \Phi^{\mathscr{g}}(z)$$

**Figure 3.1.8** Flow graph analysis of a duodesmic process.

matrix of transmissions can be written without a pause. This matrix is the matrix $\Phi^{\mathscr{G}}(z)$ of Equation 1.6.60.

**The Identity Process**

Two other three-state examples we discussed are even simpler to analyze. The first is the identity process, with transition matrix,

$$P = \begin{bmatrix} 1 & 0 & 0 \\ 0 & 1 & 0 \\ 0 & 0 & 1 \end{bmatrix} = I, \tag{3.1.5}$$

and the transition diagram of Figure 1.4.8. Its flow graph and its transmission matrix appear in Figure 3.1.9. We can obtain the transmissions directly by taking advantage of the disconnected nature of the graph. If we choose to use the inspection method, however, we get the same results. For example, the inspection method shows that the transmissions $\vec{t}_{11}$, $\vec{t}_{22}$, $\vec{t}_{33}$ are given by

$$\vec{t}_{11} = \vec{t}_{22} = \vec{t}_{33} = \frac{1 - z - z + z^2}{1 - z - z - z + z^2 + z^2 + z^2 - z^3}$$

$$= \frac{1 - 2z + z^2}{1 - 3z + 3z^2 - z^3} = \frac{(1 - z)^2}{(1 - z)^3} = \frac{1}{1 - z}, \tag{3.1.6}$$

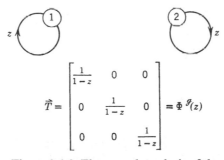

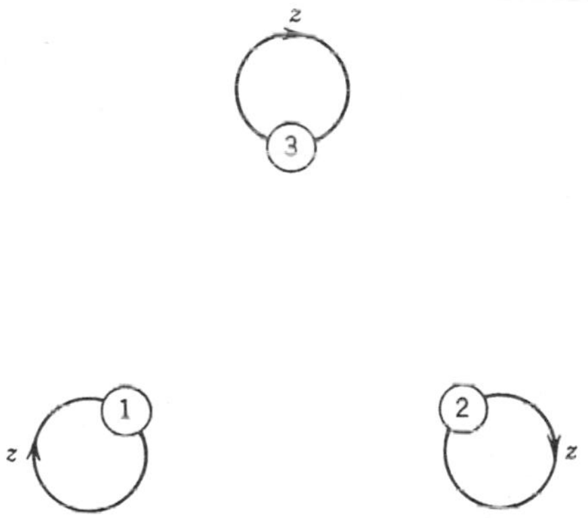


$$\vec{T} = \begin{bmatrix} \frac{1}{1-z} & 0 & 0 \\ 0 & \frac{1}{1-z} & 0 \\ 0 & 0 & \frac{1}{1-z} \end{bmatrix} = \Phi^{\mathscr{G}}(z)$$

**Figure 3.1.9** Flow graph analysis of the three-state identity process.

whereas $\vec{t}_{ij} = 0$ for $i \neq j$. The transmissions are consistent with Equation 1.6.67.

### The Periodic Process

The three-state periodic case, with transition matrix

$$P = \begin{bmatrix} 0 & 1 & 0 \\ 0 & 0 & 1 \\ 1 & 0 & 0 \end{bmatrix} \tag{3.1.7}$$

and transition diagram shown in Figure 1.4.9(a), has the flow graph of Figure 3.1.10. We can easily find every transmission of this flow graph by either simple reduction or the inspection method; in this case neither has a relative advantage. The matrix of transmissions is the $\Phi^{\mathcal{G}}(z)$ matrix of Equation 1.6.72.

### Summary

A matrix $\Phi^{\mathcal{G}}(z)$ obtained from inspection of a flow graph is thereafter treated as if it had been calculated by inversion of the matrix $[I - Pz]$. Our new procedure for calculating the multistep transition probability matrix $\Phi(n)$ thus has four steps: 1) draw the flow graph for the Markov process by multiplying each branch transmission of its transition diagram by $z$; 2) find the matrix of transmissions of the flow graph, the matrix $\Phi^{\mathcal{G}}(z)$; 3) perform a partial fraction expansion of $\Phi^{\mathcal{G}}(z)$; 4) invert the geometric transforms contained in this expansion to obtain $\Phi(n)$. If we need to find only some of the multistep transition probabilities of the process, then we can abridge this procedure by calculating only those transmissions of the flow graph.

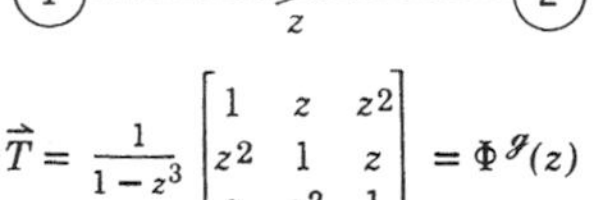


$$\vec{T} = \frac{1}{1-z^3}\begin{bmatrix} 1 & z & z^2 \\ z^2 & 1 & z \\ z & z^2 & 1 \end{bmatrix} = \Phi^{\mathcal{G}}(z)$$

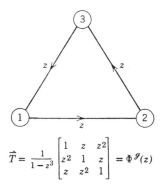

**Figure 3.1.10** Flow graph analysis of the three-state periodic process.

Whether used for part or the whole of the computational procedure, the interpretation of the Markov process as a linear system described by a flow graph provides important insight into the Markov process as a model. We can ascertain the effect of changing particular transition probabilities on the total behavior of the process. Flow graph theory makes the transition diagram an aid to computation as well as an aid to visualization.

## 3.2 MATRIX FLOW GRAPH ANALYSIS OF MARKOV PROCESSES

We have already mentioned that any Markov process can be represented by the flow graph of Figures 3.1.1 and 3.1.2. However, we often find it convenient to draw somewhat more specialized matrix flow graphs for the description, computation, and illumination of process relationships.

### Canonical Forms

For example, consider any $N$-state monodesmic Markov process that has $N_1 < N$ transient states and $N_2 = N - N_1$ recurrent states. We might find it convenient to number the states of the process so that its transition probability matrix would take the following form:

$$P = \begin{array}{c} \\ N_1 \updownarrow \\ N_2 \updownarrow \end{array} \begin{array}{c} \overset{N_1}{\longleftrightarrow} \; \overset{N_2}{\longleftrightarrow} \\ \left[\begin{array}{c|c} {}^{11}P & {}^{12}P \\ \hline 0 & {}^{22}P \end{array}\right] \end{array}. \tag{3.2.1}$$

The $N_1$ transient states are designated by $i = 1, 2, \ldots, N_1$; the $N_2$ recurrent states by $i = N_1 + 1, N_1 + 2, \ldots, N$. The quantity ${}^{11}P$ is an $N_1$ by $N_1$ matrix showing the transition probabilities that exist between transient states, and ${}^{12}P$ is an $N_1$ by $N_2$ matrix showing the transition probabilities between transient states and recurrent states. The $N_2$ by $N_2$ matrix ${}^{22}P$ is the matrix specifying transition probabilities among the recurrent states of the process; ${}^{22}P$ is, of course, a stochastic matrix that could itself govern a Markov process. We shall call a transition probability matrix that has been placed in the form of Equation 3.2.1 the transition probability matrix for the process in canonical (standard) form.

#### *Graphical representations*

We represent a process whose transition probability matrix is in canonical form by the matrix transition diagram of Figure 3.2.1(a), or by the matrix flow graph of Figure 3.2.1(b). We can perform any computation we desire on the Markov process in this form. The transient portion of the process consists of the $N_1$

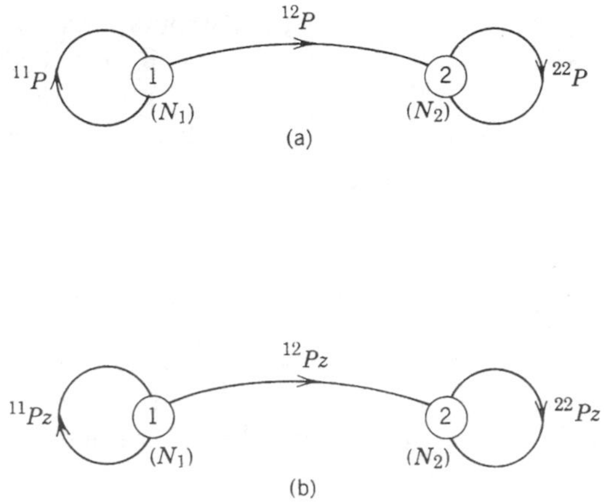


**Figure 3.2.1** Canonical form for monodesmic process. (a) matrix transition diagram. (b) Matrix flow graph.

transient states and is described by the matrices ${}^{11}P$ and ${}^{12}P$; we call this a transient process. In a transient process we are usually interested in the time for the system to leave the transient process and, hence, enter one of the $N_2$ recurrent states. In Chapter 4 we shall study such transient processes.

***Polydesmic processes***

We can extend the canonical form to the polydesmic process by placing the transition probability matrix for the process in the form

$$P = \begin{array}{c} \\ N_1 \updownarrow \\ N_2 \updownarrow \\ N_i \updownarrow \\ N_{M-1} \updownarrow \\ N_M \updownarrow \end{array} \begin{array}{c} \begin{array}{ccccc} \overset{N_1}{\longleftrightarrow} & \overset{N_2}{\longleftrightarrow} & \overset{N_i}{\longleftrightarrow} & \overset{N_{M-1}}{\longleftrightarrow} & \overset{N_M}{\longleftrightarrow} \end{array} \\ \left[\begin{array}{c:c:c:c:c} {}^{11}P & {}^{12}P & {}^{1i}P & {}^{1,M-1}P & {}^{1M}P \\ \hdashline 0 & {}^{22}P & 0 & 0 & 0 \\ \hdashline 0 & 0 & {}^{ii}P & 0 & 0 \\ \hdashline 0 & 0 & 0 & {}^{M-1,M-1}P & 0 \\ \hdashline 0 & 0 & 0 & 0 & {}^{MM}P \end{array}\right] \end{array}. \tag{3.2.2}$$

Here we have $N_1$ transient states and $M - 1$ recurrent chains with $N_2, N_3, \ldots, N_M$ states whose numbering has been arranged so as to place the transition probability matrix in the form shown. The matrices ${}^{22}P$, ${}^{33}P, \ldots, {}^{MM}P$ are all stochastic matrices and, hence, could themselves serve as transition probability matrices. In some situations we may want to consider a further partitioning of the $N_1$ transition states according to the recurrent chains that may be reached from them, but usually the canonical form of Equation 3.2.2 is sufficient. The matrix transition diagram and matrix flow graphs of Figures 3.2.2(a) and 3.2.2(b) follow from the canonical form for the $M - 1$ chain process. They are helpful in understanding and computing the properties of the process.

## General Partitioning

However, matrix flow graph techniques are not restricted to Markov processes in canonical form or to any other special type of Markov process. If we have a general $N$-state process, we can partition its states into $M$ groups indexed 1, 2, . . . ,

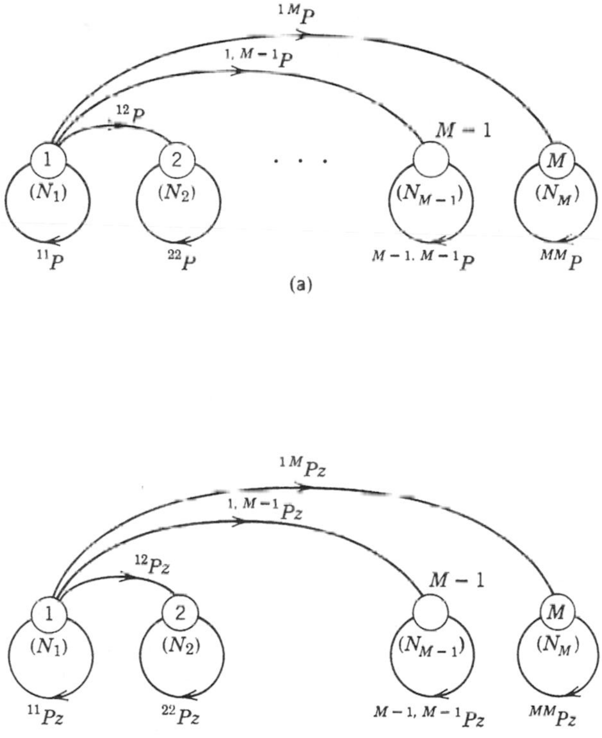


**Figure 3.2.2** Canonical form for polydesmic process. (a) Matrix transition diagram. (b) Matrix flow graph.

$M$, containing $N_1, N_2, \ldots, N_M$ states, and after a possible renumbering of states, draw its transition probability matrix in the form

$$P = \begin{array}{c} \\ N_1 \updownarrow \\ N_2 \updownarrow \\ \\ N_M \updownarrow \end{array} \begin{array}{c} \begin{array}{cccc} \overset{N_1}{\longleftrightarrow} & \overset{N_2}{\longleftrightarrow} & & \overset{N_M}{\longleftrightarrow} \end{array} \\ \left[\begin{array}{c:c:c:c} {}^{11}P & {}^{12}P & & {}^{1M}P \\ \hdashline {}^{21}P & {}^{22}P & & {}^{2M}P \\ \hdashline & & & \\ \hdashline {}^{M1}P & {}^{M2}P & & {}^{MM}P \end{array}\right] \end{array}. \qquad (3.2.3)$$

Thus, the $N$-state Markov process becomes an $M$-node matrix Markov process. The transmission from the group $i$ node to the group $j$ node in the matrix transition diagram for the process would be the $N_i$ by $N_j$ dimensional matrix ${}^{ij}P$; the corresponding transmission in the matrix flow graph would be ${}^{ij}Pz$. While this representation is possible for many processes, it will be particularly useful when many of the matrices ${}^{ij}P$ are composed entirely of zeros, for in this case considerable simplification in the computation of process relationships may result.

As a specific example, suppose that we partition an $N$-state process into two

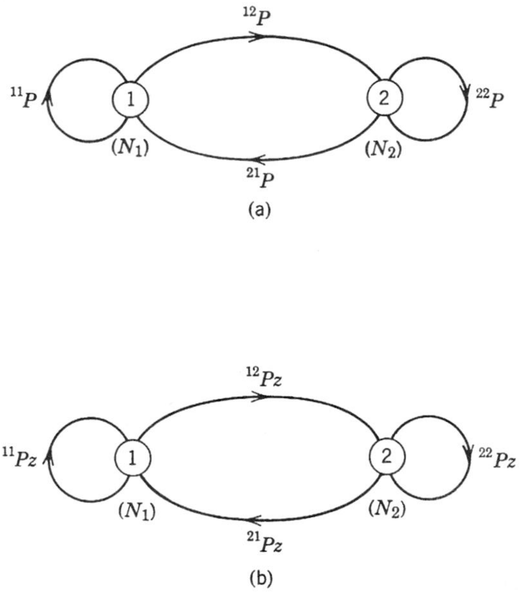


**Figure 3.2.3** Two-group matrix process. (a) Matrix transition diagram. (b) Matrix flow graph.

groups composed of its first $N_1$ states $1, 2, \ldots, N_1$ and the remaining $N_2 = N - N_1$ states $N_1 + 1, N_1 + 2, \ldots, N$. The transition probability matrix would then be

$$P = \begin{array}{c} \\ N_1 \updownarrow \\ N_2 \updownarrow \end{array} \overset{\quad N_1 \qquad N_2}{\overset{\longleftrightarrow \; \longleftrightarrow}{\left[\begin{array}{c:c} {}^{11}P & {}^{12}P \\ \hdashline {}^{21}P & {}^{22}P \end{array}\right]}} \tag{3.2.4}$$

and would be represented by the matrix transition diagram and matrix flow graph shown in Figures 3.2.3(a) and 3.2.3(b). These representations differ from the canonical form for the monodesmic process because ${}^{21}P$ is not necessarily composed of zeros.

## The Taxicab Problem

Let us now apply the matrix flow graph technique to a specific example, the three-state taxicab problem of Chapter 1 whose transition probability matrix appears in Equation 3.1.3. We partition the states into two groups: the first group consisting only of state 1, the second group consisting of states 2 and 3. Thus, $N_1 = 1$, $N_2 = 2$, and we can draw the transition probability matrix in the form

$$P = \left[\begin{array}{c:cc} 0.3 & 0.2 & 0.5 \\ \hdashline 0.1 & 0.8 & 0.1 \\ 0.4 & 0.4 & 0.2 \end{array}\right] = \left[\begin{array}{c:c} {}^{11}P & {}^{12}P \\ \hdashline {}^{21}P & {}^{22}P \end{array}\right]. \tag{3.2.5}$$

This transition probability matrix leads directly to the matrix transition diagram of Figure 3.2.4(a) and to the matrix flow graph of Figure 3.2.4(b). Suppose now that we are interested only in the multistep transition probability $\phi_{11}(n)$, the probability that the system will occupy state 1 after $n$ transitions given that it starts in state 1. We can find this probability by reducing the matrix flow graph until we obtain an expression for the transmission from state 1 to state 1.

We begin by changing the feedback matrix system ${}^{22}Pz$ surrounding node 2 in the matrix flow graph into the equivalent system with no feedback $[I - {}^{22}Pz]^{-1}$ as shown in Figure 3.2.4(c). We can evaluate $[I - {}^{22}Pz]^{-1}$ by computation of the inverse or by drawing an equivalent flow graph and then finding its transmission. Figure 3.2.4(d) shows the ordinary flow graph that is implied by the feedback system ${}^{22}Pz$. The $ij$th transmission of this flow graph is the $ij$th element of $[I - {}^{22}Pz]^{-1}$ as recorded in Figure 3.2.4(c).

Now that we have evaluated $[I - {}^{22}Pz]^{-1}$, we find the equivalent transmission of the three series systems with transmissions ${}^{12}Pz$, $[I - {}^{22}Pz]^{-1}$, ${}^{21}Pz$ in Figure 3.2.4(c) as the product of the three transmissions and record the result in Figure 3.2.4(e). The two feedback systems with transmissions ${}^{11}Pz$ and ${}^{12}Pz[I - {}^{22}Pz]^{-1}\,{}^{21}Pz$

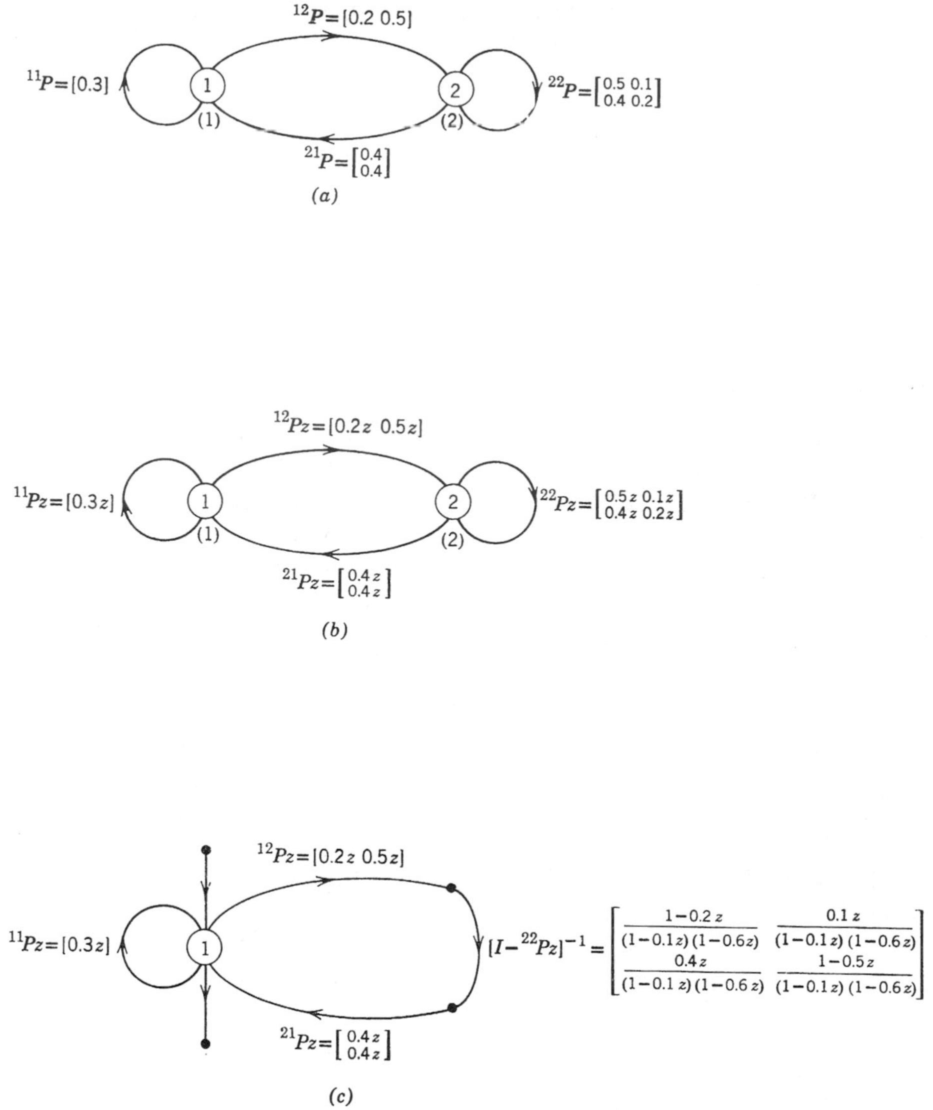


**Figure 3.2.4** The partitioned taxicab problem. (a) Matrix transistor diagram. (b) Matrix flow graph. (c) First reduction of matrix flow graph. (d) Flow graph corresponding to $[I - {}^{22}Pz]^{-1}$ (e) Second reduction of matrix flow graph. (f) Third reduction of matrix flow graph.

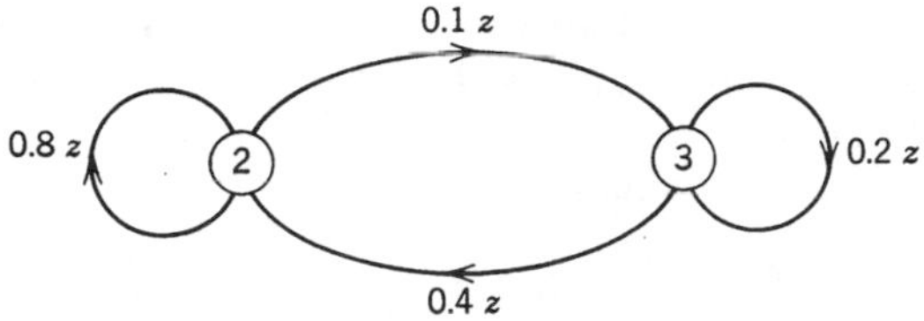


(d)

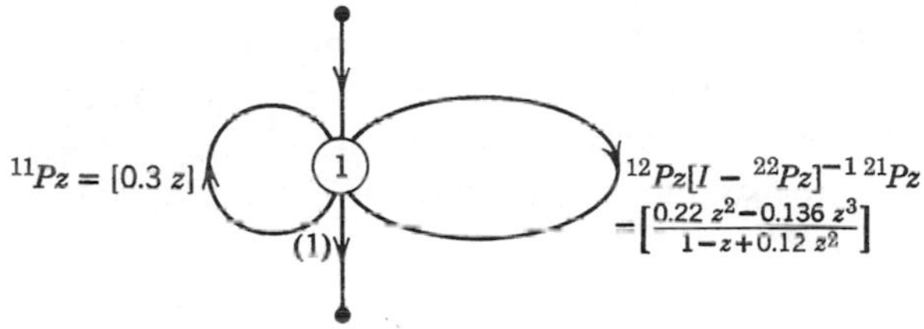


(e)

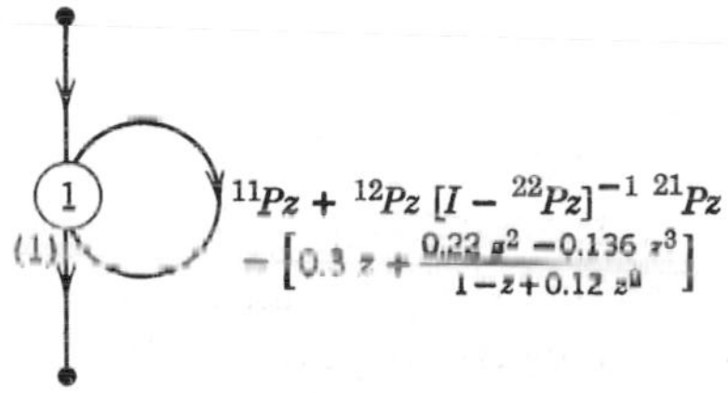


$$\vec{t}_{11} = \phi_{11}{}^{\mathcal{g}}(z) = \frac{1}{1 - \left(0.3\,z + \dfrac{0.22\,z^2 - 0.136\,z^3}{1 - z + 0.12\,z^2}\right)} = \frac{1 - z + 0.12\,z^2}{(1 - z)\,(1 - 0.5\,z)\,(1 + 0.2\,z)}$$

(f)

are in parallel and can have their transmissions added as shown in Figure 3.2.4(f). We note that this feedback system is one-dimensional, and hence that we can find the transmission of the graph as the reciprocal of one minus the feedback transmission. We obtain

$$\vec{t}_{11} = \phi_{11}{}^{g}(z) = \frac{1 - z + 0.12z^2}{(1 - z)(1 - 0.5z)(1 + 0.2z)}, \tag{3.2.6}$$

which agrees with the 11 element of $\Phi^{g}(z)$ in Equation 1.6.39 or Figure 3.1.7.

Thus, we can use matrix flow graph methods in the solution of any problem. However, the advantage will be greatest when we partition the process so that many of the matrix transmissions will be composed of zeros. Note that if we take $N_1 = 1$ and $N_2 = N - 1$ in the transition probability matrix of Equation 3.2.4, we have found a way to reduce an $N$ state problem to an $N - 1$ state problem just as we did in the example. This method of reducing the system can be very effective in special cases.

## Mergeable Processes

Suppose we can partition the $N$ states of a process into $M$ groups so that the process is mergeable. We do this by assuring that condition 1.4.40 is met by any pair of groups of states. If we say that the $i$th group of states has $N_i$ members and number states appropriately, we draw the partitioned probability matrix in the form of Equation 3.2.3. However, this form now represents not an arbitrary partitioning into $M$ groups for an $N$ state process, but rather a mergeable partitioning. This means that if we consider any off-diagonal matrix ${}^{ij}P$, $i \neq j$, in the transition probability matrix of Equation 3.2.3, the sum of the elements in each row must be the same for all rows.

Suppose that we begin to perform the matrix flow graph analysis of a process by using a partitioning of states that makes the process mergeable. Would it reveal the mergeability? Let us see. Recall the three-state mergeable process whose transition probability matrix appears in Equation 1.4.44. We use the partitioning of Equation 3.2.4 with $N_1 = 1$, $N_2 = 2$,

$$P = \left[\begin{array}{c|cc} 0.3 & 0.2 & 0.5 \\ \hline 0.4 & 0.5 & 0.1 \\ 0.4 & 0.4 & 0.2 \end{array}\right] = \left[\begin{array}{c|c} {}^{11}P & {}^{12}P \\ \hline {}^{21}P & {}^{22}P \end{array}\right] \tag{3.2.7}$$

and note that the process with this partitioning is in fact mergeable. Now we draw the corresponding matrix transition diagram and matrix flow graph of Figures 3.2.5(a) and 3.2.5(b). Suppose that we are again interested in the multistep transition probability $\phi_{11}(n)$, the probability that the system will occupy state 1 after $n$ transitions given that it starts in state 1. We perform the first step of reduction by complete analogy with Figure 3.2.4. The inverse $[I - {}^{22}Pz]^{-1}$ that appears in Figure 3.2.5(c) can be computed directly or as the transmission of the ordinary equivalent flow graph for the feedback around group 2; the appropriate flow graph for this purpose appears as Figure 3.2.5(d). Regardless of how the inverse $[I - {}^{22}Pz]^{-1}$ is obtained, the most interesting step occurs when we multiply the three transmissions ${}^{11}Pz$, $[I - {}^{22}Pz]^{-1}$, and ${}^{21}Pz$ to obtain the equivalent feedback transmission of the part of the system to the right of node 1. We find

$$
\begin{aligned}
{}^{11}Pz[I - {}^{22}Pz]^{-1}\,{}^{21}Pz &= \frac{0.28z^2(1 - 0.1z)}{(1 - 0.1z)(1 - 0.6z)} = \frac{0.28z^2}{1 - 0.6z} \\
&= \frac{(0.7z)(0.4z)}{1 - 0.6z}
\end{aligned}
\tag{3.2.8}
$$

and note that one of the denominator factors has been eliminated. The form of this equation means that the entire structure of the graph to the right of node 1 in Figure 3.2.5(c) could be replaced by a node with a feedback $0.6z$ in series with two branches with transmissions $0.7z$ and $0.4z$. We draw this form of Figure 3.2.5(c) as Figure 3.2.5(e). The resulting flow graph is an ordinary two-node flow graph and not a matrix flow graph. The transmission $\vec{t}_{11} = \phi_{11}{}^{g}(z)$ can be written by inspection of this graph as

$$
\begin{aligned}
\vec{t}_{11} = \phi_{11}{}^{g}(z) &= \frac{1 - 0.6z}{1 - 0.9z - 0.1z^2} = \frac{1 - 0.6z}{(1 - z)(1 + 0.1z)} \\
&= \frac{4/11}{1 - z} + \frac{7/11}{1 + 0.1z}.
\end{aligned}
\tag{3.2.9}
$$

Then we obtain the multistep transition probability $\phi_{11}(n)$ by inverse transformation as

$$
\phi_{11}(n) = 4/11 + (7/11)(-0.1)^n \qquad n = 0, 1, 2, \ldots. \tag{3.2.10}
$$

Thus, as far as the transitions between the two groups of states are concerned, we have been able to convert the original three-state process into a two-state process. In fact, the flow graph of Figure 3.2.5(e) is just the flow graph corresponding to the transition diagram of Figure 1.4.13 obtained in Chapter 1 for the merged

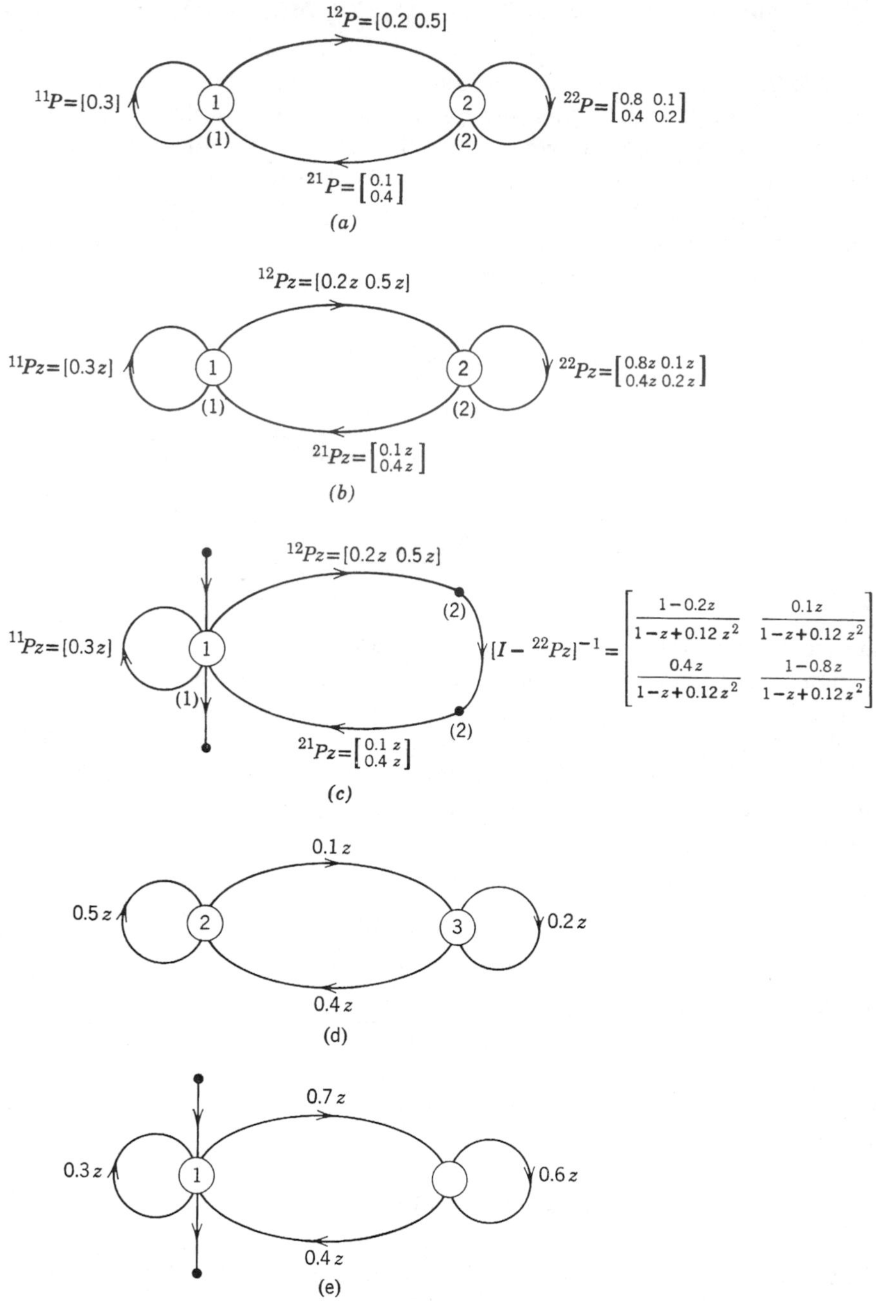


**Figure 3.2.5** A mergeable process example. (a) Matrix transition diagram. (b) Matrix flow graph. (c) First reduction of matrix flow graph. (d) Flow graph corresponding to $[I - {}^{22}Pz]^{-1}$. (e) Two-state flow graph for merged process.

process. Indeed, by referring to the solution for $\Phi(n)$ for the general two-state process found in Chapter 1, we can write the entire multistep transition probability matrix for the merged process. The transition diagram for the general two-step process appears in Figure 1.6.1, and so by comparing it with the flow graph of Figure 3.2.5(e) we take $a = 0.7$, $b = 0.4$ in the solution of Equation 1.6.23 and obtain

$$\Phi(n) = \begin{bmatrix} 4/11 & 7/11 \\ 4/11 & 7/11 \end{bmatrix} + (-0.1)^n \begin{bmatrix} 7/11 & -7/11 \\ -4/11 & 4/11 \end{bmatrix} \qquad n = 0, 1, 2, \ldots. \qquad (3.2.11)$$

The 11 element of $\Phi(n)$ is, of course, Equation 3.2.10.

In interpreting Equation 3.2.11 we must be careful to realize that it is providing multistep transition probabilities between merged groups and not between states. Thus

$$\phi_{12}(n) = 4/11 + (7/11)(-0.1)^n \qquad n = 0, 1, 2, \ldots \qquad (3.2.12)$$

is the probability that if the process starts in group 1 (state 1) at time zero, it will be in group 2 (either state 2 or state 3) after $n$ transitions. A corresponding statement applies to each of the multistep transition probabilities contained in $\Phi(n)$.

From this and our earlier discussion on mergeable processes we can draw the following conclusion. The matrix flow graph of a Markov process that is constructed according to a partition that makes the process mergeable can be replaced with an ordinary flow graph whose states represent the groups of the merged process. The transition probability from group $i$ to group $j$ is just the common row sum of ${}^{ij}P$ for all $i$ and $j$. This representation is adequate for all questions that can be phrased in terms of transitions between groups, but not for questions involving transitions within a group.

## PROBLEMS

**1.** Many of the processes in the problems of Chapter 1 can be analyzed by flow graph methods. Specifically:

a) Find $\phi_{12}(n)$ by flow graph techniques for each of the processes in Problem 1 of Chapter 1.

b) Draw the flow graph for Problem 3 of Chapter 1 and then find a new two-state flow graph for the process whose two states correspond to the generation of a 1 and a 0. The branch transmissions will be finite polynomials in $z$.

c) Solve parts a) and c) of Problem 9 of Chapter 1 using flow graphs.

d) In Problem 11 of Chapter 1 if Jones is in common market $\alpha$, what is the probability that he will be in common market $\beta$ $n$ trips from now? Find the answer using flow graphs.

e) Solve parts a) and b) of Problem 12 of Chapter 1 using flow graphs.

**2.** Find $\Phi(n)$ for the following Markov process using flow graph techniques:

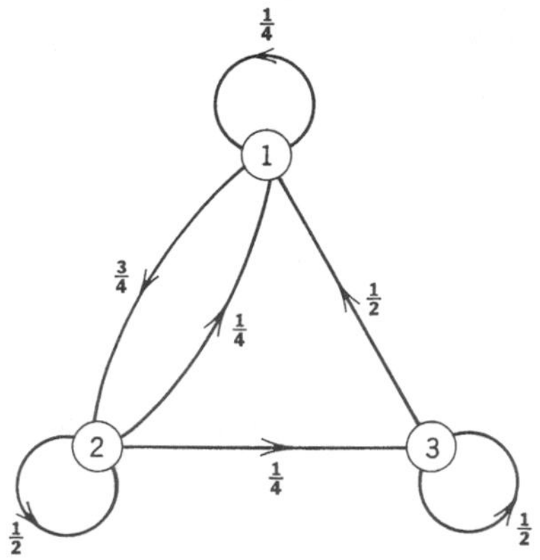


**3.** Each of the following matrices represents the transition probabilities for a Markov process.

i) $\begin{bmatrix} 2/3 & 1/3 \\ 1/2 & 1/2 \end{bmatrix}$

ii) $\begin{bmatrix} 1/3 & 1/3 & 1/3 \\ 1/3 & 1/3 & 1/3 \\ 0 & 0 & 1 \end{bmatrix}$

iii) $\begin{bmatrix} 0 & 1 & 0 \\ 1 & 0 & 0 \\ 14/37 & 13/37 & 10/37 \end{bmatrix}$

iv) $\begin{bmatrix} 1/3 & 1/3 & 1/3 \\ 0 & 0 & 1 \\ 0 & 1 & 0 \end{bmatrix}$

v) $\begin{bmatrix} 1/3 & 1/2 & 1/6 \\ 1/3 & 1/2 & 1/6 \\ 1/3 & 1/2 & 1/6 \end{bmatrix}$

vi) $\begin{bmatrix} 1/2 & 0 & 1/2 & 0 \\ 0 & 2/3 & 0 & 1/3 \\ 3/4 & 0 & 1/4 & 0 \\ 0 & 1/4 & 0 & 3/4 \end{bmatrix}$

a) Investigate each process for chain structure and periodicity.
b) Draw the flow graph for each process.
c) Using the flow graphs in b) find $\phi_{12}{}^{g}(z)$ for each process.
d) Invert the transforms in c) to find $\phi_{12}(n)$.

**4.** A certain absent-minded professor has two offices and one secretary. His secretary has a difficult time keeping track of the professor's whereabouts and has decided that his wanderings from office to office can be described by the following process: Each time the professor leaves an office he flips a weighted coin so that with probability 1/4 he chooses matrix $A$ and with probability 3/4 he chooses matrix $B$.

$$A = \begin{bmatrix} 1/3 & 2/3 \\ 5/6 & 1/6 \end{bmatrix} \qquad B = \begin{bmatrix} 8/9 & 1/9 \\ 7/18 & 11/18 \end{bmatrix}$$

The professor then uses the probabilities in the chosen matrix to determine the office he shall visit next (e.g., if he is currently in office 1 and chooses matrix $B$, his probability of staying in office 1 is 8/9, and of going to office 2 is 1/9).

a) Can the professor's transitions from office to office be described by a simple Markov process? Why? If so, what is the $P$ matrix for this process?

b) Draw the flow graph for the professor's transitions.

c) If the professor starts off in office 1, what is the probability he will be in office 2 after $n$ transitions?

d) If the professor's secretary must guess after each transition where the professor will go next, what fraction of the time will she be correct in the long run?

**5.** Describe a procedure for altering the transition diagram for a monodesmic Markov process so that the transmission to the $j$th state is the limiting state probability for that state. Illustrate:

a) For the general two-state process.

b) For the taxicab example with transition probability matrix specified by Equation 3.1.3.

**6.** The Mark V computer is a cantankerous machine. Whenever it is turned on there is a fifty fifty chance that it will be broken down at the end of the first hour of operation (thus leading to the piteous cry for the users, "Is the Mark off?"). However, if the machine survived the first hour of operation, then the probability of surviving each succeeding hour of operation is 0.8.

The maintenance crew checks the condition of the machine at the end of each hour; if the machine has just broken down, there is a 70% chance that the crew will find the fault immediately and have the machine ready to turn on again at the end of the hour. If during any one hour they do not find the fault, then there is only a probability of 0.4 that they will have the machine ready to turn on at the end of the next hour. (For some reason, the machine is only turned on at the beginning of an hour and turning the machine on after repair is the same as turning it on at the beginning of the day.)

a) Draw the transition diagram for this process when the time interval between transitions is one hour.

b) In a long time period, for what fraction of hours will the machine be operating at the beginning of the hour?

c) Draw a flow graph for this process for which the output is the transformed probability that the Mark is down at the end of the $n$th hour after being turned on.

d) Solve the flow graph to find the transformed probability in c).

**7.** In Section 1.2 we wrote a difference equation for $\phi_{ij}(n + 1)$ by considering all possible states that the process might occupy after $n$ transitions.

a) Now write a difference equation for $\phi_{ij}(n + 1)$ by considering all possible states that the process might occupy *after the first transition.*

b) Take the transform of the difference equation in part a) to yield an equation for $\phi_{ij}^{g}(z)$ in terms of the elements in the $j$th column of $\Phi^{g}(z)$.

c) From this set of transform equations, construct a flow graph whose transmission $\vec{t}_{ij}$ from $i$ to $j$ is $\phi_{ij}^{g}(z)$. Contrast it with the flow graph of Figure 3.1.2.

d) Illustrate the results of part c) with the two-state process whose transition probabilities are

$$P = \begin{bmatrix} 0.6 & 0.4 \\ 0.1 & 0.9 \end{bmatrix}.$$

**8.** Consider each branch transmission in Problem 2 of Chapter 2 to be a matrix. Find the transmission of the resulting matrix flow graph.

**9.** Place the transition probability matrices of the following problems in canonical form. Draw the relevant matrix flow graphs.

i) Problem 1, Chapter 1.

ii) Problem 3, Chapter 3.

**10.** Show that the matrix flow graph solution for the four-state mergeable process of Problem 20 of Chapter 1 produces superstate multistep transition probabilities consistent with those for the marketing example found in Sections 1.6 and 3.1.

# 4 TRANSIENT MARKOV PROCESSES

A class of processes important enough to deserve a special name is the class of transient processes. These are Markov processes containing transient states in which the time or number of transitions required to enter one of the recurrent states from each transient state is the random variable of central importance. Transient processes are significant both in their own right and as an aid to understanding more complex processes. We restrict attention initially to the most common case where there is only one recurrent state, which is, therefore, a trapping state.

## 4.1 TRANSIENT PROCESSES

We can illustrate a transient process most readily with an example. In our original marketing problem of Chapter 1 we proposed a customer who changed his allegiance between two brands $A$ and $B$ according to the transition probability matrix

$$P = \begin{bmatrix} 0.8 & 0.2 \\ 0.3 & 0.7 \end{bmatrix}. \tag{4.1.1}$$

You recall that we numbered his successive purchases $0, 1, 2, \ldots, n$. Suppose as a result of further observation of the customer's behavior we learn that a customer who originally buys brand $A$ on one or more consecutive occasions, then buys brand $B$ on one or more consecutive occasions, then buys brand $A$ again will buy brand $A$ forevermore. Such behavior could arise because brand $B$ is an inferior product—a brand $A$ customer will eventually use it, possibly for some time, but when he finally decides to switch back to brand $A$, he never leaves it again. In this model every customer sooner or later becomes a permanent brand $A$ customer. An important quantity in such a process would be the probability that a person who buys brand $A$ on purchase 0 will become a permanent brand $A$ customer by purchase $n$, that is, after he has made $n$ purchases in addition to the original one.

To treat this problem we must expand the state definition used originally to create three rather than two states. Whereas before we required only the two states of brand $A$ or brand $B$ as the customer's last purchase, now we need three states: the customer is in state 1 when his history shows only purchases of brand $A$; he is in state 2 when his latest purchase is brand $B$; and he is in state 3 when his history shows that he has bought $A$'s, then $B$'s, and finally $A$'s again. The three states thus correspond to a transient $A$ customer, a transient $B$ customer, and a permanent $A$ customer. Since we are assuming that the transition probabilities of this process are the transition probabilities of Equation 4.1.1 as modified by the trapping behavior, we can draw and label the transition diagram shown in Figure 4.1.1. A customer in state 1 has probability 0.8 of buying another brand $A$ and returning to state 1; he has a probability 0.2 of buying brand $B$ and thereby entering state 2. A customer in state 2 will repurchase brand $B$ with probability 0.7 and stay in state 2; with probability 0.3 he will buy brand $A$ and become a permanent brand $A$ customer in state 3. Because we have increased the number of states from two to three we should write a transition probability matrix for the new process. It is

$$P = \begin{bmatrix} 0.8 & 0.2 & 0 \\ 0 & 0.7 & 0.3 \\ 0 & 0 & 1 \end{bmatrix}. \tag{4.1.2}$$

We see immediately that state 3 is a trapping state of the process; moreover, by inspection of the transition matrix or by physical reasoning, we see that state 3 is the only recurrent state in the process. Notice that these comments would still apply even if $p_{13}$ or $p_{21}$ were not equal to zero as they are in this case. State 3 will be the only recurrent state in the process as long as $p_{13} + p_{23} \neq 0$ and $p_{33} = 1$.

The number of the transition on which the process enters the trapping state 3 is the random variable of current interest. Since the transition probability matrix describes a monodesmic Markov process with a single trapping state, the process is a transient process. Transient processes often arise in modeling operational problems.

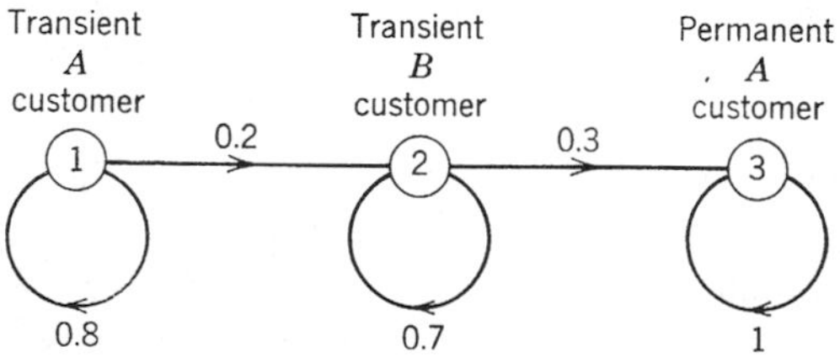


**Figure 4.1.1.** Transition diagram for a transient process.

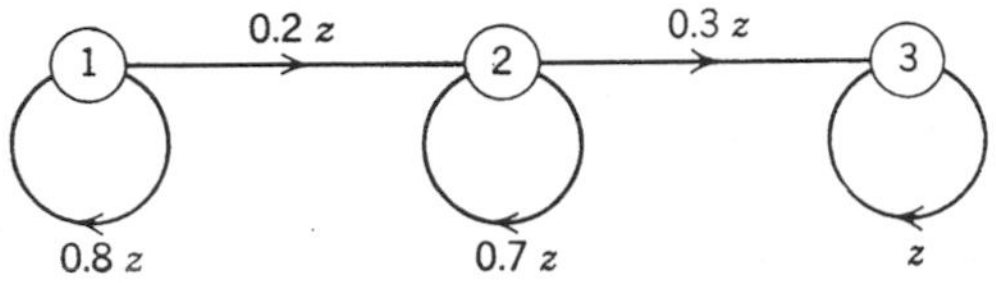


**Figure 4.1.2** Flow graph for a transient process.

## Flow Graph Analysis

The flow graph relevant to the transition diagram of Figure 4.1.1 is drawn in Figure 4.1.2. The transmission of this flow graph from state 1 to state 3 is $\phi_{13}{}^{g}(z)$, the transform of $\phi_{13}(n)$, which is the probability that a person who starts buying brand $A$ on purchase 0 will be a permanent brand $A$ customer after $n$ additional purchases. Since initial and terminal states are often understood in a transient process, we shall write this probability $\phi_{13}(n)$ in the more convenient form $\phi(n)$ and then find its transform $\phi^{g}(z) = \phi_{13}{}^{g}(z)$ by inspection of the flow graph in Figure 4.1.2. We obtain easily by either the inspection method or elementary reductions,

$$\phi^{g}(z) = \frac{0.06z^2}{(1 - 0.8z)(1 - 0.7z)(1 - z)}. \tag{4.1.3}$$

Partial fractional expansion produces

$$\phi^{g}(z) = \frac{3}{1 - 0.8z} + \frac{2}{1 - 0.7z} + \frac{1}{1 - z}. \tag{4.1.4}$$

By inverting the geometric transforms we obtain

$$\phi(n) = -3(0.8)^n + 2(0.7)^n + 1 \qquad n = 0, 1, 2, \ldots. \tag{4.1.5}$$

Notice that $\phi(0) = \phi(1) = 0$ because it is impossible to become a permanent brand $A$ customer before purchase occasion 2.

The form of the function described by Equation 4.1.5 is shown in Figure 4.1.3. The function looks like the cumulative probability distribution for a discrete random variable, and that is just what it is. The random variable is the number of transitions required to enter the trapping state, state 3. Thus $\phi(n)$ is the probability that $n$ or less transitions will be required for trapping.

## Transitions until Trapped: Modified Flow Graphs

Although we can use a cumulative probability distribution to answer all questions about the behavior of such a discrete random variable, we often find it more

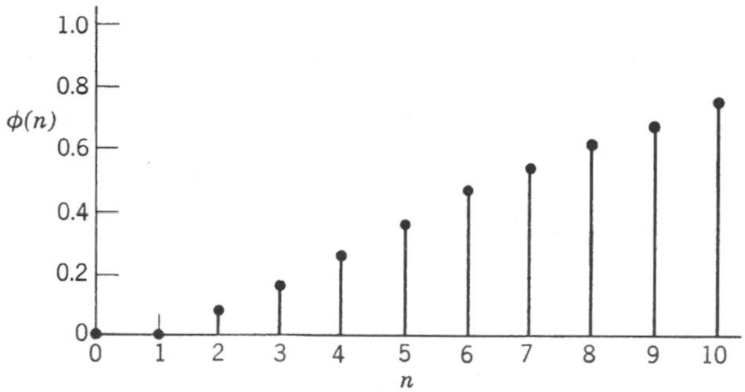


**Figure 4.1.3** Probability of being a permanent brand *A* customer by purchase occasion *n*.

convenient to work with the variable's probability mass function. If we are primarily interested in the calculation of the moments of the discrete random variable, then the mass function is particularly useful. Fortunately, in the case of transient processes we find it somewhat easier to obtain the mass function for the number of transitions until trapping than to obtain the cumulative distribution of this quantity.

To see this, let $p(n)$ be the probability that the customer *becomes* a permanent brand *A* customer on purchase occasion $n$. Since becoming a permanent brand *A* customer sooner or later is a certain event, $p(n)$ sums to one over $n$ and is therefore a probability mass function for the number of transitions required to enter the trapping state. The cumulative distribution $\phi(\cdot)$ is related to the probability mass function $p(\cdot)$ by

$$\phi(n) = \sum_{m=0}^{n} p(m) \qquad n = 0, 1, 2, \ldots. \tag{4.1.6}$$

This equation states that the probability of entering the trapping state by time $n$ is just the sum of the probabilities of entering it at times $0, 1, 2, \ldots, n$. The geometric transform of Equation 4.1.6 is

$$\phi^g(z) = \frac{1}{1-z} p^g(z), \tag{4.1.7}$$

and thus we can also write

$$p^g(z) = (1 - z)\phi^g(z). \tag{4.1.8}$$

By referring to Equation 4.1.3 we see that the effect of multiplying $\phi^g(z)$ by $1 - z$ to obtain $p^g(z)$ is to eliminate the factor $1 - z$ from the denominator of the $\phi^g(z)$

expression. Therefore we have

$$p^g(z) = \frac{0.06z^2}{(1 - 0.8z)(1 - 0.7z)}. \tag{4.1.9}$$

But a moment's reflection reminds us that the $\phi^g(z)$ expression for any transient process will have a $1 - z$ factor in its denominator because the last simple system involved in the $\phi^g(z)$ transmission is the loop with transfer function $z$ pertaining to the trapping state. Since this system has an equivalent transmission of $1/(1 - z)$, it acts as a summer to accumulate successive terms in the probability mass function. The effect of this summer is to add a $1 - z$ factor to the denominator of the transmission expression for $\phi^g(z)$. However, if once we have obtained $\phi^g(z)$ we are going to multiply it by the difference operator $1 - z$ to produce $p^g(z)$, then there is no point in having the loop around the trapping state in the original graph. We might just as well eliminate it, find the transmission of the resulting graph, and thereby generate $p^g(z)$ directly. We shall call this new flow graph the "modified" flow graph of a transient process.

The modified flow graph for the purchasing process is shown in Figure 4.1.4. It is identical with Figure 4.1.2 except for the absence of the loop around the trapping state. The transmission of this graph is just the expression for $p^g(z)$ given by Equation 4.1.9.

***Transform inversion***

We find an explicit expression for $p(n)$ by inverting the geometric transform in Equation 4.1.9. However, when we contemplate partial fraction expansion we see that the degree of $z$ is the same in the numerator and denominator. This prevents us from performing the expansion until we have placed the expression in a form that does not involve this difficulty. The simplest change is to factor a $z$ from the expression and then perform a partial fraction expansion on the remaining factor which now, of course, meets the requirement that the degree of the numerator be less than the degree of the denominator. We thus write

$$p^g(z) = \frac{0.06z^2}{(1 - 0.8z)(1 - 0.7z)} = z\left[\frac{0.06z}{(1 - 0.8z)(1 - 0.7z)}\right]$$

$$= z\left[\frac{0.6}{1 - 0.8z} + \frac{-0.6}{1 - 0.7z}\right]$$

$$= \frac{0.6z}{1 - 0.8z} + \frac{-0.6z}{1 - 0.7z}. \tag{4.1.10}$$

We invert these geometric transforms easily by recalling that a factor of $z$ in the transform represents a delay of one unit in the function,

$$p(n) = 0.6[(0.8)^{n-1} - (0.7)^{n-1}], \qquad n = 1, 2, 3, \ldots \tag{4.1.11}$$
$$p(0) = 0$$

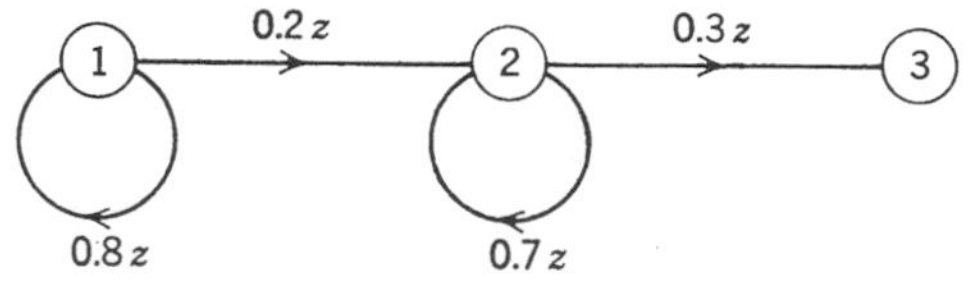


**Figure 4.1.4** Modified flow graph for a transient process.

Of course, inversion of Equation 4.1.8 shows that $p(n) = \phi(n) - \phi(n - 1)$ and thus allows us to check this result with that of Equation 4.1.5.

The form of $p(n)$ appears in Figure 4.1.5. As $n$ increases, $p(n)$ increases at first and then decreases. The maximum value of $p(n)$ is attained by a small margin with $n = 5$. Thus the customer is more likely to become a permanent brand $A$ customer on purchase occasion five than he is on any other.

Equation 4.1.11 is a closed-form expression for the probability mass function of the number of transitions to enter the trapping state, or time to absorption, as it is sometimes called. For brevity we shall often use the term "delay" to refer to the number of transitions required to enter the trapping state of a transient process. We shall use the discrete random variable $\nu$ to represent this delay.

***Difference equation solution***

We can employ Equation 4.1.11 to compute numerical values for the function shown in Figure 4.1.5, but occasionally the computational procedure introduced in Chapter 2 is more convenient. As we learned, every transform equation has buried within it a difference equation. This difference equation allows us to calculate the values of the transformed function recursively rather than in closed form.

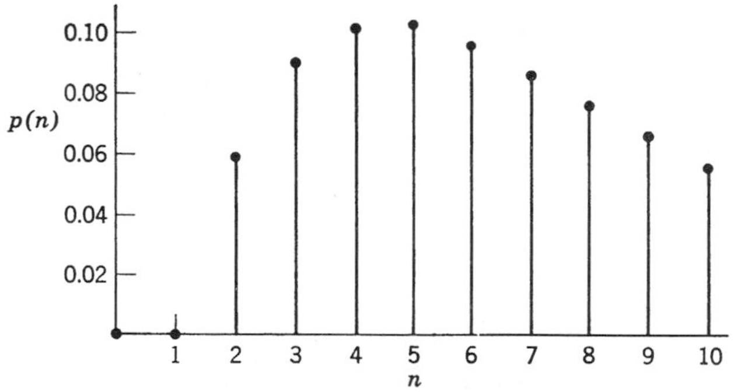


**Figure 4.1.5** Probability of becoming a permanent brand $A$ customer on purchase occasion $n$.

We can illustrate this procedure in our present example. First we manipulate $p^g(z)$ as follows:

$$p^g(z) = \frac{0.06z^2}{(1 - 0.8z)(1 - 0.7z)}$$

$$= \frac{0.06z^2}{1 - 1.5z + 0.56z^2} \tag{4.1.12}$$

or

$$\begin{aligned} p^g(z) - 1.5zp^g(z) + 0.56z^2p^g(z) &= 0.06z^2 \\ p^g(z) &= 0.06z^2 + 1.5zp^g(z) - 0.56z^2p^g(z). \end{aligned} \tag{4.1.13}$$

By taking the inverse geometric transform we obtain

$$\begin{aligned} p(n) &= 0.06\ \delta(n - 2) + 1.5p(n - 1) - 0.56p(n - 2) \qquad n = 2, 3, 4, \ldots \\ p(0) &= p(1) = 0. \end{aligned} \tag{4.1.14}$$

We can now solve for all $p(n)$'s recursively,

$$\begin{aligned} p(2) &= 0.06 \\ p(3) &= 1.5p(2) = 0.09 \\ p(4) &= 1.5p(3) - 0.56p(2) = 0.1014 \\ p(5) &= 1.5p(4) - 0.56p(3) = 0.1017 \\ p(6) &= 1.5p(5) - 0.56p(4) = 0.095766 \\ &\vdots \end{aligned} \tag{4.1.15}$$

and thus produce Figure 4.1.5. Calculations of this type are, of course, easily handled by human or electronic computers. We shall therefore often find it convenient to exploit the difference equations implied by every geometric transform.

### *A check*

Since $p^g(z)$ is the geometric transform of a probability mass function $p(\cdot)$, we should note some of its special properties. First, since

$$p^g(z) = \sum_{n=0}^{\infty} p(n)z^n, \tag{4.1.16}$$

we see immediately that

$$p^g(1) = \sum_{n=0}^{\infty} p(n) = 1. \tag{4.1.17}$$

The geometric transform of the probability mass function must be equal to one when $z = 1$. The function $p^g(z)$ given in Equation 4.1.9 meets this requirement. We can always make the evaluation easily; it serves as a useful check in this type of problem.

**Moments of Delay**

In many applications the moments $\bar{\nu}$, $\overline{\nu^2}$, etc. of the number of transitions to enter the trapping state or delay are of more immediate interest than the probability mass function itself. In these cases the transform analysis is particularly fruitful because it allows us to calculate the moments by differentiation rather than summation. If we differentiate the transform relation 4.1.16 with respect to $z$ repeatedly we obtain:

$$\begin{aligned}
\frac{d}{dz} p^{g}(z) &= \sum_{n=0}^{\infty} np(n)z^{n-1} \\
\frac{d^2}{dz^2} p^{g}(z) &= \sum_{n=0}^{\infty} n(n-1)p(n)z^{n-2} \\
\frac{d^3}{dz^3} p^{g}(z) &= \sum_{n=0}^{\infty} n(n-1)(n-2)p(n)z^{n-3} \\
&\vdots
\end{aligned} \tag{4.1.18}$$

If these derivatives are evaluated at the point $z = 1$, the relations become:

$$\begin{aligned}
\frac{d}{dz} p^{g}(z)\Big|_{z=1} &= p^{g\prime}(1) = \sum_{n=0}^{\infty} np(n) = \bar{\nu} \\
\frac{d^2}{dz^2} p^{g}(z)\Big|_{z=1} &= p^{g\prime\prime}(1) = \sum_{n=0}^{\infty} (n^2 - n)p(n) = \overline{\nu^2} - \bar{\nu} \\
\frac{d^3}{dz^3} p^{g}(z)\Big|_{z=1} &= p^{g\prime\prime\prime}(1) = \sum_{n=0}^{\infty} (n^3 - 3n^2 + 2n)p(n) = \overline{\nu^3} - 3\overline{\nu^2} + 2\bar{\nu} \\
&\vdots
\end{aligned} \tag{4.1.19}$$

Here $\overline{\nu^k}$ is the $k$th moment of the mass function. By solving these equations we can express the moments in terms of the derivatives in the form:

$$\begin{aligned}
\bar{\nu} &= p^{g\prime}(1) \\
\overline{\nu^2} &= p^{g\prime\prime}(1) + \bar{\nu} = p^{g\prime\prime}(1) + p^{g\prime}(1) \\
\overline{\nu^3} &= p^{g\prime\prime\prime}(1) + 3\overline{\nu^2} - 2\bar{\nu} = p^{g\prime\prime\prime}(1) + 3p^{g\prime\prime}(1) + p^{g\prime}(1) \\
&\vdots
\end{aligned} \tag{4.1.20}$$

Therefore the variance of the delay, $\overset{\vee}{\nu}$, is

$$\overset{\vee}{\nu} = \overline{\nu^2} - \bar{\nu}^2 = p^{g\prime\prime}(1) + p^{g\prime}(1) - [p^{g\prime}(1)]^2. \tag{4.1.21}$$

We see that the $k$th moment can be found by evaluating the 1st, 2nd, . . ., $k$th derivatives of the geometric transform at the point $z = 1$ and then combining them appropriately.

***Delay moments in the marketing problem***

We can now return to the marketing problem to use this knowledge.

**Mean.** We calculate the mean number of additional (beyond purchase 0) purchases a customer will make before becoming a permanent brand $A$ customer by differentiating $p^{\mathscr{g}}(z)$ and evaluating it at the point $z = 1$. We write

$$p^{\mathscr{g}\prime}(z) = \frac{d}{dz} p^{\mathscr{g}}(z) = \frac{d}{dz}\left[\frac{0.06z^2}{(1 - 0.8z)(1 - 0.7z)}\right]$$

$$= \frac{(1 - 0.8z)(1 - 0.7z)(0.12z) - 0.06z^2[(1 - 0.8z)(-0.7) - 0.8(1 - 0.7z)]}{(1 - 0.8z)^2(1 - 0.7z)^2}.$$

The differentiation is somewhat more easily performed on $p^{\mathscr{g}}(z)$ in the partial fraction expansion form of Equation 4.1.10. Proceeding, however, we find

$$\bar{\nu} = p^{\mathscr{g}\prime}(1) = \frac{0.0072 - 0.06[-0.14 - 0.24]}{0.0036}$$

$$= 25/3 = 8\ 1/3. \tag{4.1.22}$$

The mean number of additional purchases the customer will make before becoming a permanent brand $A$ customer is therefore 8 1/3. Note that this is considerably larger than 5, the modal number we found earlier.

It is worth pointing out that the process of finding the mean is even easier if we recognize that a factor of $z^k$ in the numerator of $p^{\mathscr{g}}(z)$ will correspond to an increase of the mean by $k$ units over the value it would have if the factor were not present. Consequently, we can always remove such a factor, differentiate, evaluate at $z = 1$, and then add $k$ units to obtain the mean. In the present example this technique is useful when $p^{\mathscr{g}}(z)$ is expressed in the form of either Equation 4.1.9 or Equation 4.1.10.

**Variance.** The variance of the number of additional purchases is computed by using Equation 4.1.21. After some labor we find

$$p^{\mathscr{g}\prime\prime}(1) = 800/9$$

and therefore

$$\overset{\mathrm{v}}{\nu} = p^{\mathscr{g}\prime\prime}(1) + p^{\mathscr{g}\prime}(1) - [p^{\mathscr{g}\prime}(1)]^2 = 800/9 + 75/9 - 625/9 = 250/9. \tag{4.1.23}$$

The standard deviation, $\overset{\mathrm{s}}{\nu} = \sqrt{\overset{\mathrm{v}}{\nu}} \approx 5.3$, turns out to be a significant fraction of the mean.

As interesting as the mean and variance of delay are to us, we would need strong motivation to differentiate the complex expressions for the transform $p^{\mathscr{g}}(z)$ that often arise in larger problems. Fortunately our flow graph methods can be of help. Before proceeding to a formal analysis, let us enjoy one of the results. First we draw a modified transition diagram for the transient process as shown in Figure 4.1.6. This is just the modified flow graph of Figure 4.1.4 with $z = 1$. Of course, it could also be written directly from the transition diagram of Figure 4.1.1 by eliminating the loop around the trapping state. Now we shall find the transmissions $\vec{t}_{11}$ and $\vec{t}_{12}$ of the modified transition diagram. They are

$$\vec{t}_{11} = \frac{1}{1 - 0.8} = 5$$

$$\vec{t}_{12} = \frac{0.2}{(1 - 0.8)(1 - 0.7)} = 10/3. \tag{4.1.24}$$

These transmissions are completely numerical because the transition diagram has only numerical transfer functions. Now we write

$$\vec{t}_{11} + \vec{t}_{12} = 25/3 = 8\ 1/3, \tag{4.1.25}$$

and we have found the mean number of transitions required to enter the trapping state by a method that is much easier than differentiation of the transform $p^{\mathscr{g}}(z)$. Why does it work? We shall now answer this question by developing relations that permit determining the mean and variance of delay in a transient process by inspection of the modified transition diagram.

## Expected Delay the Easy Way: An Indicator Variable

We begin our investigation by defining a random variable that is often useful in the analysis of Markov processes. Let $x_{ij}(n)$ equal 1 if a system started in state $i$ enters state $j$ on its $n$th transition, and let it equal zero otherwise. That is,

$$x_{ij}(n) = \begin{bmatrix} 1 & \text{if } s(n) = j \\ 0 & \text{Otherwise} \end{bmatrix} \text{ given } s(0) = i. \tag{4.1.26}$$

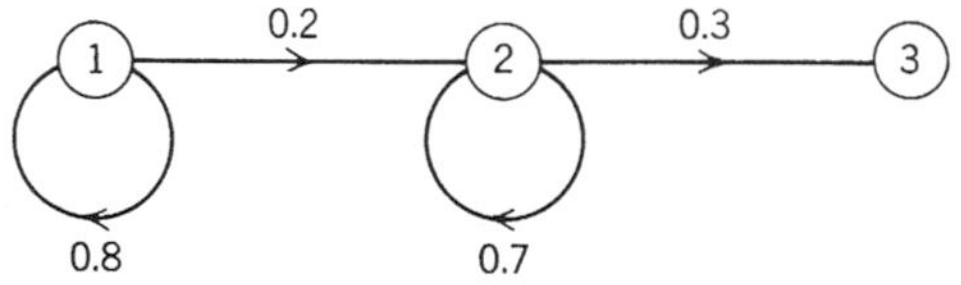


**Figure 4.1.6** Modified transition diagram for a transient process.

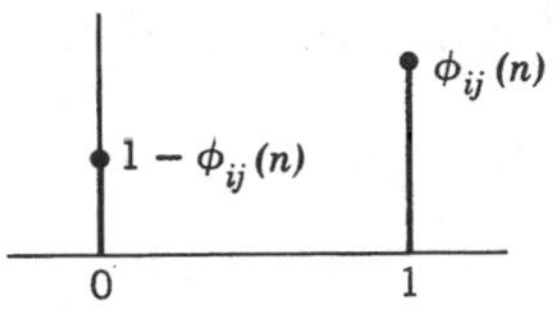


**Figure 4.1.7** Probability mass function of $X_{ij}(n)$.

The random variable $x_{ij}(n)$ thus takes on only the two discrete values 0 and 1. The probability that $x_{ij}(n)$ takes on the value 1 is just $\phi_{ij}(n)$, the multistep transition probability that we have seen many times before. The probability that $x_{ij}(n)$ equals zero is then $1 - \phi_{ij}(n)$. Therefore $x_{ij}(n)$ has the very simple probability mass function shown in Figure 4.1.7. We easily calculate the mean of the random variable, $x_{ij}(n)$, since

$$\begin{aligned} \bar{x}_{ij}(n) &= 1 \cdot \phi_{ij}(n) + 0[1 - \phi_{ij}(n)] \\ &= \phi_{ij}(n). \end{aligned} \tag{4.1.27}$$

Now we proceed to the quantity we are really interested in. Let the random variable $\nu_{ij}$ be the number of times state $j$ is entered in an infinite number of transitions if the system is started in state $i$. We can express $\nu_{ij}$ precisely in terms of $x_{ij}(n)$ by

$$\nu_{ij} = \sum_{n=0}^{\infty} x_{ij}(n). \tag{4.1.28}$$

We interpret $\nu_{ij}$ as the delay in state $j$ given that the process started in state $i$. We are generally concerned in a transient process with the moments of $\nu_{ij}$. They can be found by inspection of the transition diagram.

### Expected Delay in a Transient State

In particular, $\bar{\nu}_{ij}$, the mean number of times state $j$ will be entered in an infinite number of transitions when the system is started in state $i$, is given by

$$\bar{\nu}_{ij} = \overline{\sum_{n=0}^{\infty} x_{ij}(n)}. \tag{4.1.29}$$

Since the mean of the sum is the sum of the means we can write

$$\bar{\nu}_{ij} = \sum_{n=0}^{\infty} \bar{x}_{ij}(n). \tag{4.1.30}$$

Equation 4.1.27 now shows that $\bar{\nu}_{ij}$ is as well given by

$$\bar{\nu}_{ij} = \sum_{n=0}^{\infty} \phi_{ij}(n). \tag{4.1.31}$$

The expected delay or number of occupancies of state $j$ in an infinite number of transitions is thus equal to the sum of the multistep transition probabilities $\phi_{ij}(n)$. Of course, if state $j$ were a recurrent state, then this sum would clearly be infinite. But if state $j$ is a transient state, then the expected number of times it will be occupied in an infinite number of transitions is finite. As a matter of fact, since

$$\phi_{ij}{}^{g}(z) = \sum_{n=0}^{\infty} \phi_{ij}(n)z^n, \tag{4.1.32}$$

we find that $\bar{\nu}_{ij}$ is just the transform of $\phi_{ij}{}^{g}(z)$ evaluated at $z = 1$,

$$\bar{\nu}_{ij} = \phi_{ij}{}^{g}(1). \tag{4.1.33}$$

Now we recall that $\phi_{ij}{}^{g}(z)$ is the transfer function from node $i$ to node $j$ of the flow graph representing the process, and that the flow graph with $z = 1$ is just the transition diagram. Therefore, $\bar{\nu}_{ij}$ is the transmission $\vec{t}_{ij}$ of the transition diagram from node $i$ to node $j$:

$$\bar{\nu}_{ij} = \sum_{n=0}^{\infty} \phi_{ij}(n) = \phi_{ij}{}^{g}(1) = \vec{t}_{ij}. \tag{4.1.34}$$

This transmission will be infinite if state $j$ is a recurrent state.

We shall define a matrix $\overline{\mathrm{N}}$ with components $\bar{\nu}_{ij}$ for future convenience. We can then write immediately that this matrix is just equal to $\vec{T}$, the matrix of transmissions of the transition diagram,

$$\overline{\mathrm{N}} = \vec{T}. \tag{4.1.35}$$

### Expected Delay in a Transient Process

Suppose now that we have identified all the transient states in the process, a set we denote by $\mathscr{T}$, and that we are interested in finding the expected delay of the process, the expected number of transitions the system will make before entering a recurrent state. Let $\nu_i$ be the number of times these transient states will be occupied in an infinite number of transitions if the system is started in state $i$. This random variable is related to $\nu_{ij}$ by

$$\nu_i = \sum_{j \in \mathscr{T}} \nu_{ij}. \tag{4.1.36}$$

In this equation the summation extends over all the transient states. The mean number of transitions that will be made in the transient process is then

$$\bar{\nu}_i = \sum_{j \in \mathcal{T}} \bar{\nu}_{ij} = \sum_{j \in \mathcal{T}} \vec{t}_{ij}. \tag{4.1.37}$$

The expected delay of the transient process is the sum of the expected delays at each transient state.

Equation 4.1.37 is the relation used to write Equation 4.1.25. The process was started in state 1, so the expected delays at nodes 1 and 2 were given by the $\vec{t}_{11}$ and $\vec{t}_{12}$ computed in Equation 4.1.24. The delays at these two nodes were added to obtain the total expected delay or expected number of transitions made in the transient process. We see immediately that if the customer began by purchasing brand $B$ and entering state 2, then his expected number of purchases before becoming a permanent brand $A$ customer would be given by $\vec{t}_{22}$, which we find to be 3 1/3 by inspection of the transition diagram in Figure 4.1.1. We can also use the modified transition diagram for this calculation as long as we never try to calculate the expected delay at a recurrent state.

We can arrange the expected delays for a given starting state into a column vector $\bar{\boldsymbol{\nu}}$ with components $\bar{\nu}_i$. From Equation 4.1.37 we see that this vector is given by

$$\bar{\boldsymbol{\nu}} = \vec{T}\mathbf{s} = \bar{\mathrm{N}}\mathbf{s} \tag{4.1.38}$$

where $\mathbf{s}$ is a column vector with all components equal to one.

We have just found that the expected delay for a transient process is obtained by summing the expected delays at each transient state in the process. These delays at each node are in turn found by computing the transmission of the transition diagram from the input node to the node corresponding to the transient state. This procedure is simple and practical for small problems.

For larger problems we only need to note that since

$$\Phi^{\mathcal{T}}(z) = [I - zP]^{-1},$$

the matrix $\vec{T}$ of transmissions of the transition diagram is equal to the inverse of $[I - P]^{-1}$,

$$\vec{T} = [I - P]^{-1} = \bar{\mathrm{N}}. \tag{4.1.39}$$

Thus in a large problem we can find the transmission $\vec{t}_{ij}$ by computing the $ij$th element in the inverse of $I - P$. If we delete from $I - P$ all rows and columns corresponding to recurrent states we shall be spared the annoyance of all infinite elements.

### The Variance of Delay in a Transient State

Now that we have mastered the problem of finding expected delays in a transient process let us turn to the question of computing the variance of the delay. The procedure we shall use is identical in principle to our work on the mean delay, but a little more complicated. We shall start by calculating the variance of the time spent in each state of the transient process. In performing this calculation we find it convenient to consider the product of the random variables $\nu_{ij}$ and $\nu_{ik}$. Using Equation 4.1.28 we write

$$\nu_{ij}\nu_{ik} = \sum_{m=0}^{\infty} x_{ij}(m) \sum_{r=0}^{\infty} x_{ik}(r)$$

$$= \sum_{m=0}^{\infty} \sum_{r=0}^{\infty} x_{ij}(m)x_{ik}(r). \tag{4.1.40}$$

The expectation of this quantity is

$$\overline{\nu_{ij}\nu_{ik}} = \sum_{m=0}^{\infty} \sum_{r=0}^{\infty} \overline{x_{ij}(m)x_{ik}(r)}. \tag{4.1.41}$$

The question now is what is the expectation of $x_{ij}(m)x_{ik}(r)$ where each variable in the product is defined by Equation 4.1.26. We see immediately that the product must take on only the values 0 and 1. The product will be equal to one if the state at time $m$ is $j$ and the state at time $r$ is $k$; otherwise it will be zero. Therefore, the expectation of the product is just equal to the probability of this event:

$$\overline{x_{ij}(m)x_{ik}(r)} = \mathscr{P}\{x_{ij}(m)x_{ik}(r) = 1\} = \mathscr{P}\{s(m) = j, s(r) = k | s(0) = i\}. \tag{4.1.42}$$

This probability will occur frequently enough in our discussions to deserve a special name and symbol. We shall call it the joint transition probability and give it the symbol $\psi_{ijk}(m, r)$,

$$\psi_{ijk}(m, r) = \mathscr{P}\{s(m) = j, s(r) = k | s(0) = i\}. \tag{4.1.43}$$

We can now write

$$\overline{x_{ij}(m)x_{ik}(r)} = \psi_{ijk}(m, r) \tag{4.1.44}$$

and therefore place Equation 4.1.41 in the form

$$\overline{\nu_{ij}\nu_{ik}} = \sum_{m=0}^{\infty} \sum_{r=0}^{\infty} \psi_{ijk}(m, r). \tag{4.1.45}$$

The joint transition probability $\psi_{ijk}(m, r)$ provides the foundation for correlation studies in Markov processes. It is easily related to the interval transition probability as follows. By the definition of conditional probability,

$$\begin{aligned}
\psi_{ijk}(m, r) &\\
&= \mathscr{P}\{s(m) = j, s(r) = k | s(0) = i\} \\
&= \mathscr{P}\{s(m) = j | s(0) = i\}\mathscr{P}\{s(r) = k | s(m) = j, s(0) = i\} \qquad r \geq m \qquad (4.1.46)\\
&= \mathscr{P}\{s(r) = k | s(0) = i\}\mathscr{P}\{s(m) = j | s(r) = k, s(0) = i\} \qquad m \geq r.
\end{aligned}$$

Notice that two different expressions are required, depending on the relative sizes of $m$ and $r$. By using the basic Markovian property that only the state most recently occupied by the process affects its future behavior and the definition of the multistep transition probability $\phi_{ij}(n)$, we express $\psi_{ijk}(m, r)$ as

$$\psi_{ijk}(m, r) = \begin{cases} \phi_{ij}(m)\phi_{jk}(r - m) & r > m \\ \phi_{ik}(r)\phi_{kj}(m - r) & m > r \\ \phi_{ij}(m)\,\delta_{kj} & r = m \end{cases} . \qquad (4.1.47)$$

This equation shows that the joint transition probabilities are easily calculated from the multistep transition probabilities.

We shall now use Equation 4.1.47 to complete the calculation of Equation 4.1.45. We find

$$\begin{aligned}
\overline{\nu_{ij}\nu_{ik}} &= \sum_{m=0}^{\infty}\sum_{r=0}^{\infty} \psi_{ijk}(m, r) \\
&= \sum_{m=0}^{\infty}\sum_{r=m+1}^{\infty} \phi_{ij}(m)\phi_{jk}(r - m) + \sum_{r=0}^{\infty}\sum_{m=r+1}^{\infty} \phi_{ik}(r)\phi_{kj}(m - r) + \delta_{kj}\sum_{m=0}^{\infty} \phi_{ij}(m).
\end{aligned} \qquad (4.1.48)$$

We can simplify this result by a change of variable in two of the summations to obtain

$$\overline{\nu_{ij}\nu_{ik}} = \sum_{m=0}^{\infty} \phi_{ij}(m) \sum_{r'=1}^{\infty} \phi_{jk}(r') + \sum_{r=0}^{\infty} \phi_{ik}(r) \sum_{m'=1}^{\infty} \phi_{kj}(m') + \delta_{jk} \sum_{m=0}^{\infty} \delta_{ij}(m). \qquad (4.1.49)$$

Now we use Equation 4.1.34 to write

$$\overline{\nu_{ij}\nu_{ik}} = \vec{t}_{ij}[\vec{t}_{jk} - \phi_{jk}(0)] + \vec{t}_{ik}[\vec{t}_{kj} - \phi_{kj}(0)] + \vec{t}_{ij}\,\delta_{jk}. \qquad (4.1.50)$$

Since $\phi_{jk}(0) = \delta_{jk}$ and $\phi_{kj}(0) = \delta_{kj}$, this equation becomes

$$\begin{aligned}
\overline{\nu_{ij}\nu_{ik}} &= \vec{t}_{ij}(\vec{t}_{jk} - \delta_{jk}) + \vec{t}_{ik}(\vec{t}_{kj} - \delta_{kj}) + \vec{t}_{ij}\,\delta_{jk} \\
&= \vec{t}_{ij}\vec{t}_{jk} + \vec{t}_{ik}\vec{t}_{kj} - \vec{t}_{ik}\,\delta_{kj}.
\end{aligned} \qquad (4.1.51)$$

We shall use this result in two calculations. First we shall calculate the second moment of the time spent in each state of the transient process for each possible starting state. In our notation, the second moment of the time spent in state $j$ given that the system starts in state $i$ is $\overline{\nu_{ij}^2}$. This quantity is just the expectation of the product in Equation 4.1.51 when $j = k$,

$$\overline{\nu_{ij}^2} = \overline{\nu_{ij}\nu_{ij}} = 2\vec{t}_{ij}\vec{t}_{jj} - \vec{t}_{ij}. \tag{4.1.52}$$

This equation relates the second moment of the time spent in each state of the transient process for a given starting state to the transmissions of the transition diagram. If we define a matrix $\overline{\mathrm{N}^2}$ with components $\overline{\nu_{ij}^2}$, we can compute it from the matrix of transmissions, $\vec{T}$, using

$$\overline{\mathrm{N}^2} = 2\vec{T}\vec{T}_{\mathrm{dg}} - \vec{T} = \vec{T}(2\vec{T}_{\mathrm{dg}} - I). \tag{4.1.53}$$

The subscript dg (for "diagonal") on a square matrix means that all nondiagonal elements are to be set equal to zero.

The variance of the time spent in state $j$ of a transient process given that the process starts in state $i$, $\overset{\vee}{\nu}_{ij}$, is given by

$$\overset{\vee}{\nu}_{ij} = \overline{\nu_{ij}^2} - \overline{\nu_{ij}}^2. \tag{4.1.54}$$

We compute the matrix $\overset{\vee}{\mathrm{N}}$ with elements $\overset{\vee}{\nu}_{ij}$ by subtracting from $\overline{\mathrm{N}^2}$ a matrix whose elements are the squares of the corresponding elements in $\overline{\mathrm{N}}$.

$$\overset{\vee}{\mathrm{N}} = \overline{\mathrm{N}^2} - \overline{\mathrm{N}} \,\square\, \overline{\mathrm{N}}. \tag{4.1.55}$$

***Box notation***

In Equation 4.1.55 we have used the "box" notation for the element by element, or congruent, product of two matrices. If $A$ and $B$ are two matrices with the same dimensions, then their congruent product†

$$C = A \,\square\, B \tag{4.1.56}$$

is a matrix of the same size whose elements are computed according to the rule

$$c_{ij} = a_{ij}b_{ij}. \tag{4.1.57}$$

We shall find this operation useful in much of our future work. We see, for example, that we can express $\vec{T}_{\mathrm{dg}}$ in the box notation by

$$\vec{T}_{\mathrm{dg}} = \vec{T} \,\square\, I. \tag{4.1.58}$$

Therefore the means, second moments, and variances of the times spent in each state of a transient process as a function of starting state can be expressed com-

† The properties of congruent matrix multiplication are discussed further in Appendix A.

pactly in terms of the matrix $\vec{T}$ of transmissions of the transition diagram by the equations:

$$\overline{\mathrm{N}} = \vec{T} \tag{4.1.59}$$

$$\overline{\mathrm{N}^2} = \vec{T}[2(\vec{T} \,\square\, I) - I] = \vec{T}[(2\vec{T} - I) \,\square\, I] \tag{4.1.60}$$

$$\overset{\vee}{\mathrm{N}} = \overline{\mathrm{N}^2} - \overline{\mathrm{N}} \,\square\, \overline{\mathrm{N}}. \tag{4.1.61}$$

### *The Marketing Problem*

Let us compute these quantities for the marketing problem. Since we have already found the elements of the matrix of transmissions $\vec{T}$ for the modified transition diagram of Figure 4.1.6, we have

$$\vec{T} = \begin{bmatrix} 5 & 10/3 \\ 0 & 10/3 \end{bmatrix} = \overline{\mathrm{N}} \tag{4.1.62}$$

and then

$$\begin{aligned} \overline{\mathrm{N}^2} = \vec{T}[(2\vec{T} - I) \,\square\, I] &= \begin{bmatrix} 5 & 10/3 \\ 0 & 10/3 \end{bmatrix} \begin{bmatrix} 9 & 0 \\ 0 & 17/3 \end{bmatrix} \\ &= \begin{bmatrix} 45 & 170/9 \\ 0 & 170/9 \end{bmatrix}. \end{aligned} \tag{4.1.63}$$

We also compute

$$\overline{\mathrm{N}} \,\square\, \overline{\mathrm{N}} = \begin{bmatrix} 25 & 100/9 \\ 0 & 100/9 \end{bmatrix} \tag{4.1.64}$$

and finally write the variance matrix

$$\overset{\vee}{\mathrm{N}} = \overline{\mathrm{N}^2} - \overline{\mathrm{N}} \,\square\, \overline{\mathrm{N}} = \begin{bmatrix} 20 & 70/9 \\ 0 & 70/9 \end{bmatrix}. \tag{4.1.65}$$

We can now interpret these results. The first row of the matrix $\overline{\mathrm{N}}$ shows that if a customer originally purchases brand $A$, we expect him to buy brand $A$ 5 times and brand $B$ 3 1/3 times before the final purchase of brand $A$ that causes him to become a permanent brand $A$ customer. The expected number of additional purchases beyond the initial purchase is therefore 8 1/3. We see from the first row of the matrix $\overset{\vee}{\mathrm{N}}$ in Equation 4.1.65 that the variance of the number of brand $A$ purchases he will make under these conditions is 20, whereas the variance of the number of brand $B$ purchases is $70/9 = 7\ 7/9$.

### The Variance of Delay in a Transient Process

We have just developed methods for calculating the major statistics of the delay that a system will experience in each state of a transient process. We know, furthermore, that the mean time the system will spend in all states of the transient process is just the sum of the mean delays in each state, a result expressed by Equation 4.1.37. But how can we find the variance of the total time spent in the transient process? Is it just the sum of the variances of the times spent in each state? No, not in general, because the times spent in each state are not usually independent random variables and the variance of the sum will not ncessarily be the sum of the variances. However, we can easily find the variance of the total time in the process from the results already obtained. We first calculate the second moment of the delay in the transient process. The delay in the transient process is the quantity $\nu_i$ defined by Equation 4.1.36. The square of this delay, $\nu_i^2$, is then

$$\nu_i^2 = \sum_{j\in\mathscr{T}} \nu_{ij} \sum_{k\in\mathscr{T}} \nu_{ik} = \sum_{j\in\mathscr{T}} \sum_{k\in\mathscr{T}} \nu_{ij}\nu_{ik}. \tag{4.1.66}$$

Here, as in future equations of this section, the summation on $j$ and $k$ is over all transient states of the process.

We obtain the second moment of the delay in the transient process by taking the expectation of $\nu_i^2$

$$\overline{\nu_i^2} = \sum_{j\in\mathscr{T}} \sum_{k\in\mathscr{T}} \overline{\nu_{ij}\nu_{ik}}. \tag{4.1.67}$$

Now we use the important result of Equation 4.1.51 to write $\overline{\nu_i^2}$ in terms of the transmissions of the modified transition diagram,

$$\overline{\nu_i^2} = \sum_{j\in\mathscr{T}} \sum_{k\in\mathscr{T}} (\vec{t}_{ij}\vec{t}_{jk} + \vec{t}_{ik}\vec{t}_{kj} - \vec{t}_{ik}\,\delta_{kj}). \tag{4.1.68}$$

The symmetry of the summation permits the simplification,

$$\begin{aligned}\overline{\nu_i^2} &= \sum_{j\in\mathscr{T}} \sum_{k\in\mathscr{T}} (2\vec{t}_{ij}\vec{t}_{jk} - \vec{t}_{ik}\,\delta_{kj}) \\ &= \sum_{j\in\mathscr{T}} \vec{t}_{ij}\left(2 \sum_{k\in\mathscr{T}} \vec{t}_{jk} - 1\right) \\ &= \sum_{j\in\mathscr{T}} \vec{t}_{ij}(2\bar{\nu}_j - 1) \\ &= 2 \sum_{j\in\mathscr{T}} \vec{t}_{ij}\bar{\nu}_j - \bar{\nu}_i. \end{aligned} \tag{4.1.69}$$

We have now found for each possible starting state an expression for the second moment of the time spent in the transient process. The expression relates this quantity to the matrix of transmissions of the transition diagram $\vec{T}$ and to the row sums of that matrix, the $\bar{\nu}_j$. If we define a column vector $\overline{\boldsymbol{\nu}^2}$ whose components are the second moments $\overline{\nu_i^2}$ of the delays in the transient process for each possible starting state, then we can write Equation 4.1.69 in the form

$$\overline{\mathbf{v}^2} = 2\vec{T}\bar{\mathbf{v}} - \bar{\mathbf{v}} = (2\vec{T} - I)\bar{\mathbf{v}}. \tag{4.1.70}$$

Similarly, if we use a column vector $\check{\mathbf{v}}$ for the variance of the delay in the transient process for each starting state, then we finally obtain

$$\check{\mathbf{v}} = \overline{\mathbf{v}^2} - \bar{\mathbf{v}} \square \bar{\mathbf{v}} = (2\vec{T} - I)\bar{\mathbf{v}} - \bar{\mathbf{v}} \square \bar{\mathbf{v}}. \tag{4.1.71}$$

*The Marketing Problem*

Let us check this result for the marketing problem. The matrix of transmissions of the transition diagrams in Figures 4.1.1 or 4.1.6 when the recurrent state has been eliminated we have already found by inspection to be

$$\vec{T} = \begin{bmatrix} 5 & 10/3 \\ 0 & 10/3 \end{bmatrix}. \tag{4.1.72}$$

By Equation 4.1.38, the row sums produce $\bar{\mathbf{v}}$,

$$\bar{\mathbf{v}} = \vec{T}\mathbf{s} = \begin{bmatrix} 25/3 \\ 10/3 \end{bmatrix}, \tag{4.1.73}$$

the vector of expected delay in the transient process for each starting state. Then we use

$$\overline{\mathbf{v}^2} = (2\vec{T} - I)\bar{\mathbf{v}} \tag{4.1.74}$$

to find the vector of second moments of delay,

$$\overline{\mathbf{v}^2} = \begin{bmatrix} 9 & 20/3 \\ 0 & 17/3 \end{bmatrix}\begin{bmatrix} 25/3 \\ 10/3 \end{bmatrix} = \begin{bmatrix} 875/9 \\ 170/9 \end{bmatrix}, \tag{4.1.75}$$

compute the vector of squared expected delays,

$$\bar{\mathbf{v}} \square \bar{\mathbf{v}} = \begin{bmatrix} 625/9 \\ 100/9 \end{bmatrix}, \tag{4.1.76}$$

and finally write the vector of delay variances,

$$\check{\mathbf{v}} = \overline{\mathbf{v}^2} - \bar{\mathbf{v}} \square \bar{\mathbf{v}} = \begin{bmatrix} 250/9 \\ 70/9 \end{bmatrix} = \begin{bmatrix} 27.78 \\ 7.78 \end{bmatrix}. \tag{4.1.77}$$

Thus $\check{\nu}_1 = 250/9$ and $\check{\nu}_2 = 70/9$ are the variances of the delay in the transient process under the conditions of starting in state 1 and state 2, respectively. The variance of the delay in the transient process when the system starts in state 1, $\check{\nu}_1$, is, of course, the value found before for this quantity in Equation 4.1.23 by differentiation of the geometric transform.

By comparing Equations 4.1.77 and 4.1.65, we see that for this example the variance of the delay in the transient process is equal to the sum of the variances of the delay in each state. But this was not supposed to be true. Well, it is not true in general, but it is for this example because of the special structure of the transient process.

The transition diagram for Figure 4.1.1 shows why. Because it is impossible to return from state 2 to state 1, the total time in the process is the sum of the time to leave state 1 for the first time and the time to leave state 2 for the first time. These random variables are independent and so their variances add. In general, we shall be able to add the variances of the delay in each state to obtain the total delay in the process only when the states of the process are entered sequentially with no possibility of return. In other cases, the computation must be performed strictly in accordance with Equation 4.1.71. The method for finding the variance indicated by Equation 4.1.71 is a significant saving in effort over the previous method. It allows the variance to be found almost by inspection.

### The Case of Random Initial State

In some problems we require that the system be started in a state of a transient process that is selected according to a probability distribution rather than a state that is known. In such cases the initial state probability row vector $\boldsymbol{\pi}(0)$ specifies the probability of starting in each state of the process. We shall initially associate no delay with the placement of the system in its starting state.

We begin by studying the statistics of the total delay in the transient process under the condition of random starting. With probability $\pi_i(0)$ the system will start in state $i$ and will experience a delay $\nu_i$, the random variable defined by Equation 4.1.36. Let $\nu_{\sim}$ be the total time spent in the transient process for a given starting probability vector $\boldsymbol{\pi}(0)$. The subscript $\sim$ indicates randomization of starting state according to some known probability vector $\boldsymbol{\pi}(0)$.

We recall that the expectation of any random variable can be computed as the sum over a set of mutually exclusive and collectively exhaustive events of the expectation given one such event multiplied by its probability. We shall call this principle the expectation decomposition principle. Since the starting states of the transient Markov process are mutually exclusive and collectively exhaustive, the principle shows that the moments of $\nu_{\sim}$ are related to the moments of $\nu_i$ by

$$\bar{\nu}_{\sim} = \sum_{i \in \mathscr{T}} \pi_i(0)\bar{\nu}_i \tag{4.1.78}$$

and

$$\overline{\nu_{\sim}^{\,2}} = \sum_{i \in \mathscr{T}} \pi_i(0)\overline{\nu_i^{\,2}}. \tag{4.1.79}$$

In vector form these equations become

$$\bar{\nu}_{\sim} = \boldsymbol{\pi}(0)\bar{\boldsymbol{\nu}} \tag{4.1.80}$$

and

$$\overline{\nu_{\sim}^{2}} = \boldsymbol{\pi}(0)\overline{\boldsymbol{\nu}^{2}}, \tag{4.1.81}$$

where the vectors $\bar{\boldsymbol{\nu}}$ and $\overline{\boldsymbol{\nu}^2}$ have already been defined. Equation 4.1.70 allows us to write Equation 4.1.81 in the form

$$\overline{\nu_{\sim}^{2}} = \boldsymbol{\pi}(0)(2\vec{T} - I)\bar{\boldsymbol{\nu}}. \tag{4.1.82}$$

Finally, the variance of $\nu_{\sim}$, $\check{\nu}_{\sim}$, is written as

$$\check{\nu}_{\sim} = \overline{\nu_{\sim}^{2}} - \bar{\nu}_{\sim}^{2} = \boldsymbol{\pi}(0)(2\vec{T} - I)\bar{\boldsymbol{\nu}} - [\boldsymbol{\pi}(0)\bar{\boldsymbol{\nu}}]^{2}. \tag{4.1.83}$$

Equations 4.1.80 and 4.1.83 show that calculation of the mean and variance of the time spent in a transient process is only slightly modified by the possibility of a random starting state. We compute the same quantities as in the case of deterministic starting, but combine them with appropriate weighting by the elements of $\boldsymbol{\pi}(0)$.

### *Effect on flow graph analysis*

To incorporate the existence of random starting in the flow graph, we may consider a single state designated by a solid dot that serves as a general input state for the system. The branch that links this state to state $i$ is labeled with the starting probability $\pi_i(0)$, but with no delay $z$ if the starting is instantaneous. We denote the transmission of the flow graph from the input node to any node $j$ by $\vec{t}_{\bullet j}(z)$; for the modified transition diagram this becomes $\vec{t}_{\bullet j}$. In view of Equations 4.1.37 and 4.1.78 and the graph properties, we can write $\bar{\nu}_{\sim}$ quite simply as

$$\bar{\nu}_{\sim} = \sum_{i\in\mathcal{T}} \pi_i(0) \sum_{j\in\mathcal{T}} \vec{t}_{ij} = \sum_{j\in\mathcal{T}} \sum_{i\in\mathcal{T}} \pi_i(0)\vec{t}_{ij} = \sum_{j\in\mathcal{T}} \vec{t}_{\bullet j}. \tag{4.1.84}$$

The expected delay in the process under random starting is just the sum over all transient states of the transmission of the modified transition diagram from the general input node to each transient state.

### *The marketing problem with random starting*

We can illustrate the operations involved by returning to the marketing example and considering the situation where a customer chooses originally at random between the two brands $A$ and $B$: We shall assume that he has probability 1/2 of buying each of them on his zeroth purchase. The customer may, therefore, pass through the stages of transient $A$ customer—transient $B$ customer—permanent $A$ customer, or he may with equal probability miss the first stage. The modified transition diagram for the process appears in Figure 4.1.8. A solid dot indicates

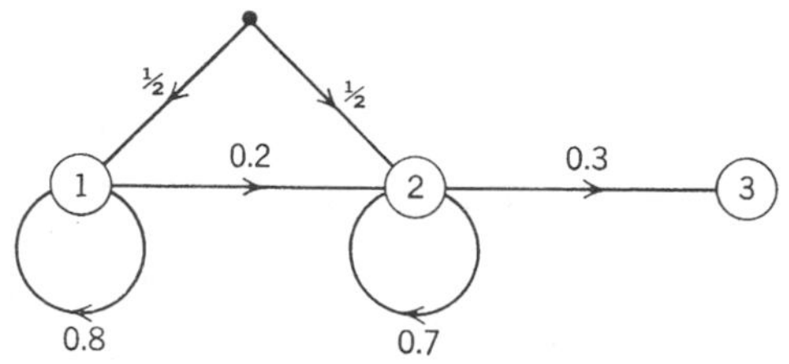


**Figure 4.1.8** Modified transition diagram for a transient process with random starting.

the general starting node from which an instantaneous transition is made with probability 1/2 to each of the states 1 and 2; all other transitions of the process require one transition time. For this starting condition the initial state probability vector is $\boldsymbol{\pi}(0) = [1/2 \quad 1/2]$. Then from Equation 4.1.80 and Equation 4.1.73 written earlier for this example we can write

$$\bar{\nu}_\sim = \boldsymbol{\pi}(0)\bar{\boldsymbol{\nu}} = [1/2 \quad 1/2]\begin{bmatrix}25/3\\10/3\end{bmatrix} = 35/6 = 5.83. \tag{4.1.85}$$

We obtain the same result immediately by using Equation 4.1.84 and computing $\bar{\nu}_\sim = \vec{t}_{\cdot 1} + \vec{t}_{\cdot 2}$. The mean number of purchases the customer will make before the purchase that makes him a permanent brand $A$ customer is therefore 35/6.

Equations 4.1.81 and 4.1.75 show how to find the second moment of the delay in the transient process. We obtain

$$\overline{\nu_\sim^{\,2}} = \boldsymbol{\pi}(0)\overline{\boldsymbol{\nu}^2} = [1/2 \quad 1/2]\begin{bmatrix}875/9\\170/9\end{bmatrix} = 1045/18. \tag{4.1.86}$$

Finally, the variance is computed from Equation 4.1.83 as

$$\check{\nu}_\sim = \overline{\nu_\sim^{\,2}} - \bar{\nu}_\sim^{\,2} = 1045/18 - (35/6)^2 = 865/36 = 24.0. \tag{4.1.87}$$

Note that although the mean delay for this random starting situation in Equation 4.1.85 is just the average of the mean delays from starting in each of the two states 1 and 2 as expressed by Equation 4.1.73, the variance found in Equation 4.1.87 is not the average of the variances in Equation 4.1.77. Errors can always be avoided by dealing with the first and second moments as a function of starting position, averaging with respect to $\boldsymbol{\pi}(0)$, and finally computing the variance of the process.

***State delay moments with random initial state***

We now apply this procedure directly to find the mean and variance of the delay in each state of a transient process for a random starting condition. To be consistent in notation, let $\nu_{\sim j}$ be the delay in state $j$ of the transient process if it is started according to a probability vector $\boldsymbol{\pi}(0)$. The random variable $\nu_{\sim j}$ will then have a mean $\bar{\nu}_{\sim j}$, a second moment $\overline{\nu_{\sim j}^2}$, and a variance $\overset{\vee}{\nu}_{\sim j}$. Since

$$\bar{\nu}_{\sim j} = \sum_{i \in \mathscr{T}} \pi_i(0) \bar{\nu}_{ij} \tag{4.1.88}$$

by the expectation decomposition principle, it follows that

$$\bar{\nu}_{\sim j} = \sum_{i \in \mathscr{T}} \pi_i(0) \vec{t}_{ij} = \vec{t}_{\bullet j}. \tag{4.1.89}$$

The expected delay in a state under random starting is just the transmission of the modified transition diagram from the general input node to that state.

The moments of state delays are readily displayed in row vectors $\bar{\boldsymbol{\nu}}_\sim$, $\overline{\boldsymbol{\nu}_\sim^2}$, and $\overset{\vee}{\boldsymbol{\nu}}_\sim$. The vectors are computed easily by using the results of Equations 4.1.59 and 4.1.60. We find

$$\begin{aligned} \bar{\boldsymbol{\nu}}_\sim &= \boldsymbol{\pi}(0)\bar{\mathrm{N}} = \boldsymbol{\pi}(0)\vec{T} \\ \overline{\boldsymbol{\nu}_\sim^2} &= \boldsymbol{\pi}(0)\overline{\mathrm{N}^2} = \pi(0)\vec{T}[(2\vec{T} - I) \square I] = \boldsymbol{\nu}_\sim[(2\vec{T} - I) \square I]. \end{aligned} \tag{4.1.90}$$

Finally we compute the variance vector,

$$\overset{\vee}{\boldsymbol{\nu}}_\sim = \overline{\boldsymbol{\nu}_\sim^2} - \bar{\boldsymbol{\nu}}_\sim \square \bar{\boldsymbol{\nu}}_\sim . \tag{4.1.91}$$

**The marketing problem with random starting.** If we want to apply these results to the marketing example with random starting, we use $\boldsymbol{\pi}(0) = [1/2 \quad 1/2]$, $\vec{T} = \bar{\mathrm{N}}$ from Equation 4.1.62, and $\overline{\mathrm{N}^2}$ from Equation 4.1.63. We obtain

$$\bar{\boldsymbol{\nu}}_\sim = \boldsymbol{\pi}(0)\bar{\mathrm{N}} = [1/2 \quad 1/2]\begin{bmatrix} 5 & 10/3 \\ 0 & 10/3 \end{bmatrix} = [5/2 \quad 10/3], \tag{4.1.92}$$

which is, of course, the vector $[\vec{t}_{\bullet 1} \quad \vec{t}_{\bullet 2}]$. Continuing, we find

$$\begin{aligned} \overline{\boldsymbol{\nu}_\sim^2} &= \boldsymbol{\pi}(0)\overline{\mathrm{N}^2} = [1/2 \quad 1/2]\begin{bmatrix} 45 & 170/9 \\ 0 & 170/9 \end{bmatrix} = [45/2 \quad 170/9], \\ \bar{\boldsymbol{\nu}}_\sim \square \bar{\boldsymbol{\nu}}_\sim &= [25/4 \quad 100/9], \end{aligned} \tag{4.1.93}$$

and

$$\overset{\vee}{\boldsymbol{\nu}}_\sim = \overline{\boldsymbol{\nu}_\sim^2} - \bar{\boldsymbol{\nu}}_\sim \square \bar{\boldsymbol{\nu}}_\sim = [65/4 \quad 70/9]. \tag{4.1.94}$$

We therefore expect a customer who makes his first purchase at random to buy brand $A$ 5/2 times with a variance of 65/4, and brand $B$ 10/3 times with a variance of 70/9 before the purchase that ultimately transforms him to a permanent brand $A$ customer. Notice that $\overset{\vee}{\boldsymbol{\nu}}_\sim$ is not equal to $\boldsymbol{\pi}(0)$ times $\overset{\vee}{\mathrm{N}}$ as given by Equation 4.1.65, even in this special example. However, the variance of the total time spent in the transient process given by Equation 4.1.87 is in this case equal to the sum of the variances of the times spent in each state, the sum of the components of $\check{\mathbf{v}}_{\sim}$. This result is a direct consequence of the sequential passage of the transient process through its states.

## The Case of Several Trapping States

Occasionally the construction of models requires that we consider a transient process that can run into a number of different trapping states. These transient processes are no different in principle from the ones we have just studied—the only distinction is that the probability "runs out of several drains" rather than just one. However, the analysis is slightly modified since we do not know for certain in which of the several trapping states the system will be captured.

### *A marketing example with trapping by both brands*

We can construct a many-outlet transient process by changing the marketing example. Let us give brand $B$ a fighting chance; we assume the same probabilities for repeating and switching brands that we used before, but now say that a customer becomes a permanent customer of a brand when he purchases it twice in succession. In this model a customer can become a permanent customer of either brand after some transient period. We assume further that a new customer is equally likely to purchase brand $A$ and brand $B$ as his zeroth purchase. The flow graph for this Markov process is shown in Figure 4.1.9. States 3 and 4 correspond to being a permanent $A$ and permanent $B$ customer. The solid dot that originates two branches with probabilities 0.5 is the starting node of the system. Transmissions from this starting node to any other node $i$ of the graph will have a dot for the initial subscript of the transmission.

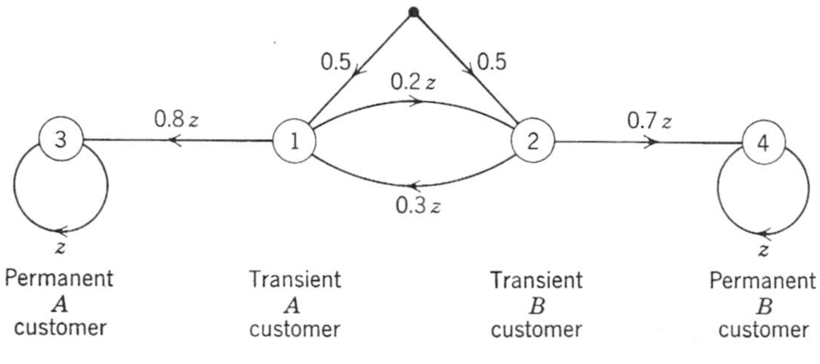


**Figure 4.1.9** Flow graph for marketing example with trapping by both brands.

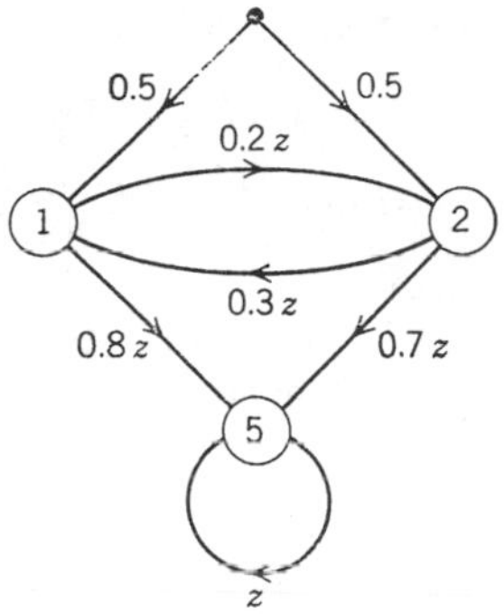


**Figure 4.1.10** Flow graph number of purchases to become permanent customer of some brand.

***Probability of delay***

We can ask many questions about this sytem. Let us begin by determining the probability of a customer's being a permanent customer of some brand by purchase $n$. We combine states 3 and 4 into a single state 5, "permanent customer," and drop the brand distinction. The flow graph for this Markov process is shown in Figure 4.1.10. This graph has only one trapping state and is the kind we have just finished studying. If we eliminate the self-loop of $z$ around state 5 we have the modified flow graph of Figure 4.1.11. The transmission of this graph from the input to node 5, $\bar{t}_{\cdot 5}$, is the geometric transform $p_5^{g}(z)$ of the probability distribution

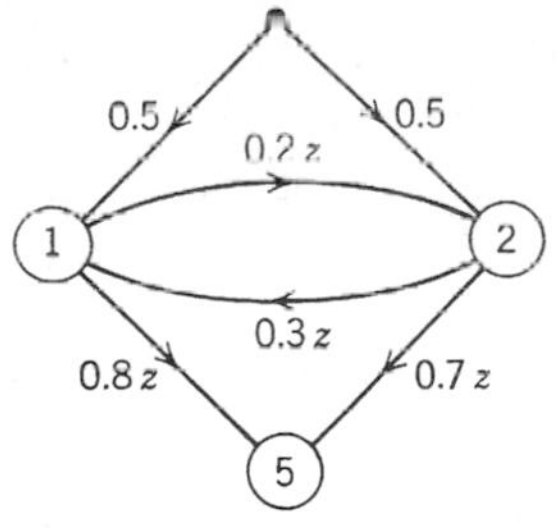


**Figure 4.1.11** Modified flow graph for number of purchases to become permanent customer of some brand.

for the number of additional purchases required to become a permanent customer of some brand. By inspection of Figure 4.1.11 we find

$$\vec{t}_{\bullet 5} = p_5{}^g(z) = \frac{0.75z + 0.19z^2}{1 - 0.06z^2}. \tag{4.1.95}$$

Evaluating this transform at $z = 1$, we confirm that ultimate arrival at node 5 is a certain event. The probability that it occurs on the $n$th additional purchase is the inverse, $p_5(n)$, of this geometric transform.

The transform

$$f^g(z) = \frac{1}{1 - 0.06z^2} \tag{4.1.96}$$

represents a time function $f(n)$ that exists only for even values of $n$ and that decays geometrically from one occurrence to the next at a rate 0.06; that is,

$$f(n) = \begin{cases} (0.06)^{n/2} & n = 0, 2, 4, \ldots \\ 0 & n = 1, 3, 5, \ldots. \end{cases} \tag{4.1.97}$$

Since the factors of $z$ and $z^2$ in the numerator of Equation 4.1.95 only delay this solution by one and two purchases, respectively, the probability distribution $p_5(n)$ is

$$p_5(n) = \begin{cases} 0 & n = 0 \\ 0.75(0.06)^{\frac{n-1}{2}} & n = 1, 3, 5, \ldots \\ 0.19(0.06)^{\frac{n-2}{2}} & n = 2, 4, 6, \ldots. \end{cases} \tag{4.1.98}$$

***Joint probability of delay and final state***

We confirm that there is indeed a marked difference in the probability of becoming a permanent customer on even and odd purchases. Yet the result is not strange when we consider that there are two different trapping mechanisms at work and that one is highly favored over the other depending on the state in which the process is started. Let us therefore reestablish the distinction between the two types of permanent customers and calculate the probability of a customer's becoming each kind on purchase $n$.

The modified flow graph suited to this computation is that of Figure 4.1.9 altered by removal of the self-loops on the trapping states; it is drawn in Figure 4.1.12. The transforms of the probabilities of becoming a permanent brand $A$ or brand $B$ customer on purchase $n$, $p_3{}^g(z)$ and $p_4{}^g(z)$, are just the transmissions from the input node to nodes 3 and 4 in this flow graph. We find

$$\vec{t}_{\bullet 3} = p_3{}^g(z) = \frac{0.4z + 0.12z^2}{1 - 0.06z^2}, \qquad \vec{t}_{\bullet 4} = p_4{}^g(z) = \frac{0.35z + 0.07z^2}{1 - 0.06z^2}. \tag{4.1.99}$$

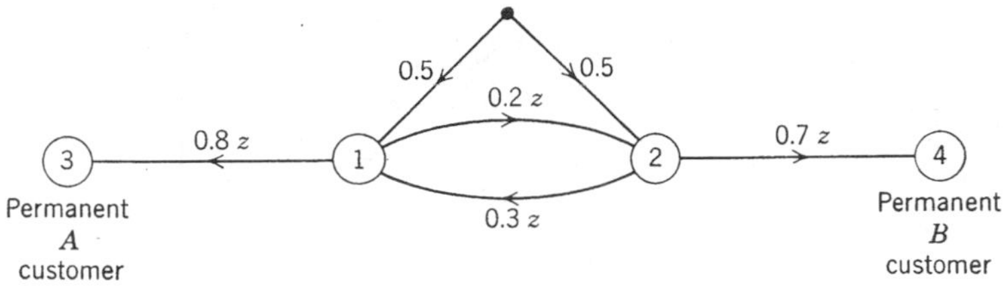


**Figure 4.1.12** Modified flow graph for time to become a permanent customer of each brand.

The sum of $p_3{}^g(z)$ and $p_4{}^g(z)$ is, of course, just the expression for $p_5{}^g(z)$ found in Equation 4.1.95. By the same inversion process used to determine $p_5(n)$ we obtain

$$p_3(n) = \begin{cases} 0 \\ 0.4(0.06)^{\frac{n-1}{2}} \\ 0.12(0.06)^{\frac{n-2}{2}} \end{cases} \qquad p_4(n) = \begin{cases} 0 & n = 0 \\ 0.35(0.06)^{\frac{n-1}{2}} & n = 1, 3, 5, \dots \\ 0.07(0.06)^{\frac{n-2}{2}} & n = 2, 4, 6, \dots . \end{cases} \tag{4.1.100}$$

These expressions must sum to the expression for $p_5(n)$ in Equation 4.1.98.

***Probability of final state***

Since we already know that $p_5(n)$ is a probability distribution, $p_3(n)$ and $p_4(n)$ cannot be. We expect this result because we know that neither becoming a permanent brand $A$ customer nor becoming a permanent brand $B$ customer is in itself a certain event. Equation 4.1.100 specifies the probability of being trapped by each brand on purchase $n$. The probability of ultimately being trapped by each of them is just the sum of $p_3(n)$ and of $p_4(n)$ over all values of $n$. But we have seen in our earlier work that there is a simple way of performing such a summation. The geometric transform evaluated at $z = 1$ is just the sum of the corresponding discrete function over all its values. Therefore, if we find $p_3{}^g(1)$ and $p_4{}^g(1)$ we shall have the probability that each brand will ultimately trap the customer. We compute from Equation 4.1.99,

$$p_3{}^g(1) = 26/47 = 0.553, \qquad p_4{}^g(1) = 21/47 = 0.447. \tag{4.1.101}$$

We have found that a new customer will eventually become a brand $A$ customer with probability 0.553. We expect capture by brand $A$ to be more probable because of the probabilistic structure of the flow graph.

***Probability of final state as transmission of modified transition diagram***

When we review the calculations we have just performed we see that we have calculated the probability of being trapped in any particular trapping state by

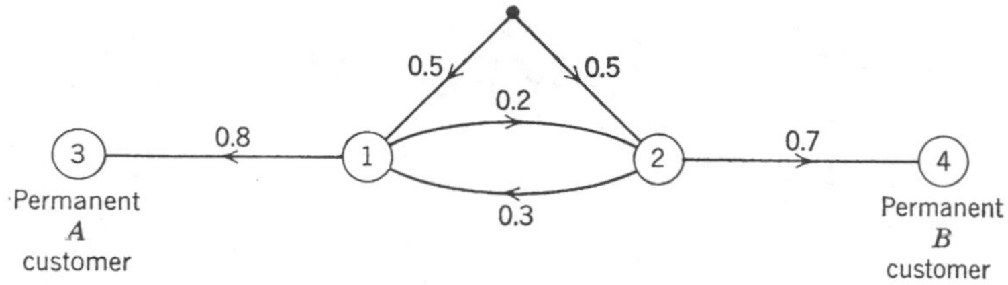


**Figure 4.1.13** Modified transition diagram used to determine ultimate probability of being trapped by each brand.

evaluating at the point $z = 1$ the transmission from the input to the node of the modified flow graph corresponding to that state. This evaluation is equivalent to simply finding the transmission from the input to that node of the modified transition diagram for the process. We draw this diagram for our example in Figure 4.1.13. Computing the transmissions $\vec{t}_{\cdot 3}$ and $\vec{t}_{\cdot 4}$ from the input to each of the nodes 3 and 4 establishes immediately the result of Equation 4.1.101. We shall often find it useful to remember that the probability that a transient process will enter any given trapping state is just the transmission of the modified transition diagram for the process from the input to the node corresponding to that state.

***Probability of delay given final state***

The quantity $p_3(n)$ given by Equation 4.1.100 is, in fact, the joint probability of becoming a permanent customer at purchase $n$ and becoming a permanent brand $A$ customer at that time. If we want the conditional probability $p_3(n|A)$ that the customer will become a permanent brand $A$ customer on his $n$th purchase given that he will ultimately become a permanent brand $A$ customer, we must divide $p_3(n)$ by $p_3^g(1)$, the ultimate probability of his becoming a brand $A$ customer. Thus,

$$p_3(n|A) = \frac{p_3(n)}{p_3^g(1)} = \begin{cases} 0 & n = 0 \\ 47/26 \times 0.4(0.06)^{\frac{n-1}{2}} & n = 1, 3, 5, \ldots \\ 47/26 \times 0.12(0.06)^{\frac{n-2}{2}} & n = 2, 4, 6, \ldots . \end{cases} \tag{4.1.102}$$

The conditional probability distribution $p_3(n|A)$ will, of course, now sum to one over all values of $n$. Its geometric transform $p_3^g(z|A)$ is just the quantity $p_3^g(z)$ appearing in Equation 4.1.99 divided by $p_3^g(1)$, or

$$p_3^g(z|A) = \frac{p_3^g(z)}{p_3^g(1)} = 47/26 \times \frac{0.4z + 0.12z^2}{1 - 0.06z^2} . \tag{4.1.103}$$

***Delay moments given final state***

The moments of the conditional distribution $p_3(n|A)$ can be obtained in the usual way by differentiating the expression for $p_3^g(z|A)$ of Equation 4.1.103 and evaluating the result when $z = 1$. In particular, we find that the mean number of purchases made when the customer becomes a permanent brand $A$ customer, $\bar{\nu}_{\sim|A}$, is given by

$$\bar{\nu}_{\sim|A} = p_3{}^{\mathscr{g}\prime}(1|A) = \frac{830}{611} = 1.358. \tag{4.1.104}$$

Naturally, a similar set of calculations would produce the corresponding quantities for the case when the customer is known to have been trapped by brand $B$.

***Expected delay given final state from the modified transition diagram***

Although we can always find the mean delay in the transient process conditioned on the state in which it is trapped by the method we have just described, we have more convenient techniques at our disposal. We already know from our earlier results in this section that the mean delay in a transient process—regardless of where the system is finally trapped—is available from inspection of the modified transition diagram. Thus the mean delay that the system will spend in state 1 of the system in Figure 4.1.13 is

$$\bar{\iota}_{\bullet 1} = \frac{0.5 + 0.5(0.3)}{1 - (0.2)(0.3)} = \frac{0.65}{0.94} = \frac{65}{94}; \tag{4.1.105}$$

the mean delay in state 2 is

$$\bar{\iota}_{\bullet 2} = \frac{0.5 + 0.5(0.2)}{1 - (0.2)(0.3)} = \frac{0.6}{0.94} = \frac{60}{94}. \tag{4.1.106}$$

Therefore, the total delay in the transient process is

$$\bar{\iota}_{\bullet 1} + \bar{\iota}_{\bullet 2} = 125/94 = 1.330. \tag{4.1.107}$$

This value for the mean delay in the transient process could also be found by differentiating $p_5{}^{\mathscr{g}}(z)$ in Equation 4.1.95 and evaluating the derivative at $z = 1$.

But we are at the moment interested in the expected delay in the transient process *given* that the process terminates in a certain state. How can we also find this quantity from the modified transition diagram of Figure 4.1.13? Let us begin by considering the amount of delay in a transient state $j$ incurred by a system that starts in state $i$ and is trapped by state $k$. This random variable could be clearly designated by $\nu_{ij|k}$.

Let $x_{ij|k}(n)$ equal one if the process started in state $i$, occupied state $j$ at time $n$, and ultimately occupied state $k$; otherwise, let it be zero.

Then we can write

$$\nu_{ij|k} = \sum_{n=0}^{\infty} x_{ij|k}(n). \tag{4.1.108}$$

The expected number of times that state $j$ will be occupied given that the system starts in state $i$ and ends in state $k$ is then

$$\bar{\nu}_{ij|k} = \overline{\sum_{n=0}^{\infty} x_{ij|k}(n)} \tag{4.1.109}$$

or

$$\bar{\nu}_{ij|k} = \sum_{n=0}^{\infty} \bar{x}_{ij|k}(n). \tag{4.1.110}$$

The random variable $x_{ij|k}(n)$ will take on the value one with probability

$$\mathscr{P}\{s(n) = j|s(0) = i, s(\infty) = k\} \tag{4.1.111}$$

and the value zero with one minus this probability. Therefore the expected value of $x_{ij}(n)$ given that $s(\infty) = k$ is

$$\bar{x}_{ij|k} = \mathscr{P}\{s(n) = j|s(0) = i, s(\infty) = k\}. \tag{4.1.112}$$

Then by substituting this result into Equation 4.1.110 we can write

$$\bar{\nu}_{ij|k} = \sum_{n=0}^{\infty} \mathscr{P}\{s(n) = j|s(0) = i, s(\infty) = k\}. \tag{4.1.113}$$

Bayes' theorem allows us to continue in the form

$$\bar{\nu}_{ij|k} = \sum_{n=0}^{\infty} \frac{\mathscr{P}\{s(n) = j|s(0) = i\}\mathscr{P}\{s(\infty) = k|s(n) = j, s(0) = i\}}{\mathscr{P}\{s(\infty) = k|s(0) = i\}}. \tag{4.1.114}$$

But, by the basic Markovian property,

$$\mathscr{P}\{s(\infty) = k|s(n) = j, s(0) = i\} = \mathscr{P}\{s(\infty) = k|s(n) = j\}. \tag{4.1.115}$$

Now Equation 4.1.114 becomes

$$\bar{\nu}_{ij|k} = \frac{\sum_{n=0}^{\infty} \mathscr{P}\{s(n) = j|s(0) = i\}\mathscr{P}\{s(\infty) = k|s(n) = j\}}{\mathscr{P}\{s(\infty) = k|s(0) = i\}}$$

$$= \frac{\sum_{n=0}^{\infty} \phi_{ij}(n)\phi_{jk}}{\phi_{ik}} = \frac{\bar{\nu}_{ij}\phi_{jk}}{\phi_{ik}} = \frac{\vec{t}_{ij}\vec{t}_{jk}}{\vec{t}_{ik}}. \tag{4.1.116}$$

Here we have used 1) the result of Equation 4.1.34 that the transmission of a modified transition diagram from a transient state $i$ to a transient state $j$ is the expected number of times state $j$ will be occupied in an infinite number of transitions if the process is started in state $i$, and 2) the result discovered in this section that the transmission of a modified transition diagram from a transient state $j$ to a trapping state $k$ is the probability that the process will be trapped by state $k$ if it is started in state $j$.

The expected total delay in the transient process is then

$$\bar{\nu}_{i|k} = \sum_{j\in\mathcal{T}} \bar{\nu}_{ij|k} = \frac{\sum_{j\in\mathcal{T}} \vec{t}_{ij}\vec{t}_{jk}}{\vec{t}_{ik}}. \tag{4.1.117}$$

If the process is started according to a random probability vector $\pi(0)$, then the expected total delay in the transient process is

$$\bar{\nu}_{\sim|k} = \frac{\sum_{j\in\mathcal{T}} \vec{t}_{\cdot j}\vec{t}_{jk}}{\vec{t}_{\cdot k}}. \tag{4.1.118}$$

Note that the expected delay in any state of the transient process and the expected total delay in the transient process conditional on being trapped by any state are computable from the transmissions of the modified transition diagram.

**Application to the marketing example.** We can use this technique to find $\bar{\nu}_{\sim|A}$ and $\bar{\nu}_{\sim|B}$, the expected delay in the transient process of Figure 4.1.13 conditional on trapping by either brand $A$ or brand $B$. We write

$$\bar{\nu}_{\sim|A} = \bar{\nu}_{\sim|3} = \frac{\vec{t}_{\cdot 1}\vec{t}_{13} + \vec{t}_{\cdot 2}\vec{t}_{23}}{\vec{t}_{\cdot 3}} \tag{4.1.119}$$

and

$$\bar{\nu}_{\sim|B} = \bar{\nu}_{\sim|4} = \frac{\vec{t}_{\cdot 1}\vec{t}_{14} + \vec{t}_{\cdot 2}\vec{t}_{24}}{\vec{t}_{\cdot 4}}. \tag{4.1.120}$$

We use Figure 4.1.13 to compute these transmissions as

$$\begin{aligned} \vec{t}_{\cdot 1} &= 65/94 & \vec{t}_{13} &= 80/94 \\ \vec{t}_{\cdot 2} &= 60/94 & \vec{t}_{23} &= 24/94 \\ \vec{t}_{\cdot 3} &= 52/94 & \vec{t}_{14} &= 14/94 \\ \vec{t}_{\cdot 4} &= 42/94 & \vec{t}_{24} &= 70/94. \end{aligned} \tag{4.1.121}$$

Now we substitute these values into Equation 4.1.119 and 4.1.120 to obtain

$$\bar{\nu}_{\sim|A} = 830/611 = 1.358 \tag{4.1.122}$$

and

$$\bar{\nu}_{\sim|B} = 365/282 = 1.294. \tag{4.1.123}$$

Equation 4.1.122 confirms Equation 4.1.104. We see that the process requires a slightly larger expected number of purchases to trap a brand $A$ customer than it does to trap a brand $B$ customer.

Of course, the expected number of purchases to trap a brand $A$ customer times the probability of becoming a permanent brand $A$ customer plus the expected number of purchases to trap a brand $B$ customer times the probability of becoming a permanent brand $B$ customer must be equal to the total expected delay in the transient process. We can verify this by referring to Equations 4.1.119 and 4.1.120 and by using the result that the sum of the transmissions from a transient state to all trapping states is one;

$$\begin{aligned}\bar{\nu}_{\sim|3}\vec{t}_{\bullet 3} + \bar{\nu}_{\sim|4}\vec{t}_{\bullet 4} &= \vec{t}_{\bullet 1}\vec{t}_{13} + \vec{t}_{\bullet 2}\vec{t}_{23} + \vec{t}_{\bullet 1}\vec{t}_{14} + \vec{t}_{\bullet 2}\vec{t}_{24} \\ &= \vec{t}_{\bullet 1}(\vec{t}_{13} + \vec{t}_{14}) + \vec{t}_{\bullet 2}(\vec{t}_{23} + \vec{t}_{24}) \\ &= \vec{t}_{\bullet 1} + \vec{t}_{\bullet 2}. \end{aligned} \tag{4.1.124}$$

We have already noted in Equation 4.1.107 that the total expected delay in the transient process is $\vec{t}_{\bullet 1} + \vec{t}_{\bullet 2}$.

We see that the existence of several trapping states can be treated by a very slight modification of the procedures for a single trapping state. If we are interested only in the delay in the transient process and not its ultimate destination, we combine all trapping states into a single trapping state and proceed as before. If we are interested in the delay in the transient process conditional on being trapped by some state, all we must do is remember that entry into any particular trapping state is no longer a certain event, and that its probability can be calculated directly from the modified transition diagram. When this is done we can easily compute the conditional probabilities for the delay—given trapping in any state—and any statistics of this distribution that we desire.

## 4.2 SOME APPLICATIONS OF TRANSIENT PROCESSES—COIN-TOSSING

The combination of the ideas of the transient process and the flow graph often allows us to formulate and solve problems almost by inspection. The method not only permits us to analyze routine problems quickly and efficiently, but also provides the insight necessary to approach problems that are not so easily understood. In this section we shall work a few simple examples to illustrate the contribution of the technique.

### Time to First Head

Let us begin with an elementary problem in probability theory. What is the probability that independent tosses of a coin whose probability of Heads is $p$ will produce a Head for the first time on the $n$th toss? To draw the flow graph we must decide the number of states that is necessary and how they should be

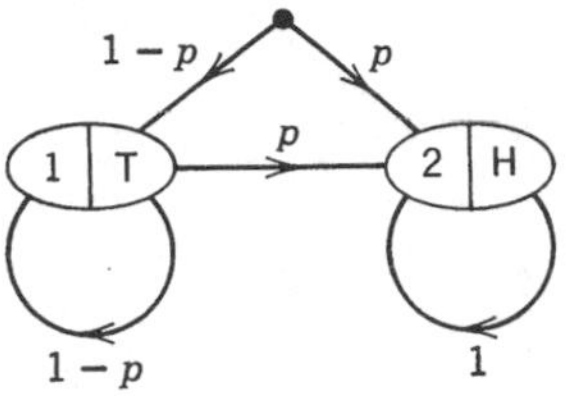


**Figure 4.2.1** Transition diagram for tosses to first Head.

described. We shall use two states: State 1 is occupied when the last toss was a Tail, state 2 when the last toss was a Head. We shall find it convenient to include a brief state description when drawing the transition diagram. Thus in the transition diagram of Figure 4.2.1 each node is not only numbered, but labeled with the first letter of the state description. The probabilities on the branches follow from the statement that the probability of Heads on any toss is $p$. Note, however, that state 2 is a trapping state; as a matter of fact, the probability of entering it after different numbers of tosses is just the problem we are trying to solve. The heavy dot is the starting node of the system.

*Flow graph*

To form the flow graph for the process we multiply each branch transmission of the transition diagram by $z$ to represent the delay of that transition. The result appears in Figure 4.2.2. Note that even the branches from the starting state are multiplied by $z$ because it will require one toss to place the system in either state 1 or state 2. Furthermore, since we want to analyze the transient process of entering state 2, the modified flow graph of Figure 4.2.3 with the transmission loop on the trapping state removed will serve us better.

Let $p(n)$ be the probability that a Head will be produced for the first time on the

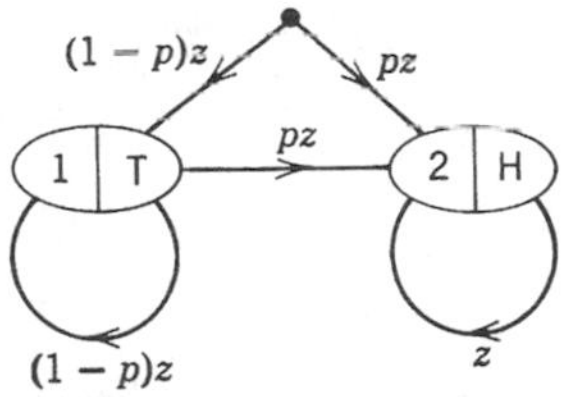


**Figure 4.2.2** Flow graph for tosses to first Head.

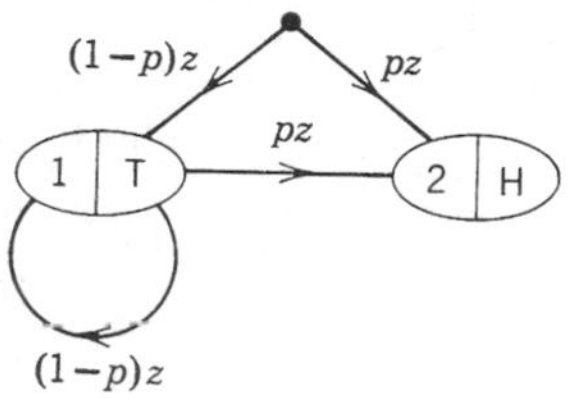


**Figure 4.2.3** Modified flow graph for tosses to first Head.

$n$th toss. Then $p^g(z)$, the geometric transform of $p(n)$, is just the transmission of the modified flow graph in Figure 4.2.3 from the input to node 2. Therefore,

$$p^g(z) = pz + \frac{(1 - p)pz^2}{1 - (1 - p)z} \tag{4.2.1}$$

or

$$p^g(z) = \frac{pz}{1 - (1 - p)z}. \tag{4.2.2}$$

We notice, however, that this would also be the transmission to node 2 of the modified flow graph in Figure 4.2.4. This flow graph is just like that of Figure 4.2.3 except for the method of starting. In the graph of Figure 4.2.4 the system is started with certainty in state 1, and no delay is associated with the starting. This method of starting is equivalent to saying that a Tail was produced on the zeroth toss. Since we are interested only in the results of physical tosses 1, 2, 3, ..., we are quite happy to accept this fiction if it will simplify our calculations. Therefore, for the purposes of answering the problem proposed, the flow graph of Figure 4.2.4 will prove to be most convenient. This is but the first of many

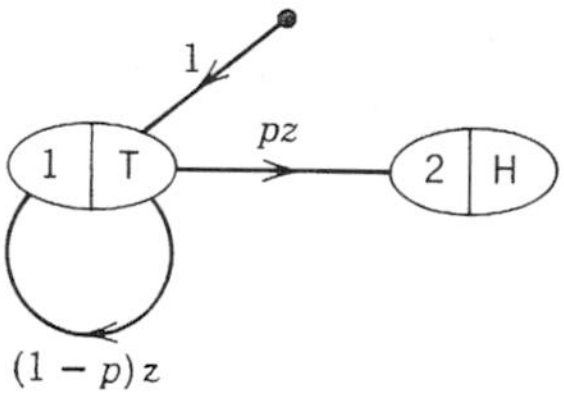


**Figure 4.2.4** Alternate modified flow graph for tosses to first Head.

examples in which the method of starting can be changed for computational simplicity without altering the answer to the question of interest.

***Probability***

We can invert the transform of Equation 4.2.2 to find the probability we seek,

$$p(0) = 0 \tag{4.2.3}$$

$$p(n) = p(1 - p)^{n-1} \qquad n = 1, 2, 3, \ldots \tag{4.2.4}$$

The probability of achieving a Head for the first time on the $n$th toss of a coin with probability of Heads $p$ is a geometric sequence with rate $1 - p$. We can interpret Equation 4.2.4 directly as the probability of a sequence of tosses in which the first $n - 1$ came up Tails and the $n$th produced a Head. Equation 4.2.4 is one form of the often-encountered "geometric" probability mass function. We shall see it frequently in our future work. Note that if the coin is fair ($p = 1/2$), then Equation 4.2.4 becomes,

$$p(n) = (1/2)^n \qquad n = 1, 2, 3, \ldots \tag{4.2.5}$$

***Moments by transform differentiation***

We find the mean and variance of the number of tosses to produce the first Head most readily by differentiating the geometric transform of Equation 4.2.2. Thus,

$$p^{g\prime}(z) = \frac{d}{dz}\left[\frac{pz}{1 - (1 - p)z}\right] = \frac{p}{[1 - (1 - p)z]^2} \tag{4.2.6}$$

and

$$p^{g\prime\prime}(z) = \frac{2p(1 - p)}{[1 - (1 - p)z]^3}\,. \tag{4.2.7}$$

Therefore, from Equation 4.1.20 we can calculate the mean $\bar{n}$ and the second moment $\overline{n^2}$ of the number of tosses to produce the first Head as

$$\bar{n} = p^{g\prime}(1) = \frac{1}{p} \tag{4.2.8}$$

and

$$\overline{n^2} = p^{g\prime\prime}(1) + p^{g\prime}(1) = \frac{2(1 - p)}{p^2} + \frac{1}{p} = \frac{2 - p}{p^2}\,. \tag{4.2.9}$$

The variance $\check{n}$ is then

$$\check{n} = \overline{n^2} - \bar{n}^2 = \frac{2 - p}{p^2} - \frac{1}{p^2} = \frac{1 - p}{p^2}\,. \tag{4.2.10}$$

These expressions for the moments of the geometric distribution will be useful in the future. Notice that if the coin is fair, both the mean and the variance of the number of tosses to produce the first Head are 2.

***Moments by transition diagram inspection***

If we only want to calculate the moments of the number of transitions to pass through the transient process, we can use inspection techniques based on the modified transition diagram of Figure 4.2.5. Since this transient process contains only one transient state, the matrix $\vec{T}$ has only one element $\vec{t}_{11}$. We find

$$\overline{\mathrm{N}} = \vec{T} = \vec{t}_{11} = \frac{1}{1-(1-p)} = \frac{1}{p}, \tag{4.2.11}$$

and this is, of course, just the mean delay in the transient process, confirming Equation 4.2.8.

The second moment matrix also has only one element given by Equation 4.1.60,

$$\overline{\mathrm{N}^2} = \vec{T}[2(\vec{T} \,\square\, I) - I] = \vec{t}_{11}(2\vec{t}_{11} - 1) = \frac{1}{p}\left(\frac{2}{p} - 1\right) = \frac{2-p}{p^2}, \tag{4.2.12}$$

as in Equation 4.2.9. The inspection methods are extremely simple for this example. For most of us it is considerably easier to find the moments by inspection of the transition diagram rather than by differentiation of the transfer function of the process, even when this transfer function has already been found.

## Three Heads in a Row

Let us now consider a more interesting transient process that is still based on the idea of coin-tossing. We shall flip a coin with probability of Heads $p$ until a sequence of three Heads in a row is produced. We can define the event $E_m$ to be the completion of a sequence of three Heads in a row for the first time on the $m$th toss of

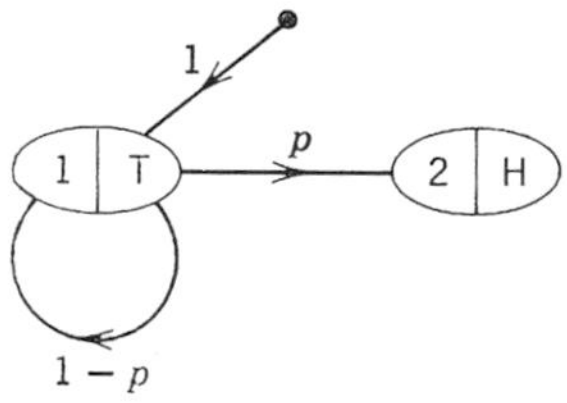


**Figure 4.2.5** Alternate modified transition diagram for tosses to first Head.

the coin. The number of tosses required to produce this sequence is a random variable $n$, where

$$\mathscr{P}\{n = m\} = \mathscr{P}\{E_m\}. \tag{4.2.13}$$

We shall select as our first task the problem of finding the mean $\bar{n}$ and variance $\breve{n}$ of the random variable $n$.

***Markov process representation***

To solve this problem with the theory we have developed we want to construct a Markov process whose statistics represent the answer to the problem. The first decision we must make is how many states to use. Some reflection reveals that the most pertinent consideration in determining whether the sequence will be completed on the next toss is the number of Heads in a row that existed up to and including the last toss; this number can be either 0, 1, or 2. We need a state to represent each possibility, plus another trapping state to represent completion of the sequence. The transition diagram for this four-state process appears in Figure 4.2.6. State 1 represents the situation where the most recent toss was a Tail; state 2, the situation where the most recent toss was a Head; state 3, the situation where the two most recent tosses were two Heads; and state 4, the situation where three Heads in a row have been obtained. Notice that this set of state descriptions has the necessary property that given any pattern of Heads and Tails up to the latest toss we can always determine a unique state for the system by considering first the three or more most recent tosses, then the two most recent tosses, etc.

The transitions and their probabilities are easy to find. If the system is in state 1 then it will return to state 1 if a Tail is produced on the next toss, and proceed to state 2 if the next toss is a Head. These events have respective probability $1 - p$ and $p$. Similarly, in state 2 a Tail will return the system to state 1, whereas a Head will advance it to state 3. A Tail in state 3 will again return the system to state 1 to begin building the sequence afresh, but a Head will complete the sequence and cause the system to enter the trapping state 4 and thus terminate the transient process.

Since assuming that a Tail occurs on a fictitious "zeroth" toss of the coin will cause no change in the statistics of interest to us, we can repeat our convention of the previous example and start the process in state 1.

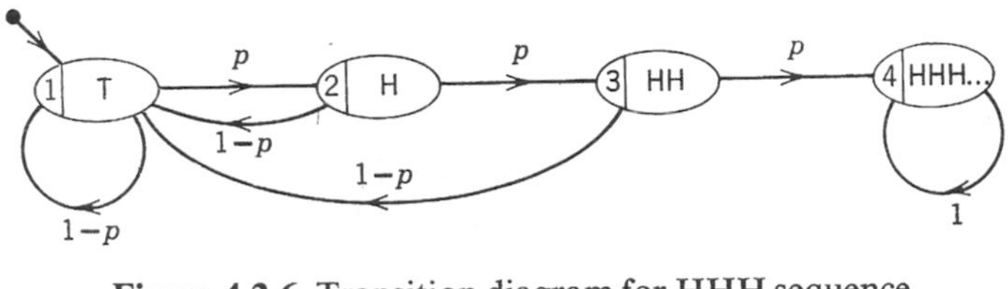


**Figure 4.2.6** Transition diagram for HHH sequence.

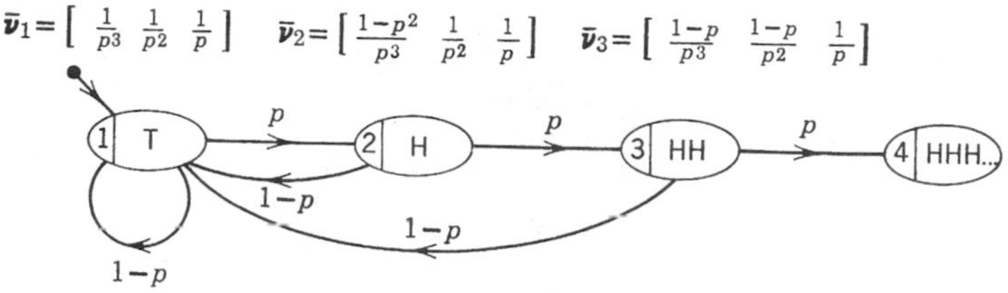


**Figure 4.2.7** Modified transition diagram for HHH sequence.

***Moments by inspection***

The delay in the transient process is the random variable $n$. We can find its moments by using the method of Section 4.1 on the modified transition diagram of the process, which is shown as Figure 4.2.7. We find the matrix of transmissions of the modified transition diagram by inspection as

$$\vec{T} = \frac{1}{p^3}\begin{bmatrix} 1 & p & p^2 \\ 1-p^2 & p & p^2 \\ 1-p & p(1-p) & p^2 \end{bmatrix} = \overline{\mathrm{N}}. \tag{4.2.14}$$

We have noted here that the matrix of transmissions is just the matrix of mean delays in each state of the transient process for each starting state, a relationship first recorded as Equation 4.1.35. Thus, the first row of $\overline{\mathrm{N}}$ represents the expected delay in each of the states if the system is started in state 1, the second row if it is started in state 2, etc. These three row vectors are drawn over the corresponding starting state in the modified transition diagram of Figure 4.2.7. This procedure is a convenient way to record the elements of $\vec{T} = \overline{\mathrm{N}}$ as they are computed.

**Mean.** For each starting state $i$, the total expected delay in the transient process $\bar{\nu}_i$ is just the sum of the expected times spent in each state, from Equation 4.1.37. The column vector representing these quantities is therefore obtained by summing across the rows of the matrix $\overline{\mathrm{N}}$,

$$\bar{\boldsymbol{\nu}} = \overline{\mathrm{N}}s = \frac{1}{p^3}\begin{bmatrix} 1+p+p^2 \\ 1+p \\ 1 \end{bmatrix} \tag{4.2.15}$$

Note that if $p = 1$,

$$\bar{\boldsymbol{\nu}} = \begin{bmatrix} 3 \\ 2 \\ 1 \end{bmatrix}, \tag{4.2.16}$$

a necessary result for a coin certain to come up Heads. If the coin is fair, $p = 1/2$, then

$$\bar{\mathbf{v}} = \begin{bmatrix} 14 \\ 12 \\ 8 \end{bmatrix} \tag{4.2.17}$$

which shows that the average number of tosses required to produce the first sequence of three Heads with a fair coin is 14. If one Head has just occurred, then the average number of tosses to complete the sequence is 12; whereas if two heads have occurred, it is 8.

The random variable $n$ in which we are interested is the delay in the transient process when it is started in state 1. For a general coin with probability of Heads $p$ the mean of $n$ is just the first component of $\bar{\mathbf{v}}$ in Equation 4.2.15,

$$\bar{n} = \bar{\nu}_1 = \frac{1}{p^3}[1 + p + p^2]. \tag{4.2.18}$$

**Variance.** To compute the variance of the delay in the transient process we use Equation 4.1.71 and recall that the $i$th component of the column vector $\overset{\vee}{\mathbf{v}}$ is the variance of the time spent in the transient process if the system is started in state $i$,

$$\overset{\vee}{\mathbf{v}} = \overline{\mathbf{v}^2} - \bar{\mathbf{v}} \,\square\, \bar{\mathbf{v}} = (2T - I)\bar{\mathbf{v}} - \bar{\mathbf{v}} \,\square\, \bar{\mathbf{v}}. \tag{4.2.19}$$

Thus we can compute $\overset{\vee}{\mathbf{v}}$ for this problem from the results of Equations 4.2.14 and 4.2.15. We find

$$\overset{\vee}{\mathbf{v}} = \frac{1}{p^6}\begin{bmatrix} 1 + 2p + 3p^2 - 3p^3 - 2p^4 - p^5 \\ 1 + 2p + 3p^2 - 3p^3 - 3p^4 \\ 1 + 2p + 2p^2 - 5p^3 \end{bmatrix}. \tag{4.2.20}$$

If $p = 1$, then

$$\overset{\vee}{\mathbf{v}} = \begin{bmatrix} 0 \\ 0 \\ 0 \end{bmatrix}, \tag{4.2.21}$$

since for a coin that is certain to produce a Head there will be no variation in the number of tosses to produce a sequence of three Heads no matter in which state of the transient process the system is started. If the coin is fair, $p = 1/2$, then

$$\overset{\vee}{\mathbf{v}} = \begin{bmatrix} 142 \\ 140 \\ 120 \end{bmatrix}. \tag{4.2.22}$$

We see that the variance in the number of tosses to produce three Heads in a row with a fair coin is decreased only slightly by the knowledge that one Head has already occurred.

The variance of the random variable $n$ for a general coin is just the first component of $\check{\mathbf{v}}$ in Equation 4.2.20,

$$\check{n} = \check{\mathbf{v}}_1 = \frac{1}{p^6}\,[1 + 2p + 3p^2 - 3p^3 - 2p^4 - p^5]. \tag{4.2.23}$$

Thus Equations 4.2.18 and 4.2.23 provide a solution to the problem we originally posed. Notice that the values of the mean and variance of the number of tosses of a fair coin to achieve a sequence of three Heads, $\bar{n} = 14$, $\check{n} = 142$, indicate that the probability mass function for the random variable $n$ is relatively broad.

***Moments by transform differentiation***

We have another method at our disposal for solving this problem. We can construct the modified flow graph for the transient process, find its transfer function, and then differentiate the transfer function to produce the moments of the random variable representing the time spent in the process. The modified flow graph of Figure 4.2.8 is just the modified transition diagram of Figure 4.2.7 with every branch transmission multiplied by $z$. We shall let $p(n)$ be the probability that $n$ tosses are required to produce the first sequence of three heads. The transmission of the modified flow graph, $p^{g}(z)$, is then equal to

$$p^{g}(z) = \frac{p^3z^3}{1 - (1 - p)z - p(1 - p)z^2 - p^2(1 - p)z^3}. \tag{4.2.24}$$

Notice that every loop in this flow graph passes through node 1; we therefore encounter no multiple loops.

To compute the moments of $n$ we must differentiate this transform and evaluate the result when $z = 1$. After much labor we find

$$p^{g\prime}(1) = \frac{1}{p^3}\,(1 + p + p^2) \tag{4.2.25}$$

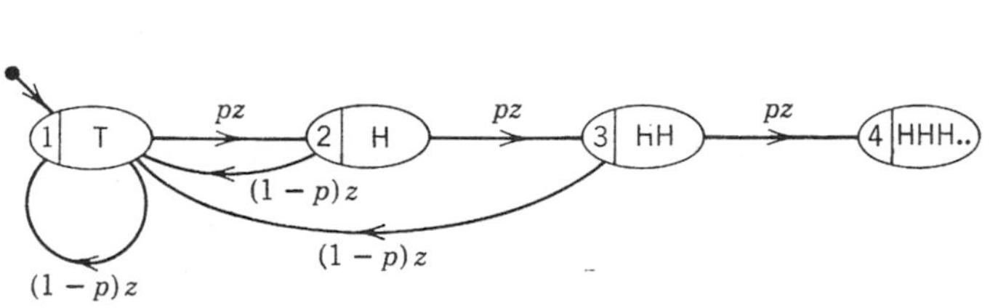


**Figure 4.2.8** Modified flow graph for HHH sequence.

and

$$p^{\mathscr{g}''}(1) = \frac{1}{p^6}(2 + 4p + 6p^2 - 2p^3 - 2p^4 - 2p^5). \tag{4.2.26}$$

The results of Section 4.1,

$$\bar{n} = p^{\mathscr{g}'}(1) \tag{4.2.27}$$

and

$$\breve{n} = p^{\mathscr{g}''}(1) + p^{\mathscr{g}'}(1) - [p^{\mathscr{g}'}(1)]^2, \tag{4.2.28}$$

produce the same values for the moments that appear in Equations 4.2.18 and 4.2.23. The tediousness of finding the moments by this approach emphasizes the value of the inspection methods suitable for transient processes.

***The probability mass function***

Suppose now that we wish to compute not just the moments of the random variable $n$, but its probability mass function $p(n)$. This problem is considerably more difficult, but of extreme importance in applications where the moments are not sufficient to describe the behavior of the random variable. Of course, we already have the transform of this probability mass function, but the problem of inverting the transform expressed in Equation 4.2.24 is indeed formidable. The denominator lacks the factor $1 - z$ because it is the expression for the transmission of a transient process rather than the transfer function of a Markov process. Factoring the denominator is therefore bound to require numerical computation.

However, we recall that every transform equation implies a difference equation. This equation allows us to solve for $p(n)$ recursively. To find this difference equation we write Equation 4.2.24 in the form

$$p^{\mathscr{g}}(z) = p^3z^3 + (1 - p)zp^{\mathscr{g}}(z) + p(1 - p)z^2p^{\mathscr{g}}(z) + p^2(1 - p)z^3p^{\mathscr{g}}(z). \tag{4.2.29}$$

From the properties of the geometric transform we know that the inverse transform of this equation is

$$p(n) = p^3\,\delta(n - 3) + (1 - p)p(n - 1) + p(1 - p)p(n - 2) + p^2(1 - p)p(n - 3)$$
$$n = 0, 1, 2, \ldots, \tag{4.2.30}$$

which implies

$$p(0) = p(1) = p(2) = 0,$$
$$p(3) = p^3,$$
$$p(n) = (1 - p)p(n - 1) + p(1 - p)p(n - 2) + p^2(1 - p)p(n - 3), \quad n > 3, \tag{4.2.31}$$

or, if we wish to continue in detail,

$$\begin{aligned} p(4) &= p^3(1 - p), \\ p(5) &= p^3(1 - p)^2 + p^4(1 - p), \\ p(6) &= p^3(1 - p)^3 + 2p^4(1 - p)^2 + p^5(1 - p), \text{ etc.} \end{aligned} \tag{4.2.32}$$

The values of $p(n)$ for a fair coin are calculated in Table 4.2.1. The most probable value of $n$ is 3; the sequence of three Heads is more likely to be completed on the third toss than on any other toss. The function $p(n)$ is plotted in Figure 4.2.9. The function seems to decrease geometrically with $n$. This tendency is confirmed by

**Table 4.2.1** Probabilities Related to Completing a Sequence of Three Heads on the $n$th Toss of a Fair Coin

| $n$ | Probability that exactly $n$ tosses are required $p(n)$ | Dominant component of $p(n)$ $0.077303(0.91964)^{n-3}$ $n \geq 3$ | Probability that more than $n$ tosses are required ${}^{>}p(n) = 1 - \sum_{j=0}^{n} p(j)$ | Probability that sequence will not be completed on $n$ + 1st toss if it has not been completed on the $n$th $\frac{{}^{>}p(n+1)}{{}^{>}p(n)}$ |
|---|---|---|---|---|
| 0 | 0 | 0 | 1.000000 | 1.000000 |
| 1 | 0 | 0 | 1.000000 | 1.000000 |
| 2 | 0 | 0 | 1.000000 | 0.875000 |
| 3 | 0.125000 | 0.077303 | 0.875000 | 0.928571 |
| 4 | 0.062500 | 0.071091 | 0.812500 | 0.923077 |
| 5 | 0.062500 | 0.065378 | 0.750000 | 0.916667 |
| 6 | 0.062500 | 0.060124 | 0.687500 | 0.920454 |
| 7 | 0.054688 | 0.055292 | 0.632812 | 0.919753 |
| 8 | 0.050781 | 0.050849 | 0.582031 | 0.919463 |
| 9 | 0.046875 | 0.046763 | 0.535156 | 0.919707 |
| 10 | 0.042969 | 0.043005 | 0.492187 | 0.919642 |
| 11 | 0.039551 | 0.039549 | 0.452636 | 0.919632 |
| 12 | 0.036377 | 0.036371 | 0.416259 | 0.919648 |
| 13 | 0.033447 | 0.033448 | 0.382812 | 0.919642 |
| 14 | 0.030762 | 0.030760 | 0.352050 | 0.919642 |
| 15 | 0.028290 | 0.028288 | 0.323760 | 0.919644 |
| 16 | 0.026016 | 0.026015 | 0.297744 | 0.919642 |
| 17 | 0.023926 | 0.023924 | 0.273818 | 0.919643 |
| 18 | 0.022003 | 0.022001 | 0.251815 | 0.919643 |
| 19 | 0.020235 | 0.020233 | 0.231580 | 0.919643 |
| 20 | 0.018609 | 0.018607 | 0.212971 | 0.919641 |
| 21 | 0.017114 | 0.017112 | 0.195857 | 0.919645 |
| 22 | 0.015738 | 0.015737 | 0.180119 | 0.919642 |
| 23 | 0.014474 | 0.014472 | 0.165645 | 0.919641 |
| 24 | 0.013311 | 0.013309 | 0.152334 | 0.919643 |
| 25 | 0.012241 | 0.012239 | 0.140093 | |

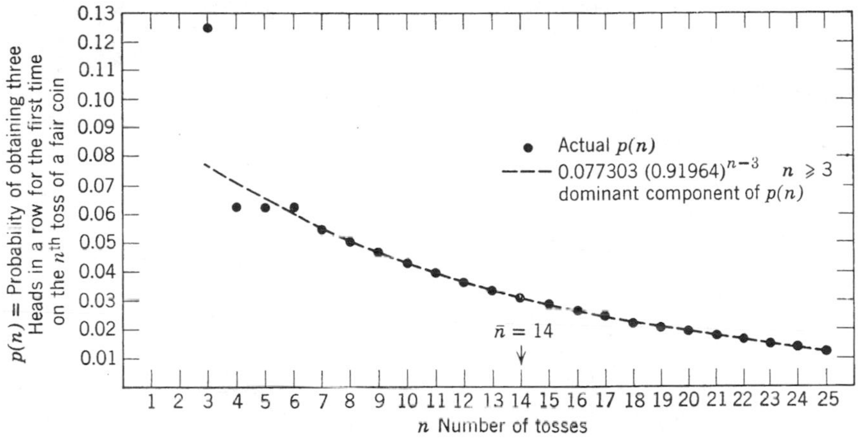


**Figure 4.2.9** Actual and dominant forms of $p(n)$.

plotting $p(n)$ on semilogarithmic paper as in Figure 4.2.10. The points almost fall on a straight line when $n$ increases beyond 7. The decay rate, as measured from Figure 4.2.10 or by calculating the ratio $p(n+1)/p(n)$ in Table 4.2.1 when $n$ is large, turns out to be approximately 0.92.

***Dominant component analysis***

Why should this be? Why does the number 0.92 appear in the calculation of the probability that a fair coin will complete a sequence of three Heads on its $n$th toss? To find this answer let us return to the transform of $p(n)$, $p^g(z)$, and factor the denominator shown in Equation 4.2.24 for the special case of $p = 1/2$. The denominator polynomial is

$$1 - (1/2)z - (1/4)z^2 - (1/8)z^3. \tag{4.2.33}$$

By any of a number of procedures we find that it may be factored in the form

$$\begin{aligned}[1 - (1/2)z - (1/4)z^2 - (1/8)z^3] \\ = (1 - 0.91964z)[1 - (-0.20982 + j0.30315)z][1 - (-0.20982 - j0.30315)z], \\ (j = \sqrt{-1}).\end{aligned} \tag{4.2.34}$$

Now we see the origin of the factor 0.92, or more precisely, 0.91964. It represents a simple geometric decay factor in the denominator of the transform. The other two factors in the denominator correspond to complex conjugate roots of the remaining quadratic part of the polynomial. The magnitude of each of these factors, about 0.37, shows that they will die out rather quickly with $n$—they contribute, in fact, only to the first few values of $p(n)$.

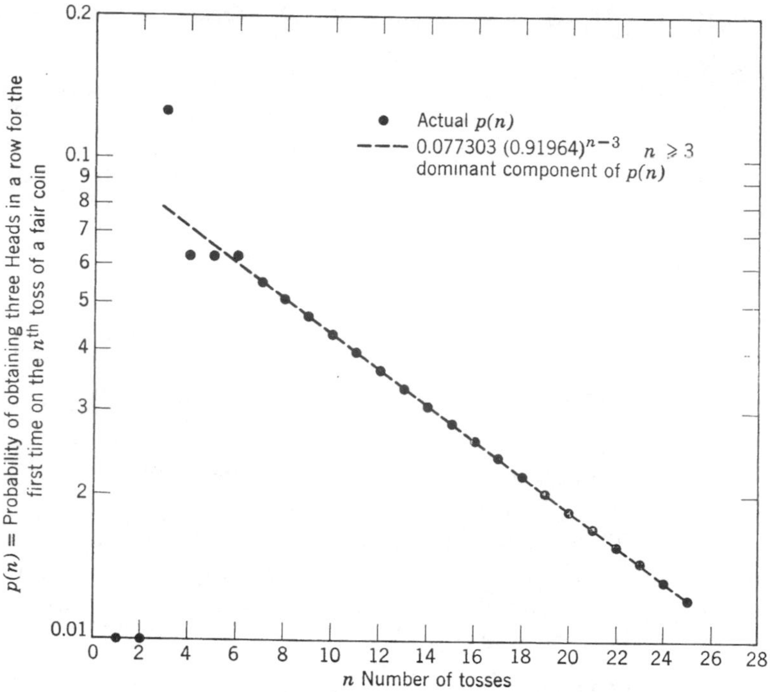


**Figure 4.2.10** Logarithmic plot of actual and dominant forms of $p(n)$.

Let us now find the part of the partial fraction expansion of $p^g(z)$ that arises from the dominant geometric component with rate 0.91964. Because the numerator and denominator of $p^g(z)$ are of the same degree in $z$, we cannot perform this expansion simply by evaluating $(1 - 0.91964z)p^g(z)$ at the point

$$z = \frac{1}{0.91964} = 1.08738. \tag{4.2.35}$$

However, we note that the factor $z^3$ in the numerator of $p^g(z)$ represents a delay of three units in $p(n)$. Therefore we can ignore the $z^3$ factor in the partial fraction expansion procedure and simply delay the final result by three units. We then compute the magnitude of the dominant component from the expression

$$(1 - 0.91964z)\frac{1/8}{1 - (1/2)z - (1/4)z^2 - (1/8)z^3}\bigg|_{z=1.08738}$$

$$= \frac{0.125}{[1 - (-0.20982 + j0.30315)z][1 - (-0.20982 - j0.30315)z]}\bigg|_{z=1.08738}$$

$$= 0.077303. \tag{4.2.36}$$

The dominant component therefore has the transform

$$\frac{0.077303z^3}{1 - 0.91964z}. \tag{4.2.37}$$

The inverse transform is

$$0.077303(0.91964)^{n-3} \qquad n = 3, 4, 5, \ldots. \tag{4.2.38}$$

We calculate immediately by evaluating Expression 4.2.37 at $z = 1$ that the sum of this component is 0.96196. Therefore it alone accounts for 96 percent of the total probability mass function.

The dominant component, Expression 4.2.38, also appears in Table 4.2.1. When $n$ is greater than 7, the difference between $p(n)$ and its dominant component is very small. The primacy of the geometric component is even more strikingly illustrated by the plots of this component in Figures 4.2.9 and 4.2.10.

If we pretend that $p^g(z)$ is equal to the transform of the geometric component in Expression 4.2.37 and then compute the moments of $n$ by differentiating with respect to $z$ and using Equations 4.2.27 and 4.2.28, we find

$$\bar{n} = 14.0087, \qquad \overset{\vee}{n} = 147.3839. \tag{4.2.39}$$

These numbers are remarkably close to the true mean, 14, and close to the true variance, 142, that we found in Equations 4.2.17 and 4.2.22. This observation is even more interesting when we recall that the geometric component fails to be a proper probability mass function because it sums only to about 0.96.

### *Principal and Dominant Components*

We shall explore the significance of this dominant component even further in what follows. However, let us now pause to consider the general implications of such components.

We know that the absolute values of the decay factors in the denominators of the transfer functions of *any* Markov process must be less than or equal to one in magnitude. Although all factors with magnitudes less than one must eventually lose their influence as the number of transitions increases, we can easily see that those decay factors that differ from one by the smallest amount will persist for the longest time. If one or two decay factors are much closer to one in magnitude than are all the others, then these components will govern most of the transient response. We shall call the contributions to the total response of the system that arise from these factors the *principal components* of the transfer function.

In transient processes, where all components are transient, the principal components play a particularly important role. Often, as in this example, only one component will dominate all others and describe the nature of the transient process in sufficient detail for many practical applications. We call this component the *dominant component*.

***Interpretation of the dominant component***

Now that we have found that the dominant component is generally interesting and important, and that it is directly related to the roots of the transfer function denominators, we may ask, does it have a physical significance? For instance, in this example, does 0.91964 represent any physical quantity, or is it just a faceless number? At first glance it is difficult to see how a number like 0.91964 is related to obtaining a sequence of Heads with a fair coin, but we shall find a direct connection.

Let us define ${}^{>}p(n)$ as the probability that more than $n$ tosses of the coin will be required to produce the sequence of three Heads. This probability is related to $p(n)$ by

$$ {}^{>}p(n) = 1 - \sum_{j=0}^{n} p(j). \tag{4.2.40} $$

The values of ${}^{>}p(n)$ are tabulated in Table 4.2.1. We note that the median value of the number of tosses to produce a sequence of three Heads is about 10—there is better than a 50 percent chance that more than 9 tosses will be required, but less than a 50 percent chance that more than 10 will be necessary.

The quantity ${}^{>}p(n)$ has been plotted in Figure 4.2.11. We observe, as we might expect, that this function appears to fall geometrically with $n$ in the same way as did $p(n)$. This tendency is confirmed by the logarithmic graph of ${}^{>}p(n)$ in Figure 4.2.12. The probability ${}^{>}p(n)$ falls geometrically with $n$ when $n$ is larger than about 6. The geometric decay rate is 0.91964, the same as for $p(n)$. This finding suggests that the ratio ${}^{>}p(n+1)/{}^{>}p(n)$ should approach the constant 0.91964 as $n$ increases. Table 4.2.1 shows how the ratio does approach this value when $n$ exceeds 7. Figure 4.2.13 shows this characteristic in graphical form.

What is the interpretation of ${}^{>}p(n+1)/{}^{>}p(n)$? It is the conditional probability that a sequence of three Heads will not be completed on the $n+1$st toss of the

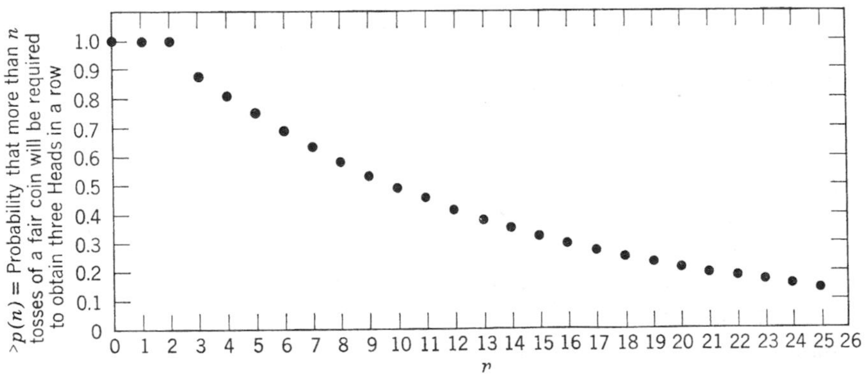


**Figure 4.2.11** Plot of ${}^{>}p(n)$.

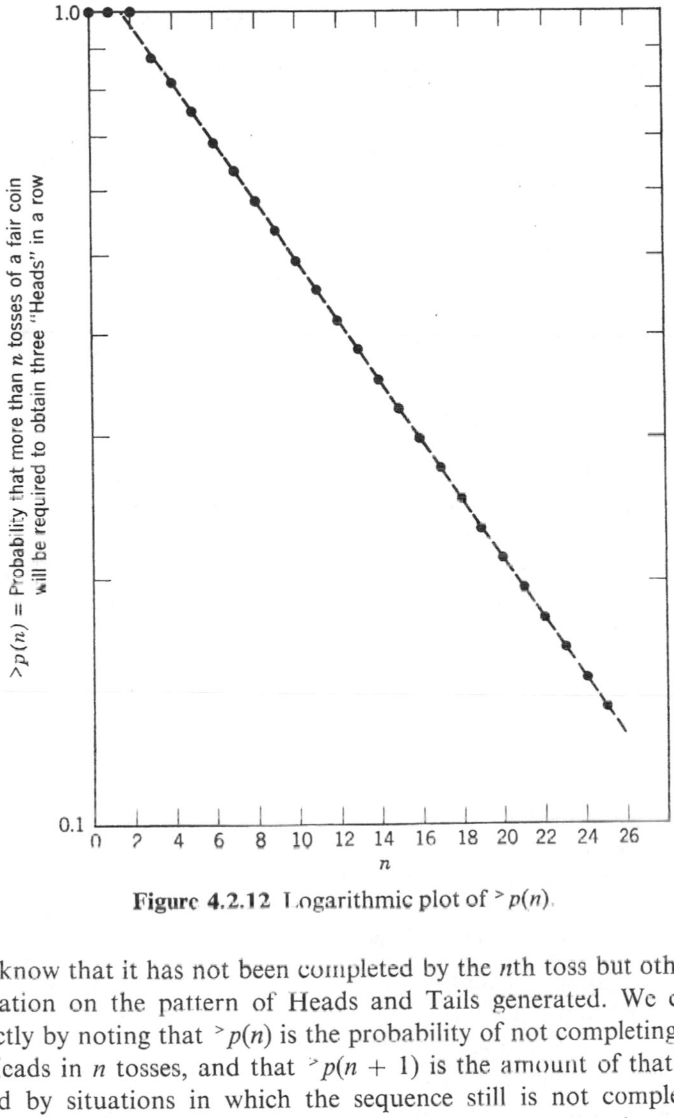


**Figure 4.2.12** Logarithmic plot of $^{>}p(n)$.

coin if we know that it has not been completed by the $n$th toss but otherwise have no information on the pattern of Heads and Tails generated. We can see this result directly by noting that $^{>}p(n)$ is the probability of not completing a sequence of three Heads in $n$ tosses, and that $^{>}p(n + 1)$ is the amount of that probability contributed by situations in which the sequence still is not completed on the $n + 1$st toss. Therefore, if you should approach someone attempting to produce a sequence of three Heads using a fair coin and learn only that he has made several tosses already, you would assign probability 0.91964 to his not completing the sequence on his next toss.

The significance of this finding is that it represents the first time we have been able to give a probabilistic interpretation for a decay factor appearing in a transfer function.

In fact, we should observe that our interpretation does not depend on the specific

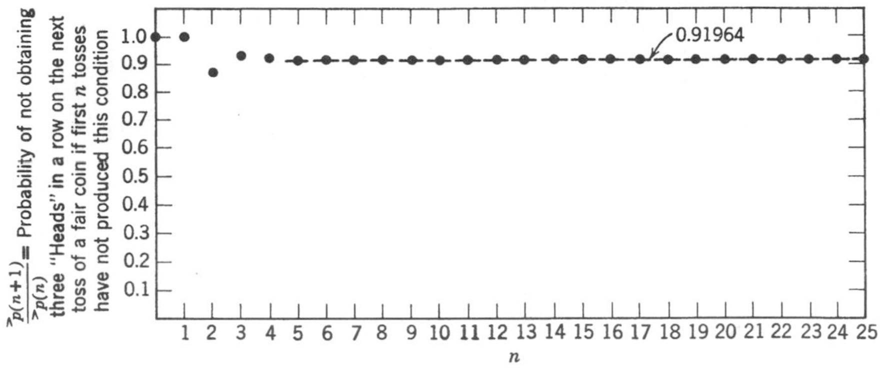


**Figure 4.2.13** Plot of $\frac{{}^{>}p(n+1)}{{}^{>}p(n)}$

example. It is always true that the probability of not terminating a transient process on the next transition, given that it has not already been terminated and that the process has made many transitions, is equal to the decay factor of the dominant component in the transfer function for the transient process. The key to the argument, which the reader may like to carry out in detail, is that ${}^{>}p(n)$ has the same dominant component as $p(n)$. We shall not always find for a general Markov process that such interpretations are possible, but they do provide considerable insight into the nature of the process when they can be made.

***Reflection***

This simple problem has revealed the power of the analytic techniques developed in Section 4.1. It is fruitful to speculate on the complications that could be introduced into the problem without requiring a corresponding change in our methods.

## Sequence HHH before Sequence HTH

Let us now consider a coin-tossing problem that produces a duodesmic Markov process. We shall find the probability that a person tossing a coin with probability of Heads $p$ will generate a sequence of three Heads in a row before generating the sequence Head-Tail-Head, or more briefly, HHH before HTH. We can imagine that the person will win the coin-tossing game if he obtains HHH before HTH and lose it if these events occur in the other order. The probability we shall calculate is therefore the probability of his winning the game.

***Markov process representation***

We construct the Markov process pertinent to this example by first deciding on the states and their descriptions. We know that we shall need all four states in the

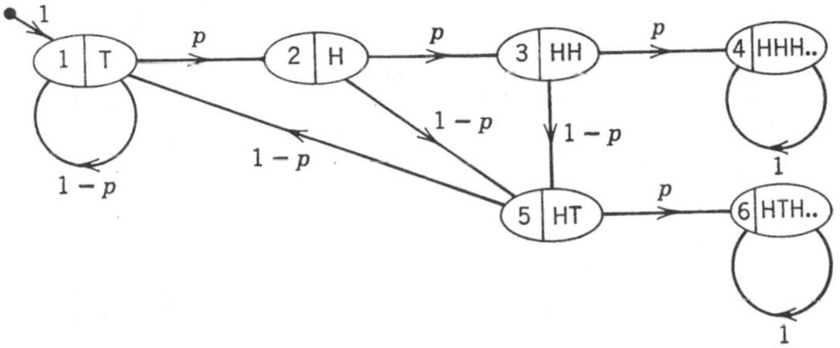


**Figure 4.2.14** Transition diagram for HHH before HTH.

example we have just completed because we are still interested in a sequence of three Heads. However, we shall need additional states to describe the building of the sequence HTH. We already have a state to represent the first Head in either sequence. We therefore have to add a state to represent the HT part of the sequence HTH and a state to represent the final completion of the sequence HTH. The total number of states we shall need is therefore 6.

The states and their descriptions appear in the transition diagram of Figure 4.2.14. Note that states 4 and 6 are both trapping states—the occurrence of either event precludes the occurrence of the other and terminates the process. Once again this set of state descriptions meets the requirement of being able to classify uniquely the current state of the system.

The reasoning underlying the transitions indicated and their accompanying probabilities is exactly the same as in the previous example. For instance, the transition from state 2 to state 5 will occur if the system is in state 2 and a Tail is produced on the next toss; the probability of a Tail is $1 - p$. Since an initial Tail does not contribute to either sequence, we shall assume that a Tail occurs on the zeroth toss and start the process in state 1.

***Probability of HHH before HTH***

The first quantity we want to find is the probability of winning, of obtaining HHH before HTH. This is just the probability that the system will be trapped by state 4. We therefore draw the modified transition diagram in Figure 4.2.15 and find its transmission from the input to state 4,

$$\begin{aligned}\mathscr{P}\{\text{Win}\} &= \mathscr{P}\{\text{HHH before HTH}\} = \vec{t}_{14} \\ &= \frac{p^3}{1 - (1 - p) - p(1 - p)^2 - p^2(1 - p)^2} \\ &= \frac{p}{1 + p - p^2}. \qquad (4.2.41)\end{aligned}$$

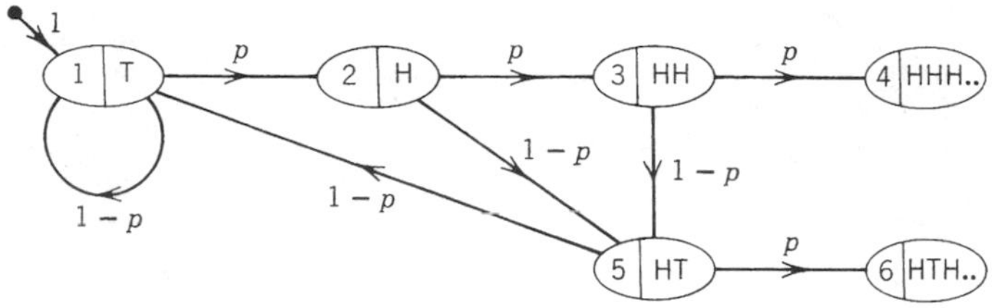


**Figure 4.2.15** Modified transition diagram for HHH before HTH.

If the coin being flipped is fair, $p = 1/2$, then we have

$$\mathscr{P}\{\text{Win}\} = \mathscr{P}\{\text{HHH before HTH}\} = 2/5. \qquad (4.2.42)$$

Thus when the coin is fair, it is more likely that the first sequence completed will be HTH.

### *Expected delay*

We can find the expected duration of the game, $\bar{\nu}_1$, by determining the transmission from 1 to each transient state including itself, and then summing. We obtain from inspection of Figure 4.2.15

$$\vec{t}_{11} = \frac{1}{p^2[1 + p - p^2]}, \qquad \vec{t}_{12} = \frac{p}{p^2(1 + p - p^2)}, \qquad \vec{t}_{13} = \frac{p^2}{p^2(1 + p - p^2)},$$

$$\vec{t}_{15} = \frac{p(1 - p^2)}{p^2(1 + p - p^2)}. \qquad (4.2.43)$$

Then

$$\bar{\nu}_1 = \vec{t}_{11} + \vec{t}_{12} + \vec{t}_{13} + \vec{t}_{15} = \frac{1 + 2p + p^2 - p^3}{p^2[1 + p - p^2]}. \qquad (4.2.44)$$

The expected duration of the game when the coin is fair is

$$\bar{\nu}_1 = 34/5 = 6.8 \qquad (4.2.45)$$

or 6.8 tosses.

### *Delay conditional on final state*

The probability that the player will win on his $n$th toss given that he will win has a transform equal to the transmission from state 1 to state 4 of the modified flow

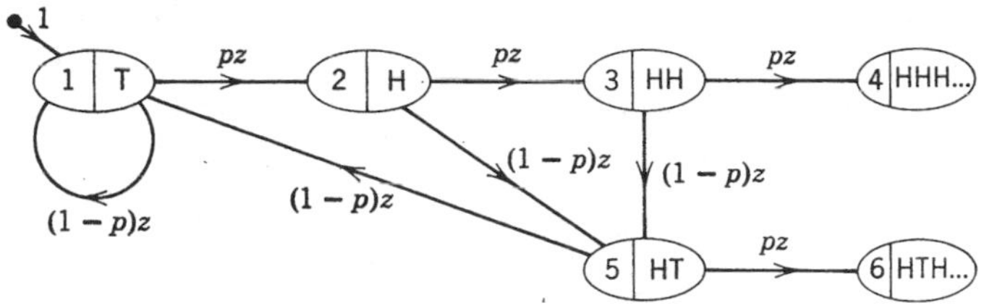


**Figure 4.2.16** Modified flow graph for HHH before HTH.

graph in Figure 4.2.16 divided by the probability of winning found in Equation 4.2.41,

$$p^{\mathscr{G}}(z|\text{Win}) = \frac{\vec{t}_{14}(z)}{\vec{t}_{14}} = \frac{(1+p-p^2)p^2z^3}{1-(1-p)z-p(1-p)^2z^3-p^2(1-p)^2z^4}. \quad (4.2.46)$$

We could invert this transform to find the probability of winning on the $n$th toss given that a win occurred for any value of $p$ by using either the transform table or the difference equation methods we have previously discussed.

We can find the mean number of tosses to win given that a win occurs, $\langle \nu_1|\text{Win}\rangle$, by differentiating the transform $p^{\mathscr{G}}(z|\text{Win})$ with respect to $z$ and evaluating the result at $z = 1$. We obtain

$$\langle \nu_1|\text{Win}\rangle = \frac{1+2p+p^2-2p^3+p^4}{p^2(1+p-p^2)} \quad (4.2.47)$$

and for $p = \frac{1}{2}$,

$$\langle \nu_1|\text{Win}\rangle = 33/5 = 6.6. \quad (4.2.48)$$

Therefore, we expect a winning game to require slightly fewer tosses than an average game.

**Inspection method.** We can also compute the expected duration of the game given a win or a loss by using the transmission of the modified transition diagram. Thus, from Equation 4.1.117,

$$\langle \nu_1|\text{Win}\rangle = \bar{\nu}_{1|4} = \frac{\vec{t}_{11}\vec{t}_{14} + \vec{t}_{12}\vec{t}_{24} + \vec{t}_{13}\vec{t}_{34} + \vec{t}_{15}\vec{t}_{54}}{\vec{t}_{14}} \quad (4.2.49)$$

and

$$\langle \nu_1|\text{Lose}\rangle = \bar{\nu}_{1|6} = \frac{\vec{t}_{11}\vec{t}_{16} + \vec{t}_{12}\vec{t}_{26} + \vec{t}_{13}\vec{t}_{36} + \vec{t}_{15}\vec{t}_{56}}{\vec{t}_{16}}. \quad (4.2.50)$$

We evaluate the additional necessary transmissions using the modified transition diagram of Figure 4.2.15,

$$\begin{aligned}
\vec{t}_{14} &= \frac{p}{1 + p - p^2} & \vec{t}_{16} &= \frac{1 - p^2}{1 + p - p^2} \\
\vec{t}_{24} &= \frac{p}{1 + p - p^2} & \vec{t}_{26} &= \frac{1 - p^2}{1 + p - p^2} \\
\vec{t}_{34} &= \frac{2p - p^2}{1 + p - p^2} & \vec{t}_{36} &= \frac{1 - p}{1 + p - p^2} \\
\vec{t}_{54} &= \frac{p - p^2}{1 + p - p^2} & \vec{t}_{56} &= \frac{1}{1 + p - p^2}.
\end{aligned} \tag{4.2.51}$$

Note that for any state $i$ in the transient process $\vec{t}_{i4} + \vec{t}_{i6} = 1$ because the system must ultimately occupy state 4 or state 6. Now we substitute the transmissions from Equations 4.2.43 and 4.2.51 into Equations 4.2.49 and 4.2.50 to obtain

$$\langle \nu_1 | \text{Win} \rangle = \frac{1 + 2p + p^2 - 2p^3 + p^4}{p^2(1 + p - p^2)} \tag{4.2.52}$$

and

$$\langle \nu_1 | \text{Lose} \rangle = \frac{1 + 3p + 3p^2}{p^2(1 + p)(1 + p - p^2)}. \tag{4.2.53}$$

Equation 4.2.52 confirms Equation 4.2.47. For a fair coin, we find

$$\langle \nu_1 | \text{Lose} \rangle = \frac{104}{15} = 6.933. \tag{4.2.54}$$

Thus we expect losing games to be slightly longer than average. We can readily verify from Equations 4.2.41, 4.2.44, 4.2.49, and 4.2.50 that

$$\bar{\nu}_1 = \langle \nu_1 | \text{Win} \rangle \mathscr{P}\{\text{Win}\} + \langle \nu_1 | \text{Lose} \rangle \mathscr{P}\{\text{Lose}\}. \tag{4.2.55}$$

Thus we have seen once more in this final example that elementary, but non-trivial problems can be solved by inspection using flow-graph techniques.

## 4.3 HOLDING-TIME STATISTICS

Since there is a possibility that a Markov process will make repeated, consecutive transitions back to the same state, we are often interested in the number of times in succession that the same state is occupied after it is first entered. This statistic we shall call the holding time of the state. The minimum holding time is one transition, corresponding to the case where the process moves directly on to another state. All states in a periodic chain would have a holding time exactly equal to one transition.

## An Equivalent Transient Process

We can consider the holding time $\tau_i$ for some state $i$ to be generated by a very simple transient process. The process continues as long as the system keeps returning to state $i$ and stops as soon as the system enters another state. The time spent in this transient process is then the holding time for state $i$. Since the probability of returning to state $i$ on the next transition is $p_{ii}$, the modified flow graph for the transient process has the simple form shown in Figure 4.3.1. The geometric transform $p^g(z)$ of the probability mass function $p(\tau_i)$ for the holding time $\tau_i$ for state $i$ is just the transmission of this modified flow graph from state 1 to state 2,

$$p^g(z) = \vec{t}_{12}(z) = \frac{(1 - p_{ii})z}{1 - p_{ii}z}. \tag{4.3.1}$$

This is the transform of a geometric distribution. It is exactly the transform in Equation 4.2.2 if we replace $p$ in that equation by $1 - p_{ii}$. Therefore we can write the inverse from Equation 4.2.4 as

$$p(\tau_i) = (1 - p_{ii})p_{ii}^{\tau_i - 1} \qquad \tau_i = 1, 2, 3, \ldots . \tag{4.3.2}$$

The mean and variance are given by Equations 4.2.8 and 4.2.10 as

$$\bar{\tau}_i = \frac{1}{1 - p_{ii}}, \qquad \overset{\vee}{\tau}_i = \frac{p_{ii}}{(1 - p_{ii})^2}. \tag{4.3.3}$$

Thus the number of transitions $\tau_i$ that a state $i$ will "hold" the process is geometrically distributed with a parameter that depends only on $p_{ii}$.

### *Examples*

**General two-state process.** For the general two-state process with transition matrix

$$P = \begin{bmatrix} 1 - a & a \\ b & 1 - b \end{bmatrix} \tag{4.3.4}$$

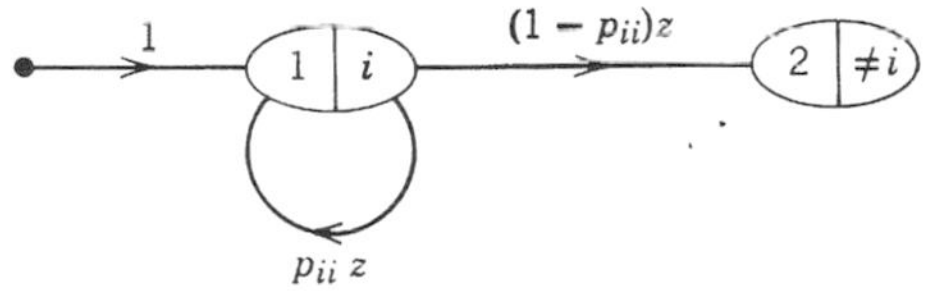


**Figure 4.3.1** Modified flow graph for holding time computation.

we have $p_{11} = 1 - a$, $p_{22} = 1 - b$. Therefore the means and variances of the holding times for the two states are

$$\bar{\tau}_1 = \frac{1}{1 - p_{11}} = \frac{1}{a}, \qquad \overset{v}{\tau}_1 = \frac{p_{11}}{(1 - p_{11})^2} = \frac{1 - a}{a^2};$$
$$\bar{\tau}_2 = \frac{1}{1 - p_{22}} = \frac{1}{b}, \qquad \overset{v}{\tau}_2 = \frac{p_{22}}{(1 - p_{22})^2} = \frac{1 - b}{b^2}. \tag{4.3.5}$$

**Marketing example.** For the two-state marketing example, we specialize the general two-state process by taking $a = 0.2$, $b = 0.3$. Then the holding time moments are

$$\bar{\tau}_1 = \frac{1}{a} = 5, \qquad \overset{v}{\tau}_1 = \frac{1 - a}{a^2} = 20;$$
$$\bar{\tau}_2 = \frac{1}{b} = \frac{10}{3} = 3\ 1/3, \qquad \overset{v}{\tau}_2 = \frac{1 - b}{b^2} = \frac{70}{9} = 7\ 7/9. \tag{4.3.6}$$

The total time spent in the transient process whose modified flow graph appears in Figure 4.1.4 is the sum of the holding times in state 1 and state 2. Therefore, the results of Equation 4.3.6 merely confirm the matrix $\bar{N}$ in Equation 4.1.62 and the matrix $\overset{v}{N}$ in Equation 4.1.65. We often find that the holding time statistics allow us to write important process statistics immediately.

**Taxicab example.** The three-state taxicab example of Chapter 1 has the transition matrix

$$P = \begin{bmatrix} 0.3 & 0.2 & 0.5 \\ 0.1 & 0.8 & 0.1 \\ 0.4 & 0.4 & 0.2 \end{bmatrix}. \tag{4.3.7}$$

with diagonal elements $p_{11} = 0.3$, $p_{22} = 0.8$, and $p_{33} = 0.2$. The holding time statistics are

$$\bar{\tau}_1 = \frac{1}{1 - p_{11}} = \frac{1}{0.7} = 1\ 3/7; \qquad \overset{v}{\tau}_1 = \frac{p_{11}}{(1 - p_{11})^2} = \frac{0.3}{0.49} = \frac{30}{49};$$
$$\bar{\tau}_2 = \frac{1}{1 - p_{22}} = \frac{1}{0.2} = 5; \qquad \overset{v}{\tau}_2 = \frac{p_{22}}{(1 - p_{22})^2} = \frac{0.8}{0.04} = 20; \tag{4.3.8}$$
$$\bar{\tau}_3 = \frac{1}{1 - p_{33}} = \frac{1}{0.8} = 1\ 1/4; \qquad \overset{v}{\tau}_3 = \frac{p_{33}}{(1 - p_{33})^2} = \frac{0.2}{0.64} = 5/16.$$

### Real and Virtual Transitions

Now let us examine more closely what we mean by a transition in a Markov process. We have always included in our concept of transition the possibility that a state will be occupied two or more times in succession. In fact, the concept of holding time would not be interesting unless our definition of transition included this situation. However, we may on occasion want to deal with processes where it is physically impossible for the system to occupy the same state for two consecutive times. Therefore, we shall find it worthwhile to expand our terminology further to include the ideas of "virtual" and "real" transitions.

A virtual transition is a transition from a state back to itself. In virtual transitions there is no actual change of state. A real transition is a transition from one state to a different state; the state indices must change. If all diagonal elements of the transition probability matrix are zero, then no virtual transitions are possible in the process—it can make only real transitions. In such a process the holding time for each state is one. Note that a periodic chain can make only real transitions while an identity process, $P = I$, can make only virtual transitions. Most processes can make both kinds, but in some problems we shall want to emphasize the difference because one or the other type will have more physical significance for us. For instance, in the marketing example we might be especially interested in the virtual transitions because they represent repeat purchases of the same brand. In the taxicab example, however, we might be more interested in the real transitions because they correspond to changes of city.

## Equivalent Real Transition Processes

If we want to examine only actual changes of state in a Markov process then we can construct the transition probability matrix $P^r$ for a process that can make only real transitions by setting the diagonal elements of the transition probability matrix $P$ equal to zero and dividing the remaining elements in each row by their sum (which will be one minus the original diagonal element) to insure that we still have a stochastic matrix. This new process will have all holding times equal to one. The $i$th limiting state probability $\pi_i^r$ of the new process will be the limiting probability that the last real transition made by the original process carried it into state $i$.

What is the relationship between the limiting state probability vector $\boldsymbol{\pi}$ of the original process and the limiting state probability vector $\boldsymbol{\pi}^r$ of the real transition process? We begin our investigation by noting that the procedure for construction of the transition probability matrix $P^r$ implies that its elements are related to those of the transition probability matrix $P$ by the equation:

$$p_{ij}^r = \begin{cases} \dfrac{p_{ij}}{1 - p_{ii}} & j \neq i \\ 0 & j = i \end{cases} \tag{4.3.9}$$

The elements of the two limiting state probability vectors $\boldsymbol{\pi}$ and $\boldsymbol{\pi}^r$ must satisfy the equations

$$\sum_{i=1}^{N} \pi_i p_{ij} = \pi_j \tag{4.3.10}$$

and

$$\sum_{i=1}^{N} \pi_i^r p_{ij}^r = \pi_j^r, \tag{4.3.11}$$

plus the requirement that the sum of the limiting state probabilities be one. If we substitute Equation 4.3.9 into Equation 4.3.11, we obtain

$$\sum_{i \neq j} \pi_i^r \frac{p_{ij}}{1 - p_{ii}} = \pi_j^r \tag{4.3.12}$$

or

$$\sum_{i=1}^{N} \pi_i^r \frac{p_{ij}}{1 - p_{ii}} - \pi_j^r \frac{p_{jj}}{1 - p_{jj}} = \pi_j^r,$$

$$\sum_{i=1}^{N} \frac{\pi_i^r}{1 - p_{ii}} p_{ij} = \frac{\pi_j^r}{1 - p_{jj}}. \tag{4.3.13}$$

Comparison of Equations 4.3.13 and 4.3.10 shows that each $\pi_i$ must be proportional to $\pi_i^r/(1 - p_{ii})$,

$$\pi_i = k\pi_i^r \cdot \frac{1}{1 - p_{ii}}, \tag{4.3.14}$$

where we have used $k$ to represent the constant of proportionality. It follows from Equation 4.3.3 that $\pi_i$ is also given by

$$\pi_i = k\pi_i^r \bar{\tau}_i. \tag{4.3.15}$$

It is indeed intuitive that the probability $\pi_i$ of finding the original process in state $i$ after many transitions should be proportional to the product of the probability $\pi_i^r$ that the original process made its last real transition to state $i$ and $\bar{\tau}_i$, the mean time the original process will spend in state $i$ when it arrives there.

The proportionality constant $k$ is determined from the requirement that

$$\sum_{i=1}^{N} \pi_i = 1, \tag{4.3.16}$$

to be

$$k = \frac{1}{\sum_{i=1}^{N} \pi_i^r \bar{\tau}_i}, \tag{4.3.17}$$

so that finally,

$$\pi_i = \frac{\pi_i{}^r \bar{\tau}_i}{\sum_{i=1}^{N} \pi_i{}^r \bar{\tau}_i} \qquad i = 1, 2, \ldots, N. \tag{4.3.18}$$

Thus the limiting state probabilities of a process are equal to the limiting state probabilities of its real transition equivalent weighted by the mean holding times in each state. This relationship provides insight that we shall find extremely important in our later work.

***Real transition equivalents of examples***

Let us confirm Equation 4.3.18 in our examples.

**General two-state process.** For the general two-state process whose transition matrix appears in Equation 4.3.4, we find that the equivalent real transition process has transition probability matrix

$$P^r = \begin{bmatrix} 0 & 1 \\ 1 & 0 \end{bmatrix}. \tag{4.3.19}$$

The equivalent real transition process is therefore the two-state periodic process. The limiting state probability vector for this process is

$$\boldsymbol{\pi}^r = [1/2 \quad 1/2]. \tag{4.3.20}$$

Now by using the results of Equations 4.3.5 and 4.3.20 in Equation 4.3.18 we can compute

$$\begin{aligned} \pi_1 &= \frac{\pi_1{}^r \bar{\tau}_1}{\pi_1{}^r \bar{\tau}_1 + \pi_2{}^r \bar{\tau}_2} = \frac{1/a}{1/a + 1/b} = \frac{b}{a+b} \\ \pi_2 &= \frac{\pi_2{}^r \bar{\tau}_2}{\pi_1{}^r \bar{\tau}_1 + \pi_2{}^r \bar{\tau}_2} = \frac{1/b}{1/a + 1/b} = \frac{a}{a+b}. \end{aligned} \tag{4.3.21}$$

We have encountered these limiting state probabilities for the general two-state Markov process on several occasions.

**Taxicab example.** We compute the real transition equivalent of the taxicab problem by deleting the diagonal elements of $P$ in Equation 4.3.7 and renormalizing to make the row sums equal to one. We obtain

$$P^r = \begin{bmatrix} 0 & 2/7 & 5/7 \\ 1/2 & 0 & 1/2 \\ 1/2 & 1/2 & 0 \end{bmatrix}. \tag{4.3.22}$$

The limiting state probability vector $\boldsymbol{\pi}^r$ we compute by our usual methods to be

$$\boldsymbol{\pi}^r = [7/21 \quad 6/21 \quad 8/21]. \tag{4.3.23}$$

Notice that the states of the original process do not vary much in the frequency with which real transitions are made into them, and that most real transitions are made into state 3. The results of Equations 4.3.8 and 4.3.23 now allow us to calculate the limiting state probabilities of the original process from Equation 4.3.18,

$$\begin{aligned}
\pi_1 &= \frac{{\pi_1}^r \bar{\tau}_1}{{\pi_1}^r \bar{\tau}_1 + {\pi_2}^r \bar{\tau}_2 + {\pi_3}^r \bar{\tau}_3} = \frac{7/21 \cdot 10/7}{7/21 \cdot 10/7 + 6/21 \cdot 5 + 8/21 \cdot 5/4} \\
&= 1/5 \\
\pi_2 &= \frac{{\pi_2}^r \bar{\tau}_2}{{\pi_1}^r \bar{\tau}_1 + {\pi_2}^r \bar{\tau}_2 + {\pi_3}^r \bar{\tau}_3} = 3/5 \\
\pi_3 &= \frac{{\pi_3}^r \bar{\tau}_3}{{\pi_1}^r \bar{\tau}_1 + {\pi_2}^r \bar{\tau}_2 + {\pi_3}^r \bar{\tau}_3} = 1/5.
\end{aligned} \tag{4.3.24}$$

These are the limiting state probabilities previously found for the taxicab problem. Notice that state 2 has a relatively high limiting state probability not because it is entered from other states with unusual frequency, but rather because it holds the process a long time when it gets it.

We shall have extensive future use for these concepts of holding time and equivalent real transition process.

## PROBLEMS

**1.** The following model of voting habits has been proposed. A voter who starts out voting for Party *A* has a probability of 0.4 of changing to Party *B* in the next election. Once a voter has voted for Party *B*, he has a probability of 0.1 of changing to Party *A* on each election, and once there he never strays from the fold again.

If a voter has a probability of 0.5 of voting for Party *A* in his first election,

a) What is the probability that a voter will be captured by Party *A* after his *n*th election?

b) What is the expected number of elections for a voter to become a Party *A* devotee?

c) What is the variance of this number?

**2.** At present Rudolph has two girl friends, Pamela and Ruth, whom he dates regularly. Every Saturday night he goes out with one or the other. The probability that he dates Pamela, given that he was out with her the previous Saturday, is 0.3, while the probability that he dates Pamela given that he was out with Ruth the previous Saturday is 0.4. Furthermore, Pamela estimates that with probability 0.4 Rudolph will pop the question while on a date with her and marry her on the following Saturday. Ruth estimates the same probability concerning herself.

a) Obviously Rudolph is eventually trapped by one of the girls. If his initial probability of dating Pamela is 0.3, what is the expected duration (in weeks) of his bachelorhood starting from the Saturday of the first date?

b) Ruth used to work for the Canadian Mounted Police and claims she always gets her man. What is the probability that she's right?

**3.** A certain taxi driver lives in town $A$. The people in town $A$ do not tip, so he decides to ply his trade only between neighboring towns $B$ and $C$. He goes where his fare demands, and fares arrive according to a Markov process with transition probability matrix

$$\begin{array}{cc} & \textit{To} \\ & \begin{array}{cc} B & C \end{array} \\ \textit{From} \begin{array}{c} B \\ C \end{array} & \begin{bmatrix} 0.4 & 0.6 \\ 0.8 & 0.2 \end{bmatrix} \end{array}.$$

But this driver does not really care for driving a cab. Every time he sees a potential fare, he must decide whether to pick up the fare or go home to his lovely young bride. Once he gets home, he stays there! The probability that he will pick up any one potential fare is 9/10. When the driver gets up in the morning, he flips a fair coin to decide whether to go to $B$ or to $C$. On this trip he does not take a passenger.

a) How many *fares* can he expect to carry *from* either $B$ or $C$ *to* $B$ before returning home? *From* $B$ or $C$ *to* $C$?

b) How many fares can he expect to carry in all before returning home?

c) How many trips does he expect to make including those to and from home?

d) Would your answers to a) and b) change if he decided to take a non-tipper from $A$ on his initial trip to the other towns? These fares are equally likely to go to towns $B$ or $C$.

**4.** Everett is a mathematician. With his birthday approaching, his wife Claudia decides to visit the local mathematics supply shop to replenish her husband's supply of means, mean squares, and variances. Just like a woman, Claudia is rather fickle. Her shopping habits may be described by

$$P = \begin{array}{c} \\ 1 \\ 2 \\ 3 \\ 4 \end{array} \begin{array}{c} \begin{array}{cccc} 1 & 2 & 3 & 4 \end{array} \\ \begin{bmatrix} 0.4 & 0.1 & 0.3 & 0.2 \\ 0.3 & 0.3 & 0.1 & 0.3 \\ 0.5 & 0.2 & 0.3 & 0 \\ 0 & 0 & 0 & 1 \end{bmatrix} \end{array}$$

where indices 1, 2, 3, 4 refer to buying a mean, a mean square, a variance, and to quitting, respectively, and

$$p_{ij} = \mathscr{P}\{\text{event } j \text{ takes place} \mid \text{event } i \text{ has just happened}\}$$

Given that Claudia starts with a variance:

a) What is the mean number of means she will buy? The mean square of the means? The variance of the means?

b) Find the mean number of items Claudia will purchase before quitting.

c) Given that Claudia has just bought a mean square what is the mean of the number of mean squares she will buy before purchasing another type of item? What is the variance?

d) What is the probability that Claudia will buy at least one variance?

**5.** A taxi driver has noticed that if he is in town $A$ he has a probability of 0.2 of going to town $B$ on his next trip and a probability of 0.8 of staying in town $A$. Similarly, if he is in town $B$, his probability of going to town $A$ is 0.25 and of staying in $B$, 0.75. Unfortunately town $B$ drivers are devotees of the Parisian school of driving so there is a probability of 7/15 that any trip within $B$ (i.e., from $B$ to $B$) will terminate prematurely in an accident severe enough to put him out of commission for the day.

a) If he starts the day in town $A$, what is the probability that he will be in $A$ after $n$ trips?

b) How many trips can he expect to complete on the average before having an accident if he starts in $A$? (Do not count the final incomplete trip.)

**6.** a) Determine the geometric transform of the probability that the sequence HHTT occurs with a fair coin for the first time on trial $n$. Find the expected number of trials to obtain this sequence using the inspection technique for a flow graph, and check your answer by taking the derivative of the appropriate transform.

b) Determine the expected number of trials to obtain the sequence HHHH for the first time.

c) Why are the answers to a) and b) different?

**7.** Smith plays the following game. A coin is flipped until it comes up Heads. The tosses are numbered in order. If the first Head occurs on an even numbered toss, Smith wins and is awarded a number of dollars equal to the number of the toss. If the coin comes up Heads on an odd numbered toss, Smith loses and receives no money. After the coin comes up Heads the game is over. The probability of Heads on a single toss is $p = 1 - q$. Successive tosses are independent.

a) What is the expected length of the game? (If the coin comes up Heads on toss number 10, the game is of length 10.)

b) What is the probability that Smith wins?

c) What are Smith's expected earnings?

**8.** Citizen Harry has decided to toss back a few ales. Harry likes to (random) walk between beverages so he operates between three bars as follows: He has his first drink in bar 1, then uses this transition probability matrix

$$P = \begin{array}{c} \\ 1 \\ 2 \\ 3 \end{array}\begin{array}{c} \begin{array}{ccc} 1 & 2 & 3 \end{array} \\ \begin{bmatrix} 0.6 & 0.1 & 0.3 \\ 0.7 & 0.2 & 0.1 \\ 0 & 0.5 & 0.5 \end{bmatrix} \end{array}$$

where $p_{ij} = \mathscr{P}\{\text{Drink } n + 1 \text{ is in bar } j | \text{Drink } n \text{ was in bar } i\}$.

a) Assuming that Harry has quaffed many, many brews, what is the probability that he passes out in bar $i$? ($i = 1, 2, 3$.) Assume that he doesn't fade between taverns.

b) Unfortunately the police do not take kindly to Harry's wanderings. Therefore, every time he changes bars there is a 1/2 probability that they will pick him up for disorderly conduct. What is the expected number of drinks that Harry downs in bar 2 before being arrested?

c) What is the mean number of drinks that Harry has in bar 1 before leaving for the first time? What is the variance of this quantity?

d) What is the probability that he will visit bar 3 at least once before being arrested?

**9.** A newly-formed record company believes that it will become an established company when it has issued a successful record in three successive months. It estimates its chances of issuing a successful record in any month as 0.5, 0.8, and 0.9 if it has issued successful records in the preceding 0, 1, and 2 months, respectively.

a) What is the expected number of months until the company becomes established?

b) What is the most probable number?

c) What is the smallest number of months such that the probability of having become established will be at least 0.5? At least 0.9?

**10.** A professor's office has three chairs, one for conference and two for waiting; there is no standing room. Before the final examination the professor's time is divided into 15-minute intervals. At each interval end, a conference in progress has a probability 3/4 of ending. During each interval there is probability 1/2 that no students will arrive for a conference and probability 1/2 that exactly one will arrive. Any student arriving when all three chairs are occupied leaves immediately.

a) Draw a flow graph to represent the number of chairs occupied at the beginning of each interval. There are 4 states: no chairs occupied, 1 chair occupied, 2 chairs occupied, and 3 chairs occupied.

b) In steady state operation, what is the expected number of occupied chairs at the beginning of an interval?

c) At the beginning of an interval an eager young student with a perplexing technical problem and a date with a beautiful coed in one hour finds himself in the third chair. What is his probability of making the date?

d) Find the probability mass function of time in the system for a student starting in the third chair. In the second chair. What is the expected time spent in the system in each case?

e) An enterprising student who arrives at the beginning of an interval finds all chairs occupied. He charmingly persuades the professor's secretary to call him the first time the professor is free. Find the probability mass function of the time until the secretary's call. What is the expected time?

**11.** The Cacophonous Symphony Orchestra has decided to hold a rehearsal in order to perfect a new symphony. The symphony has four movements. The movements require different playing times. Unfortunately, each time a movement is played there is a chance that the conductor will be dissatisfied. However, the performance of the orchestra and the conductor's feelings can both be considered independent of previous rehearsal results, i.e., any time a specific movement is played it requires a known length of time and there is a fixed probability that the conductor will be satisfied. The relevant information is as follows.

| Movement number | Time required for execution | Probability of satisfaction on each performance | Dissatisfaction implies |
|---|---|---|---|
| 1 | 10 minutes | 1/2 | Replay of movement 1 |
| 2 | 20 minutes | 3/4 | Replay of only movement 2 half of the time; Recommencement at start of movement 1 half of the time |
| 3 | 10 minutes | 3/4 | Replay of movement 3 |
| 4 | 10 minutes | 1 | — |

a) If one were interested in the playing time required to complete the entire symphony, draw the transition diagram of the appropriate Markov model.

b) If the rehearsal begins at 7:30 p.m. and there is a 15-minute intermission after the conductor is satisfied with the playing of the second movement, determine the expected time at which the rehearsal will end.

c) Harry has decided to place a telephone call during intermission in an effort to obtain a date for the weekend. He figures that if he calls before 8:25 p.m., the girl will accept with probability 0.9; however, if he phones later than 8:25 p.m., the probability drops to 0.5. Draw a relevant transition diagram for this part of the problem. What is his probability of obtaining a date? (Assume that he dashes out with the final note and is assured telephone contact in a negligible amount of time.)

**12.** A radar network has been designed to monitor the air traffic over some geographical region. The system continually scans the region with a beam of electrical pulses in the same manner that rotation of a radius vector of a circle sweeps out the area of the circle. On each scan the probability that an aircraft is detected is $p_1$ if an aircraft is actually present, while with probability $p_2$ the system falsely indicates the presence of an aircraft when none exists. To minimize the possibility of a false alarm the following strategy is adopted: The presence of an aircraft is acknowledged if it is detected by $k$ successive scans or if $k$ detections occur in $k + 1$ successive scans.

a) Determine the geometric transform of the probability that the presence of the aircraft is first acknowledged on the $n$th scan following the craft's entry into the region. You may find construction of a flow graph for the process useful.

b) Outline two methods for finding the mean acknowledgment time, illustrate for $k = 2, p_1 = 0.9$.

c) The rate at which the system can be expected to acknowledge aircraft when the sky is free of planes may be called the false alarm rate. If the system is to detect the presence of an aircraft as soon after its entry into the region as possible while maintaining a false alarm rate less than or equal to some value $r$, discuss how you could calculate the optimal value of $k$. Illustrate your results for $r = 1/50$, $p_2 = 0.1$.

**13.** We consider tossing a fair 6-sided die.

a) What is the probability the sequence 123 appears *before* the sequence 22?

b) What is the joint probability that the sequence 123 occurs for the first time on the $n$th toss (i.e., 3 on $n$, 2 on $(n-1)$, and 1 on $(n-2)$ toss), *and* the sequence 22 has not occurred within the $n$ tosses?

c) What is the probability that the sequence 123 occurs for the first time on the $n$th toss, given that the sequence 123 occurs before the sequence 22?

d) What is the expected number of tosses before either of the sequences 123 or 22 occurs?

e) What is the expected number of tosses before the sequence 123 occurs, given that 123 occurs before 22?

**14.** A spider patrols two locations numbered 1 and 2 according to the transition probability matrix

$$^{s}P = \begin{bmatrix} 0.8 & 0.2 \\ 0.3 & 0.7 \end{bmatrix}.$$

A fly wanders between the same two locations according to the transition probability matrix

$$^{f}P = \begin{bmatrix} 0.6 & 0.4 \\ 0.5 & 0.5 \end{bmatrix}.$$

Assume that the time required for all transitions is the same and that the spider starts in location 1, the fly in location 2.

Find the probability mass function for the number of transitions until the spider acquires his dinner. Evaluate the mean.

**15.** Consider an $N$-state transient process with a single trapping state, say state $N$. If the process starts in state $i$, find a simple expression (in terms of flow graph transmissions and transition probabilities) for the expected number of times that the process will make the transition from $j$ to $k$ before becoming trapped. Assume that $j \neq N$.

**16.** A certain frog has been banished to the confines of a two pad lily pond. A small boy is throwing stones at him so that his transitions from lily pad to lily pad are described by the following transition probability matrix:

$$P = \begin{bmatrix} 0.5 & 0.5 \\ 0.3 & 0.7 \end{bmatrix}.$$

a) If the frog starts off on lily pad 1, what is the geometric transform of the probability that he is on lily pad 2 after $n$ jumps?

b) Unfortunately, there is a probability $(1 - \beta)$ that, on any one jump, the frog will be knocked out of commission by the stone. When this occurs he does not complete his jump and stops jumping for the day. If the frog starts off on lily pad 1, what is the expected number of jumps that he completes to lily pad 2 before getting struck?

**17.** Consider two random variables $m$ and $n$ that can take on only the non-negative integer values 0, 1, 2, . . . .

Let

$$f(n_0) = \mathscr{P}\{n = n_0\} \qquad n_0 = 0, 1, 2, \ldots$$

$$g(m_0) = \mathscr{P}\{m = m_0\} \qquad m_0 = 0, 1, 2, \ldots$$

$$f^{g}(z) = \sum_{n_0=0}^{\infty} f(n_0) z^{n_0}$$

and

$$g^{g}(z) = \sum_{m_0=0}^{\infty} g(m_0) z^{m_0}.$$

a) Suppose that $g^{g}(z) = z^k f^{g}(z)$, where $k$ is a positive integer. By transform methods or by returning to the time domain, show that

$$\bar{m} = \bar{n} + k$$

and $\quad \check{m} = \check{n}$

where $\bar{x}$ = expected value of $x$

and $\quad \check{x}$ = variance of $x$.

b) Use the results of part a) to evaluate the mean and variance of $m$ if

$$g^{g}(z) = \frac{0.12z^3}{1 - 0.88z}.$$

c) Suppose $g^{g}(z) = az^r f^{g}(z) + bz^s f^{g}(z)$ where $r$ and $s$ are positive integers and $a, b \geq 0$. Now, how are $\bar{m}$ and $\check{m}$ related to $\bar{n}$ and $\check{n}$? What relationship must $a$ and $b$ satisfy?

d) If

$$g^{g}(z) = \frac{0.70z + 0.18z^2}{1 - 0.12z^2},$$

what is wrong with saying that

$$\bar{m} = 0.70\left(1 + \frac{d}{dz}\frac{1}{1 - 0.12z^2}\right)\bigg|_{z=1} + 0.18\left(2 + \frac{d}{dz}\frac{1}{1 - 0.12z^2}\right)\bigg|_{z=1}?$$

e) Let $f^{g}(z) = N^{g}(z)/D^{g}(z)$ where $f^{g}(1) = 1$, i.e., $f(n)$ is a true probability mass function. Show that

$$\bar{n} = f^{g\prime}(1) = \frac{N^{g\prime}(1) - D^{g\prime}(1)}{D^{g}(1)}.$$

*Note*: Use of this result often makes it as easy to obtain the mean trapping time by differentiation of the transmission of the flow graph as by using the other method involving the transmission of the transition diagram.

**18.** An $N$-state Markov process with transition probability matrix $P$ has only one trapping state $k$. Let $p_i(n)$ be the probability that the system is trapped in state $k$ on the

$n$th transition, given that is started in state $i$. The complementary cumulative distribution for the trapping time,

$$^{>}p_i(n) = \sum_{m=n+1}^{\infty} p_i(m),$$

represents the probability that the system will be trapped after time $n$ given that it is started in state $i$ at time zero.

a) Show that

$$\bar{\nu}_i = \sum_{n=0}^{\infty} {^{>}p_i(n)}.$$

b) Show that by considering the total mass of $^{>}p_i(n)$ from $n = 0$ to $\infty$, we arrive at the same method for determining mean trapping times as the one employing transition diagram transmissions.

c) Now suppose that instead of one trapping state we have a set of them. Derive an expression for the mean square time to capture in one of these states, say $j$, given that the system started in state $i$ at $n = 0$ and that it is captured by state $j$. Express your result in terms of the elements of $\Phi^{g}(1)$.

**19.** Consider a discrete time, time-invariant transient Markov process. It is known that the process was started in the $i$th state and became trapped on the $n$th transition in state $k$ (state $k$ is a trapping state).

a) Find an expression for the expected number of transitions $\bar{\nu}_{ij|k}(n)$ that the process made into a transient state $j$ conditioned on the above information.

b) Use your result from part a) to find the value of $\bar{\nu}_{12|3}(n)$ as $n \to \infty$ with the transition probability matrix:

$$P = \begin{bmatrix} 0.9 & 0.1 & 0 \\ 0 & 0.6 & 0.4 \\ 0 & 0 & 1 \end{bmatrix}$$

**20.** The Hygienic Tobacco Company of Smog Bank, Va. manufactures Long Johns and Short Ribs cigarettes. They have hired a recent graduate to conduct an analysis of the smoking habits of the U.S. addict, and our enterprising researcher has found that a smoker who last bought a pack of Long Johns will buy them again 30% of the time, and will switch to Short Ribs 50% of the time. The other 20% of the time he kicks the habit.

Similarly, the smoker who last purchased a pack of Short Ribs purchases them again 10% of the time, switches to long Johns 80% of the time, and stops smoking altogether 10% of the time.

The effect of a pack of cigarettes on the smoker has been measured and the following data are known:

$e_L(m)$ is the amount of foreign substance remaining in a smoker's lungs due to a pack of Long Johns smoked $m$ days ago.

$e_S(m)$ is the amount of foreign substance remaining in a smoker's lungs due to a pack of Short Ribs smoked $m$ days ago.

a) A certain citizen smokes a pack of cigarettes a day. If he starts smoking Long Johns on day zero, draw a flow graph whose transmission is the geometric transform of the expected total amount of foreign substance in his lungs at the end of any day. (Assume that $e_L^g(z)$ and $e_S^g(z)$ are also known.)

b) Find the transform in Part a).

[Note: You may find it useful to define $x_{ij}(\ell, n)$ as the contribution at time $n$ from a pack of cigarettes of type $j$ smoked at time $\ell$, given that $s(0) = i$.]

**21.** The Hygienic Tobacco Company of Smog Bank is still manufacturing Long Johns and Short Ribs. Our client is interested in the "loyalty" characteristics of its products and would like to know the statistics of the number of times a smoker switches brands before deciding to quit smoking altogether. Toward this end we define:

$f_i(m) = \mathscr{P}\{$a smoker currently smoking brand $i$ will change brands $m$ more times before quitting$\}$.

a) Write a set of difference equations for $f_L(m)$ and $f_S(m)$ by considering the next purchase of a smoker. Be sure your result is valid for all $m \geq 0$.

b) Take the transform of the equations and show the flow graph whose outputs are $f_L^g(z)$ and $f_S^g(z)$.

c) Find $f_L^g(z)$.

**22.** If a transient process has many trapping states and if the process will eventually be trapped in state $k$, can we describe the states of the process by a new Markov process? If so, will it be a time-invariant Markov process? What are the transition probabilities of the new process (in terms of the transition probabilities and modified transition diagram transmissions of the original process)? Illustrate with the process below when we know the process will become trapped in state 4.

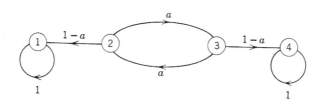

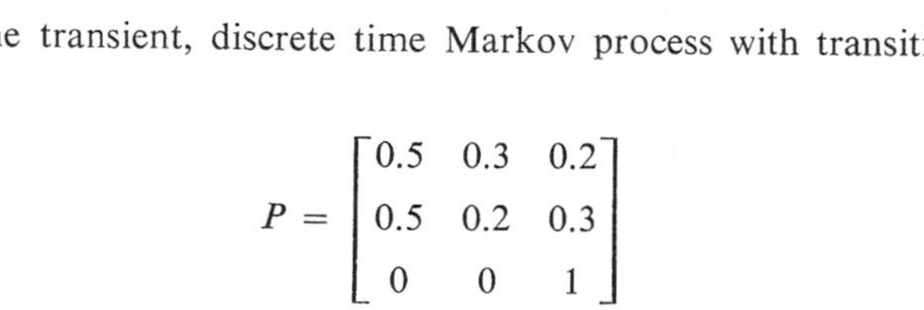


**23.** Consider the transient, discrete time Markov process with transition probability matrix:

$$P = \begin{bmatrix} 0.5 & 0.3 & 0.2 \\ 0.5 & 0.2 & 0.3 \\ 0 & 0 & 1 \end{bmatrix}$$

The random variable $\nu_{12}$ is the number of times the process has entered state 2 if it started in state 1 and has become trapped in state 3. Find $g_{12}(n) = \mathscr{P}\{\nu_{12} = n\}$.

**24.** We have shown in Section 4.1 that, for an $N$-state, discrete-time, monodesmic transient Markov process, the transmission of the modified transition diagram from node $i$ to node $j$ is equal to $\bar{\nu}_{ij}$, the expected number of times that the process enters state $j$ if it starts in $i$ and eventually becomes trapped.

a) Write the simultaneous linear equations for the $\bar{\nu}_{ij}$'s that are implied by this result.

b) For a certain three-state, transient Markov process these expected state occupancies have been measured. They are: $\bar{\nu}_{11} = 3$, $\bar{\nu}_{12} = 4$, $\bar{\nu}_{21} = 2$, $\bar{\nu}_{22} = 6$. Find the transition probability matrix $P$ for this process.

# 5 MARKOV PROCESS STATISTICS

In Chapter 4 we considered the special properties of transient Markov processes. However, the recurrent states in a Markov process have statistics that are important in many applications. The two of greatest interest are the state occupancies and first passage times. Essentially, answer the questions: "How much time did the system spend in state $i$?" and "How long did it take the system to get from state $i$ to state $j$?" Answering these questions and establishing the relationship between them is the purpose of this chapter.

## 5.1 STATE OCCUPANCY STATISTICS

We define the state occupancy random variable $\nu_{ij}(n)$ to be the number of times state $j$ is entered through time $n$ given that the system started in state $i$ at time zero. The statistics of this random variable have important interpretations in many practical problems, particularly where some reward is connected with changes of state. The probability distribution for the number of times a state is occupied in a given time will be considered in detail in later chapters (Sections 6.1 and 10.11). In the present section we shall develop methods for computing the means and variances of the state occupancies.

### Mean State Occupancies

The random variable $\nu_{ij}(n)$ is expressed in terms of the variable $x_{ij}(n)$ (defined in Equation 4.1.26) by

$$\nu_{ij}(n) = \sum_{m=0}^{n} x_{ij}(m). \tag{5.1.1}$$

The expectation of $\nu_{ij}(n)$—$\bar{\nu}_{ij}(n)$—is then equal to

$$\bar{\nu}_{ij}(n) = \sum_{m=0}^{n} \bar{x}_{ij}(m). \tag{5.1.2}$$

Now we use the result of Equation 4.1.27 to write

$$\bar{\nu}_{ij}(n) = \sum_{m=0}^{n} \phi_{ij}(m). \tag{5.1.3}$$

This equation shows that the expected number of times state $j$ is occupied after $n$ transitions starting in state $i$ is just the sum of the corresponding multistep transition probabilities for arguments zero through $n$. The initial occupancy of each state is being counted in the total number of occupancies.

We can illustrate this type of calculation for the marketing example whose transition diagram appears in Figure 1.2.1. We found the multistep transition probability matrix for this example in Equation 1.6.14,

$$\Phi(n) = P^n = \begin{bmatrix} 0.6 & 0.4 \\ 0.6 & 0.4 \end{bmatrix} + (1/2)^n \begin{bmatrix} 0.4 & -0.4 \\ -0.6 & 0.6 \end{bmatrix}. \tag{5.1.4}$$

Both states of this process are recurrent. Let us find $\bar{\nu}_{11}(n)$, the expected number of times a customer will buy brand $A$ on purchases $0, 1, \ldots, n$ given that his zeroth purchase was brand $A$. We write

$$\begin{aligned} \bar{\nu}_{11}(n) &= \sum_{m=0}^{n} \phi_{11}(m) = \sum_{m=0}^{n} [0.6 + 0.4(1/2)^m] \\ &= 0.6(n + 1) + 0.8 - 0.4(1/2)^n \qquad n = 0, 1, 2, \ldots, \end{aligned} \tag{5.1.5}$$

so that

$$\bar{\nu}_{11}(0) = 1,\ \bar{\nu}_{11}(1) = 1.8,\ \bar{\nu}_{11}(2) = 2.5,\ \bar{\nu}_{11}(3) = 3.15, \ldots. \tag{5.1.6}$$

Thus the expected number of times brand $A$ will be purchased through time 2 is 2.5—the maximum number of times it could have been bought in this interval is 3. We could, of course, repeat the calculation for any other $\bar{\nu}_{ij}(n)$ in which we were interested.

If we want to count only the entrances to each state that are internal to the process rather than including the entry at time zero, we simply begin the summation in Equation 5.1.1 at one rather than zero. We would thus construct a new variable ${}_1\nu_{ij}(n)$, the number of times the process experienced an internal transition to state $j$ in $n$ transitions given that it started in state $i$, by subtracting $\delta_{ij}$ from $\nu_{ij}(n)$,

$${}_1\nu_{ij}(n) = \nu_{ij}(n) - \delta_{ij}. \tag{5.1.7}$$

The modification will have effect only in the cases where $i = j$. For example, the values ${}_1\bar{\nu}_{11}(n)$ in the marketing example will be those in Equations 5.1.5 and 5.1.6 reduced by one. We interpret ${}_1\nu_{11}(n)$ as the expected number of times brand $A$ will be bought in $n$ purchases following a purchase of brand $A$. It is important in any model to decide whether to work with ${}_1\nu_{ij}(n)$ or $\nu_{ij}(n)$, but the relationship between these quantities is so simple that we shall confine our attention for the most part to one of them, namely, $\nu_{ij}(n)$.

### *Fractional occupancies*

The fraction of the times 0, 1, 2, . . ., $n$ that the system will occupy state $j$ given that it started in state $i$, $\zeta_{ij}(n)$, is also an interesting random variable. It is defined by

$$\zeta_{ij}(n) = \frac{1}{n+1}\,\nu_{ij}(n) \tag{5.1.8}$$

and its expectation is given by

$$\bar{\zeta}_{ij}(n) = \frac{1}{n+1}\,\bar{\nu}_{ij}(n). \tag{5.1.9}$$

By using the results of Equation 5.1.6, we find for the marketing example that

$$\bar{\zeta}_{11}(2) = (1/3)\bar{\nu}_{11}(2) = 5/6. \tag{5.1.10}$$

In other words, we expect in this interval that 5/6 of the customers' purchases will be brand $A$ purchases. More interesting is the behavior of this expected fraction when $n$ is very large, for in this case we would expect it to be the steady state probability of purchasing brand $A$. To confirm this result we use Equation 5.1.5,

$$\bar{\zeta}_{11}(n) = \frac{1}{n+1}\,\bar{\nu}_{11}(n) = 0.6 + \frac{0.8 - 0.4(1/2)^n}{n+1}. \tag{5.1.11}$$

If $n$ is very large, then

$$\bar{\zeta}_{11}(\infty) = 0.6 = \pi_1, \tag{5.1.12}$$

where $\pi_1$ is just the limiting state probability of state 1 for this process found in Equation 1.3.25. We shall later show that, in general,

$$\bar{\zeta}_{ij}(\infty) = \phi_{ij}(\infty) = \phi_{ij}. \tag{5.1.13}$$

### *Transform relations for mean occupancies*

We would find it rather tedious without a computer to evaluate $\bar{\nu}_{ij}(n)$ from the summation shown in Equation 5.1.3. We can obtain an alternative formulation by determining $\bar{\nu}_{ij}{}^g(z)$, the geometric transform of $\bar{\nu}_{ij}(n)$. If we use Table 1.5.1, we can write immediately

$$\bar{\nu}_{ij}{}^g(z) = \frac{1}{1-z}\,\phi_{ij}{}^g(z). \tag{5.1.14}$$

This result is consistent with our interpretation of $1/(1-z)$ as the transfer function of a summer. The transform of the mean number of entries to a state through time $n$ is just $1/(1 - z)$ times the transfer function of the Markov process from the input node to that state. By the final value theorem of the geometric transform,

$$\bar{\nu}_{ij}(\infty) = \bar{\nu}_{ij} = \lim_{z \to 1} (1 - z)\bar{\nu}_{ij}{}^{g}(z) = \phi_{ij}{}^{g}(1) = \vec{t}_{ij}. \tag{5.1.15}$$

Equation 5.1.15 states that the expected number of occupancies of state $j$ in an infinite number of transitions given that the system started in state $i$ is just the transmission of the transition diagram from node $i$ to node $j$, a result first expressed in Equation 4.1.34. This transmission will be finite only if state $j$ is a transient state, because any recurrent state will be occupied an unbounded number of times in an infinite number of transitions.

We can apply Equation 5.1.14 to the marketing example to find $\bar{\nu}_{11}(n)$ by this alternate method. Equation 1.6.13 shows $\Phi^{g}(z)$ for this case. We write

$$\bar{\nu}_{11}{}^{g}(z) = \frac{1}{1 - z}\phi_{11}{}^{g}(z) = \frac{1}{1 - z}\left[\frac{0.6}{1 - z} + \frac{0.4}{1 - (1/2)z}\right]$$

$$= \frac{0.6}{(1 - z)^2} + \frac{0.4}{(1 - z)(1 - (1/2)z)}$$

$$= \frac{0.6}{(1 - z)^2} + \frac{0.8}{1 - z} + \frac{-0.4}{1 - (1/2)z}. \tag{5.1.16}$$

Inversion of this transform produces

$$\bar{\nu}_{11}(n) = 0.6(n + 1) + 0.8 - 0.4(1/2)^n \qquad n = 0, 1, 2, \ldots \tag{5.1.17}$$

which is the same expression found in Equation 5.1.5 for this quantity.

We can summarize our findings up to this point by defining matrices $\bar{N}(n)$ and $\bar{N}^{g}(z)$ with elements $\bar{\nu}_{ij}(n)$ and $\bar{\nu}_{ij}{}^{g}(z)$. Equations 5.1.3 and 5.1.14 then relate these matrices to the multistep transition probabilities

$$\bar{N}(n) = \sum_{m=0}^{n} \Phi(m) = \sum_{m=0}^{n} P^m \tag{5.1.18}$$

and

$$\bar{N}^{g}(z) = \frac{1}{1 - z}\Phi^{g}(z) = \frac{1}{1 - z}[1 - Pz]^{-1}. \tag{5.1.19}$$

Fortunately in most problems we are more interested in a few selected elements of these matrices than in their complete form.

### *General two-state process*

We can easily use Equation 5.1.19 to find the mean occupancy times for the general two-state Markov process of Chapter 1. The transition matrix for this process is

$$P = \begin{bmatrix} 1-a & a \\ b & 1-b \end{bmatrix}. \tag{5.1.20}$$

From Equation 1.6.22 we see that

$$\Phi^{\mathscr{g}}(z) = \frac{1}{1-z}\begin{bmatrix} \dfrac{b}{a+b} & \dfrac{a}{a+b} \\ \dfrac{b}{a+b} & \dfrac{a}{a+b} \end{bmatrix} + \frac{1}{1-(1-a-b)z}\begin{bmatrix} \dfrac{a}{a+b} & \dfrac{-a}{a+b} \\ \dfrac{-b}{a+b} & \dfrac{b}{a+b} \end{bmatrix}. \tag{5.1.21}$$

Then we substitute this result into Equation 5.1.19 to obtain

$$\overline{\mathrm{N}}^{\mathscr{g}}(z) = \frac{1}{1-z}\Phi^{\mathscr{g}}(z) = \frac{1}{(1-z)^2}\begin{bmatrix} \dfrac{b}{a+b} & \dfrac{a}{a+b} \\ \dfrac{b}{a+b} & \dfrac{a}{a+b} \end{bmatrix}$$
$$+ \frac{1}{(1-z)[1-(1-a-b)z]}\begin{bmatrix} \dfrac{a}{a+b} & \dfrac{-a}{a+b} \\ \dfrac{-b}{a+b} & \dfrac{b}{a+b} \end{bmatrix}. \tag{5.1.22}$$

After partial fraction expansion we have

$$\overline{\mathrm{N}}^{\mathscr{g}}(z) = \frac{1}{(1-z)^2}\begin{bmatrix} \dfrac{b}{a+b} & \dfrac{a}{a+b} \\ \dfrac{b}{a+b} & \dfrac{a}{a+b} \end{bmatrix} + \frac{1}{1-z}\begin{bmatrix} \dfrac{a}{(a+b)^2} & \dfrac{-a}{(a+b)^2} \\ \dfrac{-b}{(a+b)^2} & \dfrac{b}{(a+b)^2} \end{bmatrix}$$
$$+ \frac{1}{1-(1-a-b)z}\begin{bmatrix} \dfrac{(a+b-1)a}{(a+b)^2} & \dfrac{-(a+b-1)a}{(a+b)^2} \\ \dfrac{-(a+b-1)b}{(a+b)^2} & \dfrac{(a+b-1)b}{(a+b)^2} \end{bmatrix}. \tag{5.1.23}$$

The 11 element of this matrix with $a = 0.2$, $b = 0.3$ produces Equation 5.1.16. By inverse geometric transformation of Equation 5.1.23 we can write $\overline{\mathrm{N}}(n)$,

$$\overline{\mathrm{N}}(n) = (n+1)\begin{bmatrix} \dfrac{b}{a+b} & \dfrac{a}{a+b} \\ \dfrac{b}{a+b} & \dfrac{a}{a+b} \end{bmatrix} + \begin{bmatrix} \dfrac{a}{(a+b)^2} & \dfrac{-a}{(a+b)^2} \\ \dfrac{-b}{(a+b)^2} & \dfrac{b}{(a+b)^2} \end{bmatrix}$$
$$+ (1-a-b)^n\begin{bmatrix} \dfrac{(a+b-1)a}{(a+b)^2} & \dfrac{-(a+b-1)a}{(a+b)^2} \\ \dfrac{-(a+b-1)b}{(a+b)^2} & \dfrac{(a+b-1)b}{(a+b)^2} \end{bmatrix}$$
$$n = 0, 1, 2, \ldots. \tag{5.1.24}$$

When $a = 0.2$, $b = 0.3$, the 11 element of $\overline{\mathrm{N}}(n)$ is just Equation 5.1.17.

**Second Moments of State Occupancies**

Now that we understand how to compute the mean occupancy of a state, let us proceed to calculate $\overline{\nu_{ij}^2}(n)$, the second moment of $\nu_{ij}(n)$. From Equation 5.1.1,

$$\nu_{ij}^2(n) = \sum_{m=0}^{n} x_{ij}(m) \sum_{r=0}^{n} x_{ij}(r) = \sum_{m=0}^{n} \sum_{r=0}^{n} x_{ij}(m)x_{ij}(r). \tag{5.1.25}$$

Upon taking the expectation we have

$$\overline{\nu_{ij}^2}(n) = \sum_{m=0}^{n} \sum_{r=0}^{n} \overline{x_{ij}(m)x_{ij}(r)}. \tag{5.1.26}$$

The expectation within the summation is a quantity we have encountered before; Equation 4.1.44 shows that it is the joint transition probability $\psi_{ijj}(m, r)$. Thus we have

$$\overline{\nu_{ij}^2}(n) = \sum_{m=0}^{n} \sum_{r=0}^{n} \psi_{ijj}(m, r). \tag{5.1.27}$$

We see from Equation 4.1.47 that $\psi_{ijj}(m, r)$ is given by

$$\psi_{ijj}(m, r) = \begin{cases} \phi_{ij}(m)\phi_{jj}(r - m) & r \geq m \\ \phi_{ij}(r)\phi_{jj}(m - r) & m \geq r. \end{cases} \tag{5.1.28}$$

When this expression for $\psi_{ijj}(m, r)$ is substituted into Equation 5.1.27, we obtain

$$\begin{aligned} \overline{\nu_{ij}^2}(n) &= \sum_{m=0}^{n} \sum_{r=m}^{n} \phi_{ij}(m)\phi_{jj}(r - m) \\ &\quad + \sum_{r=0}^{n} \sum_{m=r}^{n} \phi_{ij}(r)\phi_{jj}(m - r) - \sum_{m=0}^{n} \phi_{ij}(m) \\ &= 2 \sum_{m=0}^{n} \sum_{r=m}^{n} \phi_{ij}(m)\phi_{ij}(r - m) - \sum_{m=0}^{n} \phi_{ij}(m), \end{aligned} \tag{5.1.29}$$

where we have combined the two identical summations. Now we change variables in the summation, and produce

$$\overline{\nu_{ij}^2}(n) = 2 \sum_{m=0}^{n} \phi_{ij}(m) \sum_{r'=0}^{n-m} \phi_{jj}(r') - \sum_{m=0}^{n} \phi_{ij}(m) \quad n = 0, 1, 2, \ldots. \tag{5.1.30}$$

By using Equation 5.1.3 we can write

$$\overline{\nu_{ij}^2}(n) = 2 \sum_{m=0}^{n} \phi_{ij}(m)\bar{\nu}_{jj}(n - m) - \bar{\nu}_{ij}(n) \quad n = 0, 1, 2, \ldots. \tag{5.1.31}$$

We see from this equation that the second moment of the number of times state $j$ will be entered in $n$ transitions for a given starting state $i$ is equal to twice the convolution of the multistep transition probability and the mean number of entries if the starting state were $j$, less the mean number of entries of $j$ in $n$ transitions starting in state $i$. Therefore, the second moment of the state occupancy statistics is easily computable from the multistep transition probability matrix and the mean number of occupancies that was derived from it according to Equation 5.1.18. If we define a matrix $\overline{\mathrm{N}^2}(n)$ with elements $\overline{\nu_{ij}^2}(n)$, this observation may be summarized in the matrix equation

$$\overline{\mathrm{N}^2}(n) = 2\sum_{m=0}^{n} \Phi(m)[\overline{\mathrm{N}}(n-m) \square I] - \overline{\mathrm{N}}(n) \quad n = 0, 1, 2, \ldots. \tag{5.1.32}$$

**Marketing Example.** We can apply the result of Equation 5.1.31 to find $\overline{\nu_{11}^2}(n)$, the second moment of the number of times state 1 will be occupied in $n$ transitions when the system is started in state 1 for the marketing example. We use Equations 5.1.4 and 5.1.5 to write

$$\begin{aligned}\overline{\nu_{11}^2}(n) &= 2\sum_{m=0}^{n} \phi_{11}(m)\bar{\nu}_{11}(n-m) - \bar{\nu}_{11}(n)\\ &= 2\sum_{m=0}^{n} [0.6 + 0.4(1/2)^m][0.6(n-m+1) + 0.8 - 0.4(1/2)^{n-m}]\\ &\qquad - [0.6(n+1) + 0.8 - 0.4(1/2)^n]\\ &= 0.36n^2 + 2.40n + 0.60 - 0.32n(1/2)^n + 0.40(1/2)^n \quad n = 0, 1, 2, \ldots.\end{aligned} \tag{5.1.33}$$

The summations that must be computed in evaluating this quantity are elementary, but again the work is tedious.

***Transform relations for occupancy second moments***

We can exchange tedium of one sort for that of another by finding the geometric transform $\overline{\nu_{ij}^2}^g(z)$ of $\overline{\nu_{ij}^2}(n)$. By using our result that the geometric transform changes the convolution operation to multiplication of transforms, we can immediately transform Equation 5.1.31 to obtain

$$\overline{\nu_{ij}^2}^g(z) = 2\phi_{ij}^g(z)\bar{\nu}_{jj}^g(z) - \bar{\nu}_{ij}^g(z). \tag{5.1.34}$$

If we want to show the direct dependence of $\overline{\nu_{ij}^2}^g(z)$ on the transfer functions of the Markov process, we use Equation 5.1.14 and write

$$\begin{aligned}\overline{\nu_{ij}^2}^g(z) &= 2\phi_{ij}^g(z)\frac{1}{1-z}\phi_{jj}^g(z) - \frac{1}{1-z}\phi_{ij}^g(z)\\ &= \frac{1}{1-z}\phi_{ij}^g(z)[2\phi_{jj}^g(z) - 1],\end{aligned} \tag{5.1.35}$$

or even

$$\overline{\nu_{ij}{}^{2}}{}^{g}(z) = \bar{\nu}_{ij}{}^{g}(z)[2\phi_{jj}{}^{g}(z) - 1]. \tag{5.1.36}$$

We can find the limiting form of $\overline{\nu_{ij}{}^{2}}(n)$ when $n$ is large by applying the final value theorem to Equation 5.1.35. We have

$$\begin{aligned}\overline{\nu_{ij}{}^{2}}(\infty) = \overline{\nu_{ij}{}^{2}} &= \lim_{z\to1}(1 - z)\overline{\nu_{ij}{}^{2}}{}^{g}(z) = \lim_{z\to1}\phi_{ij}{}^{g}(z)[2\phi_{jj}{}^{g}(z) - 1] \\ &= \phi_{ij}{}^{g}(1)(2\phi_{jj}{}^{g}(1) - 1),\end{aligned} \tag{5.1.37}$$

or, in terms of transition diagram transmissions,

$$\overline{\nu_{ij}{}^{2}} = \vec{t}_{ij}(2\vec{t}_{jj} - 1). \tag{5.1.38}$$

We see that the second moment of the number of times each state will be entered in an infinite number of transitions is related to the transmissions of the transition diagram by an equation we first encountered as Equation 4.1.52. We shall rightly obtain an infinite value for this second moment if state $j$ is a recurrent state of the process.

**Marketing Example.** We now use Equation 5.1.36 to find $\overline{\nu_{11}{}^{2}}(n)$ for the marketing example by transform methods,

$$\overline{\nu_{11}{}^{2}}{}^{g}(z) = \bar{\nu}_{11}{}^{g}(z)[2\phi_{11}{}^{g}(z) - 1]. \tag{5.1.39}$$

From Equations 5.1.16 and 1.6.13

$$\begin{aligned}\overline{\nu_{11}{}^{2}}{}^{g}(z) &= \left[\frac{0.6}{(1 - z)^2} + \frac{0.8}{1 - z} + \frac{-0.4}{1 - (1/2)z}\right]\left[2\left(\frac{0.6}{1 - z} + \frac{0.4}{1 - (1/2)z}\right) - 1\right] \\ &= \frac{0.72}{(1 - z)^3} + \frac{0.96 - 0.6}{(1 - z)^2} + \frac{-0.48 + 0.64}{(1 - z)(1 - (1/2)z)} \\ &\quad + \frac{0.48}{(1 - z)^2(1 - (1/2)z)} + \frac{-0.32}{(1 - (1/2)z)^2} + \frac{-0.8}{1 - z} + \frac{0.4}{1 - (1/2)z} \\ &= \frac{0.72}{(1 - z)^3} + \frac{0.36}{(1 - z)^2} + 0.16\left[\frac{2}{1 - z} + \frac{-1}{1 - (1/2)z}\right] \\ &\quad + 0.48\left[\frac{2}{(1 - z)^2} + \frac{-1}{(1 - z)(1 - (1/2)z)}\right] \\ &\quad + \frac{-0.32}{(1 - (1/2)z)^2} + \frac{-0.8}{1 - z} + \frac{0.4}{1 - (1/2)z} \\ &= \frac{0.72}{(1 - z)^3} + \frac{1.32}{(1 - z)^2} + \frac{-0.48}{1 - z} + \frac{-0.32}{(1 - (1/2)z)^2} \\ &\quad + \frac{0.24}{1 - (1/2)z} - 0.48\left[\frac{2}{1 - z} + \frac{-1}{1 - (1/2)z}\right]\end{aligned} \tag{5.1.40}$$

or

$$\overline{\nu_{11}^{2}}^{g}(z) = \frac{0.72}{(1-z)^3} + \frac{1.32}{(1-z)^2} + \frac{-1.44}{1-z} + \frac{-0.32}{(1-(1/2)z)^2} + \frac{0.72}{(1-(1/2)z)}. \quad (5.1.41)$$

The reason for showing the development in detail is to illustrate that only one additional partial fraction expansion is required to establish this result, namely, the expansion of the fraction with denominator $(1 - z)(1 - (1/2)z)$. The remainder of the work is simple combination of terms. We now use the table of transforms to invert the geometric transform and find

$$\begin{aligned}\overline{\nu_{11}^{2}}(n) &= 0.72[(1/2)(n+1)(n+2)] + 1.32[n+1] - 1.44 \\ &\quad - 0.32(n+1)(1/2)^n + 0.72(1/2)^n \\ &= 0.36n^2 + 2.40n + 0.60 - 0.32n(1/2)^n + 0.40(1/2)^n \qquad n = 0, 1, 2, \ldots\end{aligned} \quad (5.1.42)$$

in agreement with Equation 5.1.33; however, it is not clear that the transform analysis has simplified the calculation very much. Finding closed-form expressions for the moments of the state occupancies is no simple task.

We place the results of Equation 5.1.36 in matrix form by using a matrix $\overline{\mathrm{N}^{2}}^{g}(z)$ for the quantities $\overline{\nu_{ij}^{2}}^{g}(z)$,

$$\overline{\mathrm{N}^{2}}^{g}(z) = \overline{\mathrm{N}}^{g}(z)[2\Phi^{g}(z) \sqcup I - I]. \quad (5.1.43)$$

### Variances of State Occupancies

Now that we have developed methods for finding the first and second moments of the state occupancies, it is easy to compute the variance. Let $\check{\nu}_{ij}(n)$ be the variance of the number of times state $j$ will be entered in $n$ transitions if the process is started in state $i$. Then,

$$\check{\nu}_{ij}(n) = \overline{\nu_{ij}^{2}}(n) - (\bar{\nu}_{ij}(n))^2, \quad (5.1.44)$$

or letting $\check{\mathrm{N}}(n)$ be the variance matrix with elements $\check{\nu}_{ij}(n)$,

$$\check{\mathrm{N}}(n) = \overline{\mathrm{N}^{2}}(n) - \overline{\mathrm{N}}(n) \sqcup \overline{\mathrm{N}}(n). \quad (5.1.45)$$

**Marketing Example.** We can find for our marketing example the variance of the number of times state 1 will be entered in $n$ transitions when the process is started in state 1 by using Equation 5.1.44 and the results in Equations 5.1.5 and 5.1.33:

$$\begin{aligned}\check{\nu}_{11}(n) &= \overline{\nu_{11}^{2}}(n) - \bar{\nu}_{11}^{2}(n) \\ &= 0.36n^2 + 2.40n + 0.60 - 0.32n(1/2)^n + 0.40(1/2)^n \\ &\quad - [0.36n^2 + 1.68n + 1.96 - 0.48n(1/2)^n - 1.12(1/2)^n + 0.16(1/4)^n] \\ &= 0.72n - 1.36 + 0.16n(1/2)^n + 1.52(1/2)^n - 0.16(1/4)^n \qquad n = 0, 1, 2, \ldots.\end{aligned} \quad (5.1.46)$$

Note for future reference that the variance grows linearly with $n$ for large $n$.

## A Coin-Tossing Example

We have now developed methods for finding the mean and variance of the number of times each state is occupied for each starting state. We can use these methods to solve many probabilistic problems that are not at first glance state occupancy problems. For example, suppose that we toss a fair coin $n$ times. What are the mean and variance of the number of occurrences of the event a Head is followed by a Tail? We can turn this problem into one of finding the occupancy statistics of an appropriate Markov process. We need to create a state that will be occupied only when a Head is followed by a Tail so that its state occupancy statistics will answer the problem. This state must be surrounded by whatever other states are required to describe the probabilistic structure governing the process.

### *Markov process representation*

The process represented by the three-state flow graph of Figure 5.1.1 meets these specifications. State 3 is occupied only when a Head-Tail sequence has been completed. State 1 is occupied if the last toss was a Tail that did not complete a Head-Tail sequence. State 2 is occupied when the last toss is a Head.

We can now follow the transitions of the process. If the process is in state 1 and a Tail is generated on the next toss it stays in state 1; if a Head is generated it proceeds to state 2. If the process is in state 2 and a Head is generated the process stays in state 2; however, if a Tail is generated, a Head-Tail sequence has been completed and the process moves to state 3. When the process is in state 3, it must move on its next transition to either state 1 or state 2 according to whether a Tail or a Head is produced—it is impossible for the process to enter state 3 again without passing through some other state. All transitions in the process have probability 1/2 because the coin is fair.

Since it does not change our statistics to say that a Tail is generated on a fictitious zeroth toss, we start the process in state 1. Then $\phi_{13}(n)$ is the probability that a Head-Tail sequence is completed on the $n$th toss, not necessarily for the first time.

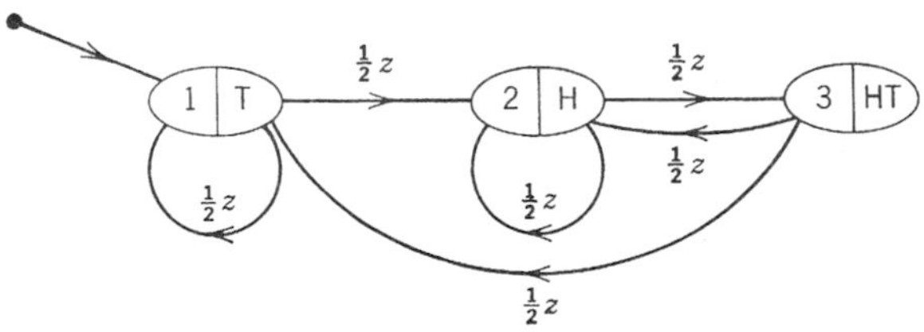


**Figure 5.1.1** Flow graph for Head-Tail sequences.

From the flow graph we have

$$\vec{t}_{13}(z) = \phi_{13}{}^g(z) = \frac{(1/4)z^2}{1 - (1/2)z - (1/2)z - (1/4)z^2 - (1/8)z^3 + (1/4)z^2 + (1/8)z^3}$$
$$= \frac{(1/4)z^2}{1 - z}. \tag{5.1.47}$$

This extremely simple transform shows that $\phi_{13}(n)$ is given by

$$\phi_{13}(n) = (1/4)u(n - 2) \qquad n = 0, 1, 2, \ldots. \tag{5.1.48}$$

Thus there is probability 1/4 of completing a Head-Tail sequence on any toss $n$ provided that $n$ is at least 2.

***Mean***

The expected number of Head-Tail sequences in $n$ tosses is $\bar{\nu}_{13}(n)$. From Equation 5.1.14,

$$\bar{\nu}_{13}{}^g(z) = \frac{1}{1 - z}\phi_{13}{}^g(z) = \frac{(1/4)z^2}{(1 - z)^2}. \tag{5.1.49}$$

Again, the inverse transform is simple. We find

$$\bar{\nu}_{13}(n) = (1/4)(n - 1) \qquad n = 2, 3, 4, \ldots$$
$$= (1/4)(n - 1)u(n - 2) \qquad n = 0, 1, 2, \ldots. \tag{5.1.50}$$

The expected number of Head-Tail sequences in $n$ tosses is thus $(1/4)(n - 1)$ if $n$ is at least 2.

***Variance***

We find the variance of the number of Head-Tail sequences in $n$ tosses by first computing the second moment $\overline{\nu_{13}{}^2}(n)$. From Equation 5.1.36,

$$\overline{\nu_{13}{}^2}{}^g(z) = \bar{\nu}_{13}{}^g(z)[2\phi_{33}{}^g(z) - 1]. \tag{5.1.51}$$

Since we have $\bar{\nu}_{13}{}^g(z)$ in Equation 5.1.49, we need only calculate $\phi_{33}{}^g(z)$ from the flow graph. We find

$$\vec{t}_{33}(z) = \phi_{33}{}^g(z) = \frac{1 - (1/2)z - (1/2)z + (1/4)z^2}{1 - z} = \frac{(1 - (1/2)z)^2}{1 - z}. \tag{5.1.52}$$

Now we substitute this result into Equation 5.1.51,

$$\overline{\nu_{13}{}^2}{}^g(z) = \frac{(1/4)z^2}{(1 - z)^2}\left[\frac{2(1 - (1/2)z)^2}{1 - z} - 1\right]$$
$$= \frac{(1/4)z^2}{(1 - z)^2}\left[\frac{1 - z + (1/2)z^2}{1 - z}\right]$$
$$= \frac{(1/4)z^2}{(1 - z)^2} + \frac{(1/8)z^4}{(1 - z)^3}. \tag{5.1.53}$$

To invert this transform we need a relation from our transform table,

$$\frac{1}{(1-z)^{k+1}} \Leftrightarrow \frac{1}{k!}(n+1)(n+2)\cdots(n+k). \tag{5.1.54}$$

Therefore,

$$\begin{aligned}
\frac{1}{(1-z)^3} &\Leftrightarrow \frac{1}{2}(n+1)(n+2)u(n) \qquad n = 0, 1, 2, \ldots \\
\frac{z^4}{(1-z)^3} &\Leftrightarrow \frac{1}{2}(n-3)(n-2)u(n-4) = \frac{1}{2}(n-3)(n-2)u(n-2) \qquad (5.1.55) \\
\frac{(1/8)z^4}{(1-z)^3} &\Leftrightarrow \frac{1}{16}(n^2-5n+6)u(n-2) \qquad n = 0, 1, 2, \ldots.
\end{aligned}$$

We can now write

$$\begin{aligned}
\overline{\nu_{13}{}^2}(n) &= [(1/4)(n-1) + (1/16)(n^2-5n+6)]u(n-2) \\
&= (1/16)(n^2-n+2)u(n-2) \qquad n = 0, 1, 2, \ldots.
\end{aligned} \tag{5.1.56}$$

By combining the results of Equations 5.1.50 and 5.1.56 we find the variance of the number of Head-Tail sequences in $n$ tosses, $\overset{\vee}{\nu}_{13}(n)$,

$$\begin{aligned}
\overset{\vee}{\nu}_{13}(n) &= \overline{\nu_{13}{}^2}(n) - (\bar{\nu}_{13}(n))^2 = 1/16[(n^2-n+2) - (n-1)^2]u(n-2) \\
&= (1/16)(n+1)u(n-2) \qquad n = 0, 1, 2, \ldots.
\end{aligned} \tag{5.1.57}$$

The variance of the number of Head-Tail sequences in $n$ tosses is therefore equal to $(1/16)(n+1)$ provided that $n$ is at least 2.

We shall usually find in analyzing a problem that can be described in Markov process terms that there are a variety of ways to create the Markov model and to define the relevant statistics. Flexibility in regarding the modeling process is a virtue often rewarded.

## Asymptotic Mean Occupancies

We see from Equations 5.1.33 and 5.1.46 that for large $n$, although the second moment increases as the square of $n$, the variance increases linearly with $n$. We shall now explore the asymptotic form of the moments for a large number of transitions. For many purposes the asymptotic forms will provide sufficient information about the behavior of the process.

We begin with the mean. We write Equation 5.1.14 using the form for $\phi_{ij}{}^{g}(z)$ given by Equation 1.6.17,

$$\begin{aligned}
\bar{\nu}_{ij}{}^{g}(z) &= \frac{1}{1-z}\phi_{ij}{}^{g}(z) = \frac{1}{(1-z)}\left(\frac{\phi_{ij}}{1-z} + t_{ij}{}^{g}(z)\right) \\
&= \frac{\phi_{ij}}{(1-z)^2} + \frac{t_{ij}{}^{g}(z)}{1-z}.
\end{aligned} \tag{5.1.58}$$

The transform $\bar{\nu}_{ij}{}^{g}(z)$ will in general have several geometric contributions because of the term with numerator $t_{ij}{}^{g}(z)$. However, the only terms in $\bar{\nu}_{ij}{}^{g}(z)$ that will persist for large $n$ are those corresponding to a step, ramp, parabolic ramp, etc.; the other terms will vanish. The term with denominator $(1 - z)^2$ corresponds to a ramp advanced by one time unit; the term with numerator $t_{ij}{}^{g}(z)$ has a step component of magnitude $t_{ij}{}^{g}(1)$ and other components that vanish. Therefore, for large $n$, $\bar{\nu}_{ij}(n)$ has the form,

$$\begin{aligned} \bar{\nu}_{ij}(n) &= (n + 1)\phi_{ij} + t_{ij}{}^{g}(1) \\ &= \phi_{ij}n + \phi_{ij} + t_{ij}{}^{g}(1), \quad \text{for large } n, \end{aligned} \tag{5.1.59}$$

or for the monodesmic case, where $\phi_{ij} = \pi_j$,

$$\bar{\nu}_{ij}(n) = \pi_j(n + 1) + t_{ij}{}^{g}(1). \tag{5.1.60}$$

This expression shows the linear rate of increase of $\bar{\nu}_{ij}(n)$ with $n$ necessary to establish Equation 5.1.13. In matrix form Equation 5.1.59 is

$$\bar{\mathrm{N}}(n) = (n + 1)\Phi + T^{g}(1), \quad \text{for large } n. \tag{5.1.61}$$

The matrix $T^{g}(1)$ has an interesting interpretation. Since the geometric transform evaluated at $z = 1$ is the sum of the values of the time function, $T^{g}(1)$ is just the sum of all the transient components $T(n) = \Phi(n) - \Phi$ of the multistep transition probabilities (defined in Equation 1.6.15). We call it the transient sum matrix; we shall see more of this quantity in the future.

### *General two-state process*

Equation 5.1.61 can be written for the general two-state Markov process whose transition probability matrix appears in Equation 5.1.20. From Equations 1.6.16 and 1.6.22 we see that

$$\Phi = \begin{bmatrix} \dfrac{b}{a + b} & \dfrac{a}{a + b} \\[2ex] \dfrac{b}{a + b} & \dfrac{a}{a + b} \end{bmatrix} \tag{5.1.62}$$

and

$$T^{g}(z) = \frac{1}{1 - (1 - a - b)z} \begin{bmatrix} \dfrac{a}{a + b} & \dfrac{-a}{a + b} \\[2ex] \dfrac{-b}{a + b} & \dfrac{b}{a + b} \end{bmatrix}. \tag{5.1.63}$$

Now we can write

$$T^{g}(1) = \begin{bmatrix} \dfrac{a}{(a + b)^2} & \dfrac{-a}{(a + b)^2} \\[2ex] \dfrac{-b}{(a + b)^2} & \dfrac{b}{(a + b)^2} \end{bmatrix} \tag{5.1.64}$$

and therefore evaluate Equation 5.1.61 for the general two-state example as

$$\overline{\mathrm{N}}(n) = (n+1)\begin{bmatrix} \dfrac{b}{a+b} & \dfrac{a}{a+b} \\ \dfrac{b}{a+b} & \dfrac{a}{a+b} \end{bmatrix} + \begin{bmatrix} \dfrac{a}{(a+b)^2} & \dfrac{-a}{(a+b)^2} \\ \dfrac{-b}{(a+b)^2} & \dfrac{b}{(a+b)^2} \end{bmatrix} \quad \text{for large } n. \tag{5.1.65}$$

***Marketing Example***

For the marketing example, where $a = 0.2$ and $b = 0.3$, we have

$$\Phi = \begin{bmatrix} 0.6 & 0.4 \\ 0.6 & 0.4 \end{bmatrix} \tag{5.1.66}$$

and

$$T^{g}(1) = \begin{bmatrix} 0.8 & -0.8 \\ -1.2 & 1.2 \end{bmatrix}. \tag{5.1.67}$$

When we substitute these values in Equation 5.1.61 we obtain

$$\overline{\mathrm{N}}(n) = (n+1)\begin{bmatrix} 0.6 & 0.4 \\ 0.6 & 0.5 \end{bmatrix} + \begin{bmatrix} 0.8 & -0.8 \\ -1.2 & 1.2 \end{bmatrix} \quad \text{for large } n. \tag{5.1.68}$$

Therefore, for the marketing example we see in particular that

$$\bar{\nu}_{11}(n) = 0.6(n+1) + 0.8, \quad \text{for large } n, \tag{5.1.69}$$

a result confirmed by Equation 5.1.5.

***Asymptotic mean occupancy differences***

Notice from Equation 5.1.61 that for a monodesmic process, where the rows of $\Phi$ are identical, the differences in the rows of $\overline{\mathrm{N}}(n)$ for large $n$ must be due to differences in the rows of $T^{g}(1)$. For example, from Equation 5.1.68,

$$\bar{\nu}_{11}(\infty) - \bar{\nu}_{21}(\infty) = 0.8 - (-1.2) = 2. \tag{5.1.70}$$

This result has a simple interpretation: a system started in state 1 is expected to visit state 1 two more times than it visits state 2 in a very large number of transitions. In the marketing example, we expect a customer to buy brand *A* an extra two times in a large number of purchases if he buys brand *A* rather than brand *B* as his initial purchase.

## Asymptotic Second Moments of Occupancies

Now let us turn to the question of developing an asymptotic expression for the second moment of the occupancies. All we have to do is use the result of Equation

1.6.17 in Equation 5.1.35,

$$\overline{\nu_{ij}^{2}}^{g}(z) = \frac{1}{1-z}\,\phi_{ij}^{g}(z)[2\phi_{jj}^{g}(z) - 1]$$

$$= \frac{1}{1-z}\left[\frac{\phi_{ij}}{1-z} + t_{ij}^{g}(z)\right]\left[2\left(\frac{\phi_{jj}}{1-z} + t_{jj}^{g}(z)\right) - 1\right]$$

$$= \frac{2\phi_{ij}\phi_{jj}}{(1-z)^3} + \frac{2t_{ij}^{g}(z)\phi_{jj} + 2\phi_{ij}t_{jj}^{g}(z) - \phi_{ij}}{(1-z)^2} + \frac{t_{ij}^{g}(z)[2t_{jj}^{g}(z) - 1]}{1-z}. \quad (5.1.71)$$

The term with denominator $(1 - z)^2$ can contain a step as well as a ramp. We shall use the differentiation method of partial fraction expansion of fractions with repeated roots in their denominators to find the magnitude of the step component. We obtain the following expansion of $\bar{\nu}_{ij}^{g}(z)$ in terms of the parabolic ramp, ramp, and step components that will exist for large $n$,

$$\overline{\nu_{ij}^{2}}^{g}(z) = \frac{2\phi_{ij}\phi_{jj}}{(1-z)^3} + \frac{2t_{ij}^{g}(1)\phi_{jj} + 2\phi_{ij}t_{jj}^{g}(1) - \phi_{ij}}{(1-z)^2}$$

$$- \frac{2t_{ij}^{g\prime}(1)\phi_{jj} + 2\phi_{ij}t_{jj}^{g\prime}(1)}{1-z} + \frac{t_{ij}^{g}(1)[2t_{jj}^{g}(1) - 1]}{1-z}$$

$$+ \text{ other terms corresponding to decaying components of } \overline{\nu_{ij}^{2}}(n). \quad (5.1.72)$$

By inverse transformation we find

$$\overline{\nu_{ij}^{2}}(n) = 2\phi_{ij}\phi_{jj}[(1/2)(n + 1)(n + 2)] + [2t_{ij}^{g}(1)\phi_{jj} + 2\phi_{ij}t_{jj}^{g}(1) - \phi_{ij}](n + 1)$$
$$- [2t_{ij}^{g\prime}(1)\phi_{jj} + 2\phi_{ij}t_{jj}^{g\prime}(1)] + t_{ij}^{g}(1)[2t_{jj}^{g}(1) - 1] \quad \text{for large } n, \quad (5.1.73)$$

or

$$\overline{\nu_{ij}^{2}}(n) = \phi_{ij}\phi_{jj}n^2 + [3\phi_{ij}\phi_{jj} + 2t_{ij}^{g}(1)\phi_{jj} + 2\phi_{ij}t_{jj}^{g}(1) - \phi_{ij}]n$$
$$+ 2\phi_{ij}\phi_{jj} + 2t_{ij}^{g}(1)\phi_{jj} + 2\phi_{ij}t_{jj}^{g}(1) - \phi_{ij} - 2t_{ij}^{g\prime}(1)\phi_{jj}$$
$$- 2\phi_{ij}t_{jj}^{g\prime}(1) + t_{ij}^{g}(1)[2t_{jj}^{g}(1) - 1] \quad \text{for large } n. \quad (5.1.74)$$

We see that $\overline{\nu_{ij}^{2}}(n)$ has terms in the zeroth, first, and second powers of $n$ when $n$ is large. This equation could be used to define the asymptotic form of $\overline{\mathrm{N}^2}(n)$, the matrix of $\overline{\nu_{ij}^{2}}(n)$, in terms of other matrices, but there is little advantage in such a formulation; we shall instead calculate the elements of $\overline{\mathrm{N}^2}(n)$ from Equation 5.1.74 whenever this matrix is required.

### *General two-state process*

We can write the asymptotic matrix $\overline{\mathrm{N}^2}(n)$ for the second moments of the occupancy times for the general two-state process whose transition matrix appears in Equation 5.1.20. We already know $\Phi$ and $T^{g}(1)$ from Equations 5.1.62 and 5.1.64.

The matrix $T^{\mathscr{g}\prime}(1)$ is found by differentiating $T^{\mathscr{g}}(z)$ in Equation 5.1.63 with respect to $z$ and evaluating the result at $z = 1$. We compute

$$\frac{d}{dz} T^{\mathscr{g}}(z) = \frac{(1 - a - b)}{[1 - (1 - a - b)z]^2} \begin{bmatrix} \dfrac{a}{a + b} & \dfrac{-a}{a + b} \\[2ex] \dfrac{-b}{a + b} & \dfrac{b}{a + b} \end{bmatrix} \tag{5.1.75}$$

and

$$T^{\mathscr{g}\prime}(1) = \frac{(1 - a - b)}{(a + b)^3} \begin{bmatrix} a & -a \\ -b & b \end{bmatrix}. \tag{5.1.76}$$

Now we use Equation 5.1.74 to write $\overline{\mathrm{N}^2}(n)$,

$$\begin{aligned} \overline{\mathrm{N}^2}(n) = {} & n^2 \begin{bmatrix} \dfrac{b^2}{(a + b)^2} & \dfrac{a^2}{(a + b)^2} \\[2ex] \dfrac{b^2}{(a + b)^2} & \dfrac{a^2}{(a + b)^2} \end{bmatrix} \\ & + n \begin{bmatrix} \dfrac{b}{a + b}\left(\dfrac{3b}{a + b} + \dfrac{4a}{(a + b)^2} - 1\right) & \dfrac{a}{a + b}\left(\dfrac{3a}{a + b} + \dfrac{2(b - a)}{(a + b)^2} - 1\right) \\[2ex] \dfrac{b}{a + b}\left(\dfrac{3b}{a + b} + \dfrac{2(a - b)}{(a + b)^2} - 1\right) & \dfrac{a}{a + b}\left(\dfrac{3a}{a + b} + \dfrac{4b}{(a + b)^2} - 1\right) \end{bmatrix} \\ & + \begin{bmatrix} \begin{aligned} & 2\frac{b^2}{(a + b)^2} + \frac{b}{a + b}\left(\frac{4a}{(a + b)^2} - 1\right) \\ & \quad - \frac{4ab(1 - a - b)}{(a + b)^4} + \frac{a}{(a + b)^2}\left(\frac{2a}{(a + b)^2} - 1\right) \end{aligned} & \begin{aligned} & 2\frac{a^2}{(a + b)^2} + \frac{a}{a + b}\left(\frac{2(b - a)}{(a + b)^2} - 1\right) \\ & \quad - \frac{2a(b - a)(1 - a - b)}{(a + b)^4} - \frac{a}{(a + b)^2}\left(\frac{2b}{(a + b)^2} - 1\right) \end{aligned} \\[4ex] \begin{aligned} & 2\frac{b^2}{(a + b)^2} + \frac{b}{a + b}\left(\frac{2(a + b)}{(a + b)^2} - 1\right) \\ & \quad - \frac{2b(a - b)(1 - a - b)}{(a + b)^4} - \frac{b}{(a + b)^2}\left(\frac{2a}{(a + b)^2} - 1\right) \end{aligned} & \begin{aligned} & 2\frac{a^2}{(a + b)^2} + \frac{a}{a + b}\left(\frac{4b}{(a + b)^2} - 1\right) \\ & \quad - \frac{4ab(1 - a - b)}{(a + b)^4} + \frac{b}{(a + b)^2}\left(\frac{2b}{(a + b)^2} - 1\right) \end{aligned} \end{bmatrix} \\ & \text{for large } n. \end{aligned} \tag{5.1.77}$$

This equation is included mainly to show the complexity of even the asymptotic form of second moments of the occupancy times.

***Marketing example***

We can use Equation 5.1.77 with $a = 0.2$ and $b = 0.3$ to find the matrix of second moments for the marketing example. It is easy, however, to use Equations 5.1.74, 5.1.66, 5.1.67, and 5.1.76 in computing $\overline{\mathrm{N}^2}(n)$ for large $n$. We first write $T^{\mathscr{g}\prime}(1)$ from Equation 5.1.76 for the marketing example,

$$T^{\mathscr{g}\prime}(1) = \begin{bmatrix} 0.8 & -0.8 \\ -1.2 & 1.2 \end{bmatrix}. \tag{5.1.78}$$

Notice that coincidentally $T^{\mathscr{g}}(1) = T^{\mathscr{g}\prime}(1)$ in this example. We finally obtain

$$\overline{\mathrm{N}^2}(n) = n^2\begin{bmatrix} 0.36 & 0.16 \\ 0.36 & 0.16 \end{bmatrix} + n\begin{bmatrix} 2.40 & 0.40 \\ 0 & 2.00 \end{bmatrix} + \begin{bmatrix} 0.60 & -1.20 \\ -0.60 & 1.60 \end{bmatrix} \quad \text{for large } n. \tag{5.1.79}$$

We see that the element $\overline{\nu_{11}{}^2}(n)$ agrees with the form of Equations 5.1.33 and 5.1.42 when $n$ is large.

### Asymptotic Occupancy Variances

The asymptotic form of the variance $\breve{\nu}_{ij}(n)$ for large $n$ is easily computed from Equation 5.1.44,

$$\breve{\nu}_{ij}(n) = \overline{\nu_{ij}{}^2}(n) - (\bar{\nu}_{ij}(n))^2, \tag{5.1.80}$$

using the asymptotic forms for $\overline{\nu_{ij}{}^2}(n)$ and $\bar{\nu}_{ij}(n)$ we have already found. We first square $\nu_{ij}(n)$ in Equation 5.1.59

$$\bar{\nu}_{ij}{}^2(n) = \phi_{ij}{}^2n^2 + 2\phi_{ij}[\phi_{ij} + t_{ij}{}^{\mathscr{g}}(1)]n + \phi_{ij}{}^2 + 2\phi_{ij}t_{ij}{}^{\mathscr{g}}(1) + [t_{ij}{}^{\mathscr{g}}(1)]^2 \quad \text{for large } n; \tag{5.1.81}$$

then we subtract this equation from Equation 5.1.74 to obtain

$$\begin{aligned} \breve{\nu}_{ij}(n) = {} & \phi_{ij}[\phi_{jj} - \phi_{ij}]n^2 \\ & + [\phi_{ij}(3\phi_{jj} - 2\phi_{ij}) + 2t_{ij}{}^{\mathscr{g}}(1)(\phi_{jj} - \phi_{ij}) + 2\phi_{ij}t_{jj}{}^{\mathscr{g}}(1) - \phi_{ij}]n \\ & + \phi_{ij}[2\phi_{jj} - \phi_{ij}] + 2t_{ij}{}^{\mathscr{g}}(1)[\phi_{jj} - \phi_{ij}] + \phi_{ij}[2t_{jj}{}^{\mathscr{g}}(1) - 1] \\ & - 2t_{ij}{}^{\mathscr{g}\prime}(1)\phi_{jj} - 2\phi_{ij}t_{jj}{}^{\mathscr{g}\prime}(1) + t_{ij}{}^{\mathscr{g}}(1)[2t_{jj}{}^{\mathscr{g}}(1) - 1 - t_{ij}{}^{\mathscr{g}}(1)]. \end{aligned} \tag{5.1.82}$$

For the important monodesmic case, where as we know $\phi_{ij} = \phi_{jj} = \pi_j$, this expression assumes the simpler form,

$$\begin{aligned} \breve{\nu}_{ij}(n) = {} & \pi_j[\pi_j + 2t_{jj}{}^{\mathscr{g}}(1) - 1]n + \pi_j{}^2 + \pi_j[2t_{jj}{}^{\mathscr{g}}(1) - 1] \\ & - 2\pi_j[t_{ij}{}^{\mathscr{g}\prime}(1) + t_{jj}{}^{\mathscr{g}\prime}(1)] + t_{ij}{}^{\mathscr{g}}(1)[2t_{jj}{}^{\mathscr{g}}(1) - 1 - t_{ij}{}^{\mathscr{g}}(1)] \quad \text{for large } n. \end{aligned} \tag{5.1.83}$$

Noteworthy here is the fact that the occupancy variances are now linear in $n$. We also point out that the difference $\overset{v}{\nu}_{ij}(n) - \overset{v}{\nu}_{kj}(n)$ in variance of occupancies of state $j$ for large $n$, due to starting in state $i$ rather than state $k$, is readily seen from the equation to be a constant.

### *General two-state process*

The asymptotic matrix of variances $\overset{v}{\mathrm{N}}(n)$ can be computed from Equation 5.1.83 or from Equation 5.1.45. We shall calculate this matrix for the general two-state example using the second procedure. We first use Equation 5.1.65 to write

$$
\begin{aligned}
\overline{\mathrm{N}}(n) \square \overline{\mathrm{N}}(n) = {} & n^2 \begin{bmatrix} \dfrac{b^2}{(a+b)^2} & \dfrac{a^2}{(a+b)^2} \\[2ex] \dfrac{b^2}{(a+b)^2} & \dfrac{a^2}{(a+b)^2} \end{bmatrix} \\
& + n \begin{bmatrix} \dfrac{b}{a+b}\left(\dfrac{2b}{a+b} + \dfrac{2a}{(a+b)^2}\right) & \dfrac{a}{a+b}\left(\dfrac{2a}{a+b} + \dfrac{-2a}{(a+b)^2}\right) \\[2ex] \dfrac{b}{a+b}\left(\dfrac{2b}{a+b} + \dfrac{-2b}{(a+b)^2}\right) & \dfrac{a}{a+b}\left(\dfrac{2a}{a+b} + \dfrac{-2b}{(a+b)^2}\right) \end{bmatrix} \\
& + \begin{bmatrix} \dfrac{b^2}{(a+b)^2} + \dfrac{b}{a+b}\left(\dfrac{2a}{(a+b)^2}\right) + \dfrac{a^2}{(a+b)^4} & \dfrac{a^2}{(a+b)^2} + \dfrac{a}{a+b}\left(\dfrac{-2a}{(a+b)^2}\right) + \dfrac{a^2}{(a+b)^4} \\[2ex] \dfrac{b^2}{(a+b)^2} + \dfrac{b}{a+b}\left(\dfrac{-2b}{(a+b)^2}\right) + \dfrac{b^2}{(a+b)^4} & \dfrac{a^2}{(a+b)^2} + \dfrac{a}{a+b}\left(\dfrac{2b}{(a+b)^2}\right) + \dfrac{b^2}{(a+b)^4} \end{bmatrix} \quad \text{for large } n.
\end{aligned} \tag{5.1.84}
$$

Then we subtract this matrix from $\overline{\mathrm{N}^2}(n)$ of Equation 5.1.77 to produce after some simplification

$$
\begin{aligned}
\overset{v}{\mathrm{N}}(n) = \overline{\mathrm{N}^2}(n) - \overline{\mathrm{N}}(n) \square \overline{\mathrm{N}}(n) = {} & n \cdot \frac{ab(2-a-b)}{(a+b)^3} \begin{bmatrix} 1 & 1 \\ 1 & 1 \end{bmatrix} \\
& + \begin{bmatrix} \dfrac{-a(1+b)}{(a+b)^2} + \dfrac{6ab}{(a+b)^3} + \dfrac{a(a-4b)}{(a+b)^4} & \dfrac{-a(1+b)}{(a+b)^2} + \dfrac{6ab}{(a+b)^3} + \dfrac{a(a-4b)}{(a+b)^4} \\[2ex] \dfrac{-b(1+a)}{(a+b)^2} + \dfrac{6ab}{(a+b)^3} + \dfrac{b(b-4a)}{(a+b)^4} & \dfrac{-b(1+a)}{(a+b)^2} + \dfrac{6ab}{(a+b)^3} + \dfrac{b(b-4a)}{(a+b)^4} \end{bmatrix} \quad \text{for large } n.
\end{aligned} \tag{5.1.85}
$$

The first thing we notice about $\overset{\vee}{\mathrm{N}}(n)$ is that its two columns are identical. We can easily show that this is a general property of the two-state Markov process for any value of $n$. The sum of the random variables $\nu_{i1}(n)$ and $\nu_{i2}(n)$ must be $n + 1$ because the system must be in one of the two states at any time. It follows immediately that the sum of the expected values of these variables $\bar{\nu}_{i1}(n)$ and $\bar{\nu}_{i2}(n)$ must equal $n + 1$ for any $n$, a fact reflected by the row sums of the matrix $\overline{\mathrm{N}}(n)$ in Equation 5.1.65. Furthermore, since the two random variables add to a constant, their variances $\overset{\vee}{\nu}_{i1}(n)$ and $\overset{\vee}{\nu}_{i2}(n)$ must be equal—$\overset{\vee}{\mathrm{N}}(n)$ must have two identical columns.

Moreover we notice in Equation 5.1.85 that the rows of the matrix multiplying $n$ are identical. This property is shown to be true for any monodesmic Markov process by the observation that the coefficient of $n$ in Equation 5.1.83 for $\overset{\vee}{\nu}_{ij}(n)$ depends only on $j$, the column index of the matrix. The linear growth of the variance with $n$ for large $n$ at a rate that is independent of the starting state is an important characteristic of monodesmic Markov processes.

### *Marketing example*

We can use either Equation 5.1.85 with $a = 0.2$ and $b = 0.3$ or the results of Equations 5.1.68 and 5.1.79 to find the asymptotic occupancy variance matrix for the marketing example. We obtain

$$\overset{\vee}{\mathrm{N}}(n) = \overline{\mathrm{N}^2}(n) - \overline{\mathrm{N}}(n) \,\square\, \overline{\mathrm{N}}(n) = n\begin{bmatrix} 0.72 & 0.72 \\ 0.72 & 0.72 \end{bmatrix} + \begin{bmatrix} -1.36 & -1.36 \\ -0.96 & -0.96 \end{bmatrix} \quad \text{for large } n. \tag{5.1.86}$$

We see that regardless of the brand the customer purchased initially, the variance of the number of times he will buy either brand in $n$ purchases grows linearly with $n$ when $n$ is large at a rate 0.72.

Let us consider how we might use these numbers. Suppose we know that a customer has made 1000 purchases during the time we have watched him. We expect that $1000\pi_1 = 1000(0.6) = 600$ of these purchases will be purchases of brand $A$ and that $1000\pi_2 = 1000(0.4) = 400$ will be purchases of brand $B$. The variance to be associated with either statistic is very close to $1000(0.72) = 720$; therefore, the standard deviation is $\sqrt{720} \approx 27$. Since we expect the central limit theorem to apply even to such a large number of dependent trials,† we can use the normal probability distribution to make estimates of the probability of various observations. For example, because 95% of the area of a normal distribution falls within 2 standard deviations of its mean, we expect with probability 0.95 to observe

† See Appendix B.

a number of brand 1 purchases between 546 and 654 and therefore a number of brand 2 purchases between 346 and 454. Rough statistical calculations of this type are often useful in determining the suitability of a particular Markov model.

## Moments of Internal Transitions

Sometimes we are interested only in the internal transitions of the process; that is, the initial placement of the process in a state will not be counted. We therefore use the modified occupancy statistic defined in Equation 5.1.7,

$$ {}_1\nu_{ij}(n) = \nu_{ij}(n) - \delta_{ij} \qquad n = 0, 1, 2, \ldots . \tag{5.1.87}$$

Our immediate task will be to show how to compute the moments of this modified occupancy statistic, either from the basic description of the process or from the moments of $\nu_{ij}(n)$.

By the properties of the expectation operator,

$$ {}_1\bar{\nu}_{ij}(n) = \bar{\nu}_{ij}(n) - \delta_{ij}. \tag{5.1.88}$$

Since $\phi_{ij}(0) = \delta_{ij}$, we can use the result of Equation 5.1.3 to write immediately

$$ {}_1\bar{\nu}_{ij}(n) = \sum_{m=1}^{n} \phi_{ij}(m) \qquad n = 1, 2, 3, \ldots . \tag{5.1.89}$$

We find the second moment of ${}_1\nu_{ij}(n)$ by squaring Equation 5.1:87 and then taking its expectation. We obtain

$$ {}_1\nu_{ij}^2(n) = \nu_{ij}^2(n) - 2\delta_{ij}\nu_{ij}(n) + \delta_{ij}^2 \tag{5.1.90}$$

and

$$ \overline{{}_1\nu_{ij}^2}(n) = \overline{\nu_{ij}^2}(n) - 2\delta_{ij}\bar{\nu}_{ij}(n) + \delta_{ij}, \tag{5.1.91}$$

where we have used the result that $\delta_{ij}^2 = \delta_{ij}$. By substituting the expression for $\overline{\nu_{ij}^2}(n)$ found in Equation 5.1.31 we find

$$ \overline{{}_1\nu_{ij}^2}(n) = 2\sum_{m=0}^{n} \phi_{ij}(m)\bar{\nu}_{jj}(n-m) - \bar{\nu}_{ij}(n) - 2\delta_{ij}\bar{\nu}_{ij}(n) + \delta_{ij}. \tag{5.1.92}$$

Since the term $-2\delta_{ij}\bar{\nu}_{ij}(n)$ can be replaced by $-2\delta_{ij}\bar{\nu}_{jj}(n)$ without changing the equation, it cancels the contribution of the summation for $m = 0$, and we have

$$ \overline{{}_1\nu_{ij}^2}(n) = 2\sum_{m=1}^{n} \phi_{ij}(m)\bar{\nu}_{jj}(n-m) - \bar{\nu}_{ij}(n) + \delta_{ij}. \tag{5.1.93}$$

Equation 5.1.88 allows us to eliminate $\bar{\nu}_{ij}(n)$ from this equation to produce

$$\overline{{}_1\nu_{ij}^2}(n) = 2\sum_{m=1}^{n} \phi_{ij}(m)[{}_1\bar{\nu}_{jj}(n-m) + 1] - {}_1\bar{\nu}_{ij}(n). \qquad (5.1.94)$$

Finally, by using Equation 5.1.89 we obtain

$$\overline{{}_1\nu_{ij}^2}(n) = 2\sum_{m=1}^{n} \phi_{ij}(m)\,{}_1\bar{\nu}_{jj}(n-m) + {}_1\bar{\nu}_{ij}(n) \qquad n = 1, 2, 3, \ldots. \qquad (5.1.95)$$

Equations 5.1.89 and 5.1.95 show us how to compute the moments of the modified occupancy statistic from the transition probability matrix of the process; Equations 5.1.88 and 5.1.91 show us how to calculate them from the moments of the original occupancy statistic.

*The multinomial process*

We shall now specialize these general results to the case of the multinomial Markov process. As defined in Chapter 1, this process has identical rows in its transition probability matrix; the probability of entering each state $j$ on the next transition is $p_j$ independent of the present state. Therefore, as we found earlier,

$$\phi_{ij}(n) = p_{ij} = p_j \qquad n = 1, 2, 3, \ldots. \qquad (5.1.96)$$

From Equations 5.1.96 and 5.1.89 we have

$${}_1\bar{\nu}_{ij}(n) = \sum_{m=1}^{n} \phi_{ij}(m) = \sum_{m=1}^{n} p_j = np_j. \qquad (5.1.97)$$

Therefore the expected number of times that state $j$ will be occupied in $n$ internal transitions if the system is started in state $i$ is $np_j$. From this result and from Equation 5.1.95 we find

$$\begin{aligned}\overline{{}_1\nu_{ij}^2}(n) &= 2\sum_{m=1}^{n} \phi_{ij}(m)\,{}_1\bar{\nu}_{jj}(n-m) + {}_1\bar{\nu}_{ij}(n)\\ &= 2\sum_{m=1}^{n} p_j(n-m)p_j + np_j\\ &= 2p_j^2\sum_{m=1}^{n}(n-m) + np_j\\ &= n(n-1)p_j^2 + np_j \qquad n \geq 1. \end{aligned} \qquad (5.1.98)$$

The variance of the number of internal transitions that will be made into state $j$ for the same starting condition, $\overset{\vee}{\nu}_{ij}(n)$, is

$$ {}_1\overset{\vee}{\nu}_{ij}(n) = {}_1\overline{\nu_{ij}^2}(n) - [{}_1\bar{\nu}_{ij}(n)]^2 = np_j - np_j^2 = np_j(1 - p_j) \qquad n \geq 1. \quad (5.1.99) $$

Equations 5.1.97 and 5.1.99 give the mean and variance of the number of successes of type $j$ in $n$ generalized Bernoulli trials where the probability of a success of type $j$ on each trial is $p_j$. This result confirms both our understanding of the multinomial process and our developments for the occupancy statistics.

For additional insight into occupancy statistics we can look at the two-state multinomial process. The two-state multinomial process is just the matrix of Equation 5.1.20 with $a = 1 - b$, or $a + b = 1$. In accordance with our work on the general multinomial process, this process must produce Bernoulli trials with a probability $a$ of being in state 2 and a probability $1 - a$ of being in state 1. The expected number of times state 2 will be occupied in $n$ trials is $na$; the variance of the number of times state 2 (or state 1) will be occupied is $a(1 - a)n$. As we see, the very important idea of Bernoulli trials is a very special case of a Markov process.

### *A symmetric two-state process—drumhead coin-tossing*

However, a more interesting special case of the two-state Markov process is the two-state symmetric process, where $a = b$ in Equation 5.1.20 and the transition matrix is

$$ P = \begin{bmatrix} 1 - a & a \\ a & 1 - a \end{bmatrix}. \qquad (5.1.100) $$

We have a very interesting interpretation for a process of this type. Suppose that a coin is placed on a horizontal drumhead, and then the drum is struck with a stick, as in Figure 5.1.2. If the drum is struck very lightly the coin is likely to face Heads if it was Heads before and Tails if it was Tails before. If the drum is struck very hard it may not be possible to predict what the coin will face even if the last facing of the coin is known; that is, a probability of 1/2 would be assigned to its coming up Heads or Tails regardless of the previous state. There may even by a striking force of such size that it is likely to change the previous facing of the coin. If we identify state 1 with Tails and state 2 with Heads then the matrix of Equation 5.1.100 will represent the process in these three cases when $a \approx 0$, $a = 1/2$, and $a > 1/2$, respectively.

Let us now find the approximate mean and variance of the number of Heads that will be observed if the drum is struck a large number of times, $n$. The asymptotic forms of the occupancy statistics are appropriate to this calculation. From Equation 5.1.60 we see that $\bar{\nu}_{ij}(n)$ grows linearly with $n$ at the rate $\pi_j$. Since the steady-state probabilities for the general two-state process are $\pi_1 = b/(a + b)$ and

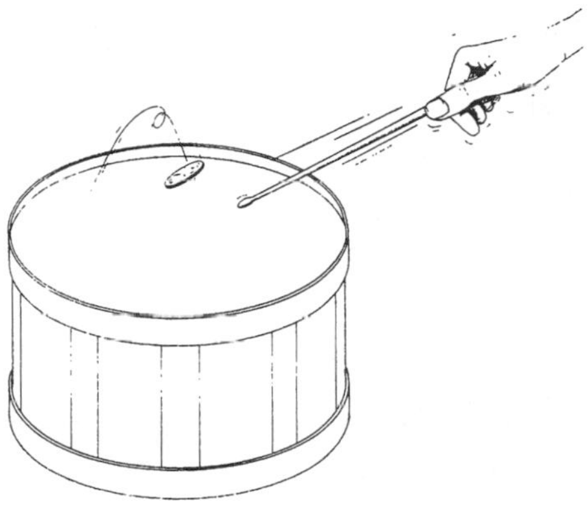

**Figure 5.1.2** Drumhead coin-tossing.

$\pi_2 = a/(a + b)$ and for this case $a = b$, we have that $\pi_1 = \pi_2 = 1/2$. Therefore, regardless of $a$, we expect that in a very large number of drumbeats, half will produce Heads and half will produce Tails.

The variance of the number of Heads in a large number $n$ of drumbeats will be equal to the variance of the number of Tails. Both grow linearly with $n$ at the rate indicated in Equation 5.1.85 with $a = b$; that is, at the rate

$$\frac{a^2(2 - 2a)}{(2a)^3} = \frac{1 - a}{4a} \qquad 0 \leq a < 1. \tag{5.1.101}$$

Notice that when $a$ is small, the variance grows very rapidly with $n$ because the system will very seldom change its original state. When $a = 1/2$ the variance grows at the rate 1/4, a result consistent with the interpretation of such a process as a succession of Bernoulli trials with probability of success 1/2. As $a$ approaches 1 the rate of growth of the variance becomes very small because the process approaches the two-state periodic case, which has a deterministic alternation of states and creates almost equal state occupancies. Expression 5.1.101 for the growth rate of the variance with $n$ thus summarizes the effect of dependence of successive tosses of a coin that is "fair" in the sense that its steady state probability of producing Heads or Tails is 1/2.

### Occupancies under Random Starting

All of our work in this section has considered situations where the starting state is known or where we are interested only in asymptotic behavior and hence the starting state does not matter. Now let us consider the question of occupancy statistics with a starting state selected at random.

Let $\nu_{\sim j}(n)$ be the number of times state $j$ is entered in $n$ transitions if the system is started according to the random starting vector $\boldsymbol{\pi}(0)$. The row vectors $\bar{\boldsymbol{\nu}}_{\sim}(n)$, $\overline{\boldsymbol{\nu}_{\sim}^{2}}(n)$, and $\check{\boldsymbol{\nu}}_{\sim}(n)$ that specify the mean, second moment, and variance of this statistic for each state under the random starting condition are then given by the expectation decomposition principle as,

$$\bar{\boldsymbol{\nu}}_{\sim}(n) = \boldsymbol{\pi}(0)\bar{\mathrm{N}}(n), \tag{5.1.102}$$

$$\overline{\boldsymbol{\nu}_{\sim}^{2}}(n) = \boldsymbol{\pi}(0)\overline{\mathrm{N}^2}(n), \tag{5.1.103}$$

and then

$$\check{\boldsymbol{\nu}}_{\sim}(n) = \overline{\boldsymbol{\nu}_{\sim}^{2}}(n) - \bar{\boldsymbol{\nu}}_{\sim}(n) \,\square\, \bar{\boldsymbol{\nu}}_{\sim}(n). \tag{5.1.104}$$

The expressions for $\bar{\mathrm{N}}(n)$ and $\overline{\mathrm{N}^2}(n)$ in these equations may be the exact quantities or they may be asymptotic forms if only the asymptotic effect of the result of random starting is needed.

An important case of random starting occurs when the system is started using the limiting state probability vector of the process; that is,

$$\boldsymbol{\pi}(0) = \boldsymbol{\pi}. \tag{5.1.105}$$

From Equations 5.1.102 and 5.1.18 we can write

$$\bar{\boldsymbol{\nu}}_{\sim}(n) = \boldsymbol{\pi}(0)\bar{\mathrm{N}}(n) = \boldsymbol{\pi}\sum_{m=0}^{n}\Phi(m) = \sum_{m=0}^{n}\boldsymbol{\pi}\Phi(m)$$

$$= \sum_{m=0}^{n}\boldsymbol{\pi} = (n+1)\boldsymbol{\pi} \qquad n = 0, 1, 2, \ldots. \tag{5.1.106}$$

In writing this equation we have been using the fact that a system started in each state with a probability given by the limiting state probability vector always has this vector for its state probability vector. We see that the expected number of times any state $j$ will be occupied in $n$ transitions is $(n+1)\pi_j$.

The second moment of the state occupancies also takes a special form for this kind of starting condition. From Equations 5.1.103 and 5.1.32 we have

$$\overline{\boldsymbol{\nu}_{\sim}^{2}}(n) = \boldsymbol{\pi}(0)\overline{\mathrm{N}^2}(n) = \boldsymbol{\pi}(0)\left[2\sum_{m=0}^{n}\Phi(m)[\bar{\mathrm{N}}(n-m)\,\square\, I] - \bar{\mathrm{N}}(n)\right]$$

$$= 2\sum_{m=0}^{n}\boldsymbol{\pi}[\bar{\mathrm{N}}(n-m)\,\square\, I] - \bar{\boldsymbol{\nu}}_{\sim}(n) \tag{5.1.107}$$

or, in terms of the elements,

$$\overline{\nu_{\sim j}^2(n)} = 2 \sum_{m=0}^{n} \pi_j \bar{v}_{jj}(n - m) - (n + 1)\pi_j$$

$$= \pi_j \left[ 2 \sum_{m=0}^{n} \bar{v}_{jj}(m) - (n + 1) \right] \qquad n = 0, 1, 2, \ldots . \qquad (5.1.108)$$

The variance of the number of state occupancies is, of course, easily derived using Equations 5.1.106 and 5.1.108.

### Computational Considerations: The Transient Sum Matrix $T^g(1)$

Let us now consider the question of computation from a practical point of view. From Equations 5.1.18 and 5.1.32 we see that the first two moments of the occupancy statistics are found by simple matrix operations on the multistep transition probability matrix, which is the heart of the Markov process. If electronic computation equipment is available, there is no difficulty even for processes with a large number of states.

Suppose, however, that we want to find the asymptotic moments using Equations 5.1.59 and 5.1.74. To calculate these quantities we have to know the limiting state probabilities of the process, and also the transform $T^g(z)$ of the transient matrix and its derivative with respect to $z$, both evaluated at $z = 1$. We already know how to find the limiting state probabilities by solving simultaneous equations, but at first glance it looks as if the transform of the transient matrix can be found only by computing the transfer function of the process $\Phi^g(z)$.

If this were so, the asymptotic results would be of little value because of the difficulty of performing the algebraic operations implied by $\Phi^g(z)$ for systems with a large number of states. Fortunately, we can compute $T^g(1)$ and $T^{g\prime}(1)$ from simple matrix operations on the transition probability matrix $P$. We shall now show how to do it.

From Equation 1.6.15 we can express the multistep transition matrix in terms of the limiting state matrix and the transient matrix by

$$\Phi(n) = P^n = \Phi + T(n) \qquad n = 0, 1, 2, \ldots \qquad (5.1.109)$$

or

$$T(n) = P^n - \Phi \qquad n = 0, 1, 2, \ldots . \qquad (5.1.110)$$

The transform $T^g(z)$ of the transient matrix is defined by

$$T^g(z) = \sum_{n=0}^{\infty} T(n) z^n . \qquad (5.1.111)$$

When we substitute Equation 5.1.110 we have

$$T^g(z) = \sum_{n=0}^{\infty} T(n)z^n = \sum_{n=0}^{\infty} (P^n - \Phi)z^n = I - \Phi + \sum_{n=1}^{\infty} (P^n - \Phi)z^n. \quad (5.1.112)$$

There are several ways we might compute $T^g(z)$. One that immediately suggests itself is direct transformation of Equation 5.1.109,

$$\Phi^g(z) = [I - Pz]^{-1} = \frac{1}{1-z}\Phi + T^g(z), \quad (5.1.113)$$

to yield

$$T^g(z) = [I - Pz]^{-1} - \frac{1}{1-z}\Phi. \quad (5.1.114)$$

However, since this representation appears on its face to become indeterminate as $z$ approaches 1, it will hardly serve as a convenient method for calculating the transient sum matrix $T^g(1)$. In fact, we shall find it much easier to compute the summation in Equation 5.1.112 directly, particularly when we establish the result that

$$P^n - \Phi = (P - \Phi)^n \qquad n = 1, 2, 3, \ldots. \quad (5.1.115)$$

To develop this relation we first note that $(P - \Phi)^n$ may be written in binomial expansion as

$$(P - \Phi)^n = \sum_{i=0}^{n} \binom{n}{i}(-1)^{n-i}P^i\Phi^{n-i}$$

$$= P^n + \sum_{i=0}^{n-1} \binom{n}{i}(-1)^{n-i}P^i\Phi^{n-i}. \quad (5.1.116)$$

In writing this expression, we have used the fact that $P$ and $\Phi$ are commutative. Any doubt on this point is rapidly dispelled by recalling that $\Phi$ is really a shorthand way of writing $P^\infty$. This observation further allows us to note that

$$P^r\Phi^s = \Phi \qquad \begin{cases} r = 0, 1, 2, \ldots \\ s = 1, 2, \ldots \end{cases} \quad (5.1.117)$$

and therefore permits writing Equation 5.1.116 in the simpler form

$$(P - \Phi)^n = P^n + \Phi \sum_{i=0}^{n-i} \binom{n}{i}(-1)^{n-i}. \quad (5.1.118)$$

Now we recognize that

$$\sum_{i=0}^{n-1} \binom{n}{i}(-1)^{n-i} = \sum_{i=0}^{n} \binom{n}{i}(-1)^{n-i} - 1 = (1-1)^n - 1 = -1 \quad (5.1.119)$$

to produce relation 5.1.115 immediately.

The establishment of Equation 5.1.115 allows us to write Equation 5.1.112 in the form

$$T^g(z) = I - \Phi + \sum_{n=1}^{\infty} (P - \Phi)^n z^n$$

$$= \sum_{n=0}^{\infty} (P - \Phi)^n z^n - \Phi. \quad (5.1.120)$$

Although $P^n$ would have a characteristic value equal to one, $(P - \Phi)^n$ does not because the steady state component has been removed. Therefore, from our table of geometric transforms we can write

$$T^g(z) = [I - zP + z\Phi]^{-1} \quad \Phi. \quad (5.1.121)$$

We have thus obtained an alternate way of expressing the transform of the transient part of the transfer function of the process. The advantage of this form is that when it is evaluated at $z = 1$ there is no problem with the non-existence of inverses and we obtain

$$T^g(1) = [I - P + \Phi]^{-1} - \Phi. \quad (5.1.122)$$

This equation shows that the transform of the transient part of the transfer function evaluated at $z = 1$ can be obtained by matrix inversion. Thus we have no difficulty in providing the $T^g(1)$ needed in Equation 5.1.59 and 5.1.74.

It is worth noting that Equation 5.1.122 for the transient sum matrix satisfies certain intuitive requirements. For example, if we write Equation 5.1.113 for the transient portion of a Markov process, we know that $\Phi = 0$ and consequently $T^g(1) = [I - P]^{-1}$. In this situation $T^g(1) = \Phi^g(1)$ is just the expected delay matrix $\overline{\mathrm{N}}$ found in Equation 4.1.39. Equation 5.1.122 does indeed show that when $\Phi = 0$, then $T^g(1) = [I - P]^{-1}$. The case $P = I$ also confirms our result, for in this case $\Phi = I$ and $T^g(1) = 0$ since there is no transient behavior. For the multinomial process, $\Phi = P$ and Equation 5.1.122 show $T^g(1) = I - P$. Since for the multinomial process the entire transient response occurs on the first transition, this transient sum matrix does indeed correspond to the behavior of the process.

### *General two-state process*

Let us illustrate the calculation of the transient sum matrix for the general two-state Markov process. The transition probability matrix $P$ appears in Equation

5.1.20; the matrix $\Phi$, in Equation 5.1.62. We write

$$I - P + \Phi = \begin{bmatrix} 1 & 0 \\ 0 & 1 \end{bmatrix} - \begin{bmatrix} 1-a & a \\ b & 1-b \end{bmatrix} + \begin{bmatrix} \dfrac{b}{a+b} & \dfrac{a}{a+b} \\ \dfrac{b}{a+b} & \dfrac{a}{a+b} \end{bmatrix}$$

$$= \begin{bmatrix} a + \dfrac{b}{a+b} & -a + \dfrac{a}{a+b} \\ -b + \dfrac{b}{a+b} & b + \dfrac{a}{a+b} \end{bmatrix}. \qquad (5.1.123)$$

The determinant of the matrix is $a + b$; therefore its inverse is

$$[I - P + \Phi]^{-1} = \frac{1}{a+b} \begin{bmatrix} b + \dfrac{a}{a+b} & a - \dfrac{a}{a+b} \\ b - \dfrac{b}{a+b} & a + \dfrac{b}{a+b} \end{bmatrix}. \qquad (5.1.124)$$

Finally we obtain $T^{\mathscr{g}}(1)$,

$$T^{\mathscr{g}}(1) = [I - P + \Phi]^{-1} - \Phi = \frac{1}{(a+b)^2} \begin{bmatrix} a & -a \\ -b & b \end{bmatrix}. \qquad (5.1.125)$$

We have previously produced this result directly from $T^{\mathscr{g}}(z)$ in Equation 5.1.64. We shall find that the transient sum matrix $T^{\mathscr{g}}(1)$ is a very important quantity in a Markov process; we are fortunate to be able to find it by the simple process of Equation 5.1.122.

### Computational Considerations: The Matrix $T^{\mathscr{g}\prime}(1)$

However, if we want to calculate the asymptotic expression for the second moment of the occupancy times given in Equation 5.1.74 we still need to find the matrix $T^{\mathscr{g}\prime}(1)$. We are again fortunate in this calculation, as we shall see. We first find $T^{\mathscr{g}\prime}(z)$ by using Equation 5.1.121,

$$T^{\mathscr{g}\prime}(z) = \frac{d}{dz} T^{\mathscr{g}}(z) = \frac{d}{dz} \{[I - zP + z\Phi]^{-1} - \Phi\}$$

$$= \frac{d}{dz} [I - zP + z\Phi]^{-1}. \qquad (5.1.126)$$

If $F(z)$ is a square matrix whose elements depend on $z$ and if $F(z)$ has an inverse, we know from Equation 1.5.24 that

$$\frac{d}{dz} F^{-1}(z) = -F^{-1}(z)\left[\frac{d}{dz} F(z)\right] F^{-1}(z). \qquad (5.1.127)$$

We apply this result to Equation 5.1.126 to obtain

$$\begin{aligned} T^{g'}(z) &= -[I - zP + z\Phi]^{-1}[-P + \Phi][I - zP + z\Phi]^{-1} \\ &= [I - zP + z\Phi]^{-1}[P - \Phi][I - zP + z\Phi]^{-1}. \end{aligned} \tag{5.1.128}$$

This equation relates $T^{g'}(z)$ to the matrix we inverted to find $T^{g}(z)$ and to $P$ and $\Phi$. It can be simplified further. First, since

$$\sum_{n=0}^{\infty} (P - \Phi)^n z^n = [I - zP + z\Phi]^{-1}, \tag{5.1.129}$$

we show that

$$\begin{aligned} (P - \Phi)[I - zP + z\Phi]^{-1} &= (P - \Phi) \sum_{n=0}^{\infty} (P - \Phi)^n z^n \\ &= \sum_{n=0}^{\infty} (P - \Phi)^{n+1} z^n \\ &= \sum_{m=1}^{\infty} (P - \Phi)^m z^{m-1} \\ &= z^{-1} \sum_{m=1}^{\infty} (P - \Phi)^m z^m \\ &= z^{-1}\left(\sum_{m=0}^{\infty} (P - \Phi)^m z^m - I\right) \\ &= z^{-1}([I - zP + \Phi]^{-1} - I). \end{aligned} \tag{5.1.130}$$

Now we use Equation 5.1.121 to place $T^{g'}(z)$ in the form

$$T^{g'}(z) = z^{-1}[T^{g}(z) + \Phi][T^{g}(z) + \Phi - I] = z^{-1}[T^{g}(z) + \Phi - I][T^{g}(z) + \Phi] \tag{5.1.131}$$

which shows how to express $T^{g'}(z)$ in terms of $T^{g}(z)$ and $\Phi$. We can, however, make the relationship even simpler. The following properties of the steady state and transient matrices are readily established from Equation 5.1.112 and the interpretation $\Phi = P^{\infty}$:

$$\Phi^2 = \Phi, \qquad T^{g}(z)\Phi = \Phi T^{g}(z) = 0. \tag{5.1.132}$$

When we multiply the matrix expressions in Equation 5.1.131 and use these relations we find the convenient result

$$T^{g'}(z) = z^{-1}T^{g}(z)(T^{g}(z) - I) = z^{-1}(T^{g}(z) - I)T^{g}(z). \tag{5.1.133}$$

Now we have expressed $T^{\mathscr{G}\prime}(z)$ in terms of $T^{\mathscr{G}}(z)$ alone. If we want to evaluate this expression at $z = 1$ we obtain

$$T^{\mathscr{G}\prime}(1) = T^{\mathscr{G}}(1)[T^{\mathscr{G}}(1) - I]. \tag{5.1.134}$$

Therefore once $T^{\mathscr{G}}(1)$ is calculated by the method of Equation 5.1.122, finding $T^{\mathscr{G}\prime}(1)$ requires just a matrix multiplication.

***General two-state process***

We shall illustrate the procedure for the general two-state Markov process. Starting with the result of Equation 5.1.125 we write

$$T^{\mathscr{G}\prime}(1) = T^{\mathscr{G}}(1)[T^{\mathscr{G}}(1) - I] = \begin{bmatrix} \dfrac{a}{(a+b)^2} & \dfrac{-a}{(a+b)^2} \\ \dfrac{-b}{(a+b)^2} & \dfrac{b}{(a+b)^2} \end{bmatrix} \begin{bmatrix} \dfrac{a}{(a+b)^2} - 1 & \dfrac{-a}{(a+b)^2} \\ \dfrac{-b}{(a+b)^2} & \dfrac{b}{(a+b)^2} - 1 \end{bmatrix}$$

$$= \frac{1-a-b}{(a+b)^3}\begin{bmatrix} a & -a \\ -b & b \end{bmatrix}. \tag{5.1.135}$$

This result was obtained before in Equation 5.1.76.

Now we have shown how the asymptotic first and second moments of the occupancy times are easily computed using matrix operations. We shall soon find that the quantities $T^{\mathscr{G}}(1)$ and $T^{\mathscr{G}\prime}(1)$ have other very important implications for Markov process statistical calculations.

## 5.2 FIRST PASSAGE TIMES

In Markov processes—as in everyday life—an important question is "How long does it take to get from here to there?". For a Markov process the question is usually phrased in the form, "How many transitions will it take to reach state $j$ for the first time if the system is in state $i$ at time zero?". Since this number of transitions is a random variable which we shall denote by $\theta_{ij}$, the answer to this question is the probability distribution of $\theta_{ij}$. We call $\theta_{ij}$ the first passage time of the system from state $i$ to state $j$ and use $f_{ij}(n)$ for the probability that $\theta_{ij}$ will be equal to $n$,

$$f_{ij}(n) = \mathscr{P}\{\theta_{ij} = n\} \qquad n = 1, 2, 3, \ldots. \tag{5.2.1}$$

Notice that in this equation $n$ takes on the integral values 1, 2, 3, . . . . Since no first passage time $\theta_{ij}$ can by definition assume the value zero, we have

$$f_{ij}(0) = 0. \tag{5.2.2}$$

If we know the probability distribution $f_{ij}(\cdot)$ for the first passage time $\theta_{ij}$, then we have all possible information on the question of how long it takes to get from one state to another in a Markov process.

The situation when $i = j$ is the case in which we are interested in how long the system will take to return to a given state when we know the time at which it left that state. Thus $\theta_{ii}$ is the number of transitions between a departure from state $i$ and the first return to $i$; it is called the first passage time from state $i$ to state $i$, or the recurrence time of state $i$. We view each occupancy of state $i$ as a time of departure, so that a virtual transition from state $i$ to state $i$ corresponds to a recurrence time $\theta_{ii}$ of one transition. The probability distribution for the recurrence time $\theta_{ii}$ is still denoted by $f_{ii}(n)$ and is defined by Equation 5.2.1.

We shall use the term "first passage time" to include the concept of recurrence time in what follows. We can then write a formal expression for the probability distribution of the first passage time in terms of our original notation as

$$f_{ij}(n) = \mathscr{P}\{\theta_{ij} = n\} = \begin{cases} \mathscr{P}\{s(n) = j, s(n-1) \neq j, \ldots, s(1) \neq j \,|\, s(0) = i\} & n > 1 \\ \mathscr{P}\{s(1) = j \,|\, s(0) = i\} = p_{ij} \quad n = 1 \end{cases} \tag{5.2.3}$$

We must now resolve the question of whether the probability distribution of Equation 5.2.3 is a probability mass function, that is, whether it sums to one. We see immediately that $f_{ij}(\cdot)$ will be a probability mass function only if the system will sooner or later enter state $j$ with certainty. If state $j$ is a recurrent state of the process, then this condition will be met. If state $j$ is a transient state and state $i$ is a recurrent state, it will not. However, if both $i$ and $j$ are transient states then the situation is not so clear. Although the condition generally will not be satisfied, it will be satisfied if, when the system leaves state $i$, it must pass through state $j$ before entering the recurrent chain. When $f_{ij}(\cdot)$ is not a probability mass function, its sum over all values of $n = 1, 2, 3, \ldots$ will be the probability that the system ever enters state $j$ in the future. Thus even when $f_{ij}(n)$ does not sum to one it provides us with information we would like to know about the process.

### Computation of First Passage Time Probability Distribution by Graph Modification

We can easily find a simple way to calculate $f_{ij}(n)$. We realize that it is the probability distribution of the time to go from state $i$ to state $j$, without reference to what happens after state $j$ is reached. But this is a problem we have already solved in Section 4.1—the problem of how long the system will spend in a transient process until it is trapped, the problem of delay computation. All we must do is make state $j$ of the Markov process a trapping state, start the system in state $i$, and observe it until trapping occurs.

To use our flow graph analysis we remove from the flow graph for the Markov process all branches that start on node $j$ and then find the transmission from node

$i$ to node $j$ of this modified flow graph. According to Section 4.1, this transmission will be $f_{ij}{}^{g}(z)$, the geometric transform of $f_{ij}(n)$ defined by

$$f_{ij}{}^{g}(z) = \sum_{n=0}^{\infty} f_{ij}(n)z^{n}. \tag{5.2.4}$$

Then we invert this geometric transform to find $f_{ij}(n)$.

We must draw a new modified flow graph for every value of $j$ we want to investigate. But the same graph will serve to find the transforms of the first passage time probability distributions from different states $i$ to the same state $j$—all that is required is that the transmission be found from different nodes on the graph to the trapping state. Therefore, in general, an $N$-state Markov process will require the construction of $N$ modified flow graphs for the purpose of investigating first passage time statistics.

***The general two-state process***

Let us now apply this technique to the general two-state process with transition matrix

$$P = \begin{bmatrix} 1 - a & a \\ b & 1 - b \end{bmatrix}. \tag{5.2.5}$$

We shall restrict $a$ and $b$ to satisfy $0 < a < 1$, $0 < b < 1$ to avoid pathological cases. The flow graph for this process is shown in Figure 5.2.1.

**First passage to state 1.** We shall first find the first passage time probability distributions to state 1. We draw the modified flow graph of Figure 5.2.2 where state 1 appears both as a transient state and as a trapping state. The graph in Figure 5.2.2 is drawn by having all branches that end on node 1 in the graph of Figure 5.2.1 end instead on a new node 1* that corresponds to state 1's role as a trapping state. The transmissions of this new graph from each node to node 1* then give the transforms of the first passage times terminating in state 1. Thus

$$f_{11}{}^{g}(z) = \vec{t}_{11*}(z) = (1 - a)z + \frac{abz^2}{1 - (1 - b)z} \tag{5.2.6}$$

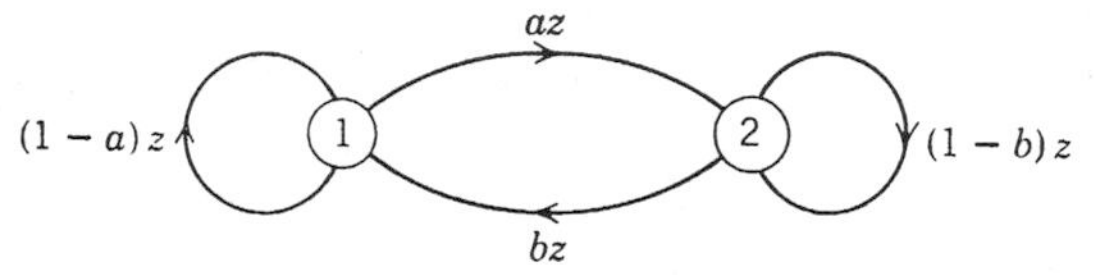


**Figure 5.2.1** Flow graph for general two-state Markov process.

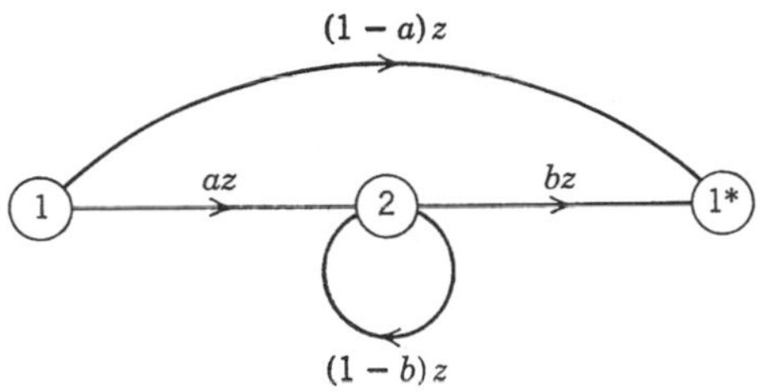


**Figure 5.2.2** Modified flow graph for first passage times to state 1.

and

$$f_{21}^{g}(z) = \vec{t}_{21^*}(z) = \frac{bz}{1 - (1 - b)z}. \tag{5.2.7}$$

By inverse transformation we then have the probability distributions of the first passage times,

$$f_{11}(n) = (1 - a)\,\delta(n - 1) + ab(1 - b)^{n-2}u(n - 2) \qquad n = 0, 1, 2, \ldots \tag{5.2.8}$$

and

$$f_{21}(n) = b(1 - b)^{n-1}u(n - 1) \qquad n = 0, 1, 2, \ldots. \tag{5.2.9}$$

We see that both probability distributions are geometric sequences with a decay rate $(1 - b)$.

If $f_{11}(n)$ and $f_{21}(n)$ are probability mass functions, then the corresponding first passages are certain events. To check that they are probability mass functions we simply evaluate their transforms at $z = 1$ and see whether we obtain 1. We find that in fact

$$f_{11}^{g}(1) = f_{21}^{g}(1) = 1. \tag{5.2.10}$$

Therefore, a return to state 1 and a first passage from state 2 to state 1 are certain events. Of course, we know this by inspection of the process transition diagram anyway.

Since $f_{11}^{g}(z)$ and $f_{21}^{g}(z)$ are the transforms of probability mass functions, we can find the moments of $\theta_{11}$ and $\theta_{21}$ by differentiating the transforms and evaluating the results at the point $z = 1$. Thus, if we let $\bar{\theta}_{ij}$, $\overline{\theta_{ij}^2}$, and $\check{\theta}_{ij}$ be the mean, second moment, and variance of the first passage time $\theta_{ij}$, we find

$$\bar{\theta}_{11} = f_{11}{}^{g\prime}(1) = \frac{d}{dz}\left[(1-a)z + \frac{abz^2}{1-(1-b)z}\right]_{z=1}$$

$$= \frac{a+b}{b}, \tag{5.2.11}$$

$$\overline{\theta_{11}{}^2} = f_{11}{}^{g\prime\prime}(1) + f_{11}{}^{g\prime}(1) = \frac{d^2}{dz^2}\left[(1-a)z + \frac{abz^2}{1-(1-b)z}\right]_{z=1} + \frac{a+b}{b}$$

$$= \frac{2a}{b^2} + \frac{a+b}{b}, \tag{5.2.12}$$

and

$$\overset{\vee}{\theta}_{11} = \overline{\theta_{11}{}^2} - \bar{\theta}_{11}{}^2 = \frac{a(2-a-b)}{b^2}. \tag{5.2.13}$$

Since $f_{21}{}^{g}(z)$ is just the geometric distribution of the holding time in state 2 found in Section 4.3, we can immediately write down its moments,

$$\bar{\theta}_{21} = \frac{1}{b}, \qquad \overline{\theta_{21}{}^2} = \frac{2-b}{b^2}, \qquad \overset{\vee}{\theta}_{21} = \frac{1-b}{b^2}. \tag{5.2.14}$$

Now we have found the probability mass functions and important moments of the statistics $\theta_{11}$ and $\theta_{21}$ by inspection of the modified flow graph of Figure 5.2.2, and operations on the transmissions thus obtained. The same techniques will allow us to find the statistical properties of $\theta_{22}$ and $\theta_{12}$.

**First passage to state 2.** We first draw the modified flow graph shown in Figure 5.2.3 by the same procedure we used in drawing Figure 5.2.2. We could then find its transmissions to obtain the transforms of the probability distributions of $\theta_{22}$ and $\theta_{12}$, invert these transforms to write the probability distributions and finally differentiate the transforms to produce the moments of $\theta_{22}$ and $\theta_{12}$ if $f_{22}(\cdot)$ and $f_{12}(\cdot)$ are probability mass functions. However, in this case we can avoid these

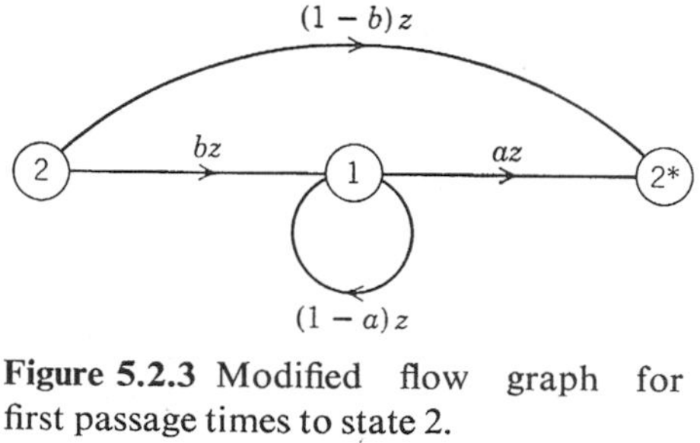


**Figure 5.2.3** Modified flow graph for first passage times to state 2.

steps by noting that the modified flow graphs of Figures 5.2.2 and 5.2.3 are identical if 1 is interchanged with 2 and $a$ is interchanged with $b$. Therefore, we can use the results of analyzing the modified flow graph of Figure 5.2.2 to write the corresponding results for Figure 5.2.3 directly. We obtain

$$f_{22}{}^{g}(z) = \vec{t}_{22\bullet}(z) = (1 - b)z + \frac{abz^2}{1 - (1 - a)z}, \tag{5.2.15}$$

$$f_{12}{}^{g}(z) = \vec{t}_{12\bullet}(z) = \frac{az}{1 - (1 - a)z}, \tag{5.2.16}$$

$$f_{22}(n) = (1 - b)\,\delta(n - 1) + ab(1 - a)^{n-2}u(n - 2) \qquad n = 0, 1, 2, \ldots, \tag{5.2.17}$$

$$f_{12}(n) = a(1 - a)^{n-1}u(n - 1) \qquad n = 0, 1, 2, \ldots, \tag{5.2.18}$$

$$\bar{\theta}_{22} = \frac{a + b}{a}, \qquad \overline{\theta_{22}{}^2} = \frac{2b}{a^2} + \frac{a + b}{a}, \qquad \check{\theta}_{22} = \frac{b(2 - a - b)}{a^2}, \tag{5.2.19}$$

$$\bar{\theta}_{12} = \frac{1}{a}, \qquad \overline{\theta_{12}{}^2} = \frac{2 - a}{a^2}, \qquad \check{\theta}_{12} = \frac{1 - a}{a^2} \tag{5.2.20}$$

The procedure for finding first passage time statistics using flow graphs would be tedious for even small problems if it were really necessary to draw each of the $N$ modified flow graphs required. Fortunately in these problems we can usually compute the transmissions for $f_{ij}{}^{g}(z)$ by making only mental changes in the basic flow graph for the Markov process. This technique would certainly be feasible in the problem we have just worked.

### Matrix Notation

The statistics of the first passage times can be placed in compact form for any problem by using matrix notation. We use square matrices $F(n)$ and $F^{g}(z)$ as the matrices of the probability distributions of first passage times $f_{ij}(n)$ and their transforms $f_{ij}{}^{g}(z)$. The first passage times $\theta_{ij}$ and their moments $\bar{\theta}_{ij}$, $\overline{\theta_{ij}{}^2}$, and $\check{\theta}_{ij}$ are expressed by square matrices $\Theta$, $\bar{\Theta}$, $\overline{\Theta^2}$, and $\check{\Theta}$. Therefore we can write

$$\bar{\Theta} = \sum_{n=0}^{\infty} nF(n) = F^{g\prime}(1) \tag{5.2.21}$$

$$\overline{\Theta^2} = \sum_{n=0}^{\infty} n^2F(n) = F^{g\prime\prime}(1) + F^{g\prime}(1) \tag{5.2.22}$$

and

$$\check{\Theta} = \overline{\Theta^2} - \bar{\Theta} \,\Box\, \bar{\Theta}. \tag{5.2.23}$$

### *General two-state process*

In this matrix notation our findings for the general two-state problem become

$$F^{\mathscr{g}}(z) = \begin{bmatrix} (1-a)z + \dfrac{abz^2}{1-(1-b)z} & \dfrac{az}{1-(1-a)z} \\ \dfrac{bz}{1-(1-b)z} & (1-b)z + \dfrac{abz^2}{1-(1-a)z} \end{bmatrix} \tag{5.2.24}$$

$$F(n) = \begin{bmatrix} (1-a)\,\delta(n-1) + ab(1-b)^{n-2}u(n-2) & a(1-a)^{n-1}u(n-1) \\ b(1-b)^{n-1}u(n-1) & (1-b)\,\delta(n-1) + ab(1-a)^{n-2}u(n-2) \end{bmatrix}$$

$$n = 0, 1, 2, \ldots \tag{5.2.25}$$

$$\bar{\Theta} = \begin{bmatrix} \dfrac{a+b}{b} & \dfrac{1}{a} \\ \dfrac{1}{b} & \dfrac{a+b}{a} \end{bmatrix} \tag{5.2.26}$$

$$\overline{\Theta^2} = \begin{bmatrix} \dfrac{2a}{b^2} + \dfrac{a+b}{b} & \dfrac{2-a}{a^2} \\ \dfrac{2-b}{b^2} & \dfrac{2b}{a^2} + \dfrac{a+b}{a} \end{bmatrix} \tag{5.2.27}$$

and

$$\overset{v}{\Theta} = \begin{bmatrix} \dfrac{a(2-a-b)}{b^2} & \dfrac{1-a}{a^2} \\ \dfrac{1-b}{b^2} & \dfrac{b(2-a-b)}{a^2} \end{bmatrix}. \tag{5.2.28}$$

### *Marketing example*

A numerical example may help in understanding the interpretation to be placed on first passage time statistics. The marketing example of Chapter 1 where customers vacillate between two brands is the special case of the general two-state Markov process where $a = 0.2$, $b = 0.3$. The means and variances of the first passage times for this example are given by Equations 5.2.26 and 5.2.28,

$$\bar{\Theta} = \begin{bmatrix} 5/3 & 5 \\ 10/3 & 5/2 \end{bmatrix} \tag{5.2.29}$$

and

$$\overset{v}{\Theta} = \begin{bmatrix} 10/3 & 20 \\ 70/9 & 45/4 \end{bmatrix}. \tag{5.2.30}$$

This numerical result shows from the statistics on $\theta_{11}$ that a customer who buys brand $A$ will on the average make 5/3 purchases before buying brand $A$ again. The variance to be associated with the number of purchases is 10/3. If a customer buys brand $A$ now, we see from the statistics on $\theta_{12}$ that the mean and variance of the number of purchases until he buys brand $B$ are 5 and 20. We would expect that since the mean recurrence time for state 1, $\bar{\theta}_{11} = 5/3$, is somewhat smaller than the mean recurrence time for state 2, $\bar{\theta}_{22} = 5/2$, the limiting state probability of state 1 should be higher than that of state 2. Indeed, we shall soon show that

$$\pi_i = \frac{1}{\bar{\theta}_{ii}} \qquad i = 1, 2, \ldots, N. \tag{5.2.31}$$

The limiting state probability is just the reciprocal of the mean recurrence time. This result is confirmed by the limiting state probability vector $\boldsymbol{\pi} = [0.6 \quad 0.4]$ for this example.

## Computation of First Passage Time Moments by Transition Diagram Inspection

Since the moments of the first passage time are of such central interest, we would like to develop more direct ways of obtaining them than by finding the transmissions of the modified flow graph and differentiating. The methods we derived for finding the moments of a transient process in Section 4.1 are directly applicable here. We know that the mean time spent in a transient process can be found by inspection of the modified transition diagram and that the second moments can be found from the means.

### *General two-state process*

Let us apply these findings to the first passage time problem of the general two-state process. The modified transition diagram appropriate to the moments of the first passage times to state 1 is just the flow graph of Figure 5.2.2 with $z$ set equal to one. We shall construct this modified transition diagram mentally. The matrix of transmission expressions for this modified transition diagram is then

$$\vec{T} = \begin{bmatrix} \vec{t}_{11} & \vec{t}_{12} \\ \vec{t}_{21} & \vec{t}_{22} \end{bmatrix} = \begin{bmatrix} 1 & \frac{a}{b} \\ 0 & \frac{1}{b} \end{bmatrix}. \tag{5.2.32}$$

In terms of our notation for transient processes we can write

$$\theta_{11} = \nu_1, \qquad \theta_{21} = \nu_2.$$

The first passage times $\theta_{11}$ and $\theta_{21}$ are just the delays in the transient process represented by the flow graph of Figure 5.2.2 under the conditions of starting in state 1 and state 2.

The vector of mean delay in the transient process is given by Equation 4.1.38,

$$\bar{\boldsymbol{\nu}} = \vec{T}\mathbf{s} = \begin{bmatrix} 1 & \frac{a}{b} \\ 0 & \frac{1}{b} \end{bmatrix} \begin{bmatrix} 1 \\ 1 \end{bmatrix} = \begin{bmatrix} \frac{a+b}{b} \\ \frac{1}{b} \end{bmatrix}. \tag{5.2.33}$$

The vector of second moments of delay in the transient process for the two starting states follows from Equation 4.1.70,

$$\overline{\boldsymbol{\nu}^2} = (2\vec{T} - I)\bar{\boldsymbol{\nu}} = \begin{bmatrix} 1 & \frac{2a}{b} \\ 0 & \frac{2}{b} - 1 \end{bmatrix} \begin{bmatrix} \frac{a+b}{b} \\ \frac{1}{b} \end{bmatrix} = \begin{bmatrix} \frac{2a}{b^2} + \frac{a+b}{b} \\ \frac{2-b}{b^2} \end{bmatrix}. \tag{5.2.34}$$

We see immediately that the vectors $\bar{\boldsymbol{\nu}}$ in Equation 5.2.33 and $\overline{\boldsymbol{\nu}^2}$ in Equation 5.2.34 are the first columns of the matrices $\bar{\Theta}$ in Equation 5.2.26 and $\overline{\Theta^2}$ in Equation 5.2.27. We compute the second column by using the symmetry argument advanced earlier. Thus we can find the first two moments of the first passage times by inspection of the modified transition diagram obtained mentally from the flow graph of the process.

But this method for finding the moments of the first passage times, although an improvement over differentiating the transform of the probability mass function, is still not practical for large problems. It would be possible to use the matrix methods of Table 4.1.1 to solve the $N$ transient-process problems implied by the passage time problem, but this approach is cumbersome. We shall return later to the question of first passage time moment computation, but first we shall investigate the problem of finding the probability distribution for a first passage time by recursive computation.

### Recursive Calculation of First Passage Time Probability Distributions

How could a process started in state $i$ at time zero arrive in state $j$ at time $n$ for the first time? If $n = 1$, this could only be accomplished if the process made its first transition to state $j$, an event with probability $p_{ij}$. If $n$ is greater than 1, however, the process must have made its first transition to some state $k$, $k \neq j$, and then have made a first passage from state $k$ to state $j$ in the remaining $n - 1$ transitions. The probability of this event is $p_{ik}f_{kj}(n-1)$, and when summed over all states $k$, $k \neq j$, we have the total probability of all such non-direct first passages. The probability $f_{ij}(n)$ that the first passage from state $i$ to state $j$ will require $n$ transitions is therefore

$$f_{ij}(n) = p_{ij}\,\delta(n-1) + \sum_{\substack{k=1 \\ k \neq j}}^{N} p_{ik}f_{kj}(n-1) \qquad n = 1, 2, 3, \ldots. \tag{5.2.35}$$

This equation provides a recursive relationship for computing the first passage time probability distributions to a state from the same distributions with smaller arguments. Note that $f_{ij}(\cdot)$ is related not only to values of this quantity with smaller arguments, but also to the first passage time probability distributions from all state $k$, $k \neq j$, to state $j$ with smaller arguments.

We note further that Equation 5.2.35 relates the off-diagonal elements of $F(n)$ (the first passage time probability distributions) to each other and does not involve the diagonal elements (the recurrence time probability distributions). Typically we use the equations first to find the first passage time probability distributions and then to write immediately the recurrence time probability distributions from them.

### *General two-state process*

To illustrate the procedure, we apply it to the general two-state process. From Equation 5.2.35 we write $f_{12}(n)$ as

$$\begin{aligned} f_{12}(n) &= p_{12}\,\delta(n-1) + p_{11}f_{12}(n-1) \\ &= a\,\delta(n-1) + (1-a)f_{12}(n-1) \qquad n = 1, 2, 3, \ldots. \end{aligned}$$

Recursive solution of this equation produces Equation 5.2.18. To find $f_{22}(n)$ we apply Equation 5.2.35 again,

$$\begin{aligned} f_{22}(n) &= p_{22}\,\delta(n-1) + p_{21}f_{12}(n-1) \\ &= (1-b)\,\delta(n-1) + bf_{12}(n-1) \qquad n = 1, 2, 3, \ldots. \end{aligned}$$

When we substitute $f_{12}(n)$ from Equation 5.2.18 we immediately obtain Equation 5.2.17. By interchanging state indices and $a$ and $b$, the same computations produce $f_{21}(n)$ and $f_{11}(n)$ as found in Equations 5.2.9 and 5.2.8.

The recursive equations provide a powerful practical method for evaluating first passage time probability distributions.

## Relationship of First Passage Times to Multistep Transition Probabilities

The multistep transition probability matrix of a Markov process $\Phi(n)$ or its transform $\Phi^{g}(z)$ provides fundamental statistics of the process. Suppose we knew them: Could we then find the first passage time statistics? The answer is yes. We shall first write an equation relating the $f_{ij}(\cdot)$'s and the $\phi_{ij}(\cdot)$'s.

How can a process that started in state $i$ at time zero be in state $j$ at time $n$? It could have entered state $j$ for the first time at a time $m \leq n$, and then somehow once more have reached state $j$ after the passage of the remaining time $n - m$. Since reaching state $j$ for the first time at time $m$ is mutually exclusive for different values of $m$, we can write the recurrence equation,

$$\phi_{ij}(n) = \sum_{m=1}^{n} f_{ij}(m)\phi_{jj}(n-m) \qquad n = 1, 2, 3, \ldots . \tag{5.2.36}$$

Although the equation relates $\phi_{ij}(\cdot)$ and $f_{ij}(\cdot)$, it does not serve well as a basis for computing the first passage time probability distributions from the multistep transition probabilities because it appears to require solving a very large set of equations in the $f_{ij}(m)$'s when $n$ is large. However, we can avoid this difficulty by writing Equation 5.2.36 in the form

$$\phi_{ij}(n) = \sum_{m=1}^{n-1} f_{ij}(m)\phi_{jj}(n-m) + f_{ij}(n) \qquad n = 2, 3, 4, \ldots \tag{5.2.37}$$

or

$$f_{ij}(n) = \begin{cases} 0 & n = 0 \\ \phi_{ij}(n) & n = 1 \\ \phi_{ij}(n) - \sum_{m=1}^{n-1} f_{ij}(m)\phi_{jj}(n-m) & n = 2, 3, 4, \ldots . \end{cases} \tag{5.2.38}$$

This recursive equation relates successive $f_{ij}(n)$'s to previously computed values of this quantity with smaller arguments and to $\phi_{ij}(n)$'s and $\phi_{jj}(n)$'s with the same or smaller arguments. Like Equation 5.2.35, it is the type of equation that is readily solved on a computer.

***Marketing example***

We shall demonstrate the use of Equation 5.2.38 to find the probability distribution $f_{12}(\cdot)$ for the first passage time $\theta_{12}$ from state 1 to state 2 in the marketing example of Chapter 1. For this case the equation assumes the form

$$f_{12}(n) = \begin{cases} 0 & n = 0 \\ \phi_{12}(n) & n = 1 \\ \phi_{12}(n) - \sum_{m=1}^{n-1} f_{12}(m)\phi_{22}(n-m) & n = 2, 3, 4, \ldots . \end{cases} \tag{5.2.39}$$

We use the values for $\phi_{12}(\cdot)$ and $\phi_{22}(\cdot)$ that appear in Equation 1.2.16 for the example.

Table 5.2.1 shows the entire calculation. The probabilities of first passage times equal to 0 through 5 are successively computed. It is clear that the first passage time probability distribution can be determined for arbitrarily large arguments by this sequential procedure.

We can verify the result of the computation in Table 5.2.1 by noting that the marketing sample is the general two-state Markov process with $a = 0.2$, $b = 0.3$ in the transition probability matrix of Equation 5.2.5. Consequently, we can

**Table 5.2.1** Computation of $f_{12}(n)$ for Marketing Example

$$f_{12}(n) = \begin{cases} 0 & n = 0 \\ \phi_{12}(1) & n = 1 \\ \phi_{12}(n) - \sum_{m=1}^{n-1} f_{12}(m)\phi_{22}(n-m) & n = 2, 3, 4, \ldots \end{cases}$$

| $n$ | 0 | 1 | 2 | 3 | 4 | 5 |
|---|---|---|---|---|---|---|
| $\phi_{12}(n)$ | 0 | 0.2 | 0.3 | 0.35 | 0.375 | 0.3875 |
| $\phi_{22}(n)$ | 1 | 0.7 | 0.55 | 0.475 | 0.4375 | 0.41875 |
| $f_{12}(n)$ | 0 | 0.2 | 0.16 | 0.128 | 0.1024 | 0.08192 |

$$f_{12}(0) = \underline{0}$$

$$f_{12}(1) = \phi_{12}(1) = \underline{0.2}$$

$$\begin{aligned} f_{12}(2) &= \phi_{12}(2) - f_{12}(1)\phi_{22}(1) \\ &= 0.3 - 0.2(0.7) \\ &= \underline{0.16} \end{aligned}$$

$$\begin{aligned} f_{12}(3) &= \phi_{12}(3) - f_{12}(1)\phi_{22}(2) - f_{12}(2)\phi_{22}(1) \\ &= 0.35 - 0.2(0.55) - 0.16(0.7) \\ &= \underline{0.128} \end{aligned}$$

$$\begin{aligned} f_{12}(4) &= \phi_{12}(4) - f_{12}(1)\phi_{22}(3) - f_{12}(2)\phi_{22}(2) - f_{12}(3)\phi_{22}(1) \\ &= 0.375 - 0.2(0.475) - 0.16(0.55) - 0.128(0.7) \\ &= \underline{0.1024} \end{aligned}$$

$$\begin{aligned} f_{12}(5) &= \phi_{12}(5) - f_{12}(1)\phi_{22}(4) - f_{12}(2)\phi_{22}(3) - f_{12}(3)\phi_{22}(2) - f_{12}(4)\phi_{22}(1) \\ &= 0.3875 - 0.2(0.4375) - 0.16(0.475) - 0.128(0.55) - 0.1024(0.7) \\ &= \underline{0.08192} \end{aligned}$$

specialize the first passage time probability distribution $f_{12}(\cdot)$ computed for the general two-state process in Equation 5.2.18 to this case by taking $a = 0.2$,

$$f_{12}(n) = a(1-a)^{n-1}u(n-1) = 0.2(0.8)^{n-1}u(n-1) \qquad n = 0, 1, 2, 3, \ldots . \tag{5.2.40}$$

Successive substitution of $n = 0, 1, 2, \ldots$ in this equation produces the values of $f_{12}(\cdot)$ that appear in Table 5.2.1.

***Transform domain relationships***

To illustrate the transform relationships between the first passage times and the multistep transition probabilities, we first realize that $f_{ij}(0) = 0$ and $\phi_{ij}(0) = \delta_{ij}$, and then write Equation 5.2.36 for the values $n = 0, 1, 2, \ldots,$

$$\phi_{ij}(n) = \delta_{ij}\,\delta(n) + \sum_{m=0}^{n} f_{ij}(m)\phi_{jj}(n-m) \qquad n = 0, 1, 2, \ldots . \tag{5.2.41}$$

We recognize the summation as a convolution. The transform of this equation is

$$\phi_{ij}^{g}(z) = \delta_{ij} + f_{ij}^{g}(z)\phi_{jj}^{g}(z). \tag{5.2.42}$$

We express the transform of the first passage time probability distributions $f_{ij}^{g}(z)$ in terms of the Markov process transfer functions as

$$f_{ij}^{g}(z) = [\phi_{ij}^{g}(z) - \delta_{ij}]\frac{1}{\phi_{jj}^{g}(z)}. \tag{5.2.43}$$

We can write Equations 5.2.36 and 5.2.43 in matrix form as

$$\Phi(n) = \sum_{m=1}^{n} F(m)[\Phi(n-m) \,\square\, I] \qquad n = 1, 2, 3, \ldots \tag{5.2.44}$$

and

$$F^{g}(z) = [\Phi^{g}(z) - I][\Phi^{g}(z) \,\square\, I]^{-1}. \tag{5.2.45}$$

**Flow graph representations.** The equations we have developed relating the first passage time statistics and the multistep transition probabilities are representative of the relations found for what are called recurrent events. We shall study recurrent events in Chapter 7 in some detail and at that time gain additional insight into the theoretical importance of first passage times. But for the moment we shall simply place the relations we have derived in a form that will make our future work easier. We first note that when $i = j$ Equation 5.2.42 takes the form

$$\phi_{jj}^{g}(z) = 1 + f_{jj}^{g}(z)\phi_{jj}^{g}(z) \tag{5.2.46}$$

or

$$\phi_{jj}^{g}(z) = \frac{1}{1 - f_{jj}^{g}(z)}. \tag{5.2.47}$$

When $i$ is not equal to $j$, we see from Equation 5.2.42 that

$$\phi_{ij}^{g}(z) = f_{ij}^{g}(z)\phi_{jj}^{g}(z) = \frac{f_{ij}^{g}(z)}{1 - f_{jj}^{g}(z)} \qquad i \neq j. \tag{5.2.48}$$

Equations 5.2.47 and 5.2.48 show that the transfer functions of a Markov process are just the transmissions of the flow graphs in Figure 5.2.4 where the branch transmissions are the first passage time transforms of the process. These flow graphs provide not only new insight into the relationship of these quantities, but also a convenient way of remembering the relationship.

**General two-state process.** We can use Equation 5.2.45 to find the transforms of the first passage time probability distributions for the general two-state process. The transfer function matrix $\Phi^{g}(z)$ for this process appears in Equation 1.6.21.

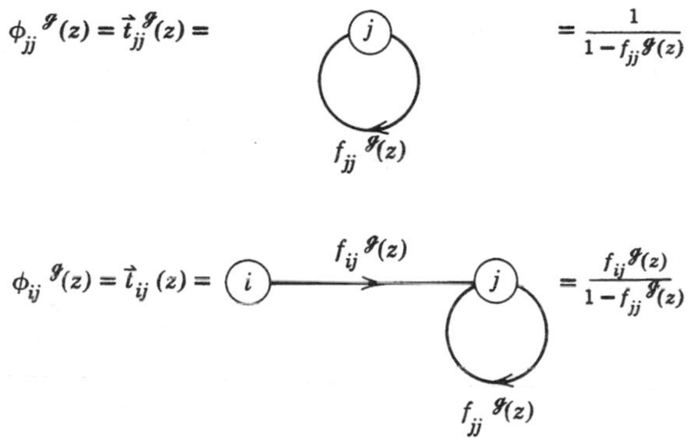


**Figure 5.2.4** Flow graphs relating the geometric transforms of first passage time probability distributions to the transfer functions of the Markov process.

We write

$$F^{\mathcal{g}}(z) = [\Phi^{\mathcal{g}}(z) - I][\Phi^{\mathcal{g}}(z) \,\square\, I]^{-1}$$

$$= \begin{bmatrix} \dfrac{1-(1-b)z}{(1-z)[1-(1-a-b)z]} - 1 & \dfrac{az}{(1-z)[1-(1-a-b)z]} \\ \dfrac{bz}{(1-z)[1-(1-a-b)z]} & \dfrac{1-(1-a)z}{(1-z)[1-(1-a-b)z]} - 1 \end{bmatrix}$$

$$\times \begin{bmatrix} \dfrac{(1-z)[1-(1-a-b)z]}{1-(1-b)z} & 0 \\ 0 & \dfrac{(1-z)[1-(1-a-b)z]}{1-(1-a)z} \end{bmatrix}$$

$$= \begin{bmatrix} 1 - \dfrac{(1-z)[1-(1-a-b)z]}{1-(1-b)z} & \dfrac{az}{1-(1-a)z} \\ \dfrac{bz}{1-(1-b)z} & 1 - \dfrac{(1-z)[1-(1-a-b)z]}{1-(1-a)z} \end{bmatrix}$$

$$= \begin{bmatrix} (1-a)z + \dfrac{abz^2}{1-(1-b)z} & \dfrac{az}{1-(1-a)z} \\ \dfrac{bz}{1-(1-b)z} & (1-b)z + \dfrac{abz^2}{1-(1-a)z} \end{bmatrix} \tag{5.2.49}$$

and obtain the same result found in Equation 5.2.24.

We have now shown how to compute first passage time statistics from the multistep transition probabilities. However, for large problems this is still not the practical method of finding the moments of first passage time distributions that we need. Fortunately, if we are interested only in the first and second moments of the first passage time for a numerical example, we can achieve a further simplification of our methods.

### Relationship of Mean First Passage Times to the Transient Sum Matrix

We shall write an equation for $\bar{\Theta}$ that will produce this matrix directly for a monodesmic process. When $i \neq j$, Equation 5.2.43 shows that

$$f_{ij}{}^{g}(z) = \frac{\phi_{ij}{}^{g}(z)}{\phi_{jj}{}^{g}(z)} \qquad i \neq j. \tag{5.2.50}$$

We know from Equation 1.6.17 that $\phi_{ij}{}^{g}(z)$ for a monodesmic process can be expressed in the form,

$$\phi_{ij}{}^{g}(z) = \frac{\pi_j}{1 - z} + t_{ij}{}^{g}(z), \tag{5.2.51}$$

where $\pi_j$ is the limiting state probability of state $j$ and $t_{ij}{}^{g}(z)$ is the transform of the transient components in $\phi_{ij}(n)$. We substitute Equation 5.2.51 into Equation 5.2.50 to produce

$$f_{ij}{}^{g}(z) = \frac{\dfrac{\pi_j}{1 - z} + t_{ij}{}^{g}(z)}{\dfrac{\pi_j}{1 - z} + t_{jj}{}^{g}(z)} = \frac{\pi_j + (1 - z)t_{ij}{}^{g}(z)}{\pi_j + (1 - z)t_{jj}{}^{g}(z)} \qquad i \neq j. \tag{5.2.52}$$

We shall now restrict our development to the case where state $j$ is a recurrent state of the process. Then we see that $f_{ij}{}^{g}(1) = 1$, and that we can find the moments of $\theta_{ij}$ by differentiating $f_{ij}{}^{g}(z)$ and evaluating the results when $z = 1$. In particular, we can find the mean first passage time $\bar{\theta}_{ij} (i \neq j)$ by differentiating Equation 5.2.52. This procedure yields

$$\bar{\theta}_{ij} = f_{ij}{}^{g\prime}(1) = [t_{jj}{}^{g}(1) - t_{ij}{}^{g}(1)]\frac{1}{\pi_j} \qquad i \neq j. \tag{5.2.53}$$

We have therefore related the mean first passage times for $i \neq j$ to the matrix $T^{g}(1)$ that first assumed prominence in Section 5.1.

When $i$ is equal to $j$ we see from Equation 5.2.42 that

$$f_{jj}{}^{g}(z) = 1 - \frac{1}{\phi_{jj}{}^{g}(z)}. \tag{5.2.54}$$

After using Equation 5.2.51 we have

$$f_{jj}^{g}(z) = 1 - \frac{1 - z}{\pi_j + (1 - z)t_{jj}^{g}(z)}. \tag{5.2.55}$$

***Interpretation of mean recurrence times***

Differentiation with respect to $z$ and evaluation at $z = 1$ produces an expression for the mean recurrence times,

$$\bar{\theta}_{jj} = f_{jj}^{g\prime}(1) = \frac{1}{\pi_j}, \tag{5.2.56}$$

and, of course,

$$\pi_j = \frac{1}{\bar{\theta}_{jj}}, \tag{5.2.57}$$

which establishes the intuitive result mentioned earlier—the mean recurrence time of a state is the reciprocal of its limiting state probability. The reason for restricting this development to recurrent states is now apparent.

***Interpretation of mean first passage times***

Equation 5.2.53 for $\bar{\theta}_{ij}$ where $i \neq j$ also has an intuitive interpretation. We write it in the form,

$$\bar{\theta}_{ij}\pi_j = \frac{\bar{\theta}_{ij}}{\bar{\theta}_{jj}} = t_{jj}^{g}(1) - t_{ij}^{g}(1). \tag{5.2.58}$$

Then we refer to Equation 5.1.59 to develop for large numbers of transitions $n$ the difference in the expected occupancies of state $j$ that results from starting the process in state $j$ rather than in state $i$,

$$\bar{\nu}_{jj}(n) - \bar{\nu}_{ij}(n) = (\phi_{jj} - \phi_{ij})(n + 1) + t_{jj}^{g}(1) - t_{ij}^{g}(1) \qquad \text{for large } n. \tag{5.2.59}$$

For a monodesmic process, where $\phi_{ij} = \phi_{jj}$, we obtain

$$\bar{\nu}_{jj}(\infty) - \bar{\nu}_{ij}(\infty) = t_{jj}^{g}(1) - t_{ij}^{g}(1). \tag{5.2.60}$$

Thus $t_{jj}^{g}(1) - t_{ij}^{g}(1)$ is just the expected difference in occupancies of state $j$ in the long run resulting from starting the system in state $i$ rather than state $j$. Consequently, $\bar{\theta}_{ij}\pi_j$ from Equation 5.2.58 has the same interpretation,

$$\bar{\theta}_{ij}\pi_j = \bar{\nu}_{jj}(\infty) - \bar{\nu}_{ij}(\infty). \tag{5.2.61}$$

The reasonableness of this interpretation is revealed when we consider that $\bar{\theta}_{ij}$ represents the expected number of transitions to pass from state $i$ to state $j$ and that on each of these transitions the expected loss of occupancy of state $j$ is

$$1 - \sum_{i \neq j} \pi_i = \pi_j.$$

### *Matrix form*

We can write Equations 5.2.53 and 5.2.56 in the combined form,

$$\bar{\theta}_{ij} = [\delta_{ij} + t_{jj}{}^{\mathscr{g}}(1) - t_{ij}{}^{\mathscr{g}}(1)]\frac{1}{\pi_j}, \qquad j \text{ a recurrent state}, \tag{5.2.62}$$

or as the matrix equation,

$$\bar{\Theta} = [I + U(T^{\mathscr{g}}(1) \square I) - T^{\mathscr{g}}(1)][\Phi \square I]^{-1}, \tag{5.2.63}$$

where $U$ is the square unity matrix composed entirely of ones.

This equation shows how, for the recurrent states of a monodesmic Markov process, the mean first passage times can be expressed in terms of the limiting state probabilities of the process and the transient sum matrix $T^{\mathscr{g}}(1)$. The computational considerations involved in finding $T^{\mathscr{g}}(1)$ were discussed at length in Section 5.1. Therefore, we have reduced to a manageable problem the computation of the mean first passage times from any state to any recurrent state.

**General two-state process.** We shall find it interesting to compute $\bar{\Theta}$ for the general two-state process using Equation 5.2.63. The matrix $\Phi$ for this example is given by Equation 5.1.62; the matrix $T^{\mathscr{g}}(1)$ by Equation 5.1.125 (or Equation 5.1.64). We write

$$\bar{\Theta} = [I + U(T^{\mathscr{g}}(1) \square I) - T^{\mathscr{g}}(1)][\Phi \square I]^{-1}$$

$$= \left\{ \begin{bmatrix} 1 & 0 \\ 0 & 1 \end{bmatrix} + \begin{bmatrix} \dfrac{a}{(a+b)^2} & \dfrac{b}{(a+b)^2} \\ \dfrac{a}{(a+b)^2} & \dfrac{b}{(a+b)^2} \end{bmatrix} - \begin{bmatrix} \dfrac{a}{(a+b)^2} & \dfrac{-a}{(a+b)^2} \\ \dfrac{-b}{(a+b)^2} & \dfrac{b}{(a+b)^2} \end{bmatrix} \right\} \begin{bmatrix} \dfrac{a+b}{b} & 0 \\ 0 & \dfrac{a+b}{a} \end{bmatrix}$$

$$= \begin{bmatrix} 1 & \dfrac{1}{a+b} \\ \dfrac{1}{a+b} & 1 \end{bmatrix} \begin{bmatrix} \dfrac{a+b}{b} & 0 \\ 0 & \dfrac{a+b}{a} \end{bmatrix}$$

$$= \begin{bmatrix} \dfrac{a+b}{b} & \dfrac{1}{a} \\ \dfrac{1}{b} & \dfrac{a+b}{a} \end{bmatrix} \tag{5.2.64}$$

in agreement with Equation 5.2.26.

**Relationship of First Passage Time Second Moments to Transient Sum Matrix**

We can derive an expression for the direct calculation of the second moments by a procedure analogous to that used for the first moments. To compute $\overline{\theta_{ij}^2}$ when $i \neq j$, we differentiate $f_{ij}^{\mathscr{g}}(z)$ in Equation 5.2.52 twice, evaluate the result when $z = 1$, and add the first moment $\bar{\theta}_{ij}$. This manipulation produces

$$\overline{\theta_{ij}^2} = \bar{\theta}_{ij}\left[\frac{2t_{jj}^{\mathscr{g}}(1)}{\pi_j} + 1\right] + 2[t_{jj}^{\mathscr{g}\prime}(1) - t_{ij}^{\mathscr{g}\prime}(1)]\frac{1}{\pi_j} \qquad i \neq j. \tag{5.2.65}$$

When $i = j$, we do the same thing, but use Equation 5.2.55 for $f_{jj}^{\mathscr{g}}(z)$. In this case we find

$$\overline{\theta_{jj}^2} = \bar{\theta}_{jj}\left[\frac{2t_{jj}^{\mathscr{g}}(1)}{\pi_j} + 1\right]. \tag{5.2.66}$$

Since the factor of Equation 5.2.65 that contains the derivatives of $t_{ij}^{\mathscr{g}}(z)$ will vanish when $i = j$, the form of Equation 5.2.65 can be used for all values of $i$ and $j$ including $i = j$. Then we can write

$$\overline{\theta_{ij}^2} = \bar{\theta}_{ij}\left[\frac{2t_{jj}^{\mathscr{g}}(1)}{\pi_j} + 1\right] + 2[t_{jj}^{\mathscr{g}\prime}(1) - t_{ij}^{\mathscr{g}\prime}(1)]\frac{1}{\pi_j} \qquad j \text{ a recurrent state}. \tag{5.2.67}$$

*Matrix form*

The somewhat elaborate matrix form of this relation is

$$\overline{\Theta^2} = \bar{\Theta}[2(T^{\mathscr{g}}(1) \square I)(\Phi \square I)^{-1} + I] + 2[U(T^{\mathscr{g}\prime}(1) \square I) - T^{\mathscr{g}\prime}(1)][\Phi \square I]^{-1}. \tag{5.2.68}$$

This equation relates the second moments of the first passage times for a recurrent state to the first moments, to the transient sum matrix $T^{\mathscr{g}}(1)$, and to $T^{\mathscr{g}\prime}(1)$. We found in Section 5.1, however, that $T^{\mathscr{g}\prime}(1)$ is easily calculated from $T^{\mathscr{g}}(1)$ according to Equation 5.1.134. Therefore, finding the second moments of the first passage times is a simple additional calculation after the first moments are computed.

**General two-state process.** We shall show that Equation 5.2.68 gives the correct result for the general two-state process. The only new quantity appearing in Equation 5.2.68 is $T^{\mathscr{g}\prime}(1)$. We calculated it from $T^{\mathscr{g}}(1)$ for this process in Equation

5.1.135. Therefore, we can immediately write

$$\overline{\Theta^2} = \begin{bmatrix} \frac{a+b}{b} & \frac{1}{a} \\ \frac{1}{b} & \frac{a+b}{a} \end{bmatrix} \left\{ 2 \begin{bmatrix} \frac{a}{(a+b)^2} & 0 \\ 0 & \frac{b}{(a+b)^2} \end{bmatrix} \begin{bmatrix} \frac{a+b}{b} & 0 \\ 0 & \frac{a+b}{a} \end{bmatrix} + \begin{bmatrix} 1 & 0 \\ 0 & 1 \end{bmatrix} \right\}$$

$$+ 2 \left\{ \begin{bmatrix} \frac{(1-a-b)a}{(a+b)^3} & \frac{(1-a-b)b}{(a+b)^3} \\ \frac{(1-a-b)a}{(a+b)^3} & \frac{(1-a-b)b}{(a+b)^3} \end{bmatrix} - \begin{bmatrix} \frac{(1-a-b)a}{(a+b)^3} & \frac{-(1-a-b)a}{(a+b)^3} \\ \frac{-(1-a-b)b}{(a+b)^3} & \frac{(1-a-b)b}{(a+b)^3} \end{bmatrix} \right\}$$

$$\times \begin{bmatrix} \frac{a+b}{b} & 0 \\ 0 & \frac{a+b}{a} \end{bmatrix}$$

$$= \begin{bmatrix} \frac{a+b}{b} & \frac{1}{a} \\ \frac{1}{b} & \frac{a+b}{a} \end{bmatrix} \begin{bmatrix} \frac{2a}{b(a+b)} + 1 & 0 \\ 0 & \frac{2b}{a(a+b)} + 1 \end{bmatrix}$$

$$+ \begin{bmatrix} 0 & \frac{2(1-a-b)}{(a+b)^2} \\ \frac{2(1-a-b)}{(a+b)^2} & 0 \end{bmatrix} \begin{bmatrix} \frac{a+b}{b} & 0 \\ 0 & \frac{a+b}{a} \end{bmatrix}$$

$$= \begin{bmatrix} \frac{2a}{b^2} + \frac{a+b}{b} & \frac{2b}{a^2(a+b)} + \frac{1}{a} \\ \frac{2a}{b^2(a+b)} + \frac{1}{b} & \frac{2b}{a^2} + \frac{a+b}{a} \end{bmatrix} + \begin{bmatrix} 0 & \frac{2(1-a-b)}{a(a+b)} \\ \frac{2(1-a-b)}{b(a+b)} & 0 \end{bmatrix}$$

$$= \begin{bmatrix} \frac{2a}{b^2} + \frac{a+b}{b} & \frac{2-a}{a^2} \\ \frac{2-b}{b^2} & \frac{2b}{a^2} + \frac{a+b}{a} \end{bmatrix} \tag{5.2.69}$$

which is the expression for $\overline{\Theta^2}$ first found in Equation 5.2.27.

### First Passage Time Moments for the Taxicab Example

We have obtained methods for calculating the mean and second moments of the first passage times from the limiting state probability vector of the process and the matrix $T^{\mathscr{g}}(1)$. We shall now apply these results to the three-state taxicab example of Chapter 1. The transition probability matrix for this example was

$$P = \begin{bmatrix} 0.3 & 0.2 & 0.5 \\ 0.1 & 0.8 & 0.1 \\ 0.4 & 0.4 & 0.2 \end{bmatrix}. \tag{5.2.70}$$

Equation 1.6.42 shows the limiting multistep transition probability matrix $\Phi$,

$$\Phi = \begin{bmatrix} 0.2 & 0.6 & 0.2 \\ 0.2 & 0.6 & 0.2 \\ 0.2 & 0.6 & 0.2 \end{bmatrix}. \tag{5.2.71}$$

Equation 1.6.43 evaluated at $z = 1$ provides $T^{\mathscr{g}}(1)$,

$$T^{\mathscr{g}}(1) = \frac{1}{30}\begin{bmatrix} 33 & -41 & 8 \\ -12 & 24 & -12 \\ 3 & -31 & 28 \end{bmatrix}. \tag{5.2.72}$$

Equation 5.1.122 indicates an alternate way of calculating $T^{\mathscr{g}}(1)$ that is more suitable for large problems,

$$T^{\mathscr{g}}(1) = [I - P + \Phi]^{-1} - \Phi. \tag{5.2.73}$$

In this problem we use the relation (5.2.73) to verify Equation 5.2.72, and it checks.

Now we have all the quantities we need to find the mean first passage times using Equation 5.2.63. We shall find it mentally easier to work with the component form of Equation 5.2.62. We compute

$$\bar{\Theta} = \frac{1}{18}\begin{bmatrix} 90 & 65 & 60 \\ 135 & 30 & 120 \\ 90 & 55 & 90 \end{bmatrix} = \begin{bmatrix} 5.00 & 3.61 & 3.33 \\ 7.50 & 1.67 & 6.67 \\ 5.00 & 3.06 & 5.00 \end{bmatrix}. \tag{5.2.74}$$

Determining the second moments of the first passage times requires calculation of $T^{\mathscr{g}\prime}(1)$. We can either differentiate $T^{\mathscr{g}}(z)$ in Equation 1.6.43 and evaluate the result at $z = 1$, or use Equation 5.1.134. We shall choose the second method and write

$$T^{\mathscr{g}\prime}(1) = T^{\mathscr{g}}(1)[T^{\mathscr{g}}(1) - I] = \frac{1}{900}\begin{bmatrix} 615 & -1355 & 740 \\ -360 & 720 & -360 \\ 465 & -805 & 340 \end{bmatrix}. \tag{5.2.75}$$

We now use Equation 5.2.67 to obtain

$$\overline{\Theta^2} = \frac{1}{54}\begin{bmatrix} 3240 & 716 & 1860 \\ 4860 & 330 & 3720 \\ 3240 & 605 & 2790 \end{bmatrix} + \frac{1}{54}\begin{bmatrix} 0 & 415 & -240 \\ 585 & 0 & 420 \\ 90 & 305 & 0 \end{bmatrix}$$

$$= \frac{1}{54}\begin{bmatrix} 3240 & 1131 & 1620 \\ 5445 & 330 & 4140 \\ 3330 & 910 & 2790 \end{bmatrix}. \tag{5.2.76}$$

We compute the matrix of first passage time variances as

$$\overset{v}{\Theta} = \overline{\Theta^2} - \bar{\Theta} \,\square\, \bar{\Theta} = \frac{1}{324}\begin{bmatrix} 11340 & 2555 & 6120 \\ 14445 & 1080 & 10440 \\ 11880 & 2435 & 8640 \end{bmatrix}$$

$$= \begin{bmatrix} 35.0000 & 7.8858 & 18.8889 \\ 44.5833 & 3.3333 & 32.2222 \\ 36.6667 & 7.5154 & 26.6667 \end{bmatrix}. \tag{5.2.77}$$

The matrix of standard deviations of the first passage times $\overset{s}{\Theta}$ is the matrix whose elements are the square roots of each $\overset{v}{\theta}_{ij}$,

$$\overset{s}{\Theta} = \begin{bmatrix} 5.92 & 2.81 & 4.35 \\ 6.68 & 1.83 & 5.68 \\ 6.06 & 2.74 & 5.16 \end{bmatrix}. \tag{5.2.78}$$

Comparison of Equations 5.2.74 and 5.2.78 for $\bar{\Theta}$ and $\overset{s}{\Theta}$ shows that the $\bar{\theta}_{ij}$ is fairly close to $\overset{s}{\theta}_{ij}$ for every $i$, $j$ pair in this example. We often find this type of numerical result in our problems.

## A Reflexive Relation for Mean First Passage Times

We shall find it useful to write a set of equations that relate the first passage time moments to themselves. These reflexive equations will provide additional insight into the meaning of first passage times and also will provide an important alternative computational procedure.

We begin by writing equations for the mean first passage times based on the same reasoning underlying Equation 5.2.35. The number of transitions required to go from state $i$ to state $j$ will be 1 if the system goes directly from $i$ to $j$ and $1 + \theta_{kj}$ if instead it goes to some state $k \neq j$ on its first transition. That is,

$$\theta_{ij} = \begin{cases} 1 & \text{with probability } p_{ij} \\ 1 + \theta_{kj} & \text{with probability } p_{ik} \quad k \neq j. \end{cases} \tag{5.2.79}$$

Therefore, the mean of $\theta_{ij}$, $\bar{\theta}_{ij}$, is simply

$$\begin{aligned} \bar{\theta}_{ij} &= p_{ij} \cdot 1 + \sum_{\substack{k=1 \\ k \neq j}}^{N} p_{ik} \overline{(1 + \theta_{kj})} \\ &= p_{ij} + \sum_{\substack{k=1 \\ k \neq j}}^{N} p_{ik}(1 + \bar{\theta}_{kj}) \\ &= \sum_{k=1}^{N} p_{ik} + \sum_{\substack{k=1 \\ k \neq j}}^{N} p_{ik} \bar{\theta}_{kj} \\ &= 1 + \sum_{\substack{k=1 \\ k \neq j}}^{N} p_{ik} \bar{\theta}_{kj} \qquad 1 \leq i, j \leq N \end{aligned} \tag{5.2.80}$$

or

$$\bar{\theta}_{ij} = 1 + \sum_{k=1}^{N} p_{ik} \bar{\theta}_{kj} - p_{ij} \bar{\theta}_{jj} \qquad 1 \leq i, j \leq N. \tag{5.2.81}$$

In matrix form this equation becomes

$$\begin{aligned} \bar{\Theta} &= U + P\bar{\Theta} - P(\bar{\Theta} \square I) \\ &= U + P[\bar{\Theta} \square (U - I)]. \end{aligned} \tag{5.2.82}$$

We can establish by matrix manipulations that are more tedious than instructive that the expression for $\bar{\Theta}$ in Equation 5.2.63 does in fact satisfy the reflexive Equation 5.2.82 when $T^{\mathscr{g}}(1)$ is given by Equation 5.1.122.

### *Computational procedure*

We proceed to the calculation of the mean first passage times by noting from Equation 5.2.80 that the mean first passage times for $i \neq j$ are related to each other by the equation

$$\bar{\theta}_{ij} = 1 + \sum_{\substack{k=1 \\ k \neq i}}^{N} p_{ik} \bar{\theta}_{kj} \qquad i \neq j,\ 1 \leq i, j \leq N. \tag{5.2.83}$$

The recurrence times $\bar{\theta}_{jj}$ do not appear in these equations at all—the equations may be solved to find the mean first passage times for $i \neq j$.

For the general two-state example, Equation 5.2.83 implies

$$\begin{aligned} \bar{\theta}_{12} &= 1 + p_{11}\bar{\theta}_{12} \\ \bar{\theta}_{21} &= 1 + p_{22}\bar{\theta}_{21}. \end{aligned} \tag{5.2.84}$$

The solution of these equations is

$$\begin{aligned} \bar{\theta}_{12} &= \frac{1}{1 - p_{11}} = \frac{1}{a} \\ \bar{\theta}_{21} &= \frac{1}{1 - p_{22}} = \frac{1}{b}, \end{aligned} \tag{5.2.85}$$

thus producing the off-diagonal elements of $\bar{\Theta}$ in Equation 5.2.26.

If we know the first passage times for $i \neq j$ and the transition probability matrix we can easily compute the mean recurrence times using Equation 5.2.80 with $i = j$,

$$\bar{\theta}_{jj} = 1 + \sum_{\substack{k=1 \\ k \neq j}}^{N} p_{jk}\bar{\theta}_{kj} \qquad j = 1, 2, \ldots, N. \tag{5.2.86}$$

For the general two-state process we obtain,

$$\begin{aligned} \bar{\theta}_{11} &= 1 + p_{12}\bar{\theta}_{21} = 1 + \frac{a}{b} \\ \bar{\theta}_{22} &= 1 + p_{21}\bar{\theta}_{12} = 1 + \frac{b}{a} \end{aligned} \tag{5.2.87}$$

which are the diagonal elements of $\bar{\Theta}$ in Equation 5.2.26.

Thus a convenient method of calculating a mean first passage time matrix from the transition probability matrix is first to calculate the off-diagonal elements from Equation 5.2.83, and then to use these results plus the transition probability matrix to find the diagonal elements from Equation 5.2.86. Of course, if the limiting state probability vector for the process has already been calculated, then we can skip the second step by capitalizing on the fact that the diagonal elements of $\bar{\Theta}$ are the reciprocals of the limiting state probabilities.

### A Reflexive Relation for First Passage Time Second Moments

Now let us develop a reflexive relation for the calculation of the second moments of the first passage times to the recurrent states of a process. Once more we can use

Equation 5.2.79 to find a relation that these second moments must satisfy. We write

$$\overline{\theta_{ij}^2} = p_{ij}\cdot 1^2 + \sum_{\substack{k=1\\k\neq j}}^{N} p_{ik}\overline{(1+\theta_{kj})^2}$$

$$= p_{ij} + \sum_{\substack{k=1\\k\neq j}}^{N} p_{ik}(1 + 2\bar{\theta}_{kj} + \overline{\theta_{kj}^2})$$

$$= \sum_{k=1}^{N} p_{ik} + 2\sum_{\substack{k=1\\k\neq j}}^{N} p_{ik}\bar{\theta}_{kj} + \sum_{\substack{k=1\\k\neq j}}^{N} p_{ik}\overline{\theta_{kj}^2}$$

$$= 1 + 2\sum_{\substack{k=1\\k\neq j}}^{N} p_{ik}\bar{\theta}_{kj} + \sum_{\substack{k=1\\k\neq j}}^{N} p_{ik}\overline{\theta_{kj}^2} \tag{5.2.88}$$

or by using Equation 5.2.80,

$$\overline{\theta_{ij}^2} = 2\bar{\theta}_{ij} - 1 + \sum_{\substack{k=1\\k\neq j}}^{N} p_{ik}\overline{\theta_{kj}^2}. \tag{5.2.89}$$

The matrix form of Equation 5.2.89 is then

$$\overline{\Theta^2} = 2\bar{\Theta} - U + P[\overline{\Theta^2} \,\square\, (U - I)]. \tag{5.2.90}$$

To prove that $\overline{\Theta^2}$ from Equation 5.2.68 does in fact satisfy this equation is merely an extended exercise in matrix algebra.

*Computational procedure*

The procedure for solving the reflexive equations to obtain the first passage time second moments parallels that for finding the mean first passage times. First, we write the Equations 5.2.89 for $i \neq j$,

$$\overline{\theta_{ij}^2} = 2\bar{\theta}_{ij} - 1 + \sum_{\substack{k=1\\k\neq j}}^{N} p_{ik}\overline{\theta_{kj}^2} \qquad i \neq j; \tag{5.2.91}$$

substitute the previously calculated off-diagonal mean first passage times $\bar{\theta}_{ij}$; and then solve for the off-diagonal first passage time second moments $\overline{\theta_{ij}^2}$.

For the general two-state example, the reflexive equations are

$$\begin{aligned}\overline{\theta_{12}^2} &= 2\bar{\theta}_{12} - 1 + p_{11}\overline{\theta_{12}^2}\\ \overline{\theta_{21}^2} &= 2\bar{\theta}_{21} - 1 + p_{22}\overline{\theta_{21}^2}.\end{aligned} \tag{5.2.92}$$

Solving for $\overline{\theta_{12}{}^2}$ and $\overline{\theta_{21}{}^2}$, using the off-diagonal $\bar{\theta}_{ij}$ from Equation 5.2.85, we find

$$\begin{aligned}\overline{\theta_{12}{}^2} &= \frac{1}{1 - p_{11}}(2\bar{\theta}_{12} - 1) = \frac{1}{a}\left(\frac{2}{a} - 1\right) = \frac{2 - a}{a^2} \\ \overline{\theta_{21}{}^2} &= \frac{1}{1 - p_{22}}(2\bar{\theta}_{21} - 1) = \frac{1}{b}\left(\frac{2}{b} - 1\right) = \frac{2 - b}{b^2}.\end{aligned} \tag{5.2.93}$$

These quantities are just the off-diagonal elements of $\overline{\Theta^2}$ in Equation 5.2.27.

Once we know the off-diagonal elements of $\overline{\Theta^2}$, we compute the diagonal elements from them by using Equation 5.2.89 once again with $i = j$,

$$\overline{\theta_{jj}{}^2} = 2\bar{\theta}_{jj} - 1 + \sum_{\substack{k=1 \\ k \neq j}}^{N} p_{jk}\overline{\theta_{kj}{}^2} \qquad j = 1, 2, \ldots, N. \tag{5.2.94}$$

These equations relate the recurrence time second moments to previously calculated mean recurrence times and off-diagonal first passage time second moments.

For the general two-state process we have

$$\begin{aligned}\overline{\theta_{11}{}^2} &= 2\bar{\theta}_{11} - 1 + p_{12}\overline{\theta_{21}{}^2} \\ \overline{\theta_{22}{}^2} &= 2\bar{\theta}_{22} - 1 + p_{21}\overline{\theta_{12}{}^2}\end{aligned} \tag{5.2.95}$$

Introducing $\bar{\theta}_{11}$ and $\bar{\theta}_{22}$ from Equation 5.2.87, and $\overline{\theta_{21}{}^2}$ and $\overline{\theta_{12}{}^2}$ from Equation 5.2.93 yields

$$\begin{aligned}\overline{\theta_{11}{}^2} &= 2\left(1 + \frac{a}{b}\right) - 1 + a\frac{2 - b}{b^2} = \frac{2a}{b^2} + \frac{a + b}{b} \\ \overline{\theta_{22}{}^2} &= 2\left(1 + \frac{b}{a}\right) - 1 + b\frac{2 - a}{a^2} = \frac{2b}{a^2} + \frac{a + b}{a}\end{aligned} \tag{5.2.96}$$

which are the diagonal elements of $\overline{\Theta^2}$ in Equation 5.2.27.

Thus the process of using the reflexive relations to compute first passage time first and second moments is straightforward and convenient. The reader can readily verify that these procedures when applied to the taxicab problem do in fact produce the moment matrices recorded for that problem in Equations 5.2.74 and 5.2.76.

### Derivation of Transition Probabilities from Mean First Passage Times

Now that we know how to compute $\bar{\Theta}$ (and $\overline{\Theta^2}$) from $P$, we might be interested in the question of whether we can calculate $P$ from $\bar{\Theta}$. Sometimes we can. We proceed by rearranging Equation 5.2.82,

$$\bar{\Theta} = U + P[\bar{\Theta} \Box (U - I)],$$

$$P[\bar{\Theta} \Box (U - I)] = \bar{\Theta} - U,$$

or

$$P = [\bar{\Theta} - U][\bar{\Theta} \Box (U - I)]^{-1}. \tag{5.2.97}$$

The inverse matrix will exist for a monodesmic process with all states recurrent; our results apply only to this case. Thus we can find the transition probability matrix for an all-recurrent-state monodesmic process if we know the mean first passage time matrix.

We confirm Equation 5.2.97 for the general two-state process by writing

$$P = \left[\begin{pmatrix} 1 + \frac{a}{b} & \frac{1}{a} \\ \frac{1}{b} & 1 + \frac{b}{a} \end{pmatrix} - \begin{pmatrix} 1 & 1 \\ 1 & 1 \end{pmatrix}\right]\begin{bmatrix} 0 & \frac{1}{a} \\ \frac{1}{b} & 0 \end{bmatrix}^{-1}$$

$$= \begin{bmatrix} \frac{a}{b} & \frac{1}{a} - 1 \\ \frac{1}{b} - 1 & \frac{b}{a} \end{bmatrix}\begin{bmatrix} 0 & b \\ a & 0 \end{bmatrix} = \begin{bmatrix} 1 - a & a \\ b & 1 - b \end{bmatrix}. \tag{5.2.98}$$

We have obtained the transition probability matrix of Equation 5.2.5, as expected.

We have now seen how to compute the transition probability matrix from the mean first passage time matrix. We know furthermore that the diagonal elements of the first passage time matrix are just the reciprocals of the limiting state probabilities, or

$$\bar{\Theta} \Box I = [\Phi \Box I]^{-1} = [(s\pi) \Box I]^{-1}. \tag{5.2.99}$$

Therefore, if we knew both the limiting state probability vector and the off-diagonal elements of the mean first passage time matrix, we could compute the mean first passage time matrix and hence the transition probability matrix. The somewhat surprising result we shall now show is that the limiting state probability vector $\pi$ can be calculated from the off-diagonal elements of the mean first passage time matrix $\bar{\Theta}$. Consequently, the off-diagonal elements of $\bar{\Theta}$ alone serve to determine both the diagonal elements of $\bar{\Theta}$ and the transition probability matrix.

***The off-diagonal mean first passage times determine the mean recurrence times***

For convenience in the development we shall define

$$D = \Phi \Box I, \qquad D^{-1} = \bar{\Theta} \Box I \tag{5.2.100}$$

and

$${}_0\bar{\Theta} = \bar{\Theta} \Box (U - I) = \bar{\Theta} - \bar{\Theta} \Box I = \bar{\Theta} - D^{-1}. \tag{5.2.101}$$

Thus $D$ is a matrix whose diagonal elements are the limiting state probabilities $\pi_i$ and whose off-diagonal elements are zero. The inverse $D^{-1}$ exists because all states are recurrent; it is a diagonal matrix whose diagonal entries are the mean recurrence times. Finally, the matrix ${}_0\bar{\Theta}$ is just the mean first passage time matrix $\bar{\Theta}$ with zero diagonal entries.

With these definitions, Equation 5.2.97 may be placed in the form

$$\begin{aligned} P &= [{}_0\bar{\Theta} + D^{-1} - U]\,{}_0\bar{\Theta}^{-1} \\ &= I + D^{-1}\,{}_0\bar{\Theta}^{-1} - U\,{}_0\bar{\Theta}^{-1}. \end{aligned} \tag{5.2.102}$$

Now we postmultiply by the summing column vector $\mathbf{s}$, all of whose components are equal to one,

$$P\mathbf{s} = \mathbf{s} = \mathbf{s} + D^{-1}\,{}_0\bar{\Theta}^{-1}\mathbf{s} - U\,{}_0\bar{\Theta}^{-1}\mathbf{s} \tag{5.2.103}$$

or

$$\begin{aligned} U\,{}_0\bar{\Theta}^{-1}\mathbf{s} &= D^{-1}\,{}_0\bar{\Theta}^{-1}\mathbf{s} \\ DU\,{}_0\bar{\Theta}^{-1}\mathbf{s} &= {}_0\bar{\Theta}^{-1}\mathbf{s}. \end{aligned} \tag{5.2.104}$$

We let $\sigma_i$ be the $i$th element of the column vector $\boldsymbol{\sigma} = {}_0\bar{\Theta}^{-1}\mathbf{s}$—it represents the sum of the elements in the $i$th row of ${}_0\bar{\Theta}^{-1}$,

$$\sigma_i = \sum_{\substack{j=1 \\ j\neq i}}^{N} \bar{\theta}_{ij}^{(-1)} = \sum_{j=1}^{N} {}_0\bar{\theta}_{ij}^{(-1)}, \tag{5.2.105}$$

and write Equation 5.2.104 as

$$DU\boldsymbol{\sigma} = \boldsymbol{\sigma}. \tag{5.2.106}$$

If we depict the matrices implied by this equation we find

$$\begin{bmatrix} \pi_1 & \pi_1 & \cdots & \pi_1 \\ \pi_2 & \pi_2 & \cdots & \pi_2 \\ \vdots & & & \\ \pi_N & \pi_N & \cdots & \pi_N \end{bmatrix} \begin{bmatrix} \sigma_1 \\ \sigma_2 \\ \vdots \\ \sigma_N \end{bmatrix} = \begin{bmatrix} \sigma_1 \\ \sigma_2 \\ \vdots \\ \sigma_N \end{bmatrix} \tag{5.2.107}$$

thereby revealing that

$$\pi_i = \frac{\sigma_i}{\sum_{i=1}^{N} \sigma_i}. \tag{5.2.108}$$

The $i$th limiting state probability is therefore proportional to the sum of the elements in the $i$th row of ${}_0\bar{\Theta}^{-1}$. The constant of proportionality is just the reciprocal of the sum of all elements of ${}_0\bar{\Theta}^{-1}$. Since the $i$th mean recurrence time $\bar{\theta}_{ii}$ is the reciprocal of $\pi_i$, we therefore have

$$\bar{\theta}_{ii} = \frac{\sum_{i=1}^{N} \sigma_i}{\sigma_i} = \frac{\sum_{i=1}^{N}\sum_{j=1}^{N} {}_0\bar{\theta}_{ij}^{(-1)}}{\sum_{j=1}^{N} {}_0\bar{\theta}_{ij}^{(-1)}}. \tag{5.2.109}$$

The mean recurrence time $\bar{\theta}_{ii}$ is the ratio of the sum of all elements of ${}_0\bar{\Theta}^{-1}$ to the sum of the elements in the $i$th row.

**The general two-state process.** To demonstrate these computations for the general two-state process, we write the off-diagonal mean first passage time matrix from Equation 5.2.26

$$ {}_0\bar{\Theta} = \begin{bmatrix} 0 & \frac{1}{a} \\ \frac{1}{b} & 0 \end{bmatrix} \tag{5.2.110} $$

and note that its inverse

$$ {}_0\bar{\Theta}^{-1} = \begin{bmatrix} 0 & b \\ a & 0 \end{bmatrix} \tag{5.2.111} $$

implies via Equation 5.2.109 that

$$ \bar{\Theta} \square I = \begin{bmatrix} \frac{a+b}{b} & 0 \\ 0 & \frac{a+b}{a} \end{bmatrix}. \tag{5.2.112} $$

Therefore,

$$ \bar{\Theta} = {}_0\bar{\Theta} + \bar{\Theta} \sqcup I = \begin{bmatrix} \frac{a+b}{b} & \frac{1}{a} \\ \frac{1}{b} & \frac{a+b}{a} \end{bmatrix} \tag{5.2.113} $$

in accordance with Equation 5.2.26.

**The taxicab example.** To illustrate the same procedure for the taxicab problem whose $\bar{\Theta}$ appears in Equation 5.2.74, we write

$$ {}_0\bar{\Theta} = \frac{1}{18} \begin{bmatrix} 0 & 65 & 60 \\ 135 & 0 & 120 \\ 90 & 55 & 0 \end{bmatrix} \tag{5.2.114} $$

and compute the inverse

$$ {}_0\bar{\Theta}^{-1} = \frac{1}{850} \begin{bmatrix} -88 & 44 & 104 \\ 144 & -72 & 108 \\ 99 & 78 & -117 \end{bmatrix}. \tag{5.2.115} $$

Then we find the row sums of the elements of the inverse,

$$\sigma_1 = {}_0\bar{\theta}_{11}^{(-1)} + {}_0\bar{\theta}_{12}^{(-1)} + {}_0\bar{\theta}_{13}^{(-1)} = \frac{60}{850}$$

$$\sigma_2 = {}_0\bar{\theta}_{21}^{(-1)} + {}_0\bar{\theta}_{22}^{(-1)} + {}_0\bar{\theta}_{23}^{(-1)} = \frac{180}{850} \tag{5.2.116}$$

$$\sigma_3 = {}_0\bar{\theta}_{31}^{(-1)} + {}_0\bar{\theta}_{32}^{(-1)} + {}_0\bar{\theta}_{33}^{(-1)} = \frac{60}{850}$$

and write

$$\bar{\theta}_{11} = \frac{\sigma_1 + \sigma_2 + \sigma_3}{\sigma_1} = \frac{300}{60} = 5$$

$$\bar{\theta}_{22} = \frac{\sigma_1 + \sigma_2 + \sigma_3}{\sigma_2} = \frac{300}{180} = \frac{5}{3} \tag{5.2.117}$$

$$\bar{\theta}_{33} = \frac{\sigma_1 + \sigma_2 + \sigma_3}{\sigma_3} = \frac{300}{60} = 5.$$

Thus,

$$\bar{\Theta} \,\square\, I = \begin{bmatrix} 5 & 0 & 0 \\ 0 & 5/3 & 0 \\ 0 & 0 & 5 \end{bmatrix} \tag{5.2.118}$$

and we have found the missing diagonal elements of $\bar{\Theta}$.

***Computation of transition probability matrix for the taxicab example***

Now that we know how to compute the entire mean first passage time matrix from its off-diagonal elements, we see that we can compute the transition probability matrix from the same fundamental information. To demonstrate this for the taxicab problem, we suppose that we start with ${}_0\bar{\Theta}$ in Equation 5.2.114, compute $\bar{\Theta} \,\square\, I$ as shown in Equation 5.2.118, and then add the two to obtain $\bar{\Theta}$,

$$\bar{\Theta} = \frac{1}{18}\begin{bmatrix} 90 & 65 & 60 \\ 135 & 30 & 120 \\ 90 & 55 & 90 \end{bmatrix}. \tag{5.2.119}$$

Now the transition probability matrix can be computed from Equation 5.2.97,

$$P = [\bar{\Theta} - U][\bar{\Theta} \,\square\, (U - I)]^{-1} = [\bar{\Theta} - U]\,{}_0\bar{\Theta}^{-1}, \tag{5.2.120}$$

where we can obtain ${}_0\bar{\Theta}^{-1}$ from Equation 5.2.115. We find

$$
\begin{aligned}
P &= [\bar{\Theta} - U]\,{}_0\bar{\Theta}^{-1} \\
&= \left(\frac{1}{18}\begin{bmatrix} 90 & 65 & 60 \\ 135 & 30 & 120 \\ 90 & 55 & 90 \end{bmatrix} - \begin{bmatrix} 1 & 1 & 1 \\ 1 & 1 & 1 \\ 1 & 1 & 1 \end{bmatrix}\right)\frac{1}{850}\begin{bmatrix} -88 & 44 & 104 \\ 144 & -72 & 108 \\ 99 & 78 & -117 \end{bmatrix} \\
&= \frac{1}{18(850)}\begin{bmatrix} 72 & 47 & 42 \\ 117 & 12 & 102 \\ 72 & 37 & 72 \end{bmatrix}\begin{bmatrix} -88 & 44 & 104 \\ 144 & -72 & 108 \\ 99 & 78 & -117 \end{bmatrix} \\
&= \frac{1}{15300}\begin{bmatrix} 4590 & 3060 & 7650 \\ 1530 & 12240 & 1530 \\ 6120 & 6120 & 3060 \end{bmatrix} \\
&= \begin{bmatrix} 0.3 & 0.2 & 0.5 \\ 0.1 & 0.8 & 0.1 \\ 0.4 & 0.4 & 0.2 \end{bmatrix}
\end{aligned}
\tag{5.2.121}
$$

which is, of course, the transition probability matrix for the taxicab problem.

At first it might seem surprising that the off-diagonal mean first passage times are supposed to determine the transition probability matrix of a Markov process; on further consideration, however, we see that it is not surprising at all. Recall that in a transition probability matrix we can specify only $N(N - 1)$ of the transition probabilities for an $N$-state process independently because of the requirement that the rows of the transition probability matrix sum to one. We must also specify $N(N - 1)$ numbers to write the off-diagonal mean first passage time matrix. It would be strange indeed if all $N^2$ entries of the entire mean first passage time matrix could be selected independently.

We have another point to note here. We have seen that the off-diagonal mean first passage times determine the diagonal mean first passage times and, in turn, that all mean first passage times determine the transition probability matrix. But, as we also know, the transition probability matrix determines all characteristics of the process, including the first passage time second moments. Consequently, the off-diagonal mean first passage times determine the first passage time second moments; ${}_0\Theta$ implies $\bar{\Theta}^2$ and hence the variance matrix $\check{\Theta}$. Therefore the means and variances of first passage times cannot be specified independently: The means determine the variances.

We see then that the off-diagonal mean first passage times are an adequate and useful alternate description of a Markov process. In some processes we find it much easier to assign mean first passage times than we do to assign transition probabilities.

**Relationship of Recurrence Time Moments to Asymptotic Occupancy Moments**

The recurrence time moments have a central importance in themselves. The moments of the recurrence times were found in Equations 5.2.56 and 5.2.66 to be

$$\bar{\theta}_{jj} = \frac{1}{\pi_j} \tag{5.2.122}$$

and

$$\overline{\theta_{jj}^{\,2}} = \bar{\theta}_{jj}\left[\frac{2t_{jj}{}^{g}(1)}{\pi_j} + 1\right] = \frac{1}{\pi_j^{\,2}}\,[2t_{jj}{}^{g}(1) + \pi_j]. \tag{5.2.123}$$

The variance of the recurrence time for state $j$ is then

$$\overset{\vee}{\theta}_{jj} = \overline{\theta_{jj}^{\,2}} - \bar{\theta}_{jj}^{\,2} = \frac{1}{\pi_j^{\,2}}\,[2t_{jj}{}^{g}(1) + \pi_j - 1]. \tag{5.2.124}$$

By referring to Equations 5.2.71 and 5.2.72 that show $\Phi$ and $T^{g}(1)$ for the taxicab problem, we confirm directly the recurrence time variances that form the diagonal elements of $\overset{\vee}{\Theta}$ found in Equation 5.2.77 for this example,

$$\begin{aligned}
\overset{\vee}{\theta}_{11} &= \frac{1}{\pi_1^{\,2}}\,[2t_{11}{}^{g}(1) + \pi_1 - 1] = \frac{1}{(0.2)^2}\left[2\left(\frac{33}{30}\right) + 0.2 - 1\right] = 35 \\
\overset{\vee}{\theta}_{22} &= \frac{1}{\pi_2^{\,2}}\,[2t_{22}{}^{g}(1) + \pi_2 - 1] = \frac{1}{(0.6)^2}\left[2\left(\frac{24}{30}\right) + 0.6 - 1\right] = 3.3333 \\
\overset{\vee}{\theta}_{33} &= \frac{1}{\pi_3^{\,2}}\,[2t_{33}{}^{g}(1) + \pi_3 - 1] = \frac{1}{(0.2)^2}\left[2\left(\frac{28}{30}\right) + 0.2 - 1\right] = 26.6667
\end{aligned} \tag{5.2.125}$$

You will recall that we first encounterd the matrix $T^{g}(1)$ in the development of asymptotic state occupancy statistics in Section 5.1. Therefore, it is not surprising to find a relationship between first passage times and limiting occupancy statistics.

***Asymptotic growth of mean state occupancies***

From Equation 5.1.60 we see that for a monodesmic process

$$\lim_{n\to\infty} \frac{1}{n}\,\bar{\nu}_{ij}(n) = \pi_j, \tag{5.2.126}$$

and then from Equation 5.2.122,

$$\lim_{n\to\infty} \frac{1}{n}\,\bar{\nu}_{ij}(n) = \frac{1}{\bar{\theta}_{jj}}. \tag{5.2.127}$$

This equation shows that the expected number of times state $j$ is occupied in $n$ transitions grows linearly with $n$ for large $n$ at a rate that is inversely proportional to the mean recurrence time of state $j$; it is, of course, independent of the starting state $i$.

### *Asymptotic growth of variance of state occupancies*

We can also relate the growth of variance of the state occupancies to the first passage time statistics. From Equation 5.1.83,

$$\lim_{n \to \infty} \frac{1}{n} \overset{\vee}{\nu}_{ij}(n) = \pi_j[\pi_j + 2t_{jj}{}^{g}(1) - 1]. \tag{5.2.128}$$

However, by using Equations 5.2.122 and 5.2.124 we can write

$$\lim_{n \to \infty} \frac{1}{n} \overset{\vee}{\nu}_{ij}(n) = \frac{\overset{\vee}{\theta}_{jj}}{\bar{\theta}_{jj}{}^{3}}. \tag{5.2.129}$$

The variance of the number of times state $j$ is occupied in $n$ transitions again grows linearly with $n$ for large $n$ at a rate equal to the ratio of the variance of the recurrence time of state $j$ to the cube of its mean recurrence time. The results of Equations 5.2.26 and 5.2.28 for the general two-state process show that this rate of growth of the occupancy variance is the same for both states and is equal to $ab(2 - a - b)/(a + b)^3$. This observation is consistent with Equation 5.1.85.

### *The taxicab example*

The asymptotic linear growth rates of occupancy means and variances for the taxicab problem appear in Table 5.2.2. In computing the linear growth rates of occupancy variances, we have used the recurrence time variances recorded in Equation 5.2.125.

These results are easily understood if we consider the case where the taxi has made 1000 trips, a case that also appears in the table. Neglecting the transient components of occupancy because of the large number of transitions, we see that we expect 200 trips to each of towns 1 and 3, and 600 trips to town 2. The variance of the number of trips to towns 1, 2, and 3 is, respectively, 280, 720, and 213 with corresponding standard deviations 16.7, 26.8, and 14.6. Finally, the standard deviation as a percentage of the expected trips to each town is 8.4% for town 1, 4.5% for town 2, and 7.3% for town 3. Consequently, we observe that there is about half the percentage fluctuation in the number of trips to town 2 that we find for the other two towns.

The lower part of the table shows the same computations if we model the taxicab operation by its multinomial projection, a multinomial process with constant probabilities $\pi_1 = 0.2$, $\pi_2 = 0.6$, $\pi_3 = 0.2$ of making a transition to each town. The computation of the variance growth rate for this case follows from Equation 5.1.99. We observe that the variances and standard deviations of trips to each town are consistently smaller for the multinomial projection than they are for the original Markov process. This behavior is consistent with our previous observation that the dependencies implicit in a Markov process have a strong effect on occupancy variances.

**Table 5.2.2** Asymptotic Linear Growth Rates of Occupancy Means and Variances for Taxicab Problem

| | $j =$ | 1 | 2 | 3 |
|---|---|---|---|---|
| $\lim\limits_{n \to \infty} \dfrac{1}{n} \bar{\nu}_{ij}(n) = \pi_j = \dfrac{1}{\bar{\theta}_{jj}}$ | | 0.2 | 0.6 | 0.2 |
| $\lim\limits_{n \to \infty} \dfrac{1}{n} \overset{\text{v}}{\nu}_{ij}(n) = \pi_j{}^3 \overset{\text{v}}{\theta}_{jj} = \dfrac{\overset{\text{v}}{\theta}_{jj}}{\bar{\theta}_{jj}{}^3}$ | | 0.28 | 0.72 | 0.21333 |

*In 1000 transitions*

| | | | |
|---|---|---|---|
| Expected occupancies | 200 | 600 | 200 |
| Variance of occupancies | 280 | 720 | 213 |
| Standard deviation of occupancies | 16.7 | 26.8 | 14.6 |
| Standard deviation as percentage of expected occupancies | 8.4% | 4.5% | 7.3% |

*Under assumption of independence*

| | | | |
|---|---|---|---|
| Expected occupancies | 200 | 600 | 200 |
| Variance of occupancies $1000\,\pi_j(1 - \pi_j)$ | 160 | 240 | 160 |
| Standard deviation of occupancies | 12.7 | 15.5 | 12.7 |
| Standard deviation as percentage of expected occupancies | 6.4% | 2.6% | 6.4% |

It is, of course, possible for a Markov process with dependencies to have occupancy variances that are lower than those for the equivalent multinomial process. Consider any periodic process, which is by its nature a case of extreme dependency of successive transitions. For such a process the recurrence time variances are always zero and in view of Equation 5.2.129, the linear growth rate of occupancy variances is zero. This fact is consistent with our intuition since there is no uncertainty in the state occupancies of a periodic process.

The relationships between the limiting occupancy statistics and the recurrence time statistics are simple and useful. They are representative of the many interconnections that exist among Markov process statistics.

## PROBLEMS

**1.** a) Joe and Sam are two drivers for the Run 'em Down Taxi Company of Problem 8, Chapter 1. If Joe starts in town $A$ and Sam starts in town $B$, and if they have both made the *same* large number of trips, how many more trips will Joe expect to make to town $A$ than Sam? [*Note:* Interpret a trip to town $A$ as one that terminates in town $A$.]

b) If a driver starts the day in town $A$ and makes $n$ trips, what is the mean and variance of the number of trips the driver will make to town $B$?

c) What are the asymptotic values for the quantities in part b)? Check their values by evaluating $T^{\mathscr{g}}(1)$ and $T^{\mathscr{g}\prime}(1)$ and using the relations in Section 5.1.

**2.** A certain machine has two critical bolts, either of which is sufficient to keep it running. If at the beginning of a week both bolts are intact, then at the end of the week exactly one will be intact with probability 0.3; neither will be intact with probability 0.4. However, if only one bolt is operative at the beginning of a week, the probability that it will still be intact at the end of the week is 0.4.

When both bolts fail, the machine is mangled. A repairman comes once a week and determines whether the machine is mangled. If not, he leaves; if so, it takes him a week to fix it.

a) If the machine starts in perfect order, how many repairs do we expect will have been performed up to and including the end of the $n$th week?

b) What fraction of the weeks is spent in repair (in the long run)?

c) While the machine is running, it yields a profit of \$5000/wk. Repairing a mangled machine costs \$9000. Suppose the repairman could also look to see how many bolts were intact, and fix the machine if only one were faulty at a cost of one week's time and \$1000. Would this be better? How much per week should we be willing to pay the repairman to perform the inspection for us?

**3.** Colonel Cathcart scores his efforts to be promoted as either feathers in his cap or black eyes. The way he sees it, his probability of earning another feather, given that he has just gotten one, is 0.6, while if he has just suffered a black eye the probability drops to 0.1.

a) If Cathcart's probability of starting his day with a black eye is 0.7, what are the mean and variance of the number of feathers he will earn in $n$ tries?

b) Given that Cathcart earns his first feather about lunchtime ($n = 3$) and that he finally gives up in despair with a black eye at time $m$ ($m > 3$), what is his mean number of feathers for the day?

**4.** It is known that a certain discrete-time, time-invariant Markov process has started in state $i$ and after $n$ transitions is currently in state $k$. Find an expression for $\bar{\nu}_{ij|k}(n)$, the expected number of times the process has entered state $j$ during these $n$ transitions. Express your answer in terms of the multistep transition probabilities.

**5.** Let $\nu_{ij|k}(m, n)$ be the number of times that an $N$-state monodesmic Markov process enters the $j$th state in the first $m$ transitions if it is known that the process started in the $i$th state, and was in the $k$th state $n$ transitions later. Assume that $n \geq m$ and define $\nu_{ij|k}(m, n) = 0$ for $m > n$. A pictorial explanation:

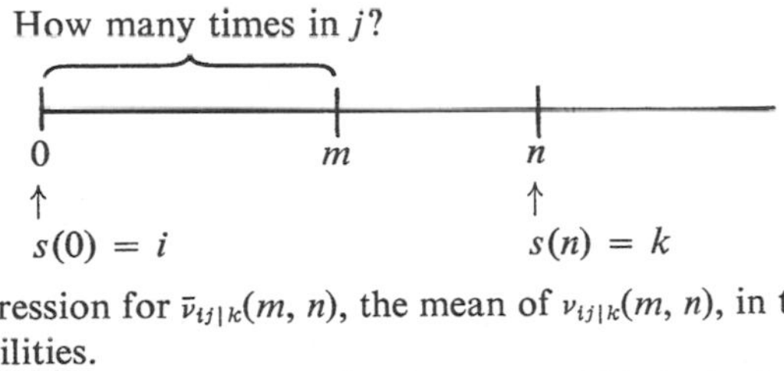


a) Find an expression for $\bar{\nu}_{ij|k}(m, n)$, the mean of $\nu_{ij|k}(m, n)$, in terms of the multistep transition probabilities.

b) Find an expression for $\bar{\nu}_{\sim j|k}(m, n)$ when probabilistic (rather than deterministic) starting is assumed, i.e., when $\pi_i(0)$ is known for all $1 \leq i \leq N$.

c) If $\pi_i(0) = \pi_i$, the limiting state probability for all $1 \le i \le N$, find the two-dimensional geometric transform of $\bar{\nu}_{\sim j|k}(m, n)$, $\bar{\nu}^{gg}_{\sim j|k}(y, z) = \sum_m \sum_n \bar{\nu}_{\sim j|k}(m, n) y^m z^n$.

**6.** Contemplate an $N$-state monodesmic Markov process with no transient states. Let $g_{ij}(k|n)$ be the probability that state $j$ is entered $k$ times in $n$ transitions if the process is started in state $i$, i.e., $g_{ij}(k|n) = \mathscr{P}\{\nu_{ij}(n) - k\}$.

a) If $f_{ij}{}^g(z)$ is the transform of the probability mass function for the first passage time from $i$ to $j$, find the geometric transform of $g_{ij}(0|n)$.

b) Find the geometric transform of $g_{ij}(k|n)$ with respect to $n$ in terms of $f_{ij}{}^g(z)$ and $f_{jj}{}^g(z)$. Do this for $i = j$, and then $i \ne j$.

c) Verify that the results of parts a) and b) are consistent with the fact that

$$\sum_{k=0}^{\infty} g_{ij}(k|n) = 1.$$

d) Use the results of parts a) and b) to find the transform of $\bar{\nu}_{ij}(n)$, the expected number of times that state $j$ is entered in $n$ transitions if the process is started in $i$. Do this in terms of $f_{ij}{}^g(z)$ and $f_{jj}{}^g(z)$ for the cases $i = j$, $i \ne j$.

e) Verify that your results are consistent with

$$\bar{\nu}_{ij}{}^g(z) = \frac{1}{1 - z} \phi_{ij}{}^g(z)$$

for the cases $i = j$, $i \ne j$.

**7.** The Drive 'n Holler Taxi Company has found that its taxicab drivers operate between towns $A$ and $B$ in such a way that the probability of going to $B$ on their next trip if they are in $A$ is 0.3 and the probability of going to $A$ if they are in $B$ is 0.6.

a) Drivers M. I. Loud and E. Z. Horn both start their days in town $B$. If they have both made twenty trips, find the expected difference of the number of trips they have made to $B$. (Include $B$ to $B$ trips. You may find an approximate answer if you desire.)

b) What is the probability mass function for the number of trips that driver Loud must take before his first arrival in $A$?

c) What are the mean and variance of the probability mass function in part b)? Check your answers by calculating at least two different ways.

**8.** If the Mark V computer in Problem 6 of Chapter 3 has just been turned on, what is the expected number of hours until the maintenance crew first find the wonderful Mark off? What is the variance?

**9.** The U-Rent-It Company has hired an analyst to describe the way its trailers move among three towns. The analyst has modeled the transitions of the trailers with a Markov process whose transition probability matrix is:

$$P = \text{From} \begin{matrix} \\ 1 \\ 2 \\ 3 \end{matrix} \overset{\textstyle \text{To}}{\begin{matrix} 1 & 2 & 3 \\ \end{matrix}} \begin{bmatrix} 1/4 & 1/4 & 1/2 \\ 1/3 & 1/3 & 1/3 \\ 1/2 & 0 & 1/2 \end{bmatrix}.$$

a) What is the probability that a trailer arrives in town 2 for the first time on its $n$th rental given that it started in town 1?

b) What is the probability that it will arrive back in town 1 for the first time $n$ rentals after leaving town 1?

c) Find $\bar{\Theta}$, the matrix of mean first passage times.

d) Find $\overline{\Theta^2}$, the matrix of mean square first passage times.

**10.** An optimistic meteorologist has found that the following Markov process transition probability matrix describes the behavior of the daily weather:

| | | *Today* | | | |
|---|---|---|---|---|---|
| | | Sunny | Cloudy | Foggy | Rainy |
| *Yesterday* | Sunny | 0.6 | 0.3 | 0.05 | 0.05 |
| | Cloudy | 0.2 | 0.5 | 0.1 | 0.2 |
| | Foggy | 0.4 | 0.2 | 0.3 | 0.1 |
| | Rainy | 0.2 | 0.2 | 0.1 | 0.5 |

where we have assumed that the states of the weather are mutually exclusive. If it is sunny today, what is the probability that the first wet weather (foggy or rainy) will occur $n$ days from today? Find the expected number of days until the next wet weather. What is the variance?

**11.** The following matrix of off-diagonal mean first passage times has been measured for a three-state Markov process,

$$\bar{\Theta} = \begin{bmatrix} - & 5 & 14/5 \\ 2 & - & 24/5 \\ 7/4 & 5 & - \end{bmatrix}$$

a) Find the transition probability matrix $P$ for this process.

b) Using $P$, find the probability distribution for the number of transitions $\theta_{12}$ for first passage from state 1 to state 2. Check to be sure that $\bar{\theta}_{12} = 5$.

**12.** In a three-state Markov process, the $\bar{\theta}_{ij}$ $(i \neq j)$ are given by the matrix:

$$\bar{\Theta} = \begin{bmatrix} - & 11/5 & 18/7 \\ 3 & - & 13/7 \\ 3 & 13/5 & - \end{bmatrix},$$

where $\bar{\theta}_{ij}$ = expected number of transitions to enter state $j$ for the first time given $s(0) = i$.

a) Find the limiting state probability vector $\pi$.

b) Find the complete $\bar{\Theta}$.

c) With the help of the result of part b), find the transition probability matrix $P$.

**13.** For the Markov process whose transition probability matrix is

$$P = \begin{bmatrix} 1/3 & 2/3 \\ 3/4 & 1/4 \end{bmatrix}:$$

a) Find the matrix of first passage time probability mass functions $F(n)$.

b) Determine $\bar{\theta}_{11}$ and $\bar{\theta}_{22}$ and check that $\bar{\theta}_{11} = 1/\pi_1$ and $\bar{\theta}_{22} = 1/\pi_2$.

c) Without resorting to the use of $\bar{\theta}_{11} = 1/\pi_1$, justify why $\bar{\theta}_{11} = 2$ for both of the following processes although they have quite different transition probability matrices. Will the variances of $\theta_{11}$ be the same for the two processes? Avoid lengthy calculations.

$$_1P = \begin{bmatrix} 0.01 & 0.99 \\ 0.99 & 0.01 \end{bmatrix} \qquad _2P = \begin{bmatrix} 0.5 & 0.5 \\ 0.5 & 0.5 \end{bmatrix}$$

**14.** An $N$-state monodesmic Markov process has no transient states.

a) Let $f_{ij}(n)$ be the probability that the first passage from $i$ to $j$ will take $n$ transitions. By expanding in terms of the first transition of the process, write a difference equation for $f_{ij}(n)$ in terms of the elements of $F(\cdot)$ and $P$. Take the transform of this set of difference equations.

b) Explain how a flow graph for this set of transform equations should be constructed so that the transmission to the $i$th state is $f_{ij}^{\mathcal{g}}(z)$. Illustrate with the process whose transition probability matrix is:

$$P = \begin{bmatrix} 0.6 & 0.4 \\ 0.1 & 0.9 \end{bmatrix}.$$

c) Explain how this flow graph may be constructed by modification of the standard flow graph for a Markov process.

d) How should the flow graph of part b) be modified so that the transmission to the $i$th state is $\bar{\theta}_{ij}$? Illustrate with the example.

e) How should the flow graph of part d) be changed so that the transmission to the $i$th state is $\overline{\theta_{ij}^2}$? (It is necessary in this part to have some branch transmissions a function of the $\bar{\theta}_{ij}$'s found in part $d$).) Illustrate with the example.

**15.** Sam has decided to model the world as a monodesmic Markov process. Unfortunately Sam has only been able to identify one state of the process but he has taken extensive statistics on the number of days it takes for the world to return to this state once it has entered it. He has concluded that the recurrence time probability mass function for this state is:

$$f_{11}(n) = \begin{cases} 0.7 & \text{for } n = 1 \\ 0.09(0.7)^{n-2} & \text{for } n \geq 2. \end{cases}$$

a) Find the long-run fraction of time Sam can expect the world to spend in the state.

b) If the world is in the state now, what is the probability it will be in the state $n$ days from now?

c) What can Sam say about how the mean and variance of the number of days the world will spend in the state will grow with time?

**16.** A consulting firm has been asked by a beer company to model the beer market. This company is planning a test advertising campaign, and they want a predictive model to measure the results of their campaign against what *would have happened* in the absence of such a campaign. It is agreed that the market can be supposed to consist of only three manufacturers; call them $A$, $B$, and $C$. The company requesting the analysis has supplied the data that follow, and suggests that a Markov model would be convenient. On the basis of the data, is such a model justified? If so, construct a reasonable one. Is there any other data you would like to see before making a final judgment? In qualitative terms, how important are the extra data you want?

a) Recently, 500 random customers were asked, "What brand did you purchase the last time you bought beer?" Unfortunately, in this survey the customers buying $A$ or $C$ (at the time of the questioning) who bought $B$ last time were not distinguished from each other. (*Hint:* It may be possible to infer from pieces of data b)–f) how these must be split.) The results were:

$$\begin{array}{c} \textit{This Purchase} \\ \textit{Previous purchase} \begin{array}{c} \\ A \\ B \\ C \end{array} \begin{array}{c} \begin{array}{ccc} A & B & C \end{array} \\ \begin{bmatrix} 56 & 0 & 24 \\ ? & 40 & ? \\ 0 & 64 & 256 \end{bmatrix} \end{array} . \end{array}$$

b) Sales data for the entire past year (*many* more customers than the 500 sampled in data a) were:

$A$: 5,325,200 bottles
$B$: 5,324,100 bottles
$C$: 15,974,920 bottles.

c) A short time ago, 60 current purchasers of brand $A$ were followed for their next 5 purchases. These purchases consisted of 115 $A$'s, 40 $B$'s, and 145 $C$'s.

d) At the same time, 146 current brand $C$ customers were followed to see how many purchases they made before again purchasing brand $C$. The average number of such purchases was 1.65.

e) Several months ago a group of 136 customers was followed for *two* transitions. Fractional results were:

$$\begin{array}{c} \textit{Second Purchase} \\ \textit{Initial Purchase} \begin{array}{c} \\ A \\ B \\ C \end{array} \begin{array}{c} \begin{array}{ccc} A & B & C \end{array} \\ \begin{bmatrix} 0.50 & 0.10 & 0.40 \\ 0.36 & 0.20 & 0.44 \\ 0.03 & 0.20 & 0.77 \end{bmatrix} \end{array} . \end{array}$$

f) Recently 5 current brand $B$ customers were followed to find the number of purchases before their first purchase of brand $C$. The average was 1.60.

# 6 TRANSITIONS: STATISTICS AND INFERENCE

We have up to this time talked about transitions in a Markov process without giving them further specification. However, in any process that models a physical system we shall be far more interested in some kinds of transitions than in others. Some system transitions may mean profits, some may mean costs, others may refer to a failure or a birth. Thus we find ourselves interested in different classes of transitions and in their statistics. We shall also study what knowledge about the process transitions allows us to infer about other process statistics.

## 6.1 TAGGING

A fundamental technique in studying transitions in Markov processes is a procedure that we call "tagging." In this section we shall develop the terminology and logic necessary to apply this procedure with efficiency and understanding.

### Transition Classes

Let us start by supposing that we divided all of the $N^2$ possible transitions that the process can make into two mutually exclusive and collectively exhaustive classes. We shall call these classes transitions of type 1 and of type 2. Then every time the process makes a transition, it makes a transition of one of the two types. As a result of this classification of transitions, we can write the transition probability matrix $P$ as the sum of two matrices ${}_1P$ and ${}_2P$. The matrix ${}_1P$ contains only the transition probabilities pertinent to type 1 transitions—those entries corresponding to type 2 transitions are equal to zero. The matrix ${}_2P$ has just the complementary composition. Thus

$$P = {}_1P + {}_2P. \tag{6.1.1}$$

We can now talk about the statistics of the number of transitions of type 1 or type 2 made in $n$ transitions. Of course, the total number of transitions that the process makes must be the sum of the numbers of each type that it makes. What then is the probability distribution for the number of transitions of each type given the total number of transitions made?

### Joint Probabilities of Transition Types

Let $\phi_{ij}(k_1, k_2|n)$ be the probability that a system which started in state $i$ and has made $n$ transitions is in state $j$ and has made $k_1$ transitions of type 1 and $k_2$ transitions of type 2. Finding this joint probability on the number of transitions of each type will certainly be the key to finding the statistics of each type of transition. For $n > 0$ we can write a recursive equation that relates $\phi_{ij}(k_1, k_2|n)$ to $\phi_{ij}(k_1, k_2|n-1)$:

$$\phi_{ij}(k_1, k_2|n) = \sum_{m=1}^{N} \phi_{im}(k_1 - 1, k_2|n-1)\, {}_1p_{mj} + \sum_{m=1}^{N} \phi_{im}(k_1, k_2 - 1|n-1)\, {}_2p_{mj}$$

$$1 \le i, j \le N;\ 0 \le k_1,\ 0 \le k_2,\ 0 < n. \qquad (6.1.2)$$

This equation holds (as indicated) even when $k_1$ or $k_2$ is equal to zero, if we understand that $\phi_{ij}(\cdot, \cdot|\cdot)$ with any of its arguments negative is equal to zero. Equation 6.1.2 states simply that the $n$th transition must have been either a type 1 or type 2 transition, and tallies it accordingly. Since no transition of either type could have been made when $n = 0$, the boundary equation is

$$\phi_{ij}(k_1, k_2|0) = \delta_{ij}\, \delta(k_1)\, \delta(k_2). \qquad (6.1.3)$$

By using once more the idea that any $\phi_{ij}$ with at least one negative argument is zero, we can write a single recursion equation that holds for all non-negative $k_1$, $k_2$, and $n$,

$$\phi_{ij}(k_1, k_2|n) = \sum_{m=1}^{N} \phi_{im}(k_1 - 1, k_2|n-1)\, {}_1p_{mj} + \sum_{m=1}^{N} \phi_{im}(k_1, k_2 - 1|n-1)\, {}_2p_{mj}$$

$$+ \delta_{ij}\, \delta(k_1)\, \delta(k_2)\, \delta(n) \qquad 1 \le i, j \le N;\ 0 \le k_1, k_2, n. \qquad (6.1.4)$$

In matrix form this equation becomes

$$\Phi(k_1, k_2|n) = \Phi(k_1 - 1, k_2|n-1)\, {}_1P + \Phi(k_1, k_2 - 1|n-1)\, {}_2P + I\, \delta(k_1)\, \delta(k_2)\, \delta(n)$$

$$0 \le k_1, k_2, n. \qquad (6.1.5)$$

Equation 6.1.5 provides a feasible computational method for calculating $\Phi(k_1, k_2|n)$ recursively for $n = 0, 1, 2, \ldots$.

### Joint Probability Transforms

However, we shall obtain more insight into the process by writing the equations after we have transformed all their variables geometrically. Let $\phi_{ij}^{ggg}(y_1, y_2|z)$ be the triple geometric transform of $\phi_{ij}(k_1, k_2|n)$ defined by

$$\phi_{ij}^{ggg}(y_1, y_2|z) = \sum_{k_1=0}^{\infty} y_1^{k_1} \sum_{k_2=0}^{\infty} y_2^{k_2} \sum_{n=0}^{\infty} z^n \phi_{ij}(k_1, k_2|n). \qquad (6.1.6)$$

The matrix of the transforms $\phi_{ij}^{ggg}(y_1, y_2|z)$ is then $\Phi^{ggg}(y_1, y_2|z)$.

The only difficulty that arises in transforming Equation 6.1.5 is finding the transforms of $\Phi(k_1 - 1, k_2|n - 1)$ and $\Phi(k_1, k_2 - 1|n - 1)$. Because of the symmetry in $k_1$ and $k_2$, if we find either one we shall be able to write the other. Therefore, let us find the transform of $\Phi(k_1 - 1, k_2|n - 1)$. We write,

$$\sum_{k_1=0}^{\infty} y_1^{k_1} \sum_{k_2=0}^{\infty} y_2^{k_2} \sum_{n=0}^{\infty} z^n \Phi(k_1 - 1, k_2|n - 1)$$

$$= \sum_{r=-1}^{\infty} y_1^{r+1} \sum_{k_2=0}^{\infty} y_2^{k_2} \sum_{s=-1}^{\infty} z^{s+1}\Phi(r, k_2|s)$$

$$= y_1 z \sum_{r=0}^{\infty} y_1^{r} \sum_{k_2=0}^{\infty} y_2^{k_2} \sum_{s=0}^{\infty} z^{s}\Phi(r, k_2|s)$$

$$= y_1 z \Phi^{ggg}(y_1, y_2|z). \tag{6.1.7}$$

The only property of $\Phi(k_1, k_2|n)$ used in this development is that it is zero when any of its arguments is zero. We see immediately, then, that the transform of $\Phi(k_1, k_2 - 1|n - 1)$ must be equal to $y_2 z \Phi^{ggg}(y_1, y_2|z)$.

Since the transform of $I\,\delta(k_1)\,\delta(k_2)\,\delta(n)$ is just $I$, the transform of Equation 6.1.5 is

$$\Phi^{ggg}(y_1, y_2|z) = y_1 z \Phi^{ggg}(y_1, y_2|z)\,{}_1P + y_2 z \Phi^{ggg}(y_1, y_2|z)\,{}_2P + I. \tag{6.1.8}$$

We can rearrange this equation to produce

$$\Phi^{ggg}(y_1, y_2|z)[I - y_1 z\,{}_1P - y_2 z\,{}_2P] = I \tag{6.1.9}$$

or

$$\Phi^{ggg}(y_1, y_2|z) = [I - y_1 z\,{}_1P - y_2 z\,{}_2P]^{-1}$$
$$= [I - z(y_1\,{}_1P + y_2\,{}_2P)]^{-1}, \tag{6.1.10}$$

which shows how to calculate the transform $\Phi^{ggg}(y_1, y_2|z)$ by inverting a matrix containing ${}_1P$ and ${}_2P$.

### Flow Graph Interpretation: Tagging

The matrix inversion can also be performed by using flow graphs. Equation 6.1.10 implies the matrix flow graph shown in Figure 6.1.1. We construct the $N$-node flow graph corresponding to this matrix flow graph by: 1) drawing the transition diagram of the process; 2) multiplying every branch transmission by $z$; and 3) multiplying each branch transmission by either $y_1$ or $y_2$ according to whether the transmission corresponds to a type 1 or type 2 transition.

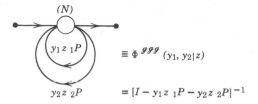

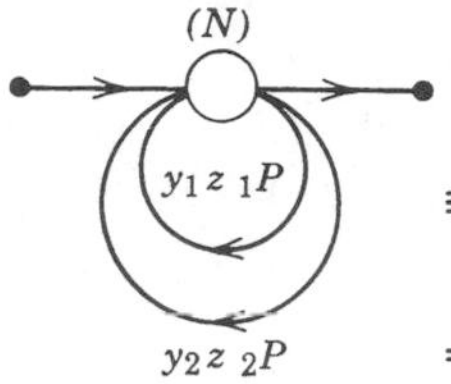


$$\equiv \Phi^{\mathcal{G}\mathcal{G}\mathcal{G}}(y_1, y_2|z)$$

$$= [I - y_1 z\, {}_1P - y_2 z\, {}_2P]^{-1}$$

**Figure 6.1.1** Matrix flow graph for transitions of two classes.

Note that if both $y_1$ and $y_2$ are set equal to one we have just the ordinary flow graph for the Markov process. This result follows directly from the observation that substituting $y_1 = y_2 = 1$ in Equation 6.1.6 implies summing $\phi_{ij}(k_1, k_2|n)$ over all values of $k_1$ and $k_2$ to obtain simply the probability $\phi_{ij}(n)$ of being in state $j$ after $n$ transitions if the system is started in state $i$, and then geometrically transforming on $n$.

We see then that we can find the statistics of transitions of two different types in a Markov process by drawing the flow graph of the process and then multiplying each branch transmission pertaining to a given type of transition by a transform variable corresponding to that type. The transmissions of the resulting flow graph are then the transforms of the probability distributions for the number of transitions of each type made in $n$ system transitions. Thus the basic computational idea is one of "tagging" each type of transition in the flow graph with a corresponding transform variable. We shall therefore refer to this solution principle as the tagging technique.

## Joint Probabilities in Closed Form

Just as

$$\Phi^{\mathcal{G}}(z) = [I - zP]^{-1} \qquad (6.1.11)$$

implies the matrix solution

$$\Phi(n) = P^n \qquad n = 0, 1, 2, \ldots, \qquad (6.1.12)$$

so does the Equation 6.1.10 imply a matrix $\Phi(k_1, k_2|n)$ in closed form. First we invert the transform on $n$ to obtain

$$\Phi^{\mathcal{G}\mathcal{G}\cdot}(y_1, y_2|n) = (y_1\, {}_1P + y_2\, {}_2P)^n \qquad n = 0, 1, 2, \ldots. \qquad (6.1.13)$$

Now the binomial expansion for matrices allows us to write this equation in the form

$$\Phi^{\mathscr{g}\mathscr{g}}(y_1, y_2|n) = \sum_{\substack{k_1=0 \\ k_1+k_2=n}}^{n} \sum_{k_2=0}^{n} y_1^{k_1} y_2^{k_2} \begin{Bmatrix} \text{Sum of products of all per-} \\ \text{mutations of } n \text{ matrices, } k_1 \\ \text{of type } {}_1P,\ k_2 \text{ of type } {}_2P \end{Bmatrix} \qquad 0 \le n; \tag{6.1.14}$$

the restriction on the summation that $k_1 + k_2 = n$ is, of course, vital. Since $\Phi(k_1, k_2|n)$ is the coefficient of $y_1^{k_1} y_2^{k_2}$ in the power series expansion of $\Phi^{\mathscr{g}\mathscr{g}}(y_1, y_2|n)$, we see immediately that $\Phi(k_1, k_2|n)$ must be given by

$$\Phi(k_1, k_2|n) = \begin{Bmatrix} \text{Sum of products of all per-} \\ \text{mutations of } n \text{ matrices, } k_1 \\ \text{of type } {}_1P,\ k_2 \text{ of type } {}_2P \end{Bmatrix}, \qquad k_1 + k_2 = n,\ 0 \le n. \tag{6.1.15}$$

This equation is computable, but probably no simpler to manage than the recursive scheme of Equation 6.1.5.

### Multiple Transition Classes

There is no need to consider only two types of transitions—we can divide each of the $N^2$ transitions that the process can make into $M \le N^2$ mutually exclusive and collectively exhaustive classes, type 1, type 2, . . . , type $M$. The matrix $P$ would then be decomposed in the form

$$P = {}_1P + {}_2P + \cdots + {}_MP \tag{6.1.16}$$

by the same procedure used before. Then we would be interested in $\phi_{ij}(k_1, k_2, \ldots, k_M|n)$, the probability that each number of transitions of each type would be made in $n$ transitions and that the process would be in state $j$ if it started in state $i$. By exactly the same arguments used previously we could write a recursive equation like Equation 6.1.4, define a transform $\phi_{ij}^{\overbrace{\mathscr{g}\mathscr{g}\cdots\mathscr{g}}^{M+1}}(y_1, y_2, \ldots, y_M|z)$, and write the matrix of the transforms $\Phi^{\overbrace{\mathscr{g}\mathscr{g}\cdots\mathscr{g}}^{M+1}}(y_1, y_2, \ldots, y_M|z)$ in the form

$$\Phi^{\overbrace{\mathscr{g}\mathscr{g}\cdots\mathscr{g}}^{M+1}}(y_1, y_2, \ldots, y_M|z) = [I - z(y_1\,{}_1P + y_2\,{}_2P + \cdots + y_M\,{}_MP)]^{-1}. \tag{6.1.17}$$

The matrix flow graph and the method of constructing the tagged flow graph are completely analogous to the two-type situation we analyzed in detail. The explicit expression for the matrix $(\Phi k_1, k_2, \ldots, k_M|n)$ is then

$$\Phi(k_1, k_2, \ldots, k_M|n) = \begin{Bmatrix} \text{Sum of products of all permutations} \\ \text{of } n \text{ matrices, } k_1 \text{ of type } {}_1P,\ k_2 \text{ of} \\ \text{type } {}_2P, \ldots, k_M \text{ of type } {}_MP \end{Bmatrix}$$

$$k_1 + k_2 + \cdots + k_M = n \tag{6.1.18}$$

and therefore represents a logical extension of our earlier results.

## Nonexclusive and Nonexhaustive Transition Classes

Although we have always considered that the transitions of the process must fall into mutually exclusive and collectively exhaustive classes, even this restriction is unnecessary. We can allow some transitions to be members of no class or of more than one class, or to be members of one or more classes as the result of a probabilistic process. Figure 6.1.2 illustrates the case where transitions can be members of no class or of more than one class. The transition from state 1 to state 2 is a member of no class and so is not tagged at all. The transition from state 2 to state 3 is a member of class 1 only, and so is tagged by $y_1$ only. The transition from state 3 to state 4 is a member of class 2 and so is tagged by $y_2$ only. The transition from state 4 to state 1 is a member of both classes 1 and 2 and so is tagged by both $y_1$ and $y_2$.

Figure 6.1.3 illustrates the case of probabilistic class membership. The transition from state 1 to state 2 is a member of class 1 with probability $a$ and a member of class 2 with probability $1 - a$. Therefore, we split the branch of the flow graph with transmission $zp_{12}$ into two branches that go from node 1 to node 2, one with transmission $azp_{12}$, the other with transmission $(1 - a)zp_{12}$. Then we tag the first with $y_1$, the second with $y_2$.

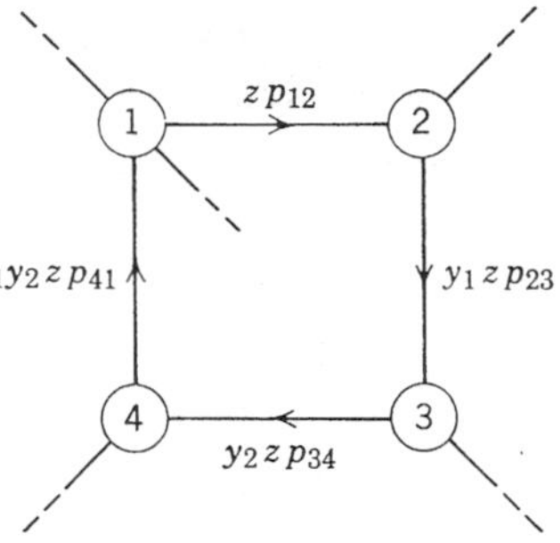


**Figure 6.1.2** Tagging with multiple classes.

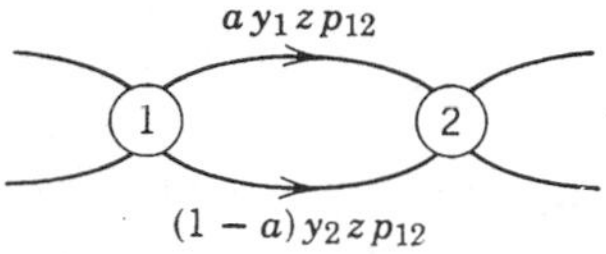


**Figure 6.1.3** Probabilistic class membership.

The flow graphs constructed without requiring mutually exclusive and collectively exhaustive classes and allowing probabilistic class membership still have transmissions that produce the proper transform $\Phi^{\overbrace{gg\cdots g}^{M}}(y_1, \ldots, y_M|z)$. The $ij$th element of the inverse transform, $\phi_{ij}(k_1, k_2, \ldots, k_M|n)$, is still the probability that in $n$ transitions a system starting in state $i$ will be in state $j$ and will have made $k_1$ type 1 transitions, $k_2$ type 2 transitions, etc. The only change is that since the classes are not mutually exclusive and collectively exhaustive there is no requirement that $k_1 + k_2 + \cdots + k_M = n$. In general, all we can say is that $k_1 \leq n$, $k_2 \leq n, \ldots, k_M \leq n$. Furthermore, the idea of splitting $P$ into $M$ parts is confused by both the multiple and probabilistic class membership. Nevertheless the tagging principle works.

The proof that it works would require creating a new expanded set of classes to represent transitions that are members of no more than one class, thus producing once again a mutually exclusive and collectively exhaustive class structure. The problem of probabilistic class membership is no problem at all—it merely means that when $P$ is decomposed into transition matrices for each of the new classes, the existence of a non-zero entry in the $ij$th element of one matrix does not imply that the $ij$th elements of all other matrices are zero. Of course, the sum of the $ij$th elements of the transition matrices for the *expanded* set of classes must be equal to $p_{ij}$, the $ij$th element of $P$. However, now that we understand how we would prove that this more general concept of tagging works, we shall not bother to do so because we shall gain no new enlightenment from carrying out the proof.

## The Tagging Philosophy

We may now state the general principle of tagging a process flow graph. We define $M$ classes of transitions that are of interest to us. There is no need for $M$ to be less than or equal to $N^2$, the number of possible system transitions. We then examine each transition of the process for membership in each of the $M$ classes. If it is a member of any class $k$, we multiply the transmission corresponding to that transition in the flow graph for the process by $y_k$. It is conceivable that a transition from state $i$ to state $j$ could belong to all of the $M$ classes. In that case the transmission of the tagged flow graph from node $i$ to node $j$ would be $y_1 y_2 \cdots y_M z p_{ij}$.

If any transition can belong to different classes with different probabilities, we split the flow graph branch transmission corresponding to that transmission into as many branches as there are possibilities for class membership and then give each branch a transition probability equal to the product of the original transition probability and the probability that the transition will belong to the class represented by the branch. Tagging each of the branches is performed according to the usual rules.

When we have constructed the tagged flow graph, we can compute its transmission from node $i$ to node $j$ and therefore obtain the transform of the probability distribution of the number of transitions of each type made, as well as the probability of the ending state conditional on the starting state and the total number of transitions made. Since inversion of this transform may not prove easy, we can write the difference equation corresponding to the transform and solve it recursively until we obtain the information we need. Thus, the tagged flow graph provides a very convenient basis for writing the difference equations of the process even though the power of its transform analysis is curtailed in even the smallest tagging problems. It has been my experience that far fewer errors result from obtaining difference equations from flow graph techniques than are produced by an attempt to write the difference equations directly without this graphical aid.

### Implications of Joint Probabilities

The probability $\phi_{ij}(k_1, k_2, \ldots, k_M|n)$ provides the basis for calculating all of the statistics of the tagging process. If we want to know the probability of each number of transitions of each type regardless of where the system is after $n$ transitions, then we must sum over $j$

$$\phi_{i\Sigma}(k_1, k_2, \ldots, k_M|n) = \sum_{j=1}^{N} \phi_{ij}(k_1, k_2, \ldots, k_M|n). \tag{6.1.19}$$

We can obtain the same effect by summing the transmissions of the tagged flow graph from node $i$ to each other node before inverse transforming.

If we want to know the probability distribution of each number of transitions of each type *conditional* on being in state $j$ at time $n$, $q_{ij}(k_1, k_2, \ldots, k_M|n)$, then we must divide the joint probability of each number of transitions and of being in state $j$, $\phi_{ij}(k_1, k_2, \ldots, k_M|n)$ by the probability of being in state $j$ at time $n$, $\phi_{ij}(n)$. Therefore,

$$q_{ij}(k_1, k_2, \ldots, k_M|n) = \frac{\phi_{ij}(k_i, k_2, \ldots, k_M|n)}{\phi_{ij}(n)}. \tag{6.1.20}$$

Through standard probabilistic methods we can calculate all moments of the number of special transitions of each type regardless of ending state by using Equation 6.1.19 and conditional on ending state by using Equation 6.1.20. However, rather than getting more involved in general results, let us turn to the case in which tagging is most frequently employed.

### Special Transitions

In most problems we are interested in only one special class of transitions, which we shall call "special" transitions. All transitions of the process are either special or nonspecial. We usually want to know the statistics of the number of special transitions made given the total number of transitions made, the starting state, and possibly the ending state. We can define $\phi_{ij}(k|n)$ to be the probability that the system has made $k$ special transitions and has ended in state $j$ given that it started in state $i$ and has made $n$ transitions. If we identify special transitions as type 1 transitions and nonspecial transitions as type 2 transitions, then we can write $\phi_{ij}(k|n)$ in terms of $\phi_{ij}(k_1, k_2|n)$ in Equation 6.1.4 as

$$\phi_{ij}(k|n) = \phi_{ij}(k, n - k|n). \tag{6.1.21}$$

Since the statistics of the number of special and nonspecial transitions are both described by $\phi_{ij}(k|n)$, this simplification is valuable and nonrestrictive.

The recursive equation corresponding to Equation 6.1.4 is then

$$\phi_{ij}(k|n) = \sum_{m=1}^{N} \phi_{im}(k - 1|n - 1)\, {}_sp_{mj} + \sum_{m=1}^{N} \phi_{im}(k|n - 1)(p_{mj} - {}_sp_{mj})$$
$$+ \delta_{ij}\, \delta(k)\, \delta(n) \qquad 1 \le i, j \le N,\ 0 \le k, n. \tag{6.1.22}$$

In this equation ${}_sp_{mj}$ is the portion of the transition probability $p_{mj}$ that is assigned to special transitions. Thus $P$ is divided into two matrices ${}_sP$ and $P - {}_sP$. The matrix form of Equation 6.1.22 is then

$$\Phi(k|n) = \Phi(k - 1|n - 1)\, {}_sP + \Phi(k|n - 1)(P - {}_sP) + I\, \delta(k)\, \delta(n) \qquad 0 \le k, n. \tag{6.1.23}$$

The probability $\phi_{ij}(k|n)$ has the transform,

$$\phi_{ij}{}^{gg}(y|z) = \sum_{k=0}^{\infty} y^k \sum_{n=0}^{\infty} z^n \phi_{ij}(k|n). \tag{6.1.24}$$

Notice that

$$\phi_{ij}{}^{gg}(y|z) = \phi_{ij}^{ggg}(y, 1|z) \tag{6.1.25}$$

in terms of the transform of Equation 6.1.6. In matrix form,

$$\Phi^{gg}(y|z) = \Phi^{ggg}(y, 1|z). \tag{6.1.26}$$

Therefore, from Equation 6.1.10 with ${}_1P = {}_sP$ and ${}_2P = P - {}_sP$ we have

$$\Phi^{gg}(y|z) = [I - yz\,{}_sP - z(P - {}_sP)]^{-1}. \tag{6.1.27}$$

Figure 6.1.4 shows the matrix flow graph corresponding to this equation. All the special transitions in the flow graph are tagged with $y$; all others are left unchanged. If some of the transitions are special with a given probability, then we have to split the corresponding transmissions of the flow graph as we discussed earlier.

By using Equations 6.1.21 and 6.1.15 we can write an explicit expression for the matrix $\Phi(k|n)$,

$$\Phi(k|n) = \left\{\begin{matrix}\text{Sum of products of all permuta-}\\ \text{tions of } n \text{ matrices, } k \text{ of type}\\ {}_sP,\ n-k \text{ of type } (P - {}_sP)\end{matrix}\right\} \quad 0 \le k \le n. \tag{6.1.28}$$

This expression is again more important for the insight it gives into the process than for any computational advantage it has over Equation 6.1.23.

***Special transition means***

Now let us turn to the question of the moments of the number of special transitions. The basic transform relation is Equation 6.1.24. If we differentiate this transform with respect to $y$, we obtain

$$\frac{d}{dy}\phi_{ij}{}^{gg}(y|z) = \sum_{n=0}^{\infty} z^n \sum_{k=0}^{\infty} ky^{k-1}\phi_{ij}(k|n). \tag{6.1.29}$$

Now if we evaluate the derivative at $y = 1$, we have

$$\frac{d}{dy}\phi_{ij}{}^{gg}(y|z)\bigg|_{y=1} = \sum_{n=0}^{\infty} z^k \sum_{n=0}^{\infty} k\phi_{ij}(k|n). \tag{6.1.30}$$

Let

$$k_{ij}(n) = \sum_{n=0}^{\infty} k\phi_{ij}(k|n), \tag{6.1.31}$$

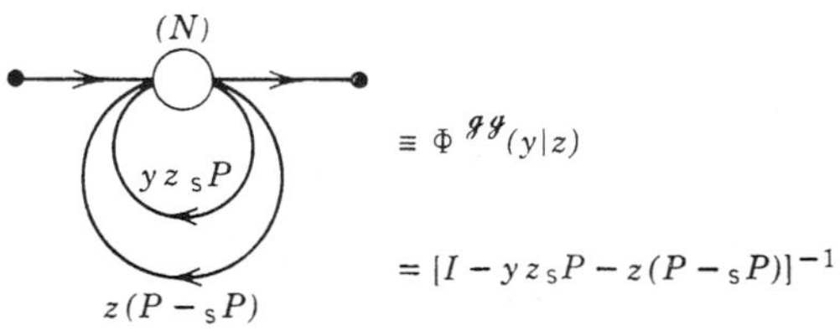


**Figure 6.1.4** Matrix flow graph for special transitions.

and let $k_{ij}(n)$ have geometric transform

$$k_{ij}{}^{g}(z) = \sum_{n=0}^{\infty} z^n k_{ij}(n). \tag{6.1.32}$$

Then Equation 6.1.30 produces the important result

$$k_{ij}{}^{g}(z) = \frac{d}{dy}\,\phi_{ij}{}^{gg}(y|z)\Big|_{y=1}. \tag{6.1.33}$$

Let $\bar{k}_{i\Sigma}(n)$ be the mean number of special transitions in time $n$ if the process starts in state $i$, but its ending state is unspecified. Then we have immediately that

$$\bar{k}_{i\Sigma}(n) = \sum_{j=1}^{N} k_{ij}(n). \tag{6.1.34}$$

Let $\bar{k}_{ij}(n)$ be the mean number of special transitions in time $n$ if the process starts in state $i$ and ends in state $j$. By the same reasoning that led to Equation 6.1.20 we have

$$\bar{k}_{ij}(n) = \frac{k_{ij}(n)}{\phi_{ij}(n)}. \tag{6.1.35}$$

Therefore the $k_{ij}(n)$ found by differentiating $\phi_{ij}{}^{gg}(y|z)$ and inverse transforming provides the basis for calculating the mean number of special transitions in $n$ transitions both regardless of ending state and conditional on ending state.

***Special transition second moments***

A similar process provides us with the second moment of the number of special transitions for these cases. If we differentiate Equation 6.1.29 again with respect to $y$, we obtain

$$\frac{d^2}{dy^2}\,\phi_{ij}{}^{gg}(y|z) = \sum_{n=0}^{\infty} z^n \sum_{k=0}^{\infty} k(k-1)y^{k-2}\phi_{ij}(k|n). \tag{6.1.36}$$

By evaluating this derivative at $y = 1$ we find

$$\frac{d^2}{dy^2}\,\phi_{ij}{}^{gg}(y|z)\Big|_{y=1} = \sum_{n=0}^{\infty} z^n\left(\sum_{k=0}^{\infty} k^2\phi_{ij}(k|n) - \sum_{k=0}^{\infty} k\phi_{ij}(k|n)\right). \tag{6.1.37}$$

Now we let

$$k_{ij}{}^{2}(n) = \sum_{k=0}^{\infty} k^2\phi_{ij}(k|n) \tag{6.1.38}$$

with geometric transform

$$k_{ij}^{2g}(z) = \sum_{n=0}^{\infty} z^n k_{ij}^2(n). \tag{6.1.39}$$

Then we can write Equation 6.1.37 in the useful form

$$\frac{d^2}{dy^2} \phi_{ij}^{gg}(y|z)\bigg|_{y=1} = k_{ij}^{2g}(z) - k_{ij}^{g}(z). \tag{6.1.40}$$

Since we already know $k_{ij}^{g}(z)$, this equation allows us to find $k_{ij}^{2g}(z)$ and then, by inverse transformation, $k_{ij}^2(n)$.

By the same reasoning as before, $k_{i\Sigma}^2(n)$, the second moment of the number of special transitions in $n$ transitions if the process starts in state $i$, regardless of ending state, is given by

$$\overline{k_{i\Sigma}^2}(n) = \sum_{j=1}^{N} k_{ij}^2(n). \tag{6.1.41}$$

Then $\overline{k_{ij}^2}(n)$, the second moment of the number of special transitions in $n$ transitions given that the process starts in state $i$ and ends in state $j$, is given by

$$\overline{k_{ij}^2}(n) = \frac{k_{ij}^2(n)}{\phi_{ij}(n)}. \tag{6.1.42}$$

The variance of either of these quantities is then computed from the first and second moments by the usual formula. Similar procedures allow calculating even higher moments of the number of special transitions if necessary.

***A cautionary note***

The most common error in calculating the moments of the number of special transitions is to omit the denominator $\phi_{ij}(n)$ in Equations 6.1.35 and 6.1.42. The reason it is required is that $\phi_{ij}(k|n)$ is the joint probability of making $k$ special transitions and ending in state $j$, not the probability of making $k$ special transitions conditional on ending in state $j$. Dividing $\phi_{ij}(k|n)$ by $\phi_{ij}(n)$, the probability of ending in $j$, creates $q_{ij}(k|n)$, the probability of making $k$ special transitions in $n$ transitions given that the process started in state $i$ and ended in state $j$. The moments of the $q_{ij}(k|n)$ distribution are the moments reported in Equations 6.1.35 and 6.1.42.

***The power of tagging***

Before we proceed to examples of these calculations there is one further point worth mentioning about the generality afforded by tagging. We have said that we can find the statistics of the number of special transitions by drawing a flow graph and tagging all branches corresponding to special transitions by a transform variable $y$. Suppose that we called all transitions special and labeled all branches with $y$. Then the number of special transitions would be equal to the number of transitions. Suppose that we labeled all transitions $y^2$. Then we would find that the number of special transitions was twice the number of transitions. Suppose now that each transition of the process requires a time $t_{ij}$, where $t_{ij}$ can be any integer 1, 2, 3, . . ., and that we tag the branch transmission from each node $i$ to each node $j$ in the flow graph by $y^{t_{ij}}$. Then the $ij$th element of $\Phi(y|z)$, the matrix of transmissions of the tagged flow graph, would be the transform of $\phi_{ij}(k|n)$, the probability that a system which started in state $i$ and made $n$ transitions would end in state $j$ and would have taken a total time $k$ to make the $n$ transitions. Thus for the first time we have a method for considering the time required to make each transition of the process. We shall not exploit this method because far more powerful techniques will be developed later. However, it does show the desirability of flexible thinking about tagging and indeed the value of the tagging concept.

### Tagging in the Marketing Example

Now let us turn to some examples of the tagging procedure. Suppose that in the marketing example we are interested in purchases of brand $A$. We therefore call any transition that represents a purchase of brand $A$ a special transition, and draw the tagged flow graph of Figure 6.1.5. Note that not only are the transitions from state 1 to state 1 and from state 2 to state 1 tagged, but the input branch to state 1 is tagged also. This tagging is a result of wanting to count even the initial purchase of brand $A$. All outputs are summed at a node because we have not yet expressed any concern about the last brand that the customer purchases.

#### *Expected number of purchases of a brand*

Let us assume that the customer buys brand $A$ originally at time zero, and that we want to find the mean number of purchases of brand $A$ through purchase time

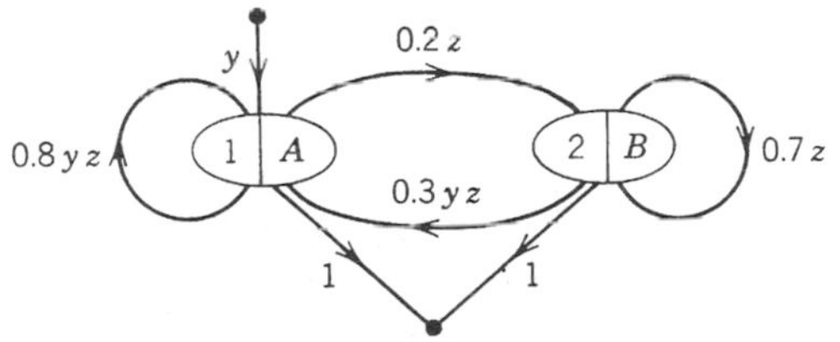


**Figure 6.1.5** Tagged flow graph for marketing example.

$n$. The transmission of the tagged flow graph of Figure 6.1.5 is then

$$\vec{t} = \phi^{gg}(y|z) = \frac{y(1 - 0.5z)}{1 - 0.7z - 0.8yz + 0.5yz^2}. \tag{6.1.43}$$

Since

$$\phi^{gg}(y|z) = \sum_{k=0}^{\infty} y^k \sum_{n=0}^{\infty} z^n \phi(k|n) \tag{6.1.44}$$

and since

$$\sum_{k=0}^{\infty} \phi(k|n) = 1 \tag{6.1.45}$$

because some number of special transitions must be made in $n$ transitions, we obtain

$$\phi^{gg}(1|z) = \sum_{n=0}^{\infty} z^n \sum_{k=0}^{\infty} \phi(k|n) = \sum_{n=0}^{\infty} z^n = \frac{1}{1 - z}. \tag{6.1.46}$$

Equation 6.1.43 passes this test.

Now we proceed to compute $k(n)$. First,

$$k^g(z) = \frac{d}{dy}\phi(y|z)\bigg|_{y=1} = \frac{1 - 0.7z}{(1 - z)^2(1 - 0.5z)} \tag{6.1.47}$$

or

$$k^g(z) = \frac{0.6}{(1 - z)^2} + \frac{0.8}{1 - z} + \frac{-0.4}{1 - 0.5z}. \tag{6.1.48}$$

Then, by inverse transformation,

$$k(n) = 0.6(n + 1) + 0.8 - 0.4(0.5)^n \qquad 0 \leq n. \tag{6.1.49}$$

Since in this case we have already summed over all states of the process, the mean number of special transitions through time $n$, $\bar{k}(n)$, is just equal to $k(n)$,

$$\bar{k}(n) = 0.6(n + 1) + 0.8 - 0.4(0.5)^n \qquad 0 \leq n. \tag{6.1.50}$$

Because in this example every special transition is associated with an occupancy of state 1, Equation 6.1.50 also gives the mean number of times state 1 will be occupied through time $n$. But this quantity was calculated for this example in Equation 5.1.5 where we discussed state occupancy. Thus tagging provides an alternate method for calculating occupancy statistics. However, a comparison of the two methods shows that the techniques of Section 5.1 will yield occupancy moments much more readily than the tagging method.

***Tagging to obtain occupancy probability distributions***

But, as you will recall, there is one result that we did not produce in Section 5.1. This was the probability distribution of the number of occupancies. Fortunately, tagging will do this analysis job in the relatively few cases where we desire this much information about occupancies. In fact, for this example the inverse transform $\phi(k|n)$ of the transform $\phi^{gg}(y|z)$ is just the probability that state 1 will be occupied $k$ times through time $n$.

Calculating the inverse transform $\phi(k|n)$ is not so simple. However, we can write the transform of Equation 6.1.43 in the form

$$\phi^{gg}(y|z) = \frac{y\left(\dfrac{1 - 0.5z}{1 - 0.7z}\right)}{1 - \left(\dfrac{0.8z - 0.5z^2}{1 - 0.7z}\right)y}. \tag{6.1.51}$$

In this form we see how to invert the transform on $k$ to produce

$$\phi^{\cdot g}(k|z) = \left(\frac{1 - 0.5z}{1 - 0.7z}\right)\left(\frac{0.8z - 0.5z^2}{1 - 0.7z}\right)^{k-1} u(k - 1) \qquad k = 0, 1, 2, \ldots. \tag{6.1.52}$$

The next step of inverting the transform on $n$ is possible, but not easy.

Therefore we shall take another tack and write the difference equation implied by Equation 6.1.43. We write

$$(1 - 0.7z - 0.8yz + 0.5yz^2)\phi^{gg}(y|z) = y - 0.5yz \tag{6.1.53}$$

or

$$\phi^{gg}(y|z) = 0.7z\phi^{gg}(y|z) + 0.8yz\phi^{gg}(y|z) - 0.5yz^2\phi^{gg}(y|z) + y - 0.5yz. \tag{6.1.54}$$

We obtain the difference equation corresponding to this equation by inspection:

$$\begin{aligned}\phi(k|n) = {} & 0.7\phi(k|n - 1) + 0.8\phi(k - 1|n - 1) - 0.5\phi(k - 1|n - 2) \\ & + \delta(k - 1)\,\delta(n) - 0.5\,\delta(k - 1)\,\delta(n - 1) \qquad 0 \le k, n.\end{aligned} \tag{6.1.55}$$

Notice that this difference equation relates $\phi(k|n)$ to the same function with arguments $n - 1$ and $n - 2$. In this respect it differs from the general form of the difference equation in Equation 6.1.22. Both equations are correct; many different forms can be used. We shall usually find that we avoid both mistakes and difficulty by writing the difference equation from the flow graph transmission, as we have done in this case.

We solve Equation 6.1.55 recursively to produce $\phi(k|n)$. Table 6.1.1 shows the calculation for the first few values of $n$. For each value of $n$, $\phi(k|n)$ is a discrete probability distribution over the number of special transitions that could have occurred. The mean of this probability distribution for each $n$ is shown in the bottom

**Table 6.1.1** Solution by Recursion for $\phi(k|n)$

| $\phi(k\|n)$ $k \backslash n$ | 0 | 1 | 2 | 3 |
|---|---|---|---|---|
| 0 | 0 | 0 | 0 | 0 |
| 1 | 1 | 0.2 | 0.14 | 0.098 |
| 2 | 0 | 0.8 | 0.22 | 0.166 |
| 3 | 0 | 0 | 0.64 | 0.224 |
| 4 | 0 | 0 | 0 | 0.512 |
| 5 | 0 | 0 | 0 | 0 |
| $\bar{k}(n) = \sum_{k=0}^{\infty} k\phi(k\|n)$ | 1 | 1.8 | 2.5 | 3.15 |

row of Table 6.1.1. In this example, these quantities are the mean number of occupancies of state 1 through time $n$. They were first reported in Equation 5.1.6; they are the expected number of purchases of brand $A$ on the purchase occasions 0 through $n$.

Thus tagging provides a method for finding the state occupancy probability distributions for any Markov process. The joint probability distribution of occupancies of all states can be found by tagging all branches that enter each state $i$ with the transform variable $y_i$. Then every branch in the process will be labeled by one of the $N$ transform variables $y_1, y_2, \ldots, y_N$. The quantity $\phi_{ij}(k_1, k_2, \ldots, k_N|n)$ obtained by inverse transforming the transmission from node $i$ to node $j$ of a flow graph tagged in this way is the joint probability that a system which starts in state $i$ and makes $n$ transitions will end in state $j$ and will occupy state 1 $k_1$ times, state 2 $k_2$ times, etc. Thus this adaptation of tagging provides complete information on the occupancy statistics of a Markov process. These statistics are computable by our difference equation methods, but the methods of Section 5.1 are preferable unless the actual probability distributions of occupancy times are required.

***Conditioning on last purchase***

Let us now return to the marketing example and retain our interest in purchases of brand $A$. However, this time let us assume that we know that the customer's first and last purchases were of brand $A$. The flow graph for this process is the same as that of Figure 6.1.5 except that the output is now taken only from node 1; the new tagged flow graph appears in Figure 6.1.6. The transmission of this flow graph is

$$\vec{t} = \phi_{11}{}^{gg}(y|z) = \frac{y(1 - 0.7z)}{1 - 0.7z - 0.8yz + 0.5z^2y}. \tag{6.1.56}$$

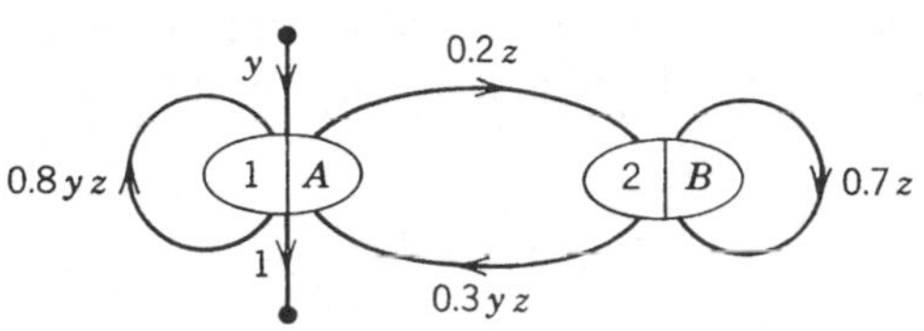


**Figure 6.1.6** Tagged flow graph for process ending in state 1.

The inverse transform of $\phi_{11}^{gg}(y|z)$ is $\phi_{11}(k|n)$, the probability that a system started in state 1 will, after $n$ transitions, be in state 1 and have made $k$ special transitions. Note that it is not the conditional probability of $k$ special transitions in $n$ transitions given that the system started and ended in state 1.

We can make the interpretation of $\phi_{11}(k|n)$ clearer by calculating it. Since the transform $\phi_{11}^{gg}(y|z)$ in Equation 6.1.56 differs only in one constant in the numerator from the transform $\phi^{gg}(y|z)$ in Equation 6.1.43, the recursive equation for $\phi_{11}^{gg}(k|n)$ differs in only one coefficient from the equation for $\phi(k|n)$ shown in Equation 6.1.55. We therefore can easily write

$$\begin{aligned}\phi_{11}(k|n) &= 0.7\phi_{11}(k|n-1) + 0.8\phi_{11}(k-1|n-1) - 0.5\phi_{11}(k-1|n-2) \\ &\quad + \delta(k-1)\,\delta(n) - 0.7\,\delta(k-1)\,\delta(n-1) \qquad 0 \le k, n. \end{aligned} \tag{6.1.57}$$

The solution to this difference equation for a few values of $k$ and $n$ appears in Table 6.1.2. Notice that the sum of $\phi_{11}(k|n)$ over $k$ is just $\phi_{11}(n)$, the $n$-step transition probability from state 1 to state 1 first found in Equation 1.2.16. This result illustrates that $\phi_{ij}(k|n)$ is a joint distribution on $j$ and $k$ which when summed over $k$ gives the marginal distribution on $j$, the state of the system at time $n$. It therefore shows the necessity of dividing $\phi_{ij}(k|n)$ by $\phi_{ij}(n)$ if we desire the conditional probability on the number of special transitions given the starting and ending state.

**Table 6.1.2** Solution by Recursion for $\phi_{11}(k|n)$

| $\phi_{11}(k\|n)$ $k \backslash n$ | 0 | 1 | 2 | 3 |
|---|---|---|---|---|
| 0 | 0 | 0 | 0 | 0 |
| 1 | 1 | 0 | 0 | 0 |
| 2 | 0 | 0.8 | 0.06 | 0.042 |
| 3 | 0 | 0 | 0.64 | 0.096 |
| 4 | 0 | 0 | 0 | 0.512 |
| 5 | 0 | 0 | 0 | 0 |
| $\phi_{11}(n) = \sum_{k=0}^{\infty} \phi_{11}(k\|n)$ | 1 | 0.8 | 0.7 | 0.65 |

**Mean purchases.** Suppose we want to find $\bar{k}_{11}(n)$, the mean number of purchases of brand $A$ through time $n$ if the first and last purchases were of brand $A$. From Equation 6.1.35,

$$\bar{k}_{11}(n) = \frac{k_{11}(n)}{\phi_{11}(n)}. \tag{6.1.58}$$

From Equation 6.1.33,

$$k_{11}{}^{g}(z) = \frac{d}{dy}\phi_{11}{}^{gg}(y|z)\bigg|_{y=1} = \frac{(1-0.7z)^2}{(1-z)^2(1-0.5z)^2}. \tag{6.1.59}$$

When we invert the geometric transform we obtain

$$k_{11}(n) = 0.36n + 1.32 + 0.16n(0.5)^n - 0.32(0.5)^n \qquad 0 \le n. \tag{6.1.60}$$

We have already found for this example that

$$\phi_{11}(n) = 0.6 + 0.4(0.5)^n. \tag{6.1.61}$$

Therefore, we can substitute the results of Equations 6.1.60 and 6.1.61 into Equation 6.1.58 to produce

$$\bar{k}_{11}(n) = \frac{0.36n + 1.32 + 0.16n(0.5)^n - 0.32(0.5)^n}{0.6 + 0.4(0.5)^n} \qquad 0 \le n. \tag{6.1.62}$$

This equation shows for any value of $n$ the expected number of purchases of brand $A$ that will be made through time $n$ given that the purchases at time zero and at time $n$ were of brand $A$. For the first few values of $n$ we find

$$\bar{k}_{11}(0) = 1, \bar{k}_{11}(1) = 2, \bar{k}_{11}(2) = 2.914, \ldots. \tag{6.1.63}$$

The value of $\bar{k}_{11}(0)$ is just what we expect because if no transitions have been made the customer must have purchased brand $A$ exactly once. Likewise, $\bar{k}_{11}(1)$ must equal 2 because the only way for a customer to have bought brand $A$ at time zero and at time 1 is to have bought brand $A$ exactly twice. Note that when $n$ is large, $\bar{k}_{11}(n) \approx 0.6n$. This is consistent with our expectation that in a large number of purchases the fraction that will be of brand $A$ is just the limiting state probability of state 1 in this process, 0.6. The influence of knowing the first and last purchases will vanish in such a situation.

### The Coin-Tossing Example

We can use tagging to investigate the problem of finding the statistics of the number of Head-Tail sequences in $n$ tosses of a fair coin, a problem first discussed in Section 5.1. We construct a two-state system that is in state 1 when the last toss was a Tail and in state 2 when the last toss was a Head. Then a Head-Tail sequence corresponds to a transition from state 2 to state 1. If we call this transition a special transition, then the statistics of the number of special transitions in $n$ transitions are the statistics we need. They can, of course, be easily found by tagging the special transitions in the flow graph and using the methods of this section.

### *Tagged flow graph*

Figure 6.1.7 shows the tagged flow graph for this system. We show the input at node 1 because creating a Tail on the zeroth toss will not affect the number of special transitions in $n$ transitions. Since we are not conditioning the statistics on the outcome of the $n$th toss, we sum all the outputs of the tagged flow graph. The transmission of this flow graph is thus the transform of the probability of each number of special transitions given the number of transitions. We find

$$\vec{t} = \phi^{gg}(y|z) = \frac{1}{1 - z + (1/4)z^2(1 - y)}. \tag{6.1.64}$$

Note that this transform meets the requirement that

$$\phi^{gg}(1|z) = 1/(1 - z),$$

a requirement developed in Equation 6.1.46 for systems that are not known to be in any particular state after $n$ transitions. The inverse $\phi(k|n)$ of this transform is the probability that $k$ Head-Tail sequences will be completed in $n$ tosses of a fair coin.

### *Mean*

We shall begin by calculating the expected number of special transitions in $n$ transitions, $\bar{k}(n)$. Because there is no conditioning on an ending state,

$$\bar{k}(n) = k(n). \tag{6.1.65}$$

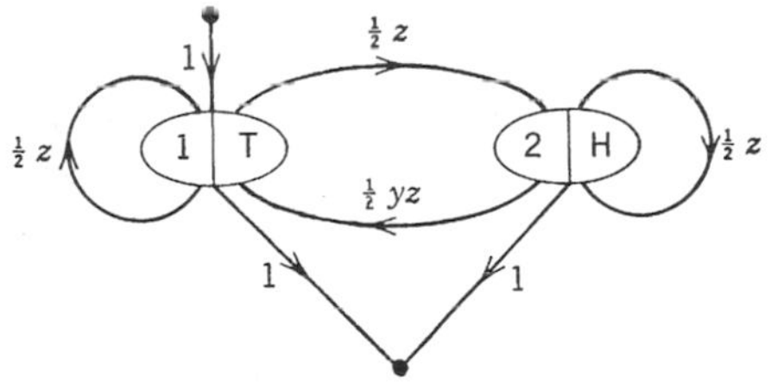


**Figure 6.1.7** Tagged flow graph for Head-Tail sequences.

We find the transform of $k(n)$ to be

$$k^{\mathscr{g}}(z) = \frac{d}{dy}\,\phi^{\mathscr{g}\mathscr{g}}(y|z)\bigg|_{y=1} = \frac{(1/4)z^2}{[1 - z + (1/4)z^2(1 - y)]^2}\bigg|_{y=1} = \frac{(1/4)z^2}{(1 - z)^2}. \quad (6.1.66)$$

Then by inverse transformation,

$$k(n) = \bar{k}(n) = (1/4)(n - 1)u(n - 2) \qquad 0 \leq n. \quad (6.1.67)$$

This result was first found in Equation 5.1.50 using occupancy statistics. The expected number of Head-Tail sequences in $n$ tosses of a fair coin is $(1/4)(n - 1)$ if $n$ is at least 2.

***Second moment***

The second moment of the number of special transitions, $\overline{k^2}(n)$, is again equal to $k^2(n)$ because there is no conditioning on an ending state,

$$\overline{k^2}(n) = k^2(n). \quad (6.1.68)$$

From Equation 6.1.40,

$$\frac{d^2}{dy^2}\,\phi^{\mathscr{g}\mathscr{g}}(y|z)\bigg|_{y=1} = k^{2\mathscr{g}}(z) - k^{\mathscr{g}}(z). \quad (6.1.69)$$

Therefore, the transform of $k^2(n)$ is

$$k^{2\mathscr{g}}(z) = k^{\mathscr{g}}(z) + \frac{d^2}{dy^2}\,\phi^{\mathscr{g}\mathscr{g}}(y|z)\bigg|_{y=1}. \quad (6.1.70)$$

We must now calculate the second derivative of $\phi^{\mathscr{g}\mathscr{g}}(y|z)$ with respect to $y$,

$$\frac{d^2}{dy^2}\,\phi^{\mathscr{g}\mathscr{g}}(y|z)\bigg|_{y=1} = \frac{(1/8)z^4}{[1 - z + (1/4)z^2(1 - y)]^3}\bigg|_{y=1} = \frac{(1/8)z^4}{(1 - z)^3}. \quad (6.1.71)$$

When we substitute the results of Equations 6.1.66 and 6.1.71 into Equation 6.1.70 we obtain

$$k^{2\mathscr{g}}(z) = \frac{(1/4)z^2}{(1 - z)^2} + \frac{(1/8)z^4}{(1 - z)^3}. \quad (6.1.72)$$

The inverse of this transform is $k^2(n)$, or equivalently, $\overline{k^2}(n)$ in view of Equation 6.1.68.

Equation 5.1.53 contains exactly this transform—its inverse was found in Equation 5.1.56. Therefore, we have the same inverse,

$$k^2(n) = \overline{k^2}(n) = (1/16)(n^2 - n + 2)u(n - 2) \qquad 0 \leq n, \quad (6.1.73)$$

and we have once again found the second moment of the number of Head-Tail sequences in $n$ tosses of a fair coin. The variance of the number of special transitions in $n$ transitions is

$$\overset{\vee}{k}(n) = \overline{k^2}(n) - (\bar{k}(n))^2 = (1/16)(n + 1)u(n - 2), \tag{6.1.74}$$

in accordance with Equation 5.1.57.

***Probability distribution***

In this problem there is little to choose between the occupancy methods of Section 5.1 and the tagging methods of this section in finding the moments. In general, the occupancy technique requires a slightly more complicated flow graph, whereas the tagging procedure involves differentiation of what can be large expressions. However, there is one area in which tagging has a unique advantage—the problem of finding the probability distribution of the number of special transitions. For example, we can use tagging to find a closed form expression for $\phi(k|n)$, the probability that $k$ Head-Tail sequences will be generated in $n$ tosses. What we must do is invert the transform of Equation 6.1.64 with respect to both $y$ and $z$.

We begin by writing $\phi^{gg}(y|z)$ in the form

$$\phi^{gg}(y|z) = \frac{1}{1 - z + (1/4)z^2 - (1/4)z^2 y} = \frac{\dfrac{1}{(1 - (1/2)z)^2}}{1 - \left(\dfrac{(1/2)z}{1 - (1/2)z}\right)^2 y}. \tag{6.1.75}$$

Now we can invert the geometric transform on the first variable by inspection to produce

$$\phi^{\cdot g}(k|z) = \frac{1}{(1 - (1/2)z)^2}\left(\frac{(1/2)z}{1 - (1/2)z}\right)^{2k} = \frac{((1/2)z)^{2k}}{(1 - (1/2)z)^{2k+2}}, \qquad k = 0, 1, 2, \ldots. \tag{6.1.76}$$

**Direct Interpretation.** This transform has a direct interpretation as the transmission of a flow graph. The flow graph appears in Figure 6.1.8. It is a flow graph with an infinite number of states, the first of this type that we have considered. As successive tosses are made, the system moves to the state to the right or stays in its present state. The first two states represent situations where no Head-Tail sequences have occurred, the next two represent situations where one Head-Tail sequence has occurred, etc. The probability $\phi(k|n)$ of making $k$ Head-Tail sequences in $n$ tosses is the probability of being in one of the two states corresponding to $k$ Head-Tail sequences after $n$ transitions. The transform on $n$ of this probability, $\phi^{\cdot g}(k|z)$, is the sum of the transmissions from the input node to each of the two states representing $k$ Head-Tail sequences in the flow graph. Since each two-state

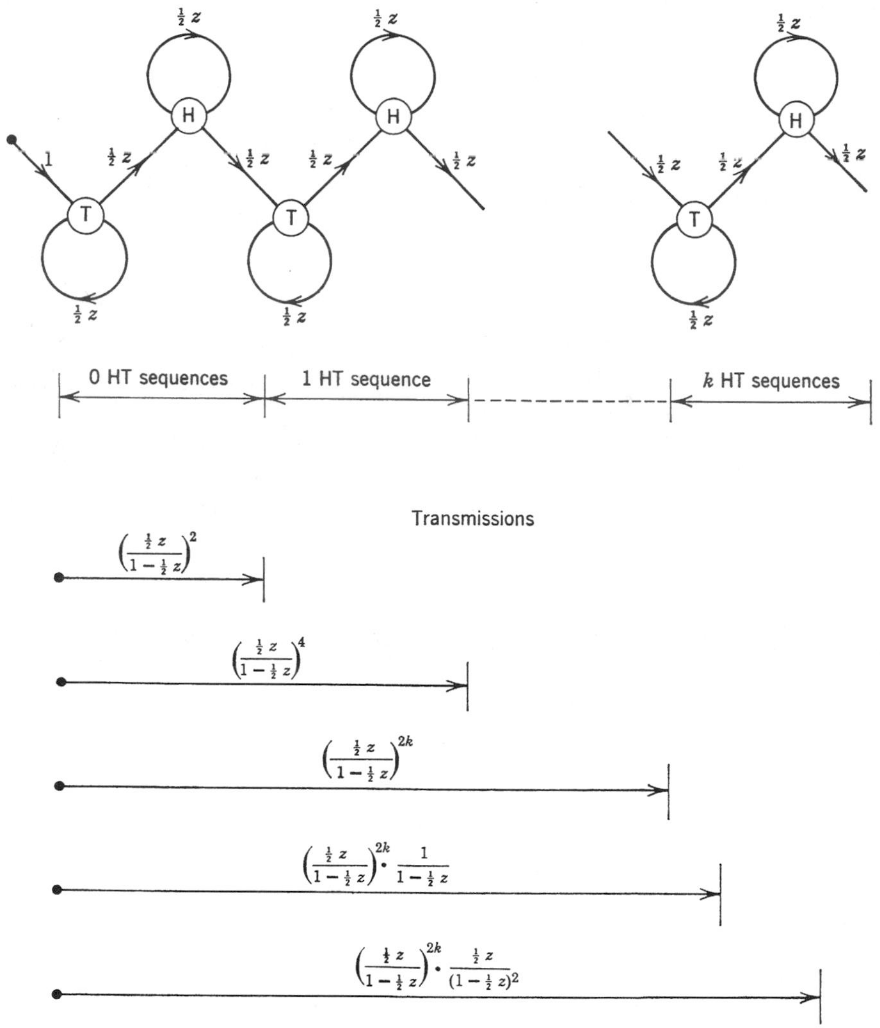


**Figure 6.1.8** Expanded flow graph for Head-Tail sequences.

section of the flow graph has a transmission $[(1/2)z/[1 - (1/2)z]]^2$, these transmissions are easy to compute. We see immediately that

$$\phi^{\mathcal{G}}(k|z) = \left(\frac{(1/2)z}{1-(1/2)z}\right)^{2k} \frac{1}{1-(1/2)z} + \left(\frac{(1/2)z}{1-(1/2)z}\right)^{2k} \frac{(1/2)z}{(1-(1/2)z)^2}$$

$$= \frac{[(1/2)z]^{2k}}{[1-(1/2)z]^{2k+2}}. \tag{6.1.77}$$

This result verifies Equation 6.1.76.

**Transform inversion.** We still face the task of inverting the remaining geometric transform in Equation 6.1.76. From our transform table relations, we find

$$\begin{aligned}
\frac{1}{(1-az)^{k+1}} &\Leftrightarrow \frac{(n+k)!}{k!\,n!}\,a^n \\
\frac{(az)^k}{(1-az)^{k+1}} &\Leftrightarrow \frac{n!}{k!\,(n-k)!}\,a^n \\
\frac{(az)^{k+1}}{(1-az)^{k+2}} &\Leftrightarrow \frac{n!}{(k+1)!\,(n-k-1)!}\,a^n \\
\frac{(az)^k}{(1-az)^{k+2}} &\Leftrightarrow \frac{(n+1)!}{(k+1)!\,(n-k)!}\,a^n \\
\frac{(az)^{2k}}{(1-az)^{2k+2}} &\Leftrightarrow \frac{(n+1)!}{(2k+1)!\,(n-2k)!}\,a^n = \binom{n+1}{2k+1}a^n \\
\frac{[(1/2)z]^{2k}}{[1-(1/2)z]^{2k+2}} &\Leftrightarrow \binom{n+1}{2k+1}(1/2)^n.
\end{aligned} \tag{6.1.78}$$

Therefore,

$$\phi(k|n) = \binom{n+1}{2k+1}(1/2)^n \qquad 0 \le k, n. \tag{6.1.79}$$

The probability of obtaining $k$ Head-Tail sequences in $n$ tosses of a fair coin is equal to the binomial coefficient $\binom{n+1}{2k+1}$ multiplied by $(1/2)^n$.

**Direct probabilistic argument.** We can confirm the result of Equation 6.1.79 by an elementary probabilistic argument. Each sequence of $n$ tosses must end with either of the two mutually exclusive outcomes Head or Tail. Therefore, the probability $\phi(k|n)$ that $n$ tosses will produce $k$ Head-Tail sequences is the sum of the probabilities of the two mutually exclusive events

$$E_1 = \{k \text{ Head-Tail sequences in } n \text{ tosses and } n\text{th toss is a Tail}\}$$

and

$$E_2 = \{k \text{ Head-Tail sequences in } n \text{ tosses and } n\text{th toss is a Head}\}.$$

Referring to Figure 6.1.8 and viewing it as a transition diagram, we see that the probability of the event $E_1$ is the probability that some subset of length $2k$ of the $n$ tosses forms the sequence

$$\overset{2k}{\overleftrightarrow{\text{HT HT} \cdots \text{HT}}}.$$

Since each sequence has probability $(1/2)^n$, and the number of ways we can select a sequence of length $2k$ out of $n$ tosses is $\binom{n}{2k}$, the probability of event $E_1$ is

$$\mathscr{P}\{E_1\} = \binom{n}{2k}(1/2)^n. \tag{6.1.80}$$

Similarly, the probability of the event $E_2$ is the probability that some subset of length $2k + 1$ of the $n$ tosses forms the sequence

$$\overset{2k+1}{\overleftrightarrow{\text{HT HT} \cdots \text{HTH}}}\,.$$

Each sequence of length $n$ has probability $(1/2)^n$, and the number of ways in which a subset of length $2k + 1$ can be selected from the $n$ tosses is $\binom{n}{2k+1}$. Therefore the probability of event $E_2$ is

$$\mathscr{P}\{E_2\} = \binom{n}{2k+1}(1/2)^n. \tag{6.1.81}$$

We can now write

$$\begin{aligned}
\phi(k|n) &= \mathscr{P}\{E_1\} + \mathscr{P}\{E_2\} = \binom{n}{2k}(1/2)^n + \binom{n}{2k+1}(1/2)^n \\
&= \left[\frac{n!}{(2k)!\,(n-2k)!} + \frac{n!}{(2k+1)!\,(n-2k-1)!}\right](1/2)^n \\
&= \frac{n!}{(2k)!\,(n-2k-1)!}\left[\frac{1}{n-2k} + \frac{1}{2k+1}\right](1/2)^n \\
&= \frac{n!}{(2k)!\,(n-2k-1)!}\left[\frac{n+1}{(n-2k)(2k+1)}\right](1/2)^n \\
&= \frac{(n+1)!}{(2k+1)!\,(n-2k)!}(1/2)^n = \binom{n+1}{2k+1}(1/2)^n.
\end{aligned} \tag{6.1.82}$$

We have thus obtained the result of Equation 6.1.79.

Table 6.1.3 shows for a few values of $k$ and $n$ the probability $\phi(k|n)$ computed using Equation 6.1.79. The means and variances appearing in this table confirm the results of Equations 6.1.67 and 6.1.74.

**Summary**

The preceding example illustrates the value of tagging and of flow graph analysis in general. Some of us may have sufficient insight to write down immediately the

**Table 6.1.3** Probability Distribution of Number of Head-Tail Sequences in $n$ Tosses of a Fair Coin

$$\phi(k|n) = \mathscr{P}\{k \text{ HT sequences in } n \text{ tosses}\} = \binom{n+1}{2k+1}(1/2)^n$$

| $\phi(k\|n)$ — $k \backslash n$ | 0 | 1 | 2 | 3 | 4 | 5 |
|---|---|---|---|---|---|---|
| 0 | 1 | 1 | 3/4 | 4/8 | 5/16 | 6/32 |
| 1 | 0 | 0 | 1/4 | 4/8 | 10/16 | 20/32 |
| 2 | 0 | 0 | 0 | 0 | 1/16 | 6/32 |
| 3 | 0 | 0 | 0 | 0 | 0 | 0 |
| 4 | 0 | 0 | 0 | 0 | 0 | 0 |
| Mean $\bar{k}(n) = \sum_{k=0}^{\infty} k\phi(k\|n)$ | 0 | 0 | 1/4 | 2/4 | 3/4 | 4/4 |
| Second moment $\overline{k^2}(n) = \sum_{k=0}^{\infty} k^2\phi(k\|n)$ | 0 | 0 | 1/4 | 1/2 | 7/8 | 11/8 |
| Variance $\overset{v}{k}(n) = \overline{k^2}(n) - (\bar{k}(n))^2$ | 0 | 0 | 3/16 | 4/16 | 5/16 | 6/16 |

solution to the Head-Tail sequence problem using elementary probabilistic reasoning—as we have done to obtain Equation 6.1.82. Anyone who can see the events that should be defined to yield the answer in almost every problem of this type has a rare gift. Most of us, however, are likely to obtain the solution in a shorter total time if we have a general structure in which to place the problem. The flow graph analysis of Markov processes is a structure that allows us to solve most problems with a minimum of ingenuity. Using ingenuity plus this structure we can treat problems that would be difficult to approach in any other way.

## 6.2 PROCESSES WITH PARTIAL INFORMATION

We have developed methods for finding the probability distribution over the states of a Markov process after $n$ transitions given its starting state. Sometimes we have additional information about what the process has been doing during the $n$ transitions. We should be able to use this information to modify the state probability distribution at time $n$. We shall call this modification procedure the analysis of processes with partial information; it is the topic of this section.

### Incorporation of Data on the Process Trajectory

We refer to the sequence of states occupied by the process during the times $0, 1, 2, \ldots, n$ as the trajectory of the process. The quantity of information we have about the trajectory of a Markov process can vary widely. At one extreme we may not know any of the activities of the process during this interval; at the other, we may know exactly which transition the process made at each transition time, i.e., the complete trajectory. These two possibilities delimit an information spectrum with which we must deal.

We shall let $\mathscr{D}$ represent whatever data are available about the process trajectory. We shall let $S_{mj}$ be the event that the system is in state $j$ at time $m$,

$$S_{mj} \equiv \{(s(m) = j)\} \tag{6.2.1}$$

and let $S_j$ be the event that the system trajectory ends in state $j$. We shall let $C_n$ be the event that $n$ transitions of the process have occurred ($C_n$ is mnenomic for $n$ changes of state). The event $C_n$ therefore implies that the system trajectory covers the times $0, 1, 2, \ldots, n$.

Then the process of incorporating the data in the probability assignment on the states at time $n$ is determined from basic probability theory by

$$\mathscr{P}\{S_j|\mathscr{D}S_{0i}C_n\} = \frac{\mathscr{P}\{S_j, \mathscr{D}|S_{0i}C_n\}}{\mathscr{P}\{\mathscr{D}|S_{0i}C_n\}} = \frac{\mathscr{P}\{S_j, \mathscr{D}|S_{0i}C_n\}}{\sum_{j=1}^{N} \mathscr{P}\{S_j, \mathscr{D}|S_{0i}C_n\}}. \tag{6.2.2}$$

The problem of using the data $\mathscr{D}$ is therefore reduced to finding $\mathscr{P}\{S_j, \mathscr{D}|S_{0i}C_n\}$, the joint probability of the data and of ending in each state after $n$ transitions given that the system started in state $i$. As we shall see, this calculation is easy when the data on the trajectory are presented in certain forms.

### *Description of trajectory data*

One of many ways to describe the information we might have on the trajectory of the process is to define a quantity $d_{ij}(m)$ by

$$d_{ij}(m) = \begin{cases} 1 & \text{if the system made its } m\text{th transition} \\ & \text{from state } i \text{ to state } j \\ 0 & \text{otherwise} \end{cases} \qquad 1 \le m \tag{6.2.3}$$

and a matrix $D(m)$ with elements $d_{ij}(m)$. Each matrix $D(m)$ will have exactly one non-zero entry corresponding to the transition made during $(m - 1, m)$. The sequence of matrices $D(1), D(2), \ldots, D(n)$ would be a rather bulky, but complete, way to specify the trajectory of the system during the interval $(0, n)$. It may help to understand this notation if we note that the probability of any trajectory can be written in the form

$$\mathscr{P}\{\text{trajectory} = D(1)D(2)\cdots D(m)\} = \prod_{m=1}^{n}\left[\sum_{i=1}^{N}\sum_{j=1}^{N} p_{ij}d_{ij}(m)\right]. \tag{6.2.4}$$

If we knew the whole trajectory of the process, there would be no problem at all. However, we usually only have partial information, if any. For example, we might know that at a particular time $m$ within the interval $(0, n)$, the system occupied a state $k$. In this case we have no difficulty in using the information because the Markovian property assures us that the trajectory in the range $(m, n)$ will depend only on the fact that the system was in state $k$ at time $m$ and not on the portion of the trajectory before time $m$. We are indeed fortunate if we have information of a kind that determines the trajectory at certain time points and thus simplifies our analysis.

### *Frequency count data*

A more common form of information arises when we know only the total number of transitions that the system has made between each pair of states, but not when those transitions occurred. Let the matrix $D^{\Sigma}$ be defined by

$$D^{\Sigma} = \sum_{m=1}^{n} D(m). \tag{6.2.5}$$

Then $d_{ij}^{\Sigma}$, the $ij$th element of $D^{\Sigma}$, is equal to the number of transitions the process has made from state $i$ to state $j$. The matrix $D^{\Sigma}$ is sometimes called the frequency count of the process. The sum of its elements must be equal to $n$, the total number of transitions the process has made. The sum of the elements of $D^{\Sigma}$ in the $i$th row gives the number of times state $i$ was entered; the sum of the elements in the $j$th column shows the number of times state $j$ was left.

If the partial information on the trajectory of the process is in the form of $D^{\Sigma}$, we can write Equation 6.2.2 in the form

$$\mathscr{P}\{S_j | D^{\Sigma} S_{0i} C_n\} = \frac{\mathscr{P}\{S_j D^{\Sigma} | S_{0i} C_n\}}{\sum_{j=1}^{N} \mathscr{P}\{S_j D^{\Sigma} | S_{0i} C_n\}}. \tag{6.2.6}$$

Now we face the problem of calculating $\mathscr{P}\{S_j D^{\Sigma} | S_{0i} C_n\}$. This is the probability of being in state $j$ at time $n$ and of having made $D^{\Sigma}$ transitions from each state to each other state. Happily, we learned how to solve problems of this type by tagging in the preceding section.

We begin by creating $N^2$ classes of special transitions, one for every type of transition the process can make, and then tag them by $N^2$ transform variables $y_1, y_2, \ldots, y_{N^2}$. By flow graph or recursive techniques we can then find the probability $\phi_{ij}(k_1, k_2, \ldots, k_{N^2} | n)$, which, when evaluated with $k_1 = d_{11}^{\Sigma}$, $k_2 = d_{12}^{\Sigma}, \ldots$, $k_{N^2} = d_{NN}^{\Sigma}$, is just $\mathscr{P}\{S_j D^{\Sigma} | S_{0i} C_n\}$. Therefore, tagging provides a complete solution to the partial information problem when the partial information is presented in the form of $D^{\Sigma}$, the frequency count of the process.†

Often the information on the process is even more limited than $D^{\Sigma}$. If only the row or column sums of $D^{\Sigma}$ are available, then we have information on state occupancies which we can incorporate by using only $N$ tagging variables. If we know only some of the elements of $D^{\Sigma}$, we can limit the number of tagging variables to the number of elements that are available—the other elements are ignored. Of course, the simplest cases from the point of view of analysis arise when we have information only on the number of transitions made between one pair of states, or when we know only the number of times one particular state was occupied. In these cases we can analyze the process with a single tagging variable and thereby simplify the calculation considerably.

It should be clear, however, that partial information $\mathscr{D}$ need not be in the form of the frequency count $D^{\Sigma}$ or its summarizations in order to write Equation 6.2.2 for $\mathscr{P}\{S_j|\mathscr{D}S_{0i}C_n\}$; all that is necessary is that we be able to compute $\mathscr{P}\{S_j, \mathscr{D}|S_{0i}C_n\}$. We learned how to perform this computation in the previous section for data $\mathscr{D}$ in the form of the number of special transitions of several classes, whether or not these classes be mutually exclusive or collectively exhaustive. Therefore, we can solve the problem of incorporating partial information on the trajectory whenever the probability of generating the partial information can be evaluated by tagging procedures.

### *Partial information in the marketing example*

Let us illustrate the procedure of analyzing partial information by returning to the marketing example. Suppose we know that the customer purchased brand $A$ on his zeroth purchase and that on the three purchase occasions $n = 1, 2, 3$ he bought brand $A$ exactly twice. What is the probability that his purchase on purchase occasion 3 was of brand $A$? We identify purchases of brand $A$ as special, tagged events. The tagged flow graph for this process is shown in Figure 6.1.6. Let $T_k$ be the event that $k$ tagged transitions have occurred. The probability we want to calculate is

$$\mathscr{P}\{S_1|S_{01}T_3C_3\} = \frac{\mathscr{P}\{S_1T_3|S_{01}C_3\}}{\mathscr{P}\{T_3|S_{01}C_3\}}. \tag{6.2.7}$$

We know

$$\phi_{ij}(k|n) = \mathscr{P}\{S_jT_k|S_{0i}C_n\}. \tag{6.2.8}$$

† In certain special cases the trajectory can be uniquely determined from the frequency count and the transition probability matrix by elementary means. However, such cases are primarily of interest to puzzle fans.

Therefore, what we must compute is

$$\mathscr{P}\{S_1|S_{01}T_3C_3\} = \frac{\phi_{11}(3|3)}{\phi_{11}(3|3) + \phi_{12}(3|3)} = \frac{\phi_{11}(3|3)}{\phi(3|3)}. \qquad (6.2.9)$$

The quantity $\phi(3|3)$ is the probability of 3 special transitions through time 3. It was calculated for this example in Table 6.1.1 as

$$\phi(3|3) = 0.224. \qquad (6.2.10)$$

The quantity $\phi_{11}(3|3)$ is the joint probability of 3 special transitions through time 3 and of ending in state 1 given that the system started in state 1. Table 6.1.2 shows that it is given by

$$\phi_{11}(3|3) = 0.096. \qquad (6.2.11)$$

Therefore, we have

$$\mathscr{P}\{S_1|S_{01}T_3C_3\} = \frac{0.096}{0.224} = 0.429 \qquad (6.2.12)$$

The probability that the customer has bought brand $A$ at time 3 given that he bought brand $A$ at time zero and at two of the three times 1, 2, 3 is 0.429. The probability that we would assign to the customer's buying brand $A$ at time 3 given only the information that the customer bought brand $A$ at time zero was computed in Equation 1.2.16 as

$$\phi_{11}(3) = \mathscr{P}\{S_1|S_{01}C_3\} = 0.65. \qquad (6.2.13)$$

Therefore, the evidence that the customer made two additional purchases of brand $A$ decreases the probability that he bought it at time 3. The decrease occurs because it is most likely that the additional purchases were made at times 1 and 2.

### Incorporation of Null Information

We have now solved, at least in principle, the problem of determining the influence of trajectory knowledge on the state probability vector at time $n$. There is one very important special case that we should explore. That is the case where we know that certain transitions of the process did not occur in the course of the trajectory. This situation arises when we have information of the form: "the rocket did not run out of fuel"; "no failures were observed"; "no purchases of brand $A$ were made."

We shall find that incorporating information of this type is very easy, a fortunate finding indeed because of the prevalence of such "null information" in practical problems.

We proceed by identifying each of the transitions that did not occur as special transitions and tagging them in the flow graph with a transform variable $y$. We are then interested in $\mathscr{P}\{S_j|T_0S_{0i}C_n\}$, the probability of being in state $j$ at time $n$ given that the system started in state $i$ at time zero and that it has made no tagged transitions. Since in this case the data $\mathscr{D}$ is just $T_0$, Equation 6.2.2 takes the form

$$\mathscr{P}\{S_j|T_0S_{0i}C_n\} = \frac{\mathscr{P}\{S_jT_0|S_{0i}C_n\}}{\sum_{j=1}^{N} \mathscr{P}\{S_jT_0|S_{0i}C_n\}}. \tag{6.2.14}$$

Thus the key to calculating the effect of knowing that no tagged transitions were made lies in evaluating $\mathscr{P}\{S_jT_0|S_{0i}C_n\}$, the joint probability of being in state $j$ at time $n$ and having made no tagged transitions. In view of Equation 6.2.8, this probability is given by

$$\mathscr{P}\{S_jT_0|S_{0i}C_n\} = \phi_{ij}(0|n), \tag{6.2.15}$$

a quantity that is readily determined by tagging. However, in this case, as we shall see, the results we obtain are particularly simple.

We first write the transform of $\phi_{ij}(k|n)$ given by Equation 6.1.24,

$$\phi_{ij}^{gg}(y|z) = \sum_{k=0}^{\infty} y^k \sum_{n=0}^{\infty} z^n\phi_{ij}(k|n). \tag{6.2.16}$$

Now we evaluate this transform at $y = 0$,

$$\phi_{ij}^{gg}(0|z) = \sum_{n=0}^{\infty} z^n\phi_{ij}(0|n) = \phi_{ij}^{\cdot g}(0|z). \tag{6.2.17}$$

We see that evaluating the transform $\phi_{ij}^{gg}(y|z)$ at $y = 0$ yields the geometric transform on $n$ of $\phi_{ij}(0|n)$. This result is really an application of the initial value theorem for the geometric transform. Therefore, we can find the matrix of transforms $\Phi^{\cdot g}(0|z)$ by evaluating $\Phi^{gg}(y|z)$ at $y = 0$.

Equation 6.1.27 shows that

$$\Phi^{gg}(y|z) = [I - yz\,{}_sP - z(P - {}_sP)]^{-1}, \tag{6.2.18}$$

where ${}_sP$ is the part of the transition probability matrix referring to special or tagged transitions. We then have

$$\Phi^{\cdot g}(0|z) = \Phi^{gg}(0|z) = [I - z(P - {}_sP)]^{-1}. \tag{6.2.19}$$

The matrix $\Phi^{\cdot g}(0|z)$ is therefore the matrix of transmissions of a flow graph from which all tagged branches have been removed. Inverse transformation then provides $\Phi(0|n)$, the matrix of quantities necessary to find $\mathscr{P}\{S_j|T_0S_{0i}C_n\}$ for any $i$ and $j$ by using Equations 6.2.14 and 6.2.15.

We have thus arrived at a particularly simple way to incorporate the knowledge that certain transitions have not occurred. We merely remove from the flow graph of the process the branches corresponding to these transitions, then find the transfer functions $\phi_{ij}{}^{\mathscr{g}}(0|z)$ and impulse responses $\phi_{ij}(0|n)$ of the remaining flow graph. The probability of being in state $j$ after $n$ transitions given that no special transitions occurred and that the process started in state $i$ is then

$$\mathscr{P}\{S_j|T_0S_{0i}C_n\} = \frac{\phi_{ij}(0|n)}{\sum_{j=1}^{N} \phi_{ij}(0|n)} = \frac{\phi_{ij}(0|n)}{\phi_{i\Sigma}(0|n)}. \qquad (6.2.20)$$

Notice that we never have to deal with the transform variable $y$ explicitly—the procedure is no more difficult than finding the multistep transition probabilities of a Markov process. In fact, by inverting the transform of Equation 6.2.19 we have immediately

$$\Phi(0|n) = (P - {}_sP)^n \qquad 0 \leq n \qquad (6.2.21)$$

so that numerical calculations are very simple. Note, however, that $\Phi(0|n)$ will not be a stochastic matrix for $n > 0$.

*Null information in the marketing example*

Let us now apply this theory to some examples. We shall begin with the marketing problem and assume that we know that the customer bought brand $A$ on his zeroth purchase and that he never bought brand $A$ two times in a row. We want to find the probability that he buys brand $B$ as his $n$th purchase. Since the customer never bought brand $A$ two times in a row, he never made a transition from state 1 to state 1. We therefore eliminate the loop around node 1 from the flow graph for the process in Figure 3.1.3 to obtain the flow graph of Figure 6.2.1.

The probability we want is $\mathscr{P}\{S_2|T_0S_{01}C_n\}$. In view of Equation 6.2.20 we have

$$\mathscr{P}\{S_2|T_0S_{01}C_n\} = \frac{\phi_{12}(0|n)}{\phi_{11}(0|n) + \phi_{12}(0|n)} = \frac{\phi_{12}(0|n)}{\phi_{1\Sigma}(0|n)}. \qquad (6.2.22)$$

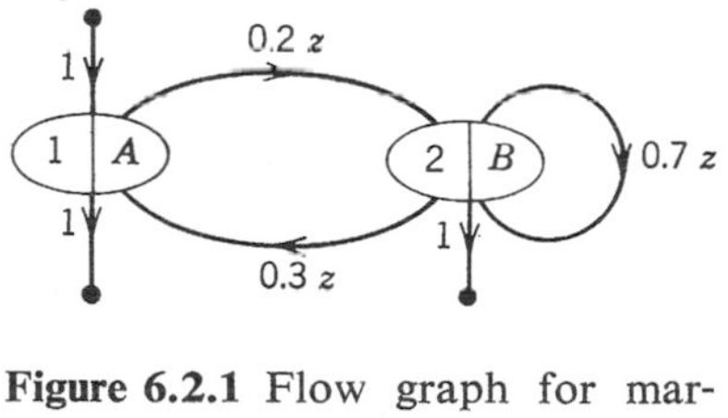


**Figure 6.2.1** Flow graph for marketing example with no transitions from state 1 to state 1.

The transforms $\phi_{11}{}^{\mathscr{g}}(0|z)$ and $\phi_{12}{}^{\mathscr{g}}(0|z)$ of $\phi_{11}(0|n)$ and $\phi_{12}(0|n)$ are equal to the transmissions $\vec{t}_{11}$ and $\vec{t}_{12}$ of the flow graph. Thus,

$$\vec{t}_{11} = \phi_{11}{}^{\mathscr{g}}(0|z) = \frac{1 - 0.7z}{1 - 0.7z - 0.06z^2}$$
$$\vec{t}_{12} = \phi_{12}{}^{\mathscr{g}}(0|z) = \frac{0.2z}{1 - 0.7z - 0.06z^2}. \tag{6.2.23}$$

Since the denominator expressions do not factor neatly, we shall write the difference equations for $\phi_{11}(0|n)$ and $\phi_{12}(0|n)$ implied by these transforms,

$$\phi_{11}(0|n) = 0.7\phi_{11}(0|n-1) + 0.06\phi_{11}(0|n-2) + \delta(n) - 0.7\,\delta(n-1) \quad 0 \le n$$
$$\phi_{12}(0|n) = 0.7\phi_{12}(0|n-1) + 0.06\phi_{12}(0|n-2) + 0.2\,\delta(n-1) \quad 0 \le n \tag{6.2.24}$$

and then solve them recursively for a few values of $n$. The result appears in Table 6.2.1.

Both $\phi_{11}(0|n)$ and $\phi_{12}(0|n)$ decrease with $n$ because the probability of never having made tagged transitions grows smaller as the number of transitions increases. The sum of the $\phi_{11}(0|n)$ and $\phi_{12}(0|n)$ is $\phi_{1\Sigma}(0|n)$, the probability of no tagged transitions through time $n$ regardless of ending state. It too must decrease with $n$ and ultimately approach zero. The ratio of $\phi_{12}(0|n)$ to $\phi_{1\Sigma}(0|n)$ is the probability we want, the probability that the customer will purchase brand $B$ at time $n$ if he bought brand $A$ at time zero and never bought brand $A$ two times in a row. Notice that this probability rapidly approaches a value near 0.72 as $n$ increases.

We might be interested in how this probability depends on the starting state, the initial brand purchased. We would expect that after a few purchases the initial state would make no difference so that even if $B$ were purchased originally, the probability that $B$ is bought at time $n$ given that brand $A$ was never bought twice

**Table 6.2.1** Marketing Example with No Transitions from State 1 to State 1

| $n$ | $\phi_{11}(0\|n)$ | $\phi_{12}(0\|n)$ | $\phi_{11}(0\|n) + \phi_{12}(0\|n) = \phi_{1\Sigma}(0\|n)$ | $\mathscr{P}\{S_2\|T_0S_{01}C_n\} = \dfrac{\phi_{12}(0\|n)}{\phi_{1\Sigma}(0\|n)}$ |
|---|---|---|---|---|
| 0 | 1 | 0 | 1 | 0 |
| 1 | 0 | 0.2 | 0.2 | 1 |
| 2 | 0.06 | 0.14 | 0.2 | 0.7 |
| 3 | 0.042 | 0.11 | 0.152 | 0.72368 |
| 4 | 0.033 | 0.0854 | 0.1184 | 0.72128 |
| 5 | 0.02562 | 0.06638 | 0.092 | 0.72152 |
| 6 | 0.019914 | 0.05159 | 0.071504 | 0.72150 |

in a row would also approach the value near 0.72. Direct calculation will affirm this result, but we shall not show the computations.

### *A tempting sophism*

Before proceeding to another example, we shall examine an alternate method, often proposed by students, for incorporating information that certain transitions have not occurred in the trajectory. The reasoning behind this procedure is this: If we know that a transition from, say, state 3 to state 4 did not occur in the trajectory, then we know that every time the system entered state 3 it behaved as if a transition to state 4 were impossible. Therefore, we shall set the transition probability from state 3 to state 4 in the flow graph equal to zero *and* multiply all the other transition probabilities out of state 3 by $1/(1 - p_{34})$ to meet the requirement that the sum of the transition probabilities out of any state be 1. We do this for every type of transition we know did not occur. The resulting flow graph then represents a Markov process whose multistep transition probabilities provide any statistics of the process that we desire.

The only thing wrong with this procedure is that it does not work. There is no reason why it should work, and yet it is so appealing to the mind that it arises spontaneously again and again. Let us put the idea to rest by showing that it gives the wrong answer for the example we have just worked.

We begin by noting that since no transition was made from state 1 to state 1 we should set $p_{11}$ equal to zero and then make $p_{12}$ equal to 1 to reflect the fact that every time the system entered state 1 it went to state 2. The result is the Markov flow graph of Figure 6.2.2. We again assume that the customer bought brand $A$ at time zero and that we want to know the probability of his buying brand $B$ at time $n$. The conjecture is that the transform of this probability is the transmission of the flow graph of Figure 6.2.2 from state 1 to state 2. We find

$$\vec{t}_{12} = \frac{z}{1 - 0.7z - 0.3z^2} = \phi_{12}{}^{g}(z) \tag{6.2.25}$$

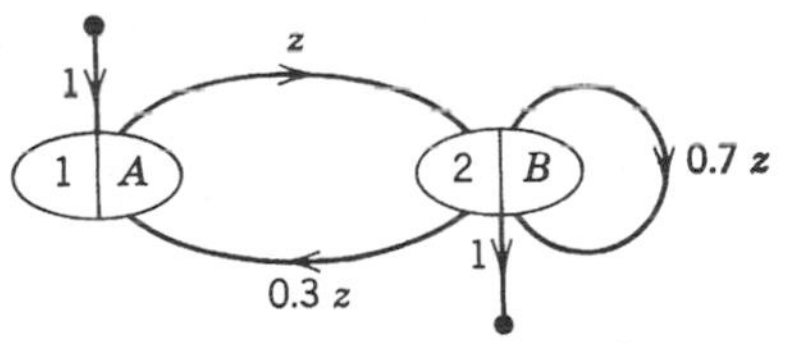


**Figure 6.2.2** A conjectured flow graph for the marketing example with no transitions from state 1 to state 2.

or

$$\phi_{12}{}^{g}(z) = \frac{z}{1 - 0.7z - 0.3z^2} = \frac{z}{(1 - z)(1 + 0.3z)}$$

$$= \frac{10/13}{1 - z} + \frac{-10/13}{1 + 0.3z}. \tag{6.2.26}$$

Then by inverse geometric transformation,

$$\phi_{12}(n) = 10/13[1 - (-0.3)^n] \qquad 0 \le n. \tag{6.2.27}$$

The numerical values for the first few $n$ are:

$$\begin{aligned}
\phi_{12}(0) &= 0\\
\phi_{12}(1) &= 1\\
\phi_{12}(2) &= 0.7\\
\phi_{12}(3) &= 0.79\\
\phi_{12}(4) &= 0.763\\
\phi_{12}(5) &= 0.7711\\
\phi_{12}(6) &= 0.76867\\
&\vdots\\
\phi_{12}(\infty) &= 10/13 = 0.76923
\end{aligned} \tag{6.2.28}$$

The values of $\phi_{12}(n)$ in Equation 6.2.28 are not the correct values for $\mathscr{P}\{S_2|T_0S_{01}C_n\}$ found in Table 6.2.1. We see in particular that the asymptotic values of the two expressions are different. Note that the two functions agree for $n = 0, 1, 2$. This initial agreement is due to the fact that we know the exact trajectory of the system through time 1. We know that the system was in state 1 when $n = 0$, and that it moved to state 2 when $n = 1$. Since it was in state 2 at $n = 1$, there is a 0.7 chance that it will stay in state 2 until time 2 and 0.3 chance that it will return to 1. Consequently, the flow graph of Figure 6.2.2 accurately describes the behavior of the system through time 2. After that time it is just not appropriate.

We must always remember that though there are many tricks we can use in flow graph analysis, each of them is based on proper logical relationships. It is the course of danger to attempt flow graph modifications without mathematical verification.

### *Null information in a coin-tossing problem*

Let us now return to a problem first encountered in Section 4.2. There we determined the probability distribution of the number of tosses of a fair coin required to produce a sequence of three Heads. Suppose it is known that $n$ tosses have been made and no such sequence has been produced. What is the probability that the last toss of the coin completed a sequence of 0, 1, or 2 Heads, the three mutually exclusive possibilities?

To answer this question we return to the transition diagram of Figure 4.2.6. Since we know that no HHH sequence has been completed, the system can have made no transitions from state 3 to state 4. This is the only transition that need be designated as special, or tagged. Therefore, in accordance with our procedure we remove this branch from the diagram. Since the coin is fair ($p = 1/2$) we obtain immediately the flow graph for the process shown in Figure 6.2.3.

We want to find $\mathscr{P}\{S_1|T_0S_{01}C_n\}$, $\mathscr{P}\{S_2|T_0S_{01}C_n\}$, and $\mathscr{P}\{S_3|T_0S_{01}C_n\}$. From Equation 6.2.20 these probabilities are equal to:

$$\mathscr{P}\{S_1|T_0S_{01}C_n\} = \frac{\phi_{11}(0|n)}{\phi_{11}(0|n) + \phi_{12}(0|n) + \phi_{13}(0|n)} = \frac{\phi_{11}(0|n)}{\phi_{1\Sigma}(0|n)}$$

$$\mathscr{P}\{S_2|T_0S_{01}C_n\} = \frac{\phi_{12}(0|n)}{\phi_{1\Sigma}(0|n)} \tag{6.2.29}$$

$$\mathscr{P}\{S_3|T_0S_{01}C_n\} = \frac{\phi_{13}(0|n)}{\phi_{1\Sigma}(0|n)}$$

The transforms of $\phi_{11}(0|n)$, $\phi_{12}(0|n)$, and $\phi_{13}(0|n)$ are the transmissions $\vec{t}_{11}$, $\vec{t}_{12}$, and $\vec{t}_{13}$ of the flow graph of Figure 6.2.3:

$$\vec{t}_{11} = \phi_{11}^{g}(0|z) = \frac{1}{1 - (1/2)z - (1/4)z^2 - (1/8)z^3}$$

$$\vec{t}_{12} = \phi_{12}^{g}(0|z) = \frac{(1/2)z}{1 - (1/2)z - (1/4)z^2 - (1/8)z^3} \tag{6.2.30}$$

$$\vec{t}_{13} = \phi_{13}^{g}(0|z) = \frac{(1/4)z^2}{1 - (1/2)z - (1/4)z^2 - (1/8)z^3}$$

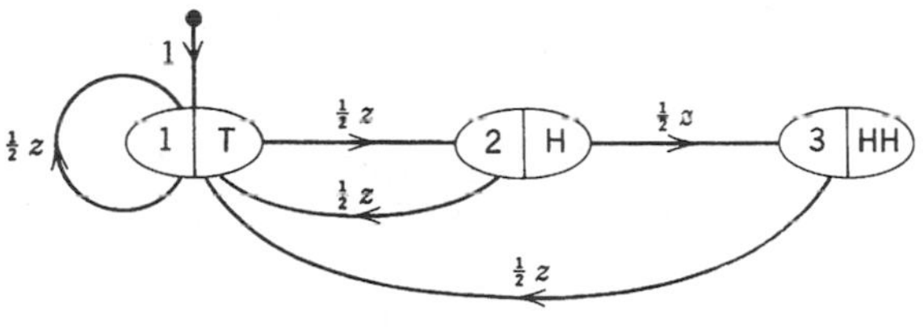


**Figure 6.2.3** Flow graph for no HHH sequence in $n$ tosses of a fair coin.

Let us now restrict our attention to the case where $n$ is large. By the same arguments used in Section 4.2, we find the dominant components of these transforms to be

$$\begin{aligned}\phi_{11}{}^{\mathscr{G}}(0|z) &\to \frac{k}{1 - 0.91964z}\\ \phi_{12}{}^{\mathscr{G}}(0|z) &\to \frac{(1/2)zk}{1 - 0.91964z}\\ \phi_{13}{}^{\mathscr{G}}(0|z) &\to \frac{(1/4)z^2k}{1 - 0.91964z}\end{aligned} \tag{6.2.31}$$

where $k = 0.61842$ is a constant that need not concern us. Then for large $n$ we can write the inverse transforms as

$$\begin{aligned}\phi_{11}(0|n) &= k(0.91964)^n\\ \phi_{12}(0|n) &= (1/2)k(0.91964)^{n-1}\\ \phi_{13}(0|n) &= (1/4)k(0.91964)^{n-2} \qquad n \text{ large.}\end{aligned} \tag{6.2.32}$$

The sum $\phi_{1\Sigma}(0|n)$ of these quantities is given by

$$\begin{aligned}\phi_{1\Sigma}(0|n) &= \phi_{11}(0|n) + \phi_{12}(0|n) + \phi_{13}(0|n)\\ &= (1/4)k(0.91964)^{n-2}[4(0.91964)^2 + 2(0.91964) + 1]\\ &= (1/4)k(0.91964)^{n-2}(6.22223) \qquad n \text{ large.}\end{aligned} \tag{6.2.33}$$

Now we use the results of Equations 6.2.32 and 6.2.33 in Equation 6.2.29 to produce the probability of being in each of the states 1, 2, 3 when the number of tosses is large and no HHH sequence has been obtained. We find

$$\begin{aligned}\mathscr{P}\{S_1|T_0S_{01}C_n\} &= \frac{\phi_{11}(0|n)}{\phi_{1\Sigma}(0|n)} = \frac{4(0.91964)^2}{6.22223} = 0.54369\\ \mathscr{P}\{S_2|T_0S_{01}C_n\} &= \frac{\phi_{12}(0|n)}{\phi_{1\Sigma}(0|n)} = \frac{2(0.91964)}{6.22223} = 0.29560\\ \mathscr{P}\{S_3|T_0S_{01}C_n\} &= \frac{\phi_{13}(0|n)}{\phi_{1\Sigma}(0|n)} = \frac{1}{6.22223} = 0.16071 \qquad n \text{ large.}\end{aligned} \tag{6.2.34}$$

We see from these results that the information that no sequence of three Heads has been completed in a large number of tosses makes it most probable that only the last toss was a Tail. There is about one chance in six that the last toss completed a sequence of two Heads. If the system is in this state, a Head on the next toss will complete the sequence of three Heads. Therefore, the probability that such a sequence will be completed on the next toss is $(0.16071)(1/2) = 0.08036$. The probability that a sequence of three Heads will not be completed on the next toss is therefore 1 − 0.08036 = 0.91964. This result was first observed in Section 4.2 in Table 4.2.1 and Figure 4.2.13. All these results hold, of course, only when the number of tosses is "large." However, we know from our work in Section 4.2 that 6 or 7 tosses are enough for the asymptotic expressions to be valid.

### Summary

Thus the problems of incorporating partial information on a trajectory are solved by simple extensions of our earlier work. The power of the tagging techniques reveals itself once more in solving the partial information problem. The case of null information on certain transitions is an important and rather easily handled special form of the theory.

## 6.3 INFERENCE

We can answer all the problems we have considered up to this time by knowing the basic probabilistic structure of the process—the transition probability matrix and the manner in which the starting state is selected. From this information we have been able to determine the probabilistic state of the system after $n$ transitions, either with or without additional information on the trajectory of the system during that interval. However, we shall now consider a situation where the number of transitions the process has made is not known, but must be inferred on the basis of information concerning the trajectory. The feature of particular interest in these problems is that they cannot be answered by using only the basic probabilistic structure of the process, i.e., the transition probability matrix and the manner of starting.

### Inference on Process Duration

We can see the issues more clearly by supposing that we have some data $\mathscr{D}$ on the trajectory, that we know the system was started in state $i$, and that we want to know the joint probability that the system has made $n$ transitions and ended in state $j$. Our notation for this probability is $\mathscr{P}\{C_nS_j|\mathscr{D}S_{0i}\}$; by elementary probability theory we can expand it into

$$\mathscr{P}\{C_nS_j|\mathscr{D}S_{0i}\} = \frac{\mathscr{P}\{C_nS_j\mathscr{D}|S_{0i}\}}{\mathscr{P}\{\mathscr{D}|S_{0i}\}} = \frac{\mathscr{P}\{C_nS_j\mathscr{D}|S_{0i}\}}{\sum_{n=0}^{\infty}\sum_{j=1}^{N}\mathscr{P}\{C_nS_j\mathscr{D}|S_{0i}\}}$$

$$= \frac{\mathscr{P}\{S_j\mathscr{D}|C_nS_{0i}\}\mathscr{P}\{C_n|S_{0i}\}}{\sum_{n=0}^{\infty}\sum_{j=1}^{N}\mathscr{P}\{S_j\mathscr{D}|C_nS_{0i}\}\mathscr{P}\{C_n|S_{0i}\}}. \tag{6.3.1}$$

Since the denominator of this expression is just the sum of the numerator over all values of $n$ and $j$, we can find the joint probability of the number of transitions and the ending state given the data and the starting state if we can find each of the numerator factors. The first numerator factor, $\mathscr{P}\{S_j\mathscr{D}|C_nS_{0i}\}$, is the joint probability of the ending state and the data given the number of transitions and the starting state. We considered this problem at length in the last two sections for the case when the data $\mathscr{D}$ are represented in a form susceptible to tagging analysis; we shall limit our present consideration to this situation. We then have no difficulty in principle in calculating the first factor of the numerator in Equation 6.3.1.

The second numerator factor, $\mathscr{P}\{C_n|S_{0i}\}$, is the probability distribution of the number of transitions given only the starting state. In some situations, as we shall see, this distribution is determinable from the probabilistic structure of the process. However, more often it must be provided externally, just like the transition probability matrix and the initial state probability vector. This situation arises when the number of transitions is not governed by the process itself, but by some other mechanism. If the number of transitions the process is allowed to make does not depend on its starting state, then we can use the simple notation $\mathscr{P}\{C_n\}$ rather than $\mathscr{P}\{C_n|S_{0i}\}$ as the probability distribution on transitions. For convenience we shall call the number of transitions the process makes the "duration" of the process. Then $\mathscr{P}\{C_n\}$ or $\mathscr{P}\{C_n|S_{0i}\}$ is the duration distribution of the process, either unconditional or conditional on starting state.

When we know both $\mathscr{P}\{S_j\mathscr{D}|C_nS_{0i}\}$ from tagging and $\mathscr{P}\{C_n|S_{0i}\}$ by specification, we can compute $\mathscr{P}\{C_nS_j|\mathscr{D}S_{0i}\}$, the joint probability of the duration and the ending state given the data and the starting state directly from Equation 6.3.1. This quantity then serves to produce a number of other interesting probabilities conditional on the data and the starting state: for example, the marginal distribution of duration,

$$\mathscr{P}\{C_n|\mathscr{D}S_{0i}\} = \sum_{j=1}^{N} \mathscr{P}\{C_nS_j|\mathscr{D}S_{0i}\}; \tag{6.3.2}$$

the marginal distribution of ending state,

$$\mathscr{P}\{S_j|\mathscr{D}S_{0i}\} = \sum_{n=0}^{\infty} \mathscr{P}\{C_nS_j|\mathscr{D}S_{0i}\}; \tag{6.3.3}$$

the conditional distribution of duration given the ending state,

$$\mathscr{P}\{C_n|S_j\mathscr{D}S_{0i}\} = \frac{\mathscr{P}\{C_nS_j|\mathscr{D}S_{0i}\}}{\mathscr{P}\{S_j|\mathscr{D}S_{0i}\}}; \tag{6.3.4}$$

and the conditional distribution of ending state given the duration,

$$\mathscr{P}\{S_j|C_n\mathscr{D}S_{0i}\} = \frac{\mathscr{P}\{C_nS_j|\mathscr{D}S_{0i}\}}{\mathscr{P}\{C_n|\mathscr{D}S_{0i}\}}. \tag{6.3.5}$$

Since this last probability is conditional on the number of transitions, it does not depend on the duration distribution, $\mathscr{P}\{C_n|S_{0i}\}$, and is given more directly by Equation 6.2.2.

Therefore, the core of the problem is to find $\mathscr{P}\{S_j\mathscr{D}|C_nS_{0i}\}$, the joint probability of ending state and data given the duration and the initial state, and to use it in Equation 6.3.1. If the data we have are that $k$ special or tagged transitions occurred in the trajectory, then our problem is to find

$$\mathscr{P}\{S_j\mathscr{D}|C_nS_{0i}\} = \mathscr{P}\{S_jT_k|C_nS_{0i}\} = \phi_{ij}(k|n), \tag{6.3.6}$$

and we can solve the problem using tagging with a single tagging variable.

***Inference on duration in the marketing example***

Let us now illustrate how to solve inference problems on process duration with an example. Suppose that in our marketing example we know that the customer bought brand $A$ at time zero and that he made $k$ changes of brand from brand $A$ to brand $B$. What is the probability that he has made $n$ purchases after his initial purchase? What is the probability that his last purchase was of brand $B$?

To answer these questions, we designate transitions from brand $A$ to brand $B$ as special transitions, and draw the tagged flow graph of Figure 6.3.1. Since the data $\mathscr{D}$ in this case are that $k$ special transmissions have been made, we can express the data simply as $T_k$ and write the fundamental inference Equation 6.3.1 in the form

$$\mathscr{P}\{C_nS_j|T_kS_{01}\} = \frac{\phi_{1j}(k|n)\mathscr{P}\{C_n\}}{\sum_{n=0}^{\infty}[\phi_{11}(k|n) + \phi_{12}(k|n)]\mathscr{P}\{C_n\}} \qquad \begin{matrix} k = 0, 1, 2, \ldots \\ n = 0, 1, 2, \ldots \\ j = 1, 2. \end{matrix} \tag{6.3.7}$$

To compute this quantity we must first specify $\mathscr{P}\{C_n\}$, the probability distribution for duration of the process before knowing of the $k$ changes from brand $A$ to brand $B$. We shall assume that the customer is equally likely to buy 1, 2, 3, 4, or 5 times after the initial purchase, and that this duration is independent of the brand purchased at time zero. Therefore,

$$\mathscr{P}\{C_n\} = 1/5 \qquad n = 1, 2, 3, 4, 5. \tag{6.3.8}$$

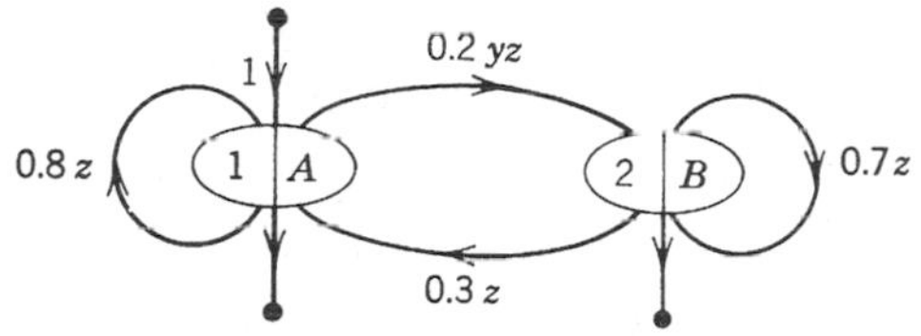


**Figure 6.3.1** Tagged flow graph for marketing inference problem.

Now we proceed to calculate the joint probability of the ending state and the data given the duration and starting state, $\phi_{ij}(k|n)$, for use in Equation 6.3.7. The transmissions $\vec{t}_{11}$ and $\vec{t}_{12}$ of the flow graph in Figure 6.3.1 are $\phi_{11}{}^{gg}(y|z)$ and $\phi_{12}{}^{gg}(y|z)$,

$$\begin{aligned} \vec{t}_{11} &= \phi_{11}{}^{gg}(y|z) = \frac{1 - 0.7z}{1 - 1.5z + 0.56z^2 - 0.06yz^2} \\ \vec{t}_{12} &= \phi_{12}{}^{gg}(y|z) = \frac{0.2yz}{1 - 1.5z + 0.56z^2 - 0.06yz^2}. \end{aligned} \tag{6.3.9}$$

The inverse transforms of these quantities are then

$$\begin{aligned} \phi_{11}(k|n) &= \mathscr{P}\{S_1 T_k | S_{01} C_n\} \\ \phi_{12}(k|n) &= \mathscr{P}\{S_2 T_k | S_{01} C_n\}. \end{aligned} \tag{6.3.10}$$

Therefore, the only remaining problem is to compute the inverse transforms of Equation 6.3.9. We do this by writing the transforms in the forms

$$\begin{aligned} \phi_{11}{}^{gg}(y|z) &= 1 - 0.7z + 1.5z\phi_{11}{}^{gg}(y|z) - 0.56z^2\phi_{11}{}^{gg}(y|z) + 0.06yz^2\phi_{11}{}^{gg}(y|z) \\ \phi_{12}{}^{gg}(y|z) &= 0.2yz + 1.5z\phi_{12}{}^{gg}(y|z) - 0.56z^2\phi_{12}{}^{gg}(y|z) + 0.06yz^2\phi_{12}{}^{gg}(y|z) \end{aligned} \tag{6.3.11}$$

and then writing the difference equations,

$$\begin{aligned} \phi_{11}(k|n) = {} & \delta(k)\,\delta(n) - 0.7\,\delta(k)\,\delta(n-1) + 1.5\phi_{11}(k|n-1) - 0.56\phi_{11}(k|n-2) \\ & + 0.06\phi_{11}(k-1|n-2) \\ \phi_{12}(k|n) = {} & 0.2\,\delta(k-1)\,\delta(n-1) + 1.5\phi_{12}(k|n-1) - 0.56\phi_{12}(k|n-2) \\ & + 0.06\phi_{12}(k-1|n-2) \qquad 0 \le k, n. \end{aligned} \tag{6.3.12}$$

Tables 6.3.1 and 6.3.2 show the values of $\phi_{11}(k|n)$ and $\phi_{12}(k|n)$ for small arguments. Note that the sum over $k$ of each of these quantities gives the corresponding multistep transition probability.

**Table 6.3.1** Computation of $\phi_{11}(k|n) = \mathscr{P}\{S_1 T_k | S_{01} C_n\}$

$\phi_{11}(k|n)$

| $k \backslash n$ | 0 | 1 | 2 | 3 | 4 | 5 | $\sum_{n=1}^{5} \phi_{11}(k\|n)$ |
|---|---|---|---|---|---|---|---|
| 0 | 1 | 0.8 | 0.64 | 0.512 | 0.4096 | 0.32768 | 2.68928 |
| 1 | 0 | 0 | 0.06 | 0.138 | 0.2118 | 0.27114 | 0.68094 |
| 2 | 0 | 0 | 0 | 0 | 0.0036 | 0.01368 | 0.01728 |
| 3 | 0 | 0 | 0 | 0 | 0 | 0 | 0 |
| 4 | 0 | 0 | 0 | 0 | 0 | 0 | 0 |
| 5 | 0 | 0 | 0 | 0 | 0 | 0 | 0 |
| $\phi_{11}(n) = \sum_{k=0}^{\infty} \phi_{11}(k\|n)$ | 1 | 0.8 | 0.7 | 0.65 | 0.625 | 0.6125 | |

**Table 6.3.2** Computation of $\phi_{12}(k|n) = \mathscr{P}\{S_2 T_k | S_{01} C_n\}$

| $\phi_{12}(k\|n)$ $k \backslash n$ | 0 | 1 | 2 | 3 | 4 | 5 | $\sum_{k=1}^{5} \phi_{12}(k\|n)$ |
|---|---|---|---|---|---|---|---|
| 0 | 0 | 0 | 0 | 0 | 0 | 0 | 0 |
| 1 | 0 | 0.2 | 0.3 | 0.338 | 0.339 | 0.31922 | 1.49622 |
| 2 | 0 | 0 | 0 | 0.012 | 0.036 | 0.06756 | 0.11556 |
| 3 | 0 | 0 | 0 | 0 | 0 | 0.00072 | 0.00072 |
| 4 | 0 | 0 | 0 | 0 | 0 | 0 | 0 |
| 5 | 0 | 0 | 0 | 0 | 0 | 0 | 0 |
| $\phi_{12}(n) = \sum_{k=0}^{\infty} \phi_{12}(k\|n)$ | 0 | 0.2 | 0.3 | 0.35 | 0.375 | 0.3875 | |

By using the distribution of Equation 6.3.8 in Equation 6.3.7 we obtain

$$\mathscr{P}\{C_n S_1 | T_k S_{01}\} = \frac{\phi_{11}(k|n)}{\sum_{n=1}^{5} \phi_{11}(k|n) + \sum_{n=1}^{5} \phi_{12}(k|n)}$$

$$\mathscr{P}\{C_n S_2 | T_k S_{01}\} = \frac{\phi_{12}(k|n)}{\sum_{n=1}^{5} \phi_{11}(k|n) + \sum_{n=1}^{5} \phi_{12}(k|n)}. \tag{6.3.13}$$

These two quantities are calculated in Table 6.3.3 for $k = 0, 1, 2, 3$ by using Tables 6.3.1 and 6.3.2. Note that when $k$ is larger than 3 it is impossible for the number of transitions to have been less than 5; therefore, the tabulated values of $k$ are the only values of interest. Each section of the table shows the joint probability of the duration and ending state for the corresponding value of $k$. The row sums of these arrays show the marginal probability of ending state, $\mathscr{P}\{S_j | T_k S_{01}\}$; the column sums are the marginal probability of duration, $\mathscr{P}\{C_n | T_k S_{01}\}$.

When $k = 0$, the customer has never made a purchase of brand $B$. Therefore, it is not surprising to see that $\mathscr{P}\{S_1 | T_0 S_{01}\} = 1$. The marginal distribution on duration for this case, $\mathscr{P}\{C_n | T_0 S_{01}\}$, shows that the customer is most likely to have made only one purchase of the product after his initial purchase.

When $k = 1$, the customer has switched from brand $A$ to brand $B$ exactly once. Since $\mathscr{P}\{S_2 | T_1 S_{01}\} = 0.68723$ we see that his ending purchase was more likely to have been of brand $B$. The duration distribution $\mathscr{P}\{C_n | T_1 S_{01}\}$ shows that the customer is now most likely to have made all five additional purchases.

When $k = 2$, the customer has twice switched from brand $A$ to brand $B$. As we would expect, the probability that his last purchase was brand $B$ is now higher; $\mathscr{P}\{S_2 | T_2 S_{01}\} = 0.86992$. From $\mathscr{P}\{C_n | T_2 S_{01}\}$ we find he could not have made fewer

**Table 6.3.3** Computation of Inference Results in Marketing Problem for $\mathscr{P}\{C_n\} = 1/5$ $n = 1, 2, 3, 4, 5$

| $\mathscr{P}\{C_nS_j \mid T_0S_{01}\}$ | $j \setminus n$ | 1 | 2 | 3 | 4 | 5 | $\mathscr{P}\{S_j \mid T_0S_{01}\}$ |
|---|---|---|---|---|---|---|---|
| $k = 0$ | 1 | 0.29748 | 0.23798 | 0.19039 | 0.15231 | 0.12185 | 1.00000 |
| | 2 | 0 | 0 | 0 | 0 | 0 | 0 |
| $\mathscr{P}\{C_n \mid T_0S_{01}\}$ | | 0.29748 | 0.23798 | 0.19039 | 0.15231 | 0.12185 | 1.00000 |

| $\mathscr{P}\{C_nS_j \mid T_1S_{01}\}$ | $j \setminus n$ | 1 | 2 | 3 | 4 | 5 | $\mathscr{P}\{S_j \mid T_1S_{01}\}$ |
|---|---|---|---|---|---|---|---|
| $k = 1$ | 1 | 0 | 0.02756 | 0.06339 | 0.09728 | 0.12454 | 0.31277 |
| | 2 | 0.09186 | 0.13779 | 0.15525 | 0.15571 | 0.14662 | 0.68723 |
| $\mathscr{P}\{C_n \mid T_1S_{01}\}$ | | 0.09186 | 0.16535 | 0.21864 | 0.25299 | 0.27116 | 1.00000 |

| $\mathscr{P}\{C_nS_j \mid T_2S_{01}\}$ | $j \setminus n$ | 1 | 2 | 3 | 4 | 5 | $\mathscr{P}\{S_j \mid T_2S_{01}\}$ |
|---|---|---|---|---|---|---|---|
| $k = 2$ | 1 | 0 | 0 | 0 | 0.02710 | 0.10298 | 0.13008 |
| | 2 | 0 | 0 | 0.09033 | 0.27100 | 0.50858 | 0.86992 |
| $\mathscr{P}\{C_n \mid T_2S_{01}\}$ | | 0 | 0 | 0.09033 | 0.29810 | 0.61156 | 1.00000 |

| $\mathscr{P}\{C_nS_j \mid T_3S_{01}\}$ | $j \setminus n$ | 1 | 2 | 3 | 4 | 5 | $\mathscr{P}\{S_j \mid T_3S_{01}\}$ |
|---|---|---|---|---|---|---|---|
| $k = 3$ | 1 | 0 | 0 | 0 | 0 | 0 | 0 |
| | 2 | 0 | 0 | 0 | 0 | 1 | 1 |
| $\mathscr{P}\{C_n \mid T_3S_{01}\}$ | 0 | 0 | 0 | 0 | 0 | 1 | 1 |

than three additional purchases and that a duration of length 5 has probability 0.61156.

When $k = 3$, we know the trajectory of the process exactly: the customer must have alternated purchases between the two brands for all five additional purchases. Therefore, we are sure that he is in state 2 and that the number of additional purchases is five.

If we knew the ending state of the trajectory, we could infer the duration probability distribution from Equation 6.3.4 with $\mathscr{D} = T_k$,

$$\mathscr{P}\{C_n|S_jT_kS_{0i}\} = \frac{\mathscr{P}\{C_nS_j|T_kS_{0i}\}}{\mathscr{P}\{S_j|T_kS_{0i}\}}. \tag{6.3.14}$$

The computation would require merely dividing each row in Table 6.3.3 by the sum of that row. Similarly, by dividing each column by the column sum we would obtain $\mathscr{P}\{S_j|C_nT_kS_{0i}\}$. However, we know this quantity does not depend on the duration distribution; we treated it separately in the last section.

### *"Noncommittal" inference in the marketing example*

Occasionally you encounter people who are reluctant to place a duration distribution $\mathscr{P}\{C_n\}$ on the process. They claim to "know nothing" about how long the process has been operating, but still want to draw inferences about the duration given the data. Without passing judgment on this point of view, let us examine its implications. One interpretation of "knowing nothing" about the duration is that the duration has the distribution

$$\mathscr{P}\{C_n\} = \frac{1}{\gamma + 1} \qquad n = 0, 1, 2, \ldots, \gamma, \tag{6.3.15}$$

where $\gamma$ is a number that is very, very large. We shall call this distribution the "maximally noncommittal" distribution for duration. With this duration distribution Equation 6.3.7 for the marketing example becomes

$$\mathscr{P}\{C_nS_j|T_kS_{01}\} = \frac{\phi_{1j}(k|n)\cdot\dfrac{1}{\gamma+1}}{\sum_{n=0}^{\gamma}[\phi_{11}(k|n) + \phi_{12}(k|n)]\dfrac{1}{\gamma+1}} = \frac{\phi_{1j}(k|n)}{\sum_{n=0}^{\infty}\phi_{11}(k|n) + \sum_{n=0}^{\infty}\phi_{12}(k|n)}. \tag{6.3.16}$$

Notice that we have used $\infty$ as the upper limit of the summation because $\gamma$ is very large by assumption. The joint probability of duration and ending state is therefore independent of $\gamma$. We have already calculated $\phi_{ij}(k|n)$ for the marketing example for a few values of $k$ and $n$, but the denominator requires the sum of these quantities for all values of $n$. Fortunately, we can find the sums in the denominator from the transforms of $\phi_{11}(k|n)$ and $\phi_{12}(k|n)$. The basic transform definition shows that

$$\begin{aligned}
\sum_{n=0}^{\infty}\phi_{11}(k|n) &= \sum_{n=0}^{\infty}\phi_{11}(k|n)z^n\Big|_{z=1} = \phi_{11}{}^{\mathscr{G}}(k|1) \\
\sum_{n=0}^{\infty}\phi_{12}(k|n) &= \sum_{n=0}^{\infty}\phi_{12}(k|n)z^n\Big|_{z=1} = \phi_{12}{}^{\mathscr{G}}(k|1).
\end{aligned} \tag{6.3.17}$$

From Equation 6.3.9,

$$\begin{aligned}\phi_{11}{}^{gg}(y|1) &= \frac{0.3}{0.06(1-y)} = \frac{5}{1-y}\\ \phi_{12}{}^{gg}(y|1) &= \frac{0.2y}{0.06(1-y)} = \frac{(10/3)y}{1-y}.\end{aligned} \tag{6.3.18}$$

Therefore, by inverting the geometric transform on $y$, we have

$$\begin{aligned}\phi_{11}{}^{\cdot g}(k|1) &= \sum_{n=0}^{\infty} \phi_{11}(k|n) = 5u(k) \qquad 0 \le k\\ \phi_{12}{}^{\cdot g}(k|1) &= \sum_{n=0}^{\infty} \phi_{12}(k|n) = \frac{10}{3}u(k-1) \qquad 0 \le k.\end{aligned} \tag{6.3.19}$$

The denominator of Equation 6.3.16 is then

$$\sum_{n=0}^{\infty} \phi_{11}(k|n) + \sum_{n=0}^{\infty} \phi_{12}(k|n) = \begin{cases} 5 & k = 0 \\ 25/3 & k = 1, 2, 3, \dots . \end{cases} \tag{6.3.20}$$

Now that we have the denominator of Equation 6.3.16 we can use this equation and the results of Tables 6.3.1 and 6.3.2 to compute the joint distributions $\mathscr{P}\{C_n S_j | T_k S_{01}\}$ for the first few values of $k$. The computations appear in Table 6.3.4. The column sums are again the marginal probability on duration. However, the row sums require summation over all values of $n$, even those not included in the table. We compute them by noting that the sum of Equation 6.3.16 over all values of $n$ shows

$$\begin{aligned}\mathscr{P}\{S_j | T_k S_{01}\} &= \sum_{n=0}^{\infty} \mathscr{P}\{C_n S_j | T_k S_{01}\} = \frac{\sum_{n=0}^{\infty} \phi_{1j}(k|n)}{\sum_{n=0}^{\infty} \phi_{11}(k|n) + \sum_{n=0}^{\infty} \phi_{12}(k|n)}\\ &= \frac{\phi_{1j}{}^{\cdot g}(k|1)}{\phi_{11}{}^{\cdot g}(k|1) + \phi_{12}{}^{\cdot g}(k|1)}.\end{aligned} \tag{6.3.21}$$

Therefore, from the results of Equation 6.3.19 the probability of having an ending state 1 is 1 if $k = 0$ and 0.6 if $k$ is larger than zero. Thus, for a maximally noncommittal duration distribution, we assign the limiting state probabilities to the ending states as long as at least one special transition has occurred.

The duration distribution $\mathscr{P}\{C_n | T_k S_{01}\}$ for the different values of $k$ shows that only when $k = 0$ do we assign high probabilities to small numbers of transitions. As $k$ increases from 1 to 3 the chances of a duration less than 5 diminish from fair (probability $\approx 1/4$) to infinitesimal (probability $= 0.000086$). Comparing Table

**Table 6.3.4** Computation of Inference Results in Marketing Problem for "Noncommittal" $\mathcal{P}(C_n)$

| $\mathcal{P}\{C_nS_j\|T_0S_{01}\}$ | $j \backslash n$ | 0 | 1 | 2 | 3 | 4 | 5 | $\mathcal{P}\{S_j\|T_0S_{01}\}$ |
|---|---|---|---|---|---|---|---|---|
| $k = 0$ | 1 | 0.2 | 0.16 | 0.128 | 0.1024 | 0.08192 | 0.065536 | 1 |
| | 2 | 0 | 0 | 0 | 0 | 0 | 0 | 0 |
| $\mathcal{P}\{C_n\|T_0S_{01}\}$ | | 0.2 | 0.16 | 0.128 | 0.1024 | 0.08192 | 0.065536 | |

| $\mathcal{P}\{C_nS_j\|T_1S_{01}\}$ | $j \backslash n$ | 0 | 1 | 2 | 3 | 4 | 5 | $\mathcal{P}\{S_j\|T_1S_{01}\}$ |
|---|---|---|---|---|---|---|---|---|
| $k = 1$ | 1 | 0 | 0 | 0.0072 | 0.01656 | 0.02542 | 0.03254 | 0.6 |
| | 2 | 0 | 0.024 | 0.0360 | 0.04056 | 0.04068 | 0.03831 | 0.4 |
| $\mathcal{P}\{C_n\|T_1S_{01}\}$ | | 0 | 0.024 | 0.0432 | 0.05712 | 0.06610 | 0.07085 | |

| $\mathcal{P}\{C_nS_j\|T_2S_{01}\}$ | $j \backslash n$ | 0 | 1 | 2 | 3 | 4 | 5 | $\mathcal{P}\{S_j\|T_2S_{01}\}$ |
|---|---|---|---|---|---|---|---|---|
| $k = 2$ | 1 | 0 | 0 | 0 | 0 | 0.000432 | 0.001642 | 0.6 |
| | 2 | 0 | 0 | 0 | 0.00144 | 0.004320 | 0.008107 | 0.4 |
| $\mathcal{P}\{C_n\|T_2S_{01}\}$ | | 0 | 0 | 0 | 0.00144 | 0.004752 | 0.009749 | |

| $\mathcal{P}\{C_nS_j\|T_3S_{01}\}$ | $j \backslash n$ | 0 | 1 | 2 | 3 | 4 | 5 | $\mathcal{P}\{S_j\|T_3S_{01}\}$ |
|---|---|---|---|---|---|---|---|---|
| $k = 3$ | 1 | 0 | 0 | 0 | 0 | 0 | 0 | 0.6 |
| | 2 | 0 | 0 | 0 | 0 | 0 | 0.000086 | 0.4 |
| $\mathcal{P}\{C_n\|T_3S_{01}\}$ | | 0 | 0 | 0 | 0 | 0 | 0.000086 | |

6.3.4 with Table 6.3.3 shows the major role played by the initial duration distribution $\mathcal{P}\{C_n\}$ in drawing inferences about the process. In most practical situations, it is not too difficult to assign a reasonable duration distribution $\mathcal{P}\{C_n\}$ and therefore to draw meaningful inferences about the process by observing the number of special transitions made.

### Inference in Processes with Natural Termination

Fortunately, in many examples the duration distribution is supplied by the process itself. In these cases we have no need for soul-searching about $\mathcal{P}\{C_n|S_{0i}\}$—

we derive it from the transition probability matrix for the process. A situation of this type arises when the process makes transitions *until* it generates the data $\mathscr{D}$. We shall call these processes, "processes with natural termination." In particular, we shall consider processes that operate until they make $k$ special or tagged transitions.

The basic problem is then to make an inference about the duration of a trajectory given its initial state and the exact number of special transitions that the trajectory was required to contain as a criterion for termination. We shall let $Q_k$ be the event that the process is stopped after $k$ special transitions ($Q$ for quit). Then the probability we are interested in is $\mathscr{P}\{C_n S_j | S_{0i} Q_k\}$, the joint probability of duration and ending state given starting state and the method of stopping.

To write this probability we shall first define two additional events. Let $T_{n,k}$ be the event that $k$ special or tagged transitions are made through $n$ transitions. Let $E_n$ be the event that the $n$th transition is a special transition ($E$ for extraordinary). Then we can write the probability we seek as

$$\mathscr{P}\{C_n S_j | S_{0i} Q_k\} = \mathscr{P}\{T_{n-1,k-1} E_n S_{nj} | S_{0i}\}. \tag{6.3.22}$$

This equation states that the probability of a duration $n$ and an ending state $j$ is equal to the joint probability of obtaining $k - 1$ special transitions on the first $n - 1$ transitions, having a special transition on the $n$th transition, and entering state $j$ on the $n$th transition. Since the system must have been in some state $r$ at time $n - 1$, we are able to write Equation 6.3.22 in the form

$$\mathscr{P}\{C_n S_j | S_{0i} Q_k\} = \sum_{r=1}^{N} \mathscr{P}\{S_{n-1,r} T_{n-1,k-1} E_n S_{nj} | S_{0i}\}. \tag{6.3.23}$$

By the Markovian property that makes the trajectory from $n - 1$ to $n$ dependent only on the state of the system at time $n - 1$, we can factor the probability in Equation 6.3.23,

$$\mathscr{P}\{C_n S_j | S_{0i} Q_k\} = \sum_{r=1}^{N} \mathscr{P}\{S_{n-1,r} T_{n-1,k-1} | S_{0i}\} \mathscr{P}\{E_n S_{nj} | S_{n-1,r}\}. \tag{6.3.24}$$

We have previously expressed the probability $\mathscr{P}\{S_{n-1,r} T_{n-1,k-1} | S_{0i}\}$ in the form,

$$\mathscr{P}\{S_{n-1,r} T_{n-1,k-1} | S_{0i}\} = \mathscr{P}\{S_r T_{k-1} | C_{n-1} S_{0i}\} = \phi_{ir}(k - 1 | n - 1). \tag{6.3.25}$$

The joint probability $\mathscr{P}\{E_n S_{nj} | S_{n-1,r}\}$ is the probability of a special transition from state $r$ to state $j$, a probability we have denoted by ${}_s p_{rj}$,

$$\mathscr{P}\{E_n S_{nj} | S_{n-1,r}\} = {}_s p_{rj}. \tag{6.3.26}$$

Therefore in the notation of Equations 6.3.25 and 6.3.26 we can write Equation 6.3.24 as

$$\mathscr{P}\{C_n S_j | S_{0i} Q_k\} = \sum_{r=1}^{N} \phi_{ir}(k - 1|n - 1)\ {}_s p_{rj}. \tag{6.3.27}$$

Thus the joint probability of duration and ending state given the starting state and the condition that the process will stop after $k$ special transitions is computable directly from the function $\phi_{ij}(k|n)$ derived from tagging and the part of the transition probability matrix corresponding to special transitions. As we would expect, the duration of the trajectory in this case is determined by the transition probability matrix for the process.

***Inference in the marketing example with natural termination***

We can apply these results to the marketing example where we tagged transitions from brand $A$ to brand $B$ as shown in Figure 6.3.1. Suppose we know that the customer bought brand $A$ initially and that the customer's behavior pattern was that he stopped buying anything when he switched from brand $A$ to brand $B$ $k$ times; that is, when he made $k$ special transitions. What is the probability that when he stopped buying he had made $n$ purchases after his initial purchase? What is the probability that when he stopped buying his last purchase was of each brand?

The answers to these questions can be determined from $\mathscr{P}\{C_n S_j | S_{01} Q_k\}$, the joint probability of duration and ending state given that the initial state was state 1 and that the process will end after $k$ special transitions. From Equation 6.3.27 we write

$$\mathscr{P}\{C_n S_j | S_{01} Q_k\} = \sum_{r=1}^{2} \phi_{1r}(k - 1|n - 1)\ {}_s p_{rj}. \tag{6.3.28}$$

Since the only transition in the process that is special is the transition from state 1 to state 2 and since this transition has probability 0.2, ${}_sP$, the part of the transition probability matrix $P$ applicable to special transitions in this example, is

$${}_sP = \begin{bmatrix} 0 & 0.2 \\ 0 & 0 \end{bmatrix}. \tag{6.3.29}$$

Because ${}_s p_{11} = {}_s p_{21} = 0$,

$$\mathscr{P}\{C_n S_1 | S_{01} Q_k\} = 0. \tag{6.3.30}$$

It is impossible for the system to be in state 1 when it stops because it must end on a special transition and special transitions end in state 2.

Therefore we can write

$$\mathscr{P}\{C_n S_2 | S_{01} Q_k\} = \mathscr{P}\{C_n | S_{01} Q_k\} = \phi_{11}(k - 1|n - 1)\ {}_s p_{12} \tag{6.3.31}$$

or

$$\mathscr{P}\{C_n|S_{01}Q_k\} = 0.2\phi_{11}(k-1|n-1). \tag{6.3.32}$$

The duration of the process has a distribution proportional to $\phi_{11}(k-1|n-1)$. Table 6.3.5 shows this duration distribution for a few values of $n$ and $k$ as computed from Equation 6.3.32 and Table 6.3.1. As we would expect, the duration distribution has its mass shifted more and more toward larger values of $n$ as $k$ increases.

**Table 6.3.5** Duration Distributions for the Marketing Problem with Natural Termination

$\mathscr{P}\{C_n|S_{01}Q_k\}$

| $k \backslash n$ | 1 | 2 | 3 | 4 | 5 | 6 |
|---|---|---|---|---|---|---|
| 1 | 0.2 | 0.16 | 0.128 | 0.1024 | 0.08192 | 0.065536 |
| 2 | 0 | 0 | 0.012 | 0.0276 | 0.04236 | 0.054228 |
| 3 | 0 | 0 | 0 | 0 | 0.00072 | 0.002736 |
| 4 | 0 | 0 | 0 | 0 | 0 | 0 |
| 5 | 0 | 0 | 0 | 0 | 0 | 0 |

We see that there is no inherent difficulty in treating processes with natural termination. The duration of these processes is not an inference problem because the duration distribution is determined from the probabilistic structure of the process. Processes with natural termination are frequently encountered as system models—their analysis yields readily to tagging methods.

## Summary

The problem of inference in Markov models as we have defined inference is basically that of using data on the process trajectory to assign probabilities to its duration and ending state. We have found that our basic tool of tagging is of both conceptual and computational advantage in obtaining a solution. However, we have also seen, in passing, the need to define carefully the events in the process to insure correct probabilistic results.

## PROBLEMS

**1.** For the taxicab driver in Problem 5 of Chapter 4:

a) What is the probability that after starting in $A$, making $n$ trips and ending in $A$, he has made exactly one trip from $B$ to $A$?

b) Find by inspection the probability that he makes exactly one trip from $B$ to $A$ in the first four trips. Assume he starts in $A$. Compare your answer to the one found in a).

c) Find by inspection the probability mass function for the number of $B$-to-$A$ trips in the first four trips. Assume he starts in $A$.

**2.** Sam, the door-to-door salesman, has observed that after making a sale his attitude improves to the point where his probability of making a sale is increased to 0.6, whereas if he fails to make a sale his discouragement reduces his probability of making a sale to 0.1.

a) If Sam's probability of making an initial sale is 0.1, what are the mean and variance of the number of sales Sam will make in $n$ customer contacts? Check your results for $n = 1, 2$.

b) Find, by tagging, the geometric transform of the mean and mean square of the number of sales in $n$ contacts and check with your results in a).

c) If Sam makes a sale on his $n$th customer contact and decides to quit for the day, what is Sam's mean number of sales for the day?

**3.** Joe Fink has been having some trouble with his wife, Mabel. In desperation, Joe has found that he can model his arguments with Mabel by a two-state Markov process. If he has had no argument with Mabel today, then his probability of having an argument tomorrow is 0.2; but if he has had an argument today, his probability of having one tomorrow is 0.6.

Joe has had no argument today. What is the probability that Joe's *second* argument with Mabel occurs $n$ days from now?

**4.** a) Find the inverse transform of $z^k/(1 - az)^{k+1}$ as follows:

i) Let $$\frac{z^k}{(1 - az)^{k+1}} = \sum_{n=0}^{\infty} f(k, n)z^n$$

ii) Find $f^{gg}(y, z)$

iii) Find $f^{g}(y, n)$

iv) Find $f(k, n)$

b) Let $q(k, n)$ = probability of getting the $k$th success on the $n$th trial in a Bernoulli process (the probability of success is $p$). Find the geometric transform $q^{g}(k, z)$.

**5.** In the two flow graphs below, $a$ and $b$ are between 0 and 1. Of what two quantities are $\vec{t}_1(z)$ and $\vec{t}_2(y, z)$ the transforms? By considering the physical similarities of these two quantities, write down the mathematical relation between $\vec{t}_1(z)$ and $\vec{t}_2(y, z)$.

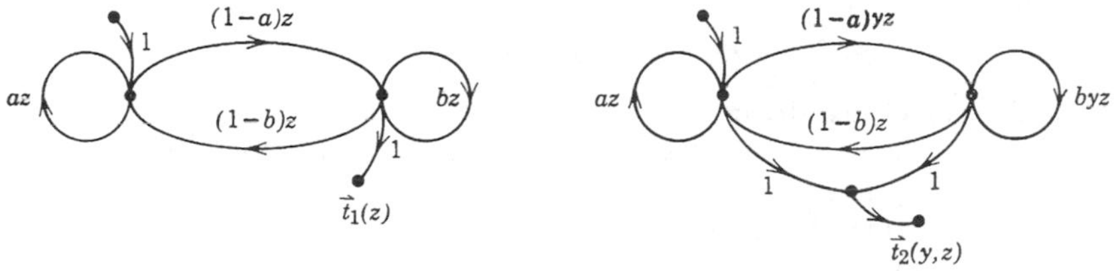

**6.** The taxi driver in Problem 3, Chapter 4, had the transition diagram:

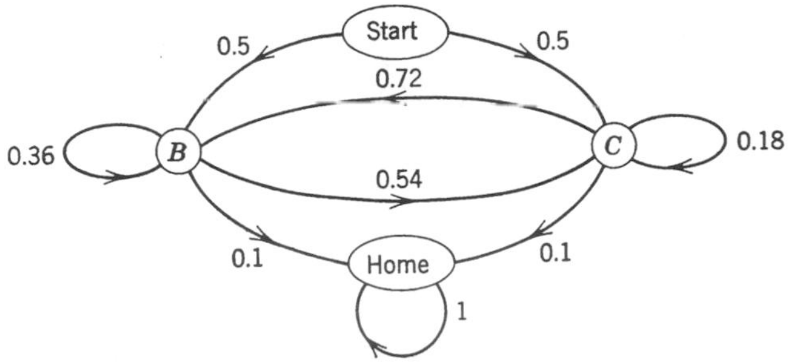


a) Find the unconditional distribution of the total number of $B$ to $C$ trips that the driver makes before going home. Prove that your method is correct.

b) Find the transform of the probability that he has carried $k$ fares from $B$ to $C$ and that he is not home yet given that he has carried $n$ fares.

c) Find a recursive relationship for your answer to part a).

d) Find the expected number of fares he carries from $B$ to $C$ given that he has carried $n$ fares and is not home yet.

**7.** a) Yogi Bear lives in Forest $A$, and spends each night there. During the day he filches picnic baskets (or parts of picnic baskets) from the tourists who frequent both Forest $A$ and Forest $B$. If he is in Forest $A$, there is a 2/3 probability that the next tourist he spots will be in $A$, and he will get one picnic basket from such a tourist. On the other hand, if he must travel to Forest $B$ (with probability 1/3) he will get 1/3 picnic basket. Similarly, if he is in Forest $B$, the next tourist he sees is in $B$ with probability 3/5, and he gets 2/3 of a picnic basket; he must travel to $A$ with probability 2/5 without receiving any picnic basket for his trouble. During the day Yogi has seen $n$ tourists, and has returned to Forest $A$ for the last tourist. Explain in detail how you would calculate the expected number of picnic baskets he has mooched during the day?

b) The average bear lives in Forest $B$ and has the same probability of spotting tourists as does our hero. He also always spends the night at home. However, he gets 1 basket if he stays in $B$, 2/3 if he stays in $A$, 1/3 if he goes from $B$ to $A$, and none if he goes from $A$ to $B$. Yogi wants to know if he really stuffs down more goodies per tourist than the average bear when both bears have seen a large number of tourists. (*Hint:* Being smarter than the average bear, Yogi figures out a quick method, which he also uses to check his answer for part a) when $n$ is large.)

**8.** An independent taxi driver plies his trade within and between towns $A$ and $B$. He has found that if he is in town $A$ his probability of going to town $B$ on his next trip is 0.5, and of staying in town $A$ is 0.5. Similarly, if he is in town $B$ his probability of going to $A$ on his next trip is 0.4 and of staying in $B$ is 0.6. He has also found that his chances of stopping for gas during a trip are:

| | |
|---|---|
| $A$ to $A$ | 0.30 |
| $A$ to $B$ | 0.75 |
| $B$ to $A$ | 0.85 |
| $B$ to $B$ | 0.25 |

a) If he starts in town $A$ and is in town $B$ $n$ trips later, what is the expected number of time he stopped for gas?

b) What is the number of trips he can expect to complete before stopping for gas if he starts in $A$? If he starts in $B$?

**9.** Mr. X operates a taxi between towns $A$ and $B$. If he has just finished a trip that terminated in town $A$, there is a probability of 0.2 that he will quit for the day, a probability of 0.3 that his next trip will be back to $A$, and a probability of 0.5 that his next trip will be to $B$. Similarly, if he is in town $B$ the probability of quitting is 0.4, of taking his next trip to $A$ is 0.2, and of taking a trip back to $B$ is 0.4.

Unfortunately, due to the Hack Law, our friend only receives a fare for $A \to A$, $A \to B$, and $B \to B$ trips.

If Mr. X has started the day in town $A$, made $k$ fare-paying trips and finally quit for the day, what is the probability that he finished the day in town $A$? (You may want to consider this as a four-state Markov process with two trapping states.)

**10.** An ore boat carries loads in and between two lakes. A trip within either lake takes 5 hours. Because of currents, trips from lake $A$ to lake $B$ take 7 hours while those in the opposite direction take only 3 hours. The boat moves according to the transition probability matrix

$$P = \text{From} \begin{matrix} & \text{To} \\ & \begin{matrix} A & B \end{matrix} \\ \begin{matrix} A \\ B \end{matrix} & \begin{bmatrix} 1/2 & 1/2 \\ 1/4 & 3/4 \end{bmatrix} \end{matrix}$$

A boat initially in lake $A$ makes $n$ trips. What is the probability distribution of the time spent in traveling?

a) Solve the problem by tagging each branch with an appropriate power of $y$.

b) Verify the limiting form of your answer by a simple argument based upon where the boat ends up plus the limiting state probabilities of the process.

**11.** Consider an $N$-state Markov process in which there are two kinds of transitions: $p$-type transitions and $q$-type transitions. If the process is currently in state $i$ then $p_{ij}$ is the probability that the next transition is a $p$-type transition to state $j$ and $q_{ij}$ is the probability that the next transition is a $q$-type transition to state $j$. Thus, since all transitions are either $p$-type or $q$-type we have $\sum_{j=1}^{N} (p_{ij} + q_{ij}) = 1$. (For convenience sake you can think of this as an $N$-city taxicab problem in which $p$-type and $q$-type transitions correspond to trips with and without stops for gas, respectively.)

Let $\omega_i$ = the number of $p$-type transitions made before the first $q$-type transition if the process starts in state $i$,

$f_i(n) = \mathscr{P}\{\omega_i = n\}$.

a) Consider all the transitions that might be made on the *first* transition and then write a difference equation for $f_i(n)$.

b) Take the geometric transform of this equation.

c) Draw the flow graph whose outputs are $f_1^g(z)$ and $f_2^g(z)$ for the following process:

$$P = \begin{bmatrix} 0.3 & 0.4 \\ 0.2 & 0.1 \end{bmatrix} \qquad Q = \begin{bmatrix} 0.2 & 0.1 \\ 0.5 & 0.2 \end{bmatrix}$$

**12.** Suppose that we are interested in the number of $(j \to k)$ transitions that a process will make if it starts in $i$, makes $n$ transitions, and ends in $l$. One method of calculating the mean number of such special transitions is to tag the $(j \to k)$ transition and then apply Equation 6.1.35:

$$\bar{k}_{il}(n) = \frac{k_{il}(n)}{\phi_{il}(n)}.$$

Without using tagging, find an expression for the geometric transform of the numerator of the equation in terms of the transition probabilities and transforms of the multistep transition probabilities. [*Hint:* Define an appropriate random variable $x_{ijkl}(m)$.]

**13.** Let $\tau_{ij}$ equal the number of time intervals required for a Markov process to make a direct transition from state $i$ to state $j$ (for the processes we have considered we have always had $\tau_{ij} = 1$). Let

$\phi_{ij}(k|n) = \mathscr{P}\{k$ time intervals have elapsed and the process is in state $j$ after $n$ transitions given that the process was initially in state $i\}$.

a) Use the principles of tagging to describe the general matrix and flow graph solutions for

$$\phi_{ij}^{gg}(y|z) = \sum_{k=0}^{\infty} y^k \sum_{n=0}^{\infty} z^n \phi_{ij}(k|n).$$

b) Let $f_{ij}(k, n) = \mathscr{P}\{$the first passage from state $i$ to state $j$ requires $k$ time intervals and $n$ transitions given that the process was initially in state $i\}$. Write a two-dimensional difference equation relating $\phi_{ij}(k|n)$ and $f_{ij}(k, n)$. Let $f_{ij}^{gg}(y, z)$ be the two-dimensional geometric transform of $f_{ij}(k, n)$. Solve the transformed difference equation for $f_{ij}^{gg}(y, z)$ in terms of $\phi_{ij}^{gg}(y|z)$.

c) Let $f_{ij}(k)$ be the usual first passage time density function, that is, the probability that the first passage from $i$ to $j$ requires $k$ time intervals, and let $f_{ij}^g(y)$ be its transform. Find $f_{ij}^g(y)$ in terms of $f_{ij}^{gg}(y, z)$, and then in terms of $\phi_{ij}^{gg}(y, z)$.

d) Write a difference equation for $f_{ij}(k)$ directly and use the transform of this equation to develop a method for modifying the flow graph of the process so that the transmission to node $i$ is $f_{ij}^g(y)$.

e) Show that the transformed difference equation of part d) is consistent with the results of part c).

**14.** A certain Markov process is known to have started in state $i$ and stopped $n$ transitions later in state $k$. What is the probability that the process was in state $j$ after transition $n - 1$? Express your answer in terms of transition probabilities and multistep transition probabilities.

**15.** A certain taxicab company operates between two towns according to the two-state Markov process with transition probability matrix,

$$P = \textit{From town} \begin{matrix} \\ 1 \\ 2 \end{matrix} \overset{\textit{To town}}{\begin{matrix} 1 & 2 \\ \end{matrix}} \begin{bmatrix} 0.6 & 0.4 \\ 0.4 & 0.6 \end{bmatrix}.$$

Driver Sam Clod lives in Town 1 and has found through experience that the probability of making $n$ trips during any one day is $(1/2)^{n+1}u(n)$. After finishing the day, Sam reveals that he made *at least one* trip from *Town 1* to *Town 1*, and ended the day in Town 2. What is the posterior mass function for the number of trips made? Helpful hint: $\mathscr{P}\{A\} = \mathscr{P}\{A, B\} + \mathscr{P}\{A, \text{not } B\}$.

**16.** A certain taxi company operates among three towns according to the transition probability matrix,

$$P = \textit{From town} \begin{matrix} \\ 1 \\ 2 \\ 3 \end{matrix} \overset{\textit{To town}}{\begin{matrix} 1 & 2 & 3 \end{matrix}} \begin{bmatrix} 0.4 & 0.4 & 0.2 \\ 0.8 & 0 & 0.2 \\ 0.5 & 0.5 & 0 \end{bmatrix}.$$

One of the drivers has studied his past record and found that the probability mass function for the number of trips he makes per day is uniform over a very large range. This same driver starts off in town 1 and upon arriving home announces that he made no trips from 2 to 1 or from 1 to 3. What is the probability that he ended his day in town 1? In town 2? Solve this problem by expressing your answer in terms of flow graph transmissions, and then substitute the appropriate quantities.

**17.** A local taxi company has expanded its area of operations from two to three towns and has found that the trips of any one driver can be described by the following transition diagram:

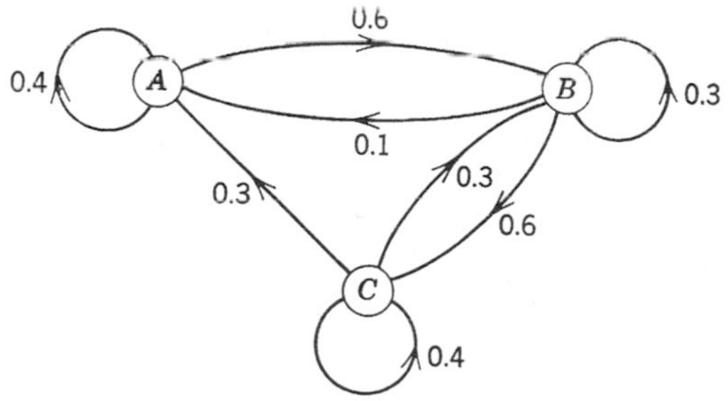

Sam Hacker starts his day in $A$ and has found over the years that his chances of making any particular number of trips during the day are approximately constant over a very wide range. If Sam's first trip to $C$ turns out to be his last trip of the day, what is the probability that he made $n$ trips during the day?

**18.** The Run 'em Down Taxi Company of Problem 8, Chapter 1, continues to operate between towns $A$ and $B$ in such a way that the probability of a driver going to $B$ on his next trip if he is in $A$ is 0.4 and the probability of going to $A$ if he is in $B$ is 0.1.

a) Upon arriving home at night, driver T. D. Fast tells his wife that after starting in $A$ he made $n$ trips, *at least* one of which was from $A$ to $B$. What is the probability that his final trip was to $A$? You may find the following helpful:

$$\mathcal{P}\{C, D\} + \mathcal{P}\{C, \text{not } D\} = \mathcal{P}\{C\}.$$

b) Driver Gus Aristotle also works for the Run 'em Down Taxi Company. One night he tells his wife that after starting in $A$ he made at least one trip from $A$ to $B$ and ended his day in $A$. His wife's feelings about the number of trips Gus makes in a day, *prior* to this information, are independent of his starting point and uniform (equally likely) over the range from 1 to 3 trips inclusive. What is her posterior distribution for the number of trips Gus has made? What is the mean of this distribution?

**19.** An independent taxi driver plies his trade within and between towns $A$ and $B$. He has found that if he is in town $A$ his probability of going to town $B$ on his next trip is 0.5 and of staying in $A$ is 0.5. Similarly, if he is in town $B$ his probability of going to town $A$ on his next trip is 0.4 and of staying in $B$ is 0.6. He has also found that the number of trips made in a day can be described by a random process whose probability mass function, $p(n)$, is

$$p(n) = \mathcal{P}\{n \text{ trips in a day}\} = (\tfrac{1}{2})^n u(n - 1).$$

a) At the end of a day the driver finds that he has made 2 trips from $A$ to $B$. We wish to find the posterior mass function for the number of trips he has made if he started in $A$. By application of Bayes' rule find an expression for the posterior in terms of $\phi_{A\Sigma}(2|n)$, the probability that he makes 2 $A \rightarrow B$ trips if he starts in $A$ and makes a total of $n$ trips. Then find the geometric transform of $\phi_{A\Sigma}(2|n)$. Don't bother to invert it. Explain how you would make use of the inverted expression to find the mean of the posterior.

b) To the extent of the analysis of part a), determine the posterior distribution for each of the following cases:

i) The driver has started in $A$, made 2 trips from $A$ to $B$, and ended in $B$.
ii) The driver has started in $A$, made 2 trips from $A$ to $B$ and his final trip was from $A$ to $B$.
iii) The driver started in $A$ and decided to quit after his second trip from $A$ to $B$.
For what prior distributions are the answers to ii) and iii) the same?

# 7 RECURRENT EVENTS AND RANDOM WALKS

We often encounter situations in which events occur over and over again according to the same probabilistic mechanisms. These situations are of such practical importance that they are studied as a special case of Markov process theory.

## 7.1 RECURRENT EVENTS—THE RENEWAL PROCESS

The basic idea of recurrent events is that the times between successive occurences of some event are all selected independently from the same probability distribution, which we shall call the "lifetime distribution." For the present we shall consider those situations where the event can occur only at integral, non-zero, multiples of some fundamental time unit; the lifetime distribution is therefore a probability mass function over the number of these time units separating successive events on the time axis. We shall represent the lifetime distribution by $p(l)$, the probability that the time between successive events is exactly $l$ time units. We require, of course, that $p(0) = 0$.

We can think of many applications of this model. Failures of complex equipment, purchases of a particular brand of product, or the counting of atomic particles might all be modeled by appropriate recurrent events. We shall see that the general Markov process has embedded within it a number of recurrent event processes.

The actual recurrence of such an event is often called a "renewal." For this reason recurrent event processes and renewal processes are both terms for the same kind of probabilistic mechanism; we shall use them interchangeably.

### Recurrent and Transient Renewal Processes

Now we shall proceed to analyze the renewal process. The lifetime distribution $p(l)$ has a geometric transform $p^g(z)$. The sum of $p(l)$ over all values of $l(l = 1, 2, 3, \ldots)$, however, may not be one because in some examples there is a finite probability that a renewal will not have occurred, even in an infinite time.

We shall define this sum to be $p_R$, the probability of eventual renewal. When $p_R = 1$, as we have in most cases, the renewal process is said to be recurrent. When $p_R < 1$, then the renewal process is said to be transient—there may be no recurrences. Since the geometric transform evaluated at $z = 1$ is the sum over all the values of the function it transforms, we have immediately

$$p_R = \sum_{l=1}^{\infty} p(l) = p^{g}(1). \tag{7.1.1}$$

We can explore the question of eventual renewal further by defining $q(r, n)$ to be the probability that on the $n$th time unit following some renewal designated as the zeroth renewal we shall see the $r$th renewal of the process following the zeroth renewal. The probability of seeing the $r$th renewal at time $n$ is just the probability that $r$ independent samples from the lifetime distribution have the sum $n$. Therefore, the geometric transform on $n$ of $q(r, n)$ is just the $r$th power of the transform of the lifetime distribution,

$$q^{\cdot g}(r, z) = [p^{g}(z)]^r. \tag{7.1.2}$$

The probability that the event will recur on at least $r$ occasions in an infinite time is then

$$\sum_{n=1}^{\infty} q(r, n) = q^{\cdot g}(r, 1) = [p^{g}(1)]^r = p_R{}^r \qquad r = 0, 1, 2, \ldots. \tag{7.1.3}$$

The probability of at least $r$ renewals in an infinite time is just the probability of eventual renewal raised to the $r$th power. Since

$$\lim_{r\to\infty} p_R{}^r = \begin{cases} 1 & \text{if } p_R = 1 \\ 0 & \text{if } p_R < 1, \end{cases} \tag{7.1.4}$$

we see immediately that renewals in recurrent renewal processes occur infinitely often in infinite time while renewals in transient renewal processes have a zero probability of occurring infinitely often.

Let $g(r)$ be the probability that the event will occur exactly $r$ times in an infinite number of trials. This probability is the probability that the event will occur at least $r$ times in an infinite time less the probability that it will occur at least $r + 1$ times in an infinite time; thus,

$$g(r) = p_R{}^r - p_R^{r+1} = (1 - p_R)p_R{}^r \qquad r = 0, 1, 2, \ldots. \tag{7.1.5}$$

The probability of exactly $r$ recurrences in an infinite time is therefore a geometric distribution with parameter $p_R$. However, unlike the geometric distribution we discussed in Section 4.1, the domain of this distribution is 0, 1, 2, . . . rather than

1, 2, 3, . . . . The mean value of $r$, $\bar{r}$, is therefore one less than the mean of the geometric distribution of Section 4.1,

$$\bar{r} = \frac{1}{1 - p_R} - 1 = \frac{p_R}{1 - p_R}. \qquad (7.1.6)$$

However, the variance $\overset{v}{r}$ is the same as before,

$$\overset{v}{r} = \frac{p_R}{(1 - p_R)^2}. \qquad (7.1.7)$$

Thus it is very simple to compute the probability distribution for the number of renewals in an infinite time and its moments from the probability of eventual renewal $p_R$ for a transient renewal process—no other property of the lifetime distribution is required.

Unless we specify otherwise, the renewal processes which we shall treat are recurrent; that is $p_R = p^g(1) = 1$. Although we deal in practice more often with recurrent renewal processes, the case of transient renewal processes is of great importance theoretically, as we shall see later in this chapter.

When $p_R = 1$, the lifetime distribution $p(l)$ is a probability mass function since it sums to one. Therefore we can find the moments of $l$ either from the probability mass function or from the geometric transform of the lifetime distribution. The mean lifetime $\bar{l}$ is

$$\bar{l} = \sum_{l=1}^{\infty} l p(l) = \frac{d}{dz} p^g(z) \Big|_{z=1}, \qquad (7.1.8)$$

and the second moment of $l$ is

$$\overline{l^2} = \sum_{l=1}^{\infty} l^2 p(l) = \frac{d^2}{dz^2} p^g(z) \Big|_{z=1} + \frac{d}{dz} p^g(z) \Big|_{z=1}. \qquad (7.1.9)$$

As usual, we find the variance from

$$\overset{v}{l} = \overline{l^2} - \bar{l}^2. \qquad (7.1.10)$$

### The Probability of Renewal at Time $n$

Probably the most important calculation in renewal processes is the calculation of the probability that a renewal (maybe not the first) will occur $n$ time units after a time zero specified in some way. The most convenient way to specify time zero is to say as we did above that a renewal, the zeroth renewal, is known to have occurred at time zero. We shall define $\phi(n)$ to be the probability of a renewal at time $n$ (not necessarily the first) given that a renewal occurred at time zero; consequently $\phi(0) = 1$. As we shall see, the function $\phi(n)$ is the impulse response of a renewal process, and is also the multistep transition probability from and to the renewal state. Most other statistics of the process are derivable from it, including those based on other specifications of the starting condition.

We can relate $\phi(\cdot)$ and $p(\cdot)$ in a difference equation by enumerating all the ways to produce a renewal at time $n$. When $n \geq 1$, we could have a first renewal at time $n$; this event has probability $p(n)$. We could also have the first renewal at time $m < n$ and then somehow have obtained another renewal (maybe the last of several) at the end of an interval of length $n - m$ that begins with the first renewal. The probability of this circumstance is $p(m)\phi(n - m)$; this quantity must be summed over all the mutually exclusive values of $m$ from 1 to $n - 1$. Thus we have

$$\phi(n) = p(n) + \sum_{m=1}^{n-1} p(m)\phi(n - m) \qquad n = 1, 2, 3, \ldots. \tag{7.1.11}$$

However, since $\phi(0) = 1$, we can use the more compact form

$$\phi(n) = \sum_{m=1}^{n} p(m)\phi(n - m) \qquad n = 1, 2, 3, \ldots. \tag{7.1.12}$$

To extend the domain of the equation to $n = 0$ and thereby make explicit the definition that $\phi(0) = 1$, we recall that $p(0) = 0$ and then write

$$\phi(n) = \delta(n) + \sum_{m=0}^{n} p(m)\phi(n - m), \qquad n = 0, 1, 2, \ldots. \tag{7.1.13}$$

The first few values of $\phi(\cdot)$ are:

$$\begin{aligned}
\phi(0) &= 1 \\
\phi(1) &= p(1) \\
\phi(2) &= p(1)\phi(1) + p(2) = [p(1)]^2 + p(2) \\
\phi(3) &= p(1)\phi(2) + p(2)\phi(1) + p(3) = [p(1)]^3 + 2p(1)p(2) + p(3) \\
&\vdots
\end{aligned} \tag{7.1.14}$$

Of course, $\phi(n)$ depends only on the lifetime distribution values $p(1), p(2), \ldots, p(n)$; the individual terms in $\phi(n)$ are just the probabilities of the different ways of obtaining a renewal at time $n$.

***Transform analysis***

If we take the geometric transform of Equation 7.1.13, we have

$$\phi^{g}(z) = 1 + p^{g}(z)\phi^{g}(z) \tag{7.1.15}$$

or

$$\phi^g(z) = \frac{1}{1 - p^g(z)}. \tag{7.1.16}$$

The transform of the probability of recurrence at any time is just the reciprocal of one minus the transform of the lifetime distribution. This result becomes more meaningful if we expand the transform in series form,

$$\phi^g(z) = \frac{1}{1 - p^g(z)} = 1 + p^g(z) + [p^g(z)]^2 + [p^g(z)]^3 + \cdots. \tag{7.1.17}$$

From Equation 7.1.2 we have the alternate form

$$\phi^g(z) = 1 + q^{g}(1, z) + q^{g}(2, z) + q^{g}(3, z) + \cdots \tag{7.1.18}$$

with inverse transform,

$$\phi(n) = \delta(n) + q(1, n) + q(2, n) + q(3, n) + \cdots \tag{7.1.19}$$

This equation states that the probability of a renewal at time $n$ is the sum of the probabilities of a zeroth renewal at time $n$ (zero except when $n = 0$), a first renewal at time $n$, a second renewal at time $n$, etc. The logic of this statement is unassailable.

***Flow graph representation***

The transform of Equation 7.1.16 implies the simple flow graph shown in Figure 7.1.1. The flow graph makes it easy to visualize the process returning again and again to the renewal state through the branch whose transmission is the transform of the lifetime distribution. The flow graph is unusual for us because we are accustomed to having branch transmissions that are simply multiples of $z$. However, our basic theory of Chapter 2 considered the case of more general systems in detail. In fact, one of the motivating reasons for using transforms in the first place was to have the ability to solve simple feedback systems like that of Figure 7.1.1. We shall exploit the form of the flow graph further, but let us first use our results to solve a simple example.

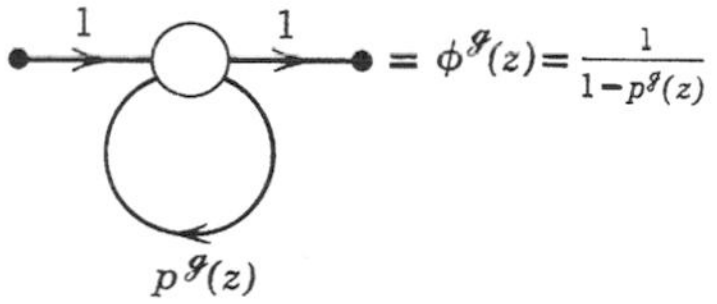


**Figure 7.1.1** Flow graph for the basic renewal process.

### *A renewal process example*

Consider a renewal process whose lifetime distribution shows that the time between renewals takes on the values 1 and 2 with equal probability; that is,

$$p(l) = \begin{cases} 0 & l = 0 \\ 1/2 & l = 1 \\ 1/2 & l = 2 \\ 0 & l = 3, 4, 5, \ldots . \end{cases} \tag{7.1.20}$$

Let us find $\phi(n)$, the probability of a renewal at time $n$ given a renewal at time zero. Working first with the difference equation, Equation 7.1.13, we have

$$\begin{aligned} \phi(n) &= \delta(n) + \sum_{m=0}^{n} p(m)\phi(n - m) \qquad n = 0, 1, 2, \ldots \\ &= \delta(n) + p(1)\phi(n - 1) + p(2)\phi(n - 2) \\ &= \delta(n) + 1/2[\phi(n - 1) + \phi(n - 2)]. \end{aligned} \tag{7.1.21}$$

Then we can write

$$\begin{aligned} &\phi(0) = 1 && \phi(1) = 1/2 = 0.5 \\ &\phi(2) = 3/4 = 0.75 && \phi(3) = 5/8 = 0.625 \\ &\phi(4) = 11/16 = 0.6875 && \phi(5) = 21/32 = 0.65625 \\ &\phi(6) = 43/64 = 0.671875 && \phi(7) = 85/128 = 0.6640625. \end{aligned} \tag{7.1.22}$$

Note that $\phi(n)$ seems to be approaching a limit in the vicinity of 0.66. The exact nature of this behavior is better revealed by solving the same problem using geometric transforms.

First we write the geometric transform of the lifetime distribution,

$$p^{g}(z) = \sum_{l=0}^{\infty} p(l)z^{l} = (1/2)z + (1/2)z^{2}. \tag{7.1.23}$$

Then we use Equation 7.1.16 to find the geometric transform of the probability of renewal at time $n$,

$$\begin{aligned} \phi^{g}(z) &= \frac{1}{1 - p^{g}(z)} = \frac{1}{1 - (1/2)z - (1/2)z^{2}} \\ &= \frac{1}{(1 - z)[1 + (1/2)z]} \\ &= \frac{2/3}{1 - z} + \frac{1/3}{1 + (1/2)z}. \end{aligned} \tag{7.1.24}$$

When we invert this transform, we have

$$\phi(n) = 2/3 + (1/3)(-1/2)^n \qquad n = 0, 1, 2, \dots \tag{7.1.25}$$

which agrees completely with the results of Equation 7.1.22 obtained by solving the difference equation. The probability $\phi(n)$ has an oscillatory component that dies away at rate $-1/2$ leaving only the constant component of 2/3. Therefore, when $n$ is large there is a probability 2/3 of a replacement at any particular time.

Of course, in general it will not be so easy to solve the transform Equation 7.1.16. If $p(l)$ is non-zero for large values of $l$ then the denominator polynomial in $z$ in the equation will be of high degree, making partial fraction expansion extremely difficult. In such cases the only practical alternative is to solve the difference equation 7.1.13.

### *Limiting probability of renewal*

However, regardless of the complexity of $p^g(z)$ there is one result that is very easy to calculate. That is the value of $\phi(n)$ when $n$ is large, 2/3 in our example. We call this quantity $\phi$ and find it by applying the final value theorem of geometric transforms to Equation 7.1.16. We write

$$\phi = \phi(\infty) = \lim_{n\to\infty} \phi(n) = \lim_{z\to 1} (1 - z)\phi^g(z)$$

$$= \lim_{z\to 1} \frac{1 - z}{1 - p^g(z)}. \tag{7.1.26}$$

However, this last limit is indeterminate, so we apply l'Hôpital's rule and find

$$\phi = \lim_{z\to 1} \frac{1}{\dfrac{d}{dz}p^g(z)} = \frac{1}{\bar{l}} \tag{7.1.27}$$

The limiting value of $\phi(n)$ for large $n$ is just the reciprocal of the mean lifetime. This result says that if the mean time between events is, say, 7 units then the probability of an event at any particular time after the process has been running for a while is 1/7. In our example the mean lifetime is $(1/2)(1) + (1/2)(2) = 3/2$; therefore we found $\phi = 2/3$.

If these results strike a familiar chord it is not surprising. In Section 5.1 we found that the limiting state probability of a recurrent state in a monodesmic Markov process was the reciprocal of the mean recurrence time of that state. We can consider the result of Equation 7.1.27 as a special case of this earlier result. Entries into a recurrent state of a Markov process are renewal processes with a lifetime distribution equal to the recurrence time distribution for the state. Entries into a transient state of a Markov process are transient events ($p_R < 1$). Consequently, all the results of this section have an implication for recurrence times in Markov processes.

## Transient Markov Process Representation

If we can treat some aspects of Markov processes as renewal processes, we expect that we would also be able to solve renewal process problems using Markov process theory. The first problem is one of representing the lifetime distribution by a suitable Markov process.

We begin by defining ${}^{>}p(m)$ as the probability that the lifetime exceeds $m$ units,

$$ {}^{>}p(m) = \sum_{l=m+1}^{\infty} p(l). \tag{7.1.28} $$

We shall call ${}^{>}p(m)$ the complementary cumulative probability distribution on lifetime. Since $p(l)$ is a probability mass function,

$$ {}^{>}p(m) = 1 - {}^{\leq}p(m) = 1 - \sum_{l=0}^{m} p(l). \tag{7.1.29} $$

We can therefore write the geometric transform ${}^{>}p^{g}(z)$ for future reference directly from relations 3 and 25 of transform Table 1.5.1,

$$ {}^{>}p^{g}(z) = \frac{1}{1-z} - \frac{1}{1-z}p^{g}(z) $$

$$ = \frac{1 - p^{g}(z)}{1-z}. \tag{7.1.30} $$

This result for the geometric transform of the complementary cumulative probability distribution holds for any non-negative, discrete random variable.

### *Flow graph representation*

Now we observe that we can interpret $p^{g}(z)$ as the transform of the probability mass function of the time to complete the transient process whose flow graph appears in Figure 7.1.2. Notice that the transmission of this flow graph is just the sum of the path transmissions, or

$$ \vec{t} = p(1)z + p(2)z^2 + p(3)z^3 + \cdots $$

$$ = p^{g}(z). \tag{7.1.31} $$

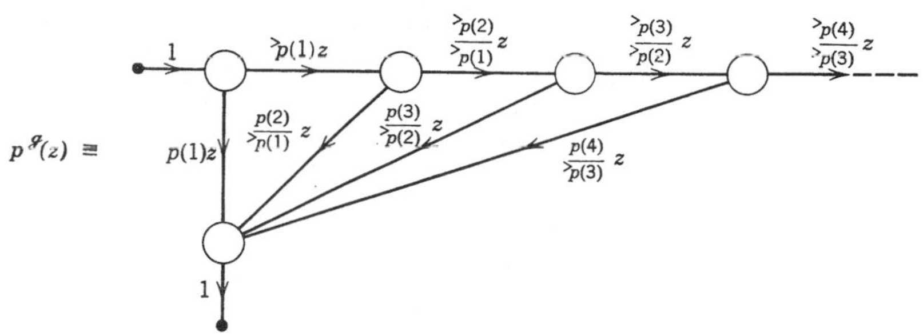


**Figure 7.1.2** Representation of a general lifetime distribution as a transient process.

The flow graph of Figure 7.1.2 therefore allows us to represent the loop transmission $p^g(z)$ of the basic renewal process of Figure 7.1.1 as a transient Markov process wherein each transition requires exactly one time unit. For our example, Equation 7.1.28 yields

$$ {}^{>}p(m) = \begin{cases} 1 & m = 0 \\ 1/2 & m = 1 \\ 0 & m \geq 2. \end{cases} \tag{7.1.32} $$

Figure 7.1.3 shows the transient Markov process equivalent of the lifetime distribution transform for this example.

The next step is to join the input and output nodes of the transient Markov process so that we represent the renewal process of Figure 7.1.1. We accomplish this in the present example by designating both input and output nodes of Figure 7.1.3 as state 1, and the remaining node as state 2. The equivalent Markov process

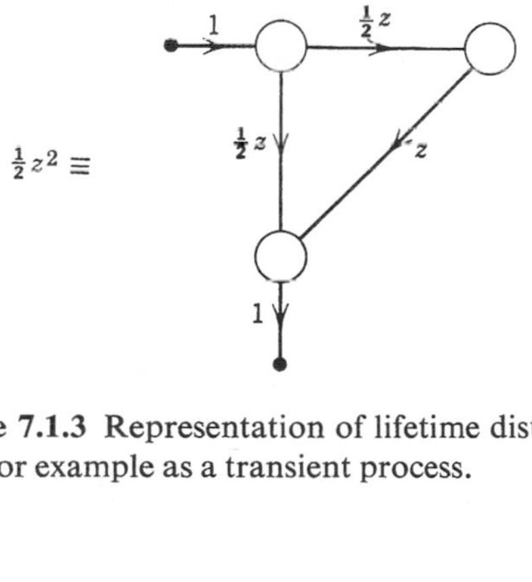


**Figure 7.1.3** Representation of lifetime distribution for example as a transient process.

flow graph then appears as shown in Figure 7.1.4. The transition probability matrix for this two-state process is

$$P = \begin{bmatrix} 1/2 & 1/2 \\ 1 & 0 \end{bmatrix}. \tag{7.1.33}$$

The multistep transition probability $\phi_{11}(n)$ of occupying state 1 at time $n$ given that state 1 was occupied at time zero is just the renewal probability $\phi(n)$. The transmission $\vec{t}_{11} = \phi_{11}{}^{g}(z)$ of this flow graph is therefore the transform $\phi^{g}(z)$ of the renewal probability developed in Equation 7.1.24.

There is a potential catch in performing such a representation, however. The flow graph of Figure 7.1.2 will require an infinite number of nodes unless the lifetime distribution has a finite domain. Infinite-node flow graphs mean infinite-state Markov processes. This is a complication we are not yet prepared to handle.

### *Geometric lifetime example*

The difficulty is a real one because we can easily construct a physically interesting lifetime distribution with an infinite domain. For example, consider the geometric distribution,

$$p(l) = p(1-p)^{l-1} \qquad 0 < p < 1,\ l = 1, 2, \ldots, \tag{7.1.34}$$

with transform

$$p^{g}(z) = \frac{pz}{1-(1-p)z}. \tag{7.1.35}$$

For this distribution the transform of the recurrence probability is

$$\phi^{g}(z) = \frac{1}{1-p^{g}(z)} = \frac{1-(1-p)z}{1-z} = 1 + \frac{pz}{1-z}, \tag{7.1.36}$$

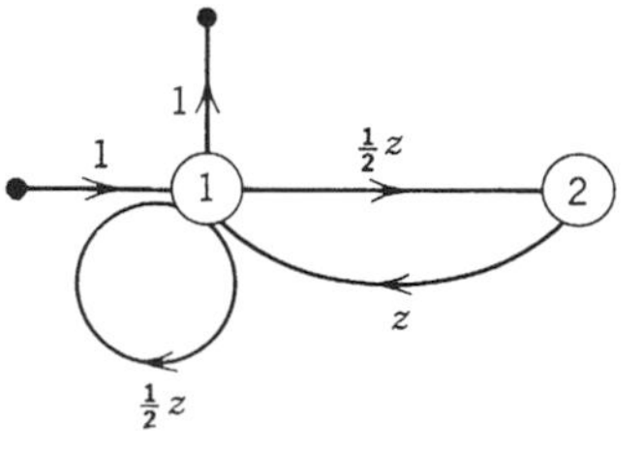


$$\vec{t}_{11} = \phi_{11}{}^{g}(z) = \frac{1}{1-\frac{1}{2}z-\frac{1}{2}z^2} = \phi^{g}(z)$$

**Figure 7.1.4** Markov process equivalent of a renewal process.

and the recurrence probability is therefore

$$\phi(n) = \delta(n) + pu(n-1) \qquad n = 0, 1, 2, \ldots \tag{7.1.37}$$

or

$$\phi(n) = \begin{cases} 1 & n = 0 \\ p & n = 1, 2, 3, \ldots. \end{cases} \tag{7.1.38}$$

Although the domain of the lifetime distribution is infinite, the recurrence probability is simply the constant $p$ at all future times. Can we construct a finite-state Markov process whose multistep transition probability is precisely $\phi(n)$?

We begin by computing the complementary cumulative probability distribution ${}^{>}p(m)$ corresponding to the geometric distribution,

$${}^{>}p(m) = \sum_{l=m+1}^{\infty} p(l) = p \sum_{l=m+1}^{\infty} (1-p)^{l-1} = (1-p)^m \qquad m = 0, 1, 2, \ldots. \tag{7.1.39}$$

The flow graph of Figure 7.1.2 now takes the form shown in Figure 7.1.5, an infinite-node flow graph of rather simple structure. The transmission of the graph is

$$p^{g}(z) = pz + pz(1-p)z + pz(1-p)^2z^2 + \cdots$$

$$= \frac{pz}{1-(1-p)z}. \tag{7.1.40}$$

We might be tempted to use the finite-state transient Markov process of Figure 7.1.6 to represent $p^{g}(z)$, for it has the proper transmission. However, a moment's reflection reveals that when we join input and output nodes to draw the Markov process in the next step, we shall lose the internal structure and obtain the wrong answer.

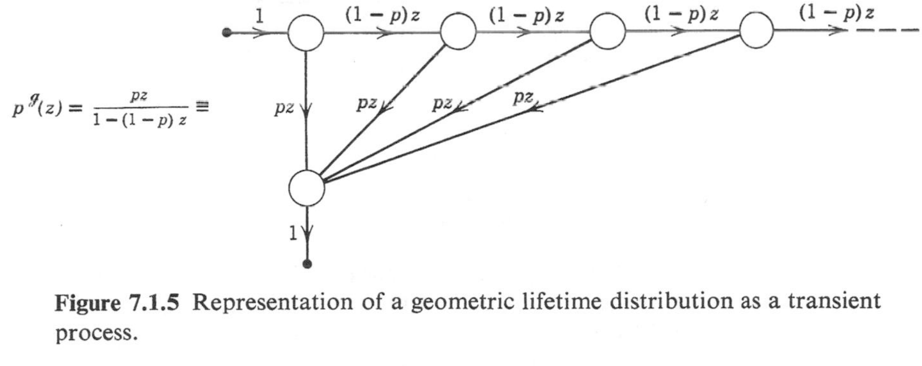


**Figure 7.1.5** Representation of a geometric lifetime distribution as a transient process.

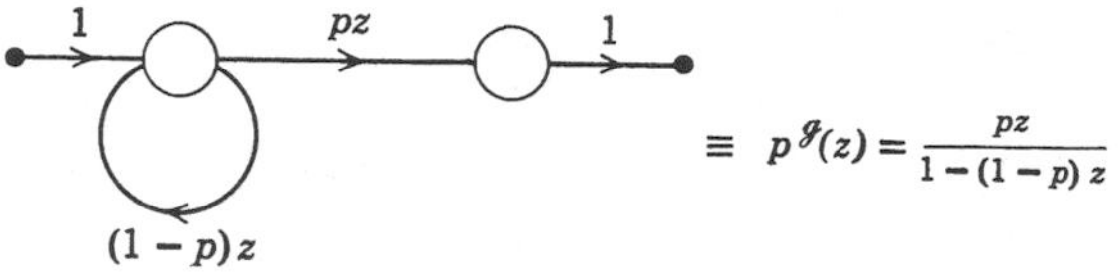


**Figure 7.1.6** Finite-state representation of a geometric lifetime distribution as a transient process.

Therefore we choose to represent $p^{\mathscr{g}}(z)$ by the finite-state transient Markov process of Figure 7.1.7. It also has the proper transmission, but avoids the node identity difficulty. If we designate both input and output nodes of the process as state 1 and the remaining node as state 2, we obtain the finite-state Markov process representation whose flow graph appears as Figure 7.1.8 and whose transition probability matrix is

$$P = \begin{bmatrix} p & 1-p \\ p & 1-p \end{bmatrix}. \tag{7.1.41}$$

The transmission $\vec{t}_{11}$ of the flow graph is then $\phi_{11}{}^{\mathscr{g}}(z)$ and, in turn, the renewal probability transform $\phi^{\mathscr{g}}(z)$ found in Equation 7.1.36.

We were fortunate in this case that the simple structure of the infinite-state graph allowed us to characterize the process by using feedback. In general, we shall be able to do this only when $p^{\mathscr{g}}(z)$ has the form of the ratio of two polynomials in $z$. Since there exist infinite domain lifetime distributions whose transforms cannot be placed in this form, not all discrete-time renewal processes have finite-state Markov process equivalents.

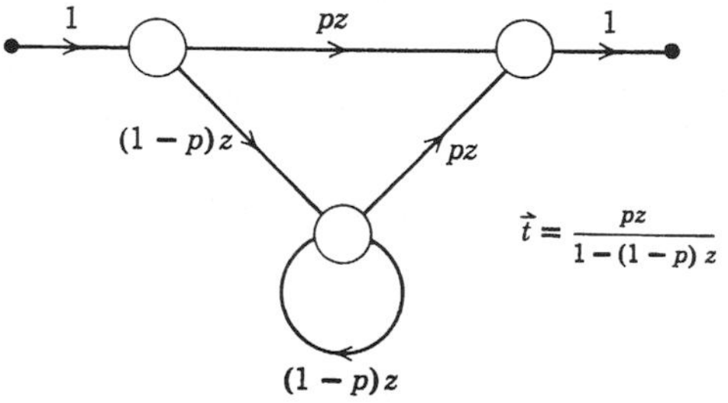


**Figure 7.1.7** Finite-state transient Markov process representation of geometric lifetime distribution.

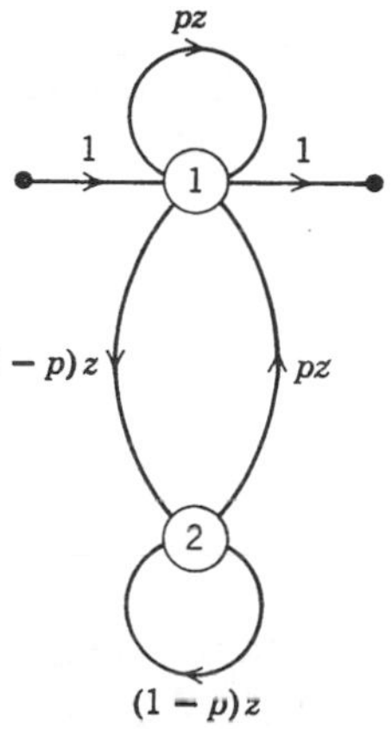


$$\vec{t}_{11} = \phi_{11}{}^{\mathscr{g}}(z) = 1 + \frac{pz}{1-z} = \phi\,{}^{\mathscr{g}}(z)$$

**Figure 7.1.8** Finite-state Markov process representation of geometric lifetime renewal process.

### The General Renewal Process

Up to this time we have considered the probability of renewal at time $n$ given that a renewal occurred at time zero. Suppose now that instead of specifying what happened at time zero we assign a probability mass function ${}_{f}p(\cdot)$ for the time of the first renewal after time zero. Once this first renewal occurs, all future renewals are separated by times selected according to the lifetime distribution $p(\cdot)$. This generalization of starting condition allows us to analyze processes for which the time to the first renewal may be statistically different from the time between subsequent renewals. The domain of ${}_{f}p(m)$ is $m = 1, 2, 3, \ldots$ since we can think of it as the lifetime distribution for the first renewal only. Thus

$${}_{f}p(m) = p(m) \qquad m = 1, 2, 3, \ldots \tag{7.1.42}$$

is the condition for which the earlier results of this section were computed.

Since $\phi(n)$ was based on the assumption that a renewal occurred at time zero, we must define a new quantity ${}_{f}\phi(n)$ to be the probability that a renewal occurs at time $n \geq 1$ given that the first renewal after time zero has a probability mass function ${}_{f}p(\cdot)$. The probability of a renewal at time $n$ under this starting condition is just the probability that the first renewal occurred at time $m$, and another renewal (perhaps the last of many) occurred after $n - m$ given that one occurred at $m$, summed over all values of $m$ from 1 to $n$. Thus we obtain the equation

$${}_{f}\phi(n) = \sum_{m=1}^{n} {}_{f}p(m)\phi(n-m) \qquad n = 1, 2, 3, \ldots. \tag{7.1.43}$$

Note that the quantity on the right of this equation is $\phi(\cdot)$, not ${}_f\phi(\cdot)$, because the interval $n - m$ is known to begin with a renewal. The equation relates the probability of renewal under the more general starting condition to the results we obtained earlier.

If we transform Equation 7.1.43, we find directly

$$ {}_f\phi^g(z) = {}_fp^g(z)\phi^g(z). \tag{7.1.44} $$

Therefore, from Equation 7.1.16 we write

$$ {}_f\phi^g(z) = \frac{{}_fp^g(z)}{1 - p^g(z)} \tag{7.1.45} $$

which states that the geometric transform of the probability of recurrence at time $n$, given a general starting condition, is the ratio of the geometric transform of the time to the first renewal divided by one minus the transform of the lifetime distribution.

***Flow graph representation***

The transform Equation 7.1.45 implies the flow graph of Figure 7.1.9 for this more general renewal process. The form of the flow graph indicates very clearly the difference between the time to the first renewal and the time between successive renewals. When we compare this flow graph with the flow graph in Figure 5.2.4 for the relationships between state probabilities, first passage times, and recurrence times, we see that the graphs are identical if we interpret ${}_fp(\cdot)$ as a first passage time distribution to some recurrent state of a Markov process and $p(\cdot)$ as the recurrence time distribution of that state. Thus we can consider the problem of finding the multistep transition probability to any particular recurrent state in a Markov process as a general renewal process problem. We therefore expect that state occupancies, etc. will be computable from renewal process statistics, and vice versa. The connection between these concepts gives us valuable, exploitable flexibility in modeling problems.

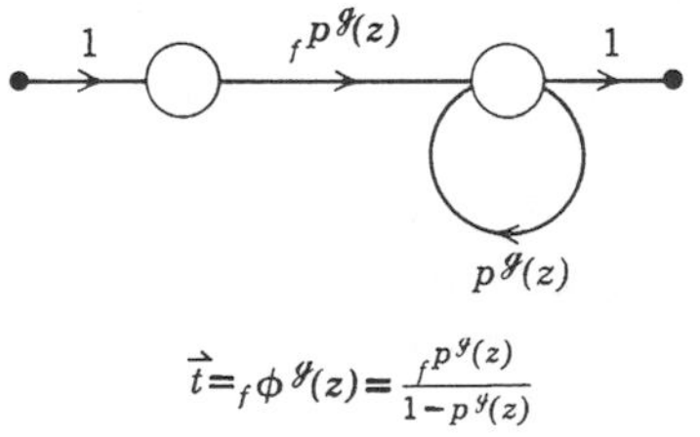


**Figure 7.1.9** Flow graph for the general renewal process.

***The two-lifetime example***

Let us now return to the two-lifetime example and introduce a distribution for the time to first occurrence. We shall assume that there is a probability 2/3 of observing the first renewal at time 1 and a probability 1/3 of having to wait until time 2; there is, therefore, no chance of having to wait more than two time units for a renewal. Thus we have:

$$ {}_{f}p(m) = \begin{cases} 0 & m = 0 \\ 2/3 & m = 1 \\ 1/3 & m = 2 \\ 0 & m \geq 3 \end{cases} \tag{7.1.46} $$

For this choice of ${}_{f}p(\cdot)$ Equation 7.1.43 becomes

$$ {}_{f}\phi(n) = \sum_{m=1}^{n} {}_{f}p(m)\phi(n - m) = (2/3)\phi(n - 1) + (1/3)\phi(n - 2) \qquad n = 1, 2, 3, \ldots \tag{7.1.47} $$

with our usual understanding that functions with negative arguments are zero. Now we can write ${}_{f}\phi(n)$ by applying this equation to the values of $\phi(n)$ we recorded in Equation 7.1.22. We find

$$ {}_{f}\phi(1) = 2/3 \qquad {}_{f}\phi(2) = 2/3 \qquad {}_{f}\phi(3) = 2/3 \qquad {}_{f}\phi(4) = 2/3 \text{ etc.} \tag{7.1.48} $$

Apparently, for this starting condition, ${}_{f}\phi(n) = 2/3$ for $n = 1, 2, 3, \ldots$.

We can confirm this result by computing the transform ${}_{f}\phi^{g}(z)$. First we write the transform of ${}_{f}p(\cdot)$,

$$ {}_{f}p^{g}(z) = \sum_{m=0}^{\infty} {}_{f}p(m)z^{m} = (2/3)z + (1/3)z^{2}. \tag{7.1.49} $$

Then we substitute this result and that of Equation 7.1.24 into Equation 7.1.44. We obtain

$$ {}_{f}\phi^{g}(z) = {}_{f}p^{g}(z)\phi^{g}(z) = \frac{(2/3)z + (1/3)z^{2}}{(1 - z)[1 + (1/2)z]} = \frac{(2/3)z}{1 - z}. \tag{7.1.50} $$

Therefore, by inverse transformation we verify that, in fact,

$$ {}_{f}\phi(n) = 2/3 \qquad n = 1, 2, 3, \ldots. \tag{7.1.51} $$

But what is so special about this starting condition that causes ${}_{f}\phi(n)$ to be a constant? We shall soon see.

## Random Starting

Suppose that we choose time zero by entering the process at a time selected completely independent of the operation of the process. We shall call this "entering at random." What then will be the probability distribution for the time to the first renewal that we see? We shall let ${}_{\mathrm{r}}p(m)$ be the probability that the first renewal will occur at time $m$, $m = 1, 2, 3, \ldots$, for a process entered at random. We shall have to wait a time $m$ for the next renewal if and only if a renewal occurs at time $m$ and no renewal occurs at the times $1, 2, \ldots, m - 1$. The probability that a renewal will take place at any particular time in a renewal process, when we have no information about when the process started, we have found previously to be the reciprocal of the mean lifetime of the process; therefore, the probability of a renewal at time $m$ is $1/\bar{l}$. No renewal will occur at times $1, 2, \ldots, m - 1$ when the time since the last renewal before the one at time $m$ is at least $m$; the probability of this event is ${}^{>}p(m - 1)$. Therefore, the probability that the first renewal will be observed at time $m$ after the random starting point is just

$$ {}_{\mathrm{r}}p(m) = \frac{1}{\bar{l}}\,{}^{>}p(m - 1) \qquad m = 1, 2, 3, \ldots. \tag{7.1.52} $$

The probability of waiting time $m$ for the next renewal after random entry is just the probability that the lifetime is greater than $m - 1$ divided by the mean lifetime. We can and will derive this relationship in other ways in our future work.

The geometric transform ${}_{\mathrm{r}}p^{g}(z)$ of the time to the first renewal for this case we find directly from the geometric transform definition to be

$$ \begin{aligned} {}_{\mathrm{r}}p^{g}(z) &= \frac{1}{\bar{l}} \sum_{m=1}^{\infty} z^{m}\,{}^{>}p(m - 1) = \frac{1}{\bar{l}} \sum_{r=0}^{\infty} z^{r+1}\,{}^{>}p(r) \\ &= \frac{z}{\bar{l}}\,{}^{>}p^{g}(z), \end{aligned} \tag{7.1.53} $$

or using Equation 7.1.30,

$$ {}_{\mathrm{r}}p^{g}(z) = \frac{z}{\bar{l}} \cdot \frac{1 - p^{g}(z)}{1 - z}. \tag{7.1.54} $$

***Specialization of the general renewal process to the case of random starting***

If a process is entered at random, the probability distribution of the time to the first renewal is just

$$ {}_{\mathrm{f}}p(m) = {}_{\mathrm{r}}p(m) \qquad m = 1, 2, 3, \ldots. \tag{7.1.55} $$

We can therefore use all the results of the general renewal process for analyzing the case of random entering just by substituting the result of Equation 7.1.52 for

${}_{f}p(m)$ in all of our earlier expressions. For example, let ${}_{r}\phi(n)$ be the probability of a renewal at time $n$ for a process entered at random. Then we can use Equations 7.1.52 and 7.1.54 to write the difference equation and transform of this quantity directly from Equations 7.1.43 and 7.1.45. We obtain

$${}_{r}\phi(n) = \frac{1}{\bar{l}} \sum_{m=1}^{\infty} {}^{>}p(m-1)\phi(n-m) \qquad n = 1, 2, 3, \ldots \qquad (7.1.56)$$

and

$$\begin{aligned} {}_{r}\phi^{g}(z) &= \frac{{}_{r}p^{g}(z)}{1 - p^{g}(z)} \\ &= \frac{1}{\bar{l}} \frac{z(1 - p^{g}(z))}{(1 - z)(1 - p^{g}(z))} \\ &= \frac{1}{\bar{l}} \cdot \frac{z}{1 - z}. \end{aligned} \qquad (7.1.57)$$

Although Equation 7.1.56 may not be very revealing in its present form, Equation 7.1.57 for the transform of ${}_{r}\phi(n)$ shows immediately that ${}_{r}\phi(n)$ is just the reciprocal of the mean lifetime for all $n$,

$${}_{r}\phi(n) = \frac{1}{\bar{l}} \qquad n = 1, 2, 3, \ldots. \qquad (7.1.58)$$

This equation states that if we enter a renewal process at random we have a probability of $1/\bar{l}$ of seeing a renewal at every future time point. This result is just what we would expect from our original statement that the time of entering is in no way dependent on the behavior of the process.

The situation is analogous to that of entering a Markov process according to the probabilities in the limiting state probability vector. If you enter the process in such a way that you have only steady state knowledge about its present state, then you can never change this state of knowledge until you are allowed to observe the process.

### *The two-lifetime example*

You will recall that in our example of the general renewal process we used the distribution of Equation 7.1.46 for the time to the first renewal, and found in Equation 7.1.51 that there was a constant probability of 2/3 of seeing a renewal at any time. The results of random starting then lead us to believe that the distribution of Equation 7.1.46 must be just the distribution ${}_{r}p(\cdot)$ for the time to first renewal in a process entered at random. To confirm this conjecture we obtain ${}_{r}p(m)$ by dividing ${}^{>}p(m-1)$ from Equation 7.1.32 by $\bar{l} = 2/3$,

$$ {}_{r}p(m) = \frac{1}{\bar{l}}\,{}^{>}p(m-1) = \begin{cases} 0 & m = 0 \\ 2/3 & m = 1 \\ 1/3 & m = 2 \\ 0 & m \geq 3 \end{cases} \tag{7.1.59} $$

which specifies the distribution we used in Equation 7.1.46 for the example. The distribution for time to the first renewal that we assumed for the example is just the distribution that would result if the process were entered at random. Therefore, the observation that there was a constant probability of renewal at all times was to be expected.

## 7.2 RENEWAL STATISTICS

One of the most important statistics of a renewal process is the number of renewals that will occur in any time period after a time zero selected in a specified way. We might be interested in the moments of this statistic or in its complete probability distribution.

### Counting Renewals

We shall let $\phi(k|n)$ be the probability that $k$ renewals will take place at times 1 through $n$ if it is known that a renewal occurred at time 0; we define $\phi(k|0) = \delta(k)$.

We can have $k \geq 1$ renewals in the interval $(1, n)$ if the first renewal after time zero occurs at time $m < n$ and then $k-1$ renewals occur in the interval $(m+1, n)$, an interval that is immediately preceded by a renewal. The probability of these events, $p(m)\phi(k-1|n-m)$, must be summed over all values of $m$ from 1 to $n-1$. When $k = 1$, the first renewal could also occur at time $n$—an event with probability $p(n)$. We include this possibility by extending the sum to include $m = n$. Of course, including $m = 0$ does not affect the sum since $p(0) = 0$. Therefore we obtain

$$ \phi(k|n) = \sum_{m=0}^{n} p(m)\phi(k-1|n-m) \qquad \begin{matrix} k = 1, 2, 3, \ldots \\ n = 0, 1, 2, \ldots . \end{matrix} \tag{7.2.1} $$

To extend this equation to $k = 0$, we recognize that the only way to have no renewals in the interval $(1, n)$ occurs when the time to the first renewal is greater than $n$, an event with probability ${}^{>}p(n)$. Under our convention that discrete functions with negative arguments are interpreted to be zero, we can then write

$$ \phi(k|n) = \delta(k)\,{}^{>}p(n) + \sum_{m=0}^{n} p(m)\phi(k-1|n-m) \qquad \begin{matrix} k = 0, 1, 2, \ldots \\ n = 0, 1, 2, \ldots . \end{matrix} \tag{7.2.2} $$

This equation allows us to calculate recursively the probability distribution of the number of renewals in any period following a renewal.

If we geometrically transform on both variables $k$ and $n$ we obtain

$$\phi^{gg}(y|z) = {}^{>}p^{g}(z) + yp^{g}(z)\phi^{gg}(y|z) \tag{7.2.3}$$

or

$$\phi^{gg}(y|z) = \frac{{}^{>}p^{g}(z)}{1 - yp^{g}(z)}. \tag{7.2.4}$$

This transform has a very simple interpretation as the transmission of the flow graph in Figure 7.2.1. We see that this is just like the flow graph for the basic renewal process with the addition of a tagging variable $y$ to count renewals and a branch ${}^{>}p^{g}(z)$ to account for the possibility of no further renewals in the interval.

Since the geometric transform ${}^{>}p^{g}(z)$ is related to $p^{g}(z)$ by Equation 7.1.30, we can write Equation 7.2.4 in the alternate form

$$\phi^{gg}(y|z) = \frac{1 - p^{g}(z)}{(1 - z)[1 - yp^{g}(z)]} \tag{7.2.5}$$

to show the direct relationship of $\phi^{gg}(y|z)$ to the transform of the lifetime distribution. Note that the requirement that $\sum_{k=0}^{\infty} \phi(k|n) = 1$ ensures that the transform will always specialize to $\phi^{gg}(1|z) = 1/(1 - z)$.

We can apply these results to find the probability distribution of the number of renewals through time $n$ after a renewal for the example considered earlier whose lifetime distribution appears in Equation 7.1.20. For this case, the difference Equation 7.2.2 becomes

$$\phi(k|n) = \delta(k)[\delta(n) + (1/2)\,\delta(n-1)] + (1/2)\phi(k-1|n-1) + (1/2)\phi(k-1|n-2). \tag{7.2.6}$$

The solution to this equation by recursion is shown for a few values of $k$ and $n$ in Table 7.2.1.

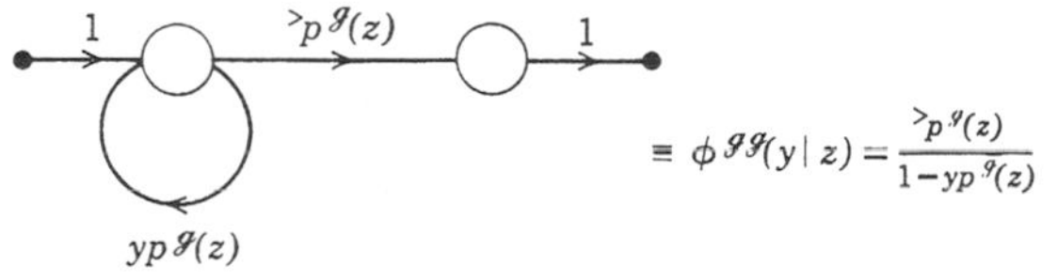


**Figure 7.2.1** Tagged flow graph for counting renewals.

**Table 7.2.1** Probability of $k$ Renewals in Time $n$ Following a Renewal

$\phi(k|n)$

| $k \backslash n$ | 0 | 1 | 2 | 3 | 4 | 5 |
|---|---|---|---|---|---|---|
| 0 | 1 | 1/2 | 0 | 0 | 0 | 0 |
| 1 | 0 | 1/2 | 3/4 | 1/4 | 0 | 0 |
| 2 | 0 | 0 | 1/4 | 5/8 | 1/2 | 1/8 |
| 3 | 0 | 0 | 0 | 1/8 | 7/16 | 9/16 |
| 4 | 0 | 0 | 0 | 0 | 1/16 | 9/32 |
| 5 | 0 | 0 | 0 | 0 | 0 | 1/32 |
| 6 | 0 | 0 | 0 | 0 | 0 | 0 |
| $\bar{k}(n)$ | | 1/2 | 5/4 | 15/8 | 41/16 | 103/32 |
| $\overline{k^2}(n)$ | | 1/2 | 7/4 | 31/8 | 111/16 | 347/32 |
| $\check{k}(n)$ | | 1/4 | 3/16 | 23/64 | 95/256 | 495/1024 |

***Renewal moments***

We can use the function $\phi(k|n)$ to compute the mean $\bar{k}(n)$, second moment $\overline{k^2}(n)$, and variance $\check{k}(n)$ of the number of renewals through time $n$ according to the equations,

$$\bar{k}(n) = \sum_{k=0}^{n} k\phi(k|n), \tag{7.2.7}$$

$$\overline{k^2}(n) = \sum_{k=0}^{n} k^2\phi(k|n), \tag{7.2.8}$$

$$\check{k}(n) = \overline{k^2}(n) - [\bar{k}(n)]^2. \tag{7.2.9}$$

Typical computations of moments for the example are presented in Table 7.2.1. Note the interesting result that as $n$ takes on the values 1, 2, 3, . . . the variance of the number of renewals in the period $(1, n)$, $\check{k}(n)$, passes through a minimum when $n = 2$.

The results of Table 7.2.1 are also computable from transform operations. For our example, we find the transform $\phi^{gg}(y|z)$ from Equation 7.2.5 to be

$$\phi^{gg}(y|z) = \frac{1 - (1/2)z - (1/2)z^2}{(1 - z)[1 - (1/2)yz - (1/2)yz^2]}. \tag{7.2.10}$$

Since inverting such transforms is usually neither easy nor illuminating, we prefer to find numerical values by the recursive methods based on the difference Equation

7.2.2. We can, however, use the transform expression $\phi^{gg}(y|z)$ to derive transform expressions for the moments of the number of renewals through time $n$. These transform expressions can be the source of analytical closed-form expressions for the moments in problems where we place a high value on closed-form results.

**Mean.** The geometric transform $\bar{k}^g(z)$ of $\bar{k}(n)$, the expected number of renewals through time $n$, is the derivative of $\phi^{gg}(y|z)$ with respect to $y$ evaluated at $y = 1$,

$$\bar{k}^g(z) = \frac{d}{dy}\phi^{gg}(y|z)\Big|_{y=1}. \tag{7.2.11}$$

From Equation 7.2.4 we find

$$\bar{k}^g(z) = \frac{d}{dy}\frac{{}^{>}p^g(z)}{1 - yp^g(z)}\Big|_{y=1} = \frac{{}^{>}p^g(z)p^g(z)}{[1 - yp^g(z)]^2}\Big|_{y=1}$$

$$= \frac{{}^{>}p^g(z)p^g(z)}{[1 - p^g(z)]^2} \tag{7.2.12}$$

or, alternately,

$$\bar{k}^g(z) = \frac{p^g(z)}{(1 - z)[1 - p^g(z)]}. \tag{7.2.13}$$

The transform of the expected number of renewals through time $n$ is therefore simply related to the transform of the lifetime distribution.

**Second moment.** Similarly, the second moment of the number of renewals through time $n$, $\overline{k^2}(n)$, has a geometric transform $\overline{k^2}^g(z)$ obtained as

$$\overline{k^2}^g(z) = \frac{d^2}{dy^2}\phi^{gg}(y|z)\Big|_{y=1} + \bar{k}^g(z) = \frac{2\,{}^{>}p^g(z)[p^g(z)]^2}{[1 - yp^g(z)]^3}\Big|_{y=1} + \bar{k}^g(z)$$

$$= \frac{2\,{}^{>}p^g(z)[p^g(z)]^2}{[1 - p^g(z)]^3} + \frac{{}^{>}p^g(z)p^g(z)}{[1 - p^g(z)]^2}$$

$$= \frac{{}^{>}p^g(z)p^g(z)[1 + p^g(z)]}{[1 - p^g(z)]^3}. \tag{7.2.14}$$

The alternate form in terms of $p^g(z)$ alone is then

$$\overline{k^2}^g(z) = \frac{p^g(z)[1 + p^g(z)]}{(1 - z)[1 - p^g(z)]^2}. \tag{7.2.15}$$

**The two-lifetime example.** For the majority of interesting lifetime distributions the transforms for the first and second moments obtained from these expressions are too complicated to be practical, although they may be of theoretical value. However, we can calculate a closed-form expression for the mean in our example by simply substituting the transform of the lifetime distribution into Equation 7.2.13,

$$\bar{k}^g(z) = \frac{(1/2)z + (1/2)z^2}{(1 - z)[1 - (1/2)z - (1/2)z^2]} = \frac{(1/2)z(1 + z)}{(1 - z)^2[1 + (1/2)z]}$$

$$= \frac{2/3}{(1 - z)^2} + \frac{-7/9}{1 - z} + \frac{1/9}{1 + (1/2)z}. \tag{7.2.16}$$

From inverse transformation, we find

$$\begin{aligned}\bar{k}(n) &= (2/3)(n + 1) - 7/9 + 1/9(-1/2)^n \\ &= (2/3)n - 1/9[1 - (-1/2)^n] \qquad n = 1, 2, 3, \ldots .\end{aligned} \tag{7.2.17}$$

This equation, of course, yields the values of $\bar{k}(n)$ that appear in Table 7.2.1.

**Computational considerations.** The problem of factoring the denominator of $\bar{k}^g(z)$ inhibits finding analytic expressions for even the mean if several different lifetimes are possible. One would need to be grimly determined to have an analytical expression for moments in order to go to the trouble. Of course, other methods are available for finding numerical values. Direct computation from $\phi(k|n)$ was the method used in Table 7.2.1. We could also use the dominant component approach of Section 4.2 to solve equations like 7.2.13 and 7.2.15 in approximate form. Still another alternative is to write the difference equations implied by these transform equations and to solve for the moments alone by recursion. The method to be used will depend mostly on the type of answer desired.

**Relationship of expected number of renewals to renewal probabilities.** We can sometimes capitalize on the fact that the impulse response $\phi(n)$ of a renewal process determines all its statistics. For example, let us put Equation 7.2.13 in the form

$$\bar{k}^g(z) = \frac{p^g(z)}{(1 - z)[1 - p^g(z)]} = \frac{1}{1 - z}\left[\frac{1}{1 - p^g(z)} - 1\right]. \tag{7.2.18}$$

Then from Equation 7.1.16,

$$\bar{k}^g(z) = \frac{1}{1 - z}[\phi^g(z) - 1] \tag{7.2.19}$$

and by inverse transformation,

$$\begin{aligned}\bar{k}(n) &= \sum_{m=0}^{n} \phi(m) - 1 \\ &= \sum_{m=1}^{n} \phi(m).\end{aligned} \tag{7.2.20}$$

This equation shows that the expected number of renewals in the interval $(1, n)$ is just the sum of the probabilities of renewal at each time in the interval. This is merely the renewal process special case of the occupancy statistic result of Equation 5.1.89 for the general Markov process. When we calculate $\bar{k}(n)$ for the example using Equation 7.2.20 and the results of Equation 7.1.22, we obtain the values of $\bar{k}(n)$ computed in Table 7.2.1.

### *Bernoulli trials*

Finding the probability of $k$ renewals in time $n$ and the moments of $k$ are difficult transform operations for even simple problems. However, we can perform all computations very easily for the case where a renewal occurs at each time point with a probability $p$, independent of previous renewals. In this case we can think of renewals as successes in Bernoulli trials with a probability of success $p$. The time between successes in Bernoulli trials is a random variable we first considered in Section 4.2. We found the time between successes to be given by the geometric distribution we have seen most recently as Equation 7.1.34, and with the transform of Equation 7.1.35. This geometric distribution is the lifetime distribution for the renewal process where a renewal is considered to be a success in a Bernoulli trial. The probability of a lifetime exceeding any given number of time units is shown in Equation 7.1.39, whose geometric transform is

$$ {}^{>}p^{g}(z) = \frac{1}{1 - (1 - p)z}. \tag{7.2.21} $$

For this lifetime distribution we can then draw the flow graph of Figure 7.2.1 in the form of Figure 7.2.2. We find from the flow graph or from Equation 7.2.4 that the transform of the number of renewals in any time period is

$$ \phi^{gg}(y|z) = \frac{1}{1 - (1 - p)z - pyz}. \tag{7.2.22} $$

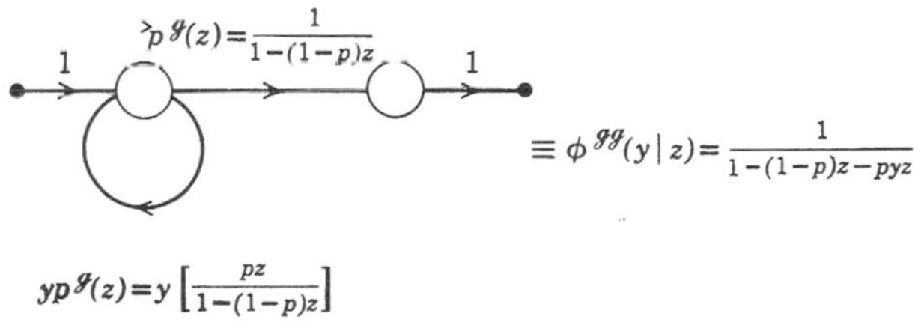


**Figure 7.2.2** Flow graph for number of successes in $n$ Bernoulli trails.

We invert first the transform on $n$,

$$\phi^{\mathcal{g}}(y|n) = [(1 - p) + py]^n = \sum_{k=0}^{n} \binom{n}{k} (py)^k(1 - p)^{n-k}, \tag{7.2.23}$$

and then the transform on $k$,

$$\phi(k|n) = \binom{n}{k} p^k(1 - p)^{n-k}. \tag{7.2.24}$$

This is, of course, just the Bernoulli binomial distribution for the probability of $k$ successes in $n$ Bernoulli trials with probability of success $p$.

An alternate way to solve this same problem would be to model the process as a one-node Markov process where the transitions from the node to itself may be of two types: successes or failures. The probabilities of each of these types are $p$ and $1 - p$. We could then tag every success type transition with a transform variable $y$ and draw the flow graph of Figure 7.2.3. The transmission of this tagged flow graph is just the transform given by Equation 7.2.22, thus illustrating the variety of the modeling methods at our disposal.

**Moments.** We can compute the mean and variance of the number of renewals in time $n$ or, equivalently, the number of successes in $n$ trials from the moment transform Equations 7.2.13 and 7.2.15. Proceeding first with the mean, we write

$$\bar{k}^{\mathcal{g}}(z) = \frac{p^{\mathcal{g}}(z)}{(1 - z)[1 - p^{\mathcal{g}}(z)]} = \frac{\dfrac{pz}{1 - (1 - p)z}}{(1 - z)\left[1 - \dfrac{pz}{1 - (1 - p)z}\right]} = \frac{pz}{(1 - z)^2}. \tag{7.2.25}$$

Therefore,

$$\bar{k}(n) = pn \qquad n = 1, 2, 3, \ldots. \tag{7.2.26}$$

The expected number of renewals, or successes, in $n$ trials is just the probability of success on each trial times the number of trials, a well-known property of the binomial distribution.

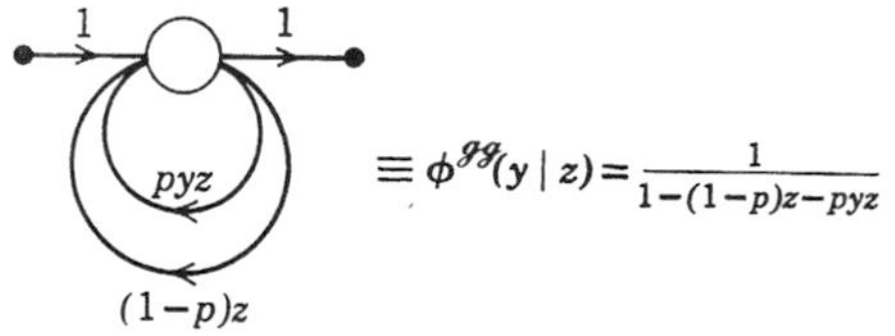


**Figure 7.2.3** Tagged flow graph for successes in Bernoulli trials.

For the second moment, we write

$$\overline{k^2}(z) = \frac{p^g(z)[1 + p^g(z)]}{(1 - z)[1 - p^g(z)]^2} = \frac{\dfrac{pz}{1-(1-p)z}\left[1 + \dfrac{pz}{1-(1-p)z}\right]}{(1-z)\left[1 - \dfrac{pz}{1-(1-p)z}\right]^2}$$

$$= \frac{pz}{(1-z)^2} + \frac{2p^2z^2}{(1-z)^3}. \tag{7.2.27}$$

Then by inverse transformation,

$$\overline{k^2}(n) = pn + 2p^2[(1/2)(n-1)n] = pn + p^2n^2 - p^2n \qquad n = 1, 2, 3, \ldots. \tag{7.2.28}$$

Finally, we compute the variance $\overset{\vee}{k}(n)$ from

$$\overset{\vee}{k}(n) = \overline{k^2}(n) - [\bar{k}(n)]^2 = pn - p^2n = p(1-p)n. \tag{7.2.29}$$

We have obtained the well-known property that the variance of the binomial distribution is the probability of success times 1 minus the probability of success times the number of trials.

As in this case, many of the results of elementary probability theory can be easily derived as special cases of Markov process theory. However, these results are more satisfying from the point of view of integration of our knowledge than they are important in themselves.

**Counting Renewals in the General Renewal Process**

We often want to count renewals in a situation where the time to the first renewal is selected from a different distribution from that used for subsequent renewals. Thus we define ${}_f\phi(k|n)$ to be the probability that $k$ renewals will take place in the interval $(1, n)$ when the time to the first renewal is selected from the distribution ${}_fp(\cdot)$ defined in the previous section and the time between all further renewals is selected from the lifetime distribution $p(\cdot)$. The event of $k$ renewals in the interval $(1, n)$ can happen in a number of ways. Let us categorize them by the time of the first renewal. With probability ${}_f\overset{>}{p}(n)$, the first renewal will occur after time $n$; then $k$ renewals in the interval $(1, n)$ can occur only when $k = 0$. With probability ${}_fp(n)$, the first renewal will occur at time $n$; then $k$ renewals in $(1, n)$ can occur only when $k = 1$. With probability ${}_fp(m)$, the first renewal will occur at some time $m = 1, 2, \ldots, n - 1$. In this case $1 \leq k \leq n$ renewals will occur in $(1, n)$ only if $k - 1$ additional renewals follow in the interval $(m + 1, n)$, an event with probability $\phi(k-1|n-m)$. When we sum the contribution for each $k$ over all possible times for the first renewal, we obtain

$$ {}_f\phi(k|n) = {}_f^{>}p(n)\,\delta(k) + {}_fp(n)\,\delta(k-1) + \sum_{m=1}^{n-1} {}_fp(m)\phi(k-1|n-m)\,u(n-1) \qquad \begin{array}{l} k = 0, 1, 2, \ldots \\ n = 0, 1, 2, \ldots . \end{array} \quad (7.2.30) $$

With the definitions that ${}_f\phi(k|0) = \phi(k|0) = \delta(k)$ and that functions with negative arguments are assumed to be zero, we can write this expression in the form,

$$ {}_f\phi(k|n) = \delta(k)\,{}_f^{>}p(n) + \sum_{m=0}^{n} {}_fp(m)\phi(k-1|n-m) \qquad \begin{array}{l} k = 0, 1, 2, \ldots \\ n = 0, 1, 2, \ldots , \end{array} \quad (7.2.31) $$

which shows how to build a table for ${}_f\phi(\cdot|\cdot)$ from a table for $\phi(\cdot|\cdot)$ when the starting distribution ${}_fp(\cdot)$ is specified.

The double geometric transform of Equation 7.2.31 we obtain immediately as

$$ {}_f\phi^{gg}(y|z) = {}_f^{>}p^{g}(z) + y\,{}_fp^{g}(z)\phi^{gg}(y|z). \quad (7.2.32) $$

We can write this transform in terms of the transform expression $\phi^{gg}(y|z)$ from Equation 7.2.4 as

$$ {}_f\phi^{gg}(y|z) = {}_f^{>}p^{g}(z) + \frac{y\,{}_fp^{g}(z)\,{}^{>}p^{g}(z)}{1 - yp^{g}(z)}. \quad (7.2.33) $$

***Flow graph representation***

The transform ${}_f\phi^{gg}(y|z)$ can now be interpreted as the transmission of the flow graph in Figure 7.2.4, a major part of which is the flow graph of Figure 7.2.1. The differences between the two graphs are due only to the difference in the probabilistic mechanism for the first renewal. If we know that the process had a renewal at time zero, then the distribution of time to the first renewal ${}_fp(\cdot)$ is just the same as the lifetime distribution $p(\cdot)$. Therefore, if we replace ${}_fp(\cdot)$ in every expression

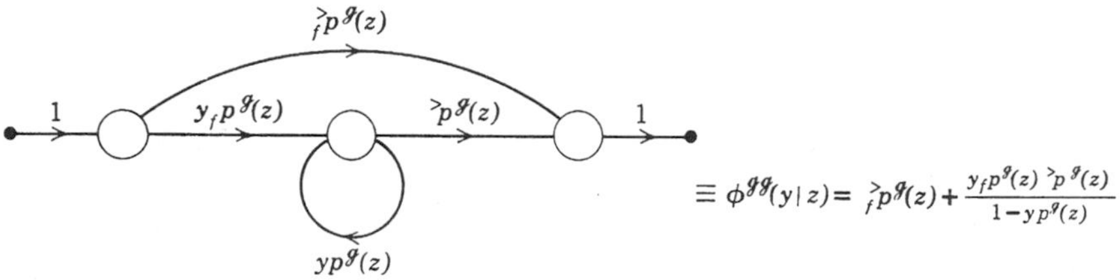


**Figure 7.2.4** Flow graph for counting transitions in general renewal process.

by $p(\cdot)$, all the results of the present discussion must simplify to those developed for counting after a renewal is known to have just occurred. We see, for example, that if we substitute $p^g(z)$ and ${}^{>}p^g(z)$ for ${}_{\mathrm{f}}p^g(z)$ and ${}^{>}_{\mathrm{f}}p^g(z)$ in Equation 7.2.33 we obtain immediately Equation 7.2.4. The equivalence of the flow graphs of Figures 7.2.4 and 7.2.1 for this case is most clearly evidenced by the identity of their transmission expressions.

***Random starting***

These results also provide the answer to the problem of counting renewals under random starting. We simply define ${}_{\mathrm{r}}\phi(k|n)$ to be the probability of observing $k$ renewals in the interval $(1, n)$ following a time zero selected at random. Then the probability mass function ${}_{\mathrm{f}}p(\cdot)$ becomes the probability mass function ${}_{\mathrm{r}}p(\cdot)$ defined by Equation 7.1.52. To write the equations for counting renewals after random starting we therefore change the subscript f to r in our expressions. The recursive Equation 7.2.31 becomes

$$ {}_{\mathrm{r}}\phi(k|n) = \delta(k)\,{}^{>}_{\mathrm{r}}p(n) + \sum_{m=0}^{n} {}_{\mathrm{r}}p(m)\phi(k-1|n-m) \qquad \begin{matrix} k = 0, 1, 2, \dots \\ n = 0, 1, 2, \dots \end{matrix} \tag{7.2.34} $$

with transform

$$ {}_{\mathrm{r}}\phi^{gg}(y|z) = {}^{>}_{\mathrm{r}}p^g(z) + \frac{y\,{}_{\mathrm{r}}p^g(z)\,{}^{>}p^g(z)}{1 - yp^g(z)}, \tag{7.2.35} $$

corresponding to Equation 7.2.33.

***The two-lifetime example***

To illustrate the procedure for counting renewals when we have a starting distribution ${}_{\mathrm{f}}p(\ )$, let us assume that the first renewal will occur with certainty at time 2 and that all further renewals will have the lifetime distribution we defined in Equation 7.1.20 for our example. Since ${}_{\mathrm{f}}p(m) = \delta(m-2)$, Equation 7.2.31 becomes

$$ {}_{\mathrm{f}}\phi(k|n) = \delta(k)[\delta(n) + \delta(n-1)] + \phi(k-1|n-2) \qquad \begin{matrix} k = 0, 1, 2, \dots \\ n = 0, 1, 2, \dots . \end{matrix} \tag{7.2.36} $$

The equation states that ${}_{\mathrm{f}}\phi(k|1) = \delta(k)$ and that ${}_{\mathrm{f}}\phi(k|n) = \phi(k-1|n-2)$, $k \geq 1$, $n \geq 2$; therefore, ${}_{\mathrm{f}}\phi(k|n)$ in this case can be read directly from Table 7.2.1. For example, the probability of seeing three renewals through time 5 is ${}_{\mathrm{f}}\phi(3|5) = \phi(2|3) = 5/8$.

If the time to the first renewal could take on more than one value, then we would have to compute a linear combination of entries from the table rather than read a single number. For example, let us consider the case where the time to the first

**Table 7.2.2** Probability of $k$ Renewals in Time $n$ Starting at Random

${}_r\phi(k|n)$

| $k \backslash n$ | 0 | 1 | 2 | 3 | 4 | 5 |
|---|---|---|---|---|---|---|
| 0 | 1 | 1/3 | 0 | 0 | 0 | 0 |
| 1 | 0 | 2/3 | 2/3 | 1/6 | 0 | 0 |
| 2 | 0 | 0 | 1/3 | 2/3 | 5/12 | 1/12 |
| 3 | 0 | 0 | 0 | 1/6 | 1/2 | 13/24 |
| 4 | 0 | 0 | 0 | 0 | 1/12 | 1/3 |
| 5 | 0 | 0 | 0 | 0 | 0 | 1/24 |
| 6 | 0 | 0 | 0 | 0 | 0 | 0 |
| ${}_r\bar{k}(n)$ | | 2/3 | 4/3 | 6/3 | 8/3 | 10/3 |
| ${}_r\overline{k^2}(n)$ | | 2/3 | 2 | 13/3 | 15/2 | 139/12 |
| ${}_r\overset{\vee}{k}(n)$ | | 2/9 | 2/9 | 1/3 | 7/18 | 17/36 |

event is that due to entering the process at random. We use ${}_rp(\cdot)$ from Equation 7.1.59 to write the recursive Equation 7.2.34 in the form

$${}_r\phi(k|n) = \delta(k)[\delta(n) + (1/3)\,\delta(n-1)] + (2/3)\phi(k-1|n-1) + (1/3)\phi(k-1|n-2)$$
$$k = 0, 1, 2, \ldots \qquad n = 0, 1, 2, \ldots . \tag{7.2.37}$$

Table 7.2.2 shows the result of applying this difference equation to the results of Table 7.2.1 for a few values of $k$ and $n$. The moments of the number of renewals in time $n$ are also computed in the table from the ${}_r\phi(k|n)$ distributions. Note that the mean ${}_r\bar{k}(n)$ grows linearly in $n$—we shall soon verify this characteristic.

***Moments of renewals in the general renewal process***

Although we could always compute the moments of the number of renewals in any time from tables of the probability distribution, moments play such an important role in our understanding of processes that we make a special effort to find direct and approximate methods for calculating them. The most important moments for the general renewal process are the mean ${}_f\bar{k}(n)$, second moment ${}_f\overline{k^2}(n)$, and variance ${}_f\overset{\vee}{k}(n)$ of the number of renewals in the interval $(1, n)$. These moments are defined by

$${}_f\bar{k}(n) = \sum_{k=0}^{n} k\ {}_f\phi(k|n) \tag{7.2.38}$$

$${}_f\overline{k^2}(n) = \sum_{k=0}^{n} k^2\ {}_f\phi(k|n) \tag{7.2.39}$$

and, of course,

$$ {}_f\overset{\vee}{k}(n) = {}_f\overline{k^2}(n) - [{}_f\bar{k}(n)]^2 . \tag{7.2.40} $$

When the time to the first event is determined by entering the process at random, we replace *f* by *r* in the notation as we did in Table 7.2.2.

**Mean.** The moments ${}_f\bar{k}(n)$ and ${}_f\overline{k^2}(n)$ have geometric transforms ${}_f\bar{k}^g(z)$ and ${}_f\overline{k^2}{}^g(z)$ that we can calculate from the transform ${}_f\phi^{gg}(y|z)$ in Equation 7.2.33 by the differentiation methods of Section 6.1. To find the transform of the mean, we write

$$ {}_f\bar{k}^g(z) = \frac{d}{dy}\,{}_f\phi^{gg}(y|z) = \frac{d}{dy}\left[{}^{>}_{f}p^g(z) + \frac{y\,{}_fp^g(z)\,{}^{>}p^g(z)}{1 - yp^g(z)}\right] $$

$$ = \left\{\frac{[1 - yp^g(z)]\,{}_fp^g(z)\,{}^{>}p^g(z) + y\,{}_fp^g(z)\,{}^{>}p^g(z)p^g(z)}{[1 - yp^g(z)]^2}\right\}_{y=1} $$

$$ = \left\{\frac{{}_fp^g(z)\,{}^{>}p^g(z)}{[1 - yp^g(z)]^2}\right\}_{y=1} $$

$$ = \frac{{}_fp^g(z)\,{}^{>}p^g(z)}{[1 - p^g(z)]^2} . \tag{7.2.41} $$

Now we use Equation 7.1.30 to obtain

$$ {}_f\bar{k}^g(z) = \frac{{}_fp^g(z)}{(1 - z)[1 - p^g(z)]} . \tag{7.2.42} $$

**Second moment.** For the transform of the second moment, ${}_f\overline{k^2}{}^g(z)$, we have

$$ {}_f\overline{k^2}{}^g(z) = \frac{d^2}{dy^2}\,{}_f\phi^{gg}(y|z)\Big|_{y=1} + {}_f\bar{k}^g(z) $$

$$ = \left\{\frac{d}{dy}\frac{{}_fp^g(z)\,{}^{>}p^g(z)}{[1 - yp^g(z)]^2}\right\}_{y=1} + {}_f\bar{k}^g(z) $$

$$ = \left\{\frac{2\,{}_fp^g(z)\,{}^{>}p^g(z)p^g(z)}{[1 - yp^g(z)]^3}\right\}_{y=1} + {}_f\bar{k}^g(z) $$

$$ = \frac{2\,{}_fp^g(z)\,{}^{>}p^g(z)p^g(z)}{[1 - p^g(z)]^3} + {}_f\bar{k}^g(z) . \tag{7.2.43} $$

With ${}_f\bar{k}^g(z)$ from Equation 7.2.41, we write

$$ {}_f\overline{k^2}{}^g(z) = \frac{2\,{}_fp^g(z)\,{}^{>}p^g(z)p^g(z)}{[1 - p^g(z)]^3} + \frac{{}_fp^g(z)\,{}^{>}p^g(z)}{[1 - p^g(z)]^2} $$

$$ = \frac{{}_fp^g(z)[1 + p^g(z)]\,{}^{>}p^g(z)}{[1 - p^g(z)]^3} . \tag{7.2.44} $$

If we now use Equation 7.1.30 for ${}^{>}p^{g}(z)$, we obtain

$$ {}_{f}\overline{k^2}^{g}(z) = \frac{{}_{f}p^{g}(z)[1 + p^{g}(z)]}{(1 - z)[1 - p^{g}(z)]^2}. \tag{7.2.45}$$

Note also that since from Equations 7.2.12 and 7.2.14 we have

$$ \overline{k^2}^{g}(z) + \bar{k}^{g}(z) = \frac{2\ {}^{>}p^{g}(z)p^{g}(z)}{[1 - p^{g}(z)]^3}, \tag{7.2.46}$$

Equation 7.2.43 can be written in the form

$$ {}_{f}\overline{k^2}^{g}(z) = {}_{f}p^{g}(z)[\overline{k^2}^{g}(z) + \bar{k}^{g}(z)] + {}_{f}\bar{k}^{g}(z), \tag{7.2.47}$$

which relates the transform of the second moment of renewals for the general process directly to the other renewal moment transforms.

**Moments under random starting.** For the special case of starting at random we find the moment transform equations by substituting ${}_{r}p^{g}(z)$ from Equation 7.1.54 for ${}_{f}p^{g}(z)$ in Equations 7.2.42 and 7.2.45. The results are

$$ {}_{r}\bar{k}^{g}(z) = \frac{z}{\bar{l}(1 - z)^2} \tag{7.2.48}$$

and

$$ {}_{r}\overline{k^2}^{g}(z) = \frac{z[1 + p^{g}(z)]}{\bar{l}(1 - z)^2[1 - p^{g}(z)]}. \tag{7.2.49}$$

The expression for ${}_{r}\bar{k}^{g}(z)$ shows that ${}_{r}\bar{k}(n)$ is given by

$$ {}_{r}\bar{k}(n) = \frac{1}{\bar{l}}\,n, \qquad n = 1, 2, 3, \ldots, \tag{7.2.50}$$

that is, the expected number of renewals at the $n$ time-points following a starting point picked at random is just $n$ times the reciprocal of the mean lifetime. It is therefore equal to the steady state probability of renewal at each time-point multiplied by the number of time-points. This result satisfies our intuition about the meaning of random starting and, of course, explains the results we obtained for ${}_{r}\bar{k}(n)$ in Table 7.2.2.

### *The Bernoulli process with random starting*

The Bernoulli process has the interesting and important property that its renewal statistics are the same whether we start examining the process after an event or we enter it at random. We illustrate this property by calculating ${}_{r}p^{g}(z)$ for the Bernoulli process from Equations 7.1.54 and 7.1.40:

$$ {}_{r}p^{g}(z) = \frac{z[1 - p^{g}(z)]}{\bar{l}(1 - z)} = \frac{z\left[1 - \dfrac{pz}{1 - (1 - p)z}\right]}{\dfrac{1}{p}(1 - z)} = \frac{pz}{1 - (1 - p)z} = p^{g}(z). \quad (7.2.51) $$

Thus the probability distribution for the time to the first renewal under random starting is just the lifetime distribution itself. When the time to the first renewal is the lifetime distribution, the probability of $k$ renewals in $(1, n)$ is given by $\phi(k|n)$ defined in Equation 7.2.2, and therefore this must also be the distribution for random starting. Thus, for the Bernoulli process,

$$ {}_{r}\phi(k|n) = \phi(k|n). \quad (7.2.52) $$

All statistics, including the moments, will be the same for a Bernoulli process whether we enter it after a success or enter it at random. This result implies, correctly, that information on the time of the last renewal is not helpful in determining when the next renewal will occur.

Interestingly enough, the Bernoulli process is the only one in which we do not prejudice the statistics by starting after an event rather than at random. We can see this by requiring that

$$ {}_{r}p^{g}(z) = p^{g}(z) = \frac{z[1 - p^{g}(z)]}{\bar{l}(1 - z)} \quad (7.2.53) $$

and solving for $p^{g}(z)$. We find

$$ \begin{aligned} \bar{l}(1 - z)p^{g}(z) &= z - zp^{g}(z), \\ p^{g}(z)[z + \bar{l}(1 - z)] &= z \end{aligned} \quad (7.2.54) $$

or

$$ p^{g}(z) = \frac{\dfrac{1}{\bar{l}}z}{1 - \left(1 - \dfrac{1}{\bar{l}}\right)z}. \quad (7.2.55) $$

Thus the lifetime distribution must be geometric: The renewals must be generated by Bernoulli trials with $p = 1/\bar{l}$.

### Asymptotic Moments

Although the moment transforms are not often useful for practical calculations, they have theoretical uses in deriving asymptotic forms for the moments that in turn have great practical value.

***Properties of the lifetime distribution transform***

Before we develop these results, let us note that we can use Equation 7.1.30 to write one minus the transform of the lifetime distribution in the form

$$1 - p^g(z) = (1 - z)\,{}^{>}p^g(z). \tag{7.2.56}$$

Since ${}^{>}p(\cdot)$ is a complementary cumulative probability distribution, it must approach zero for large values of its arguments. Therefore, ${}^{>}p^g(z)$ contains only transient components. We shall assume that every lifetime distribution we consider has a finite mean and variance.

We can find relationships between the derivatives of $p^g(z)$ and ${}^{>}p^g(z)$ at the point $z = 1$ by differentiating Equation 7.2.56 with respect to $z$. Our first differentiation produces

$$-p^{g\prime}(z) = -\,{}^{>}p^g(z) + (1 - z)\,{}^{>}p^{g\prime}(z) \tag{7.2.57}$$

or

$$-p^{g\prime}(1) = -\,{}^{>}p^g(1). \tag{7.2.58}$$

Then,

$${}^{>}p^g(1) = p^{g\prime}(1) = \bar{l}. \tag{7.2.59}$$

This equation establishes the well-known result that the area under the complementary cumulative probability mass function for a non-negative random variable is equal to its mean.

Another differentiation with respect to $z$ of Equation 7.2.57 yields

$$-p^{g\prime\prime}(z) = -2\,{}^{>}p^{g\prime}(z) + (1 - z)\,{}^{>}p^{g\prime\prime}(z) \tag{7.2.60}$$

or

$$-p^{g\prime\prime}(1) = -2\,{}^{>}p^{g\prime}(1). \tag{7.2.61}$$

Finally, we have

$${}^{>}p^{g\prime}(1) = (1/2)p^{g\prime\prime}(1) = (1/2)(\overline{l^2} - \bar{l}). \tag{7.2.62}$$

***Asymptotic first moment of renewals***

Now we proceed to find asymptotic expressions for the moments of the number of renewals in the interval $(1, n)$ when $n$ is large. We begin by using Equation 7.2.56 to write Equation 7.2.41 as

$${}_{f}\bar{k}^g(z) = \frac{{}_{f}p^g(z)\,{}^{>}p^g(z)}{[1 - p^g(z)]^2} = \frac{{}_{f}p^g(z)}{(1 - z)^2\,{}^{>}p^g(z)}\cdot \tag{7.2.63}$$

From the form of the denominator in this expression we know that the only terms in ${}_f\bar{k}(n)$ that will persist for large $n$ are the ramp and step components. Therefore, we write the approximate expression

$$ {}_f\bar{k}^g(z) = \frac{a_1}{(1-z)^2} + \frac{a_2}{1-z} \tag{7.2.64}$$

with inverse,

$$ {}_f\bar{k}(n) = a_1(n+1) + a_2 = a_1 n + a_1 + a_2 \qquad \text{for large } n. \tag{7.2.65}$$

Then we solve for $a_1$ and $a_2$ by using the differentiation method for partial fraction expansion with repeated roots on Equation 7.2.63. We obtain

$$ a_1 = \lim_{z\to 1} (1-z)^2 \, {}_f\bar{k}^g(z) = \lim_{z\to 1} \frac{{}_fp^g(z)}{{}^{>}p^g(z)} = \frac{{}_fp^g(1)}{{}^{>}p^g(1)} = \frac{1}{\bar{l}}; \tag{7.2.66}$$

and

$$ a_2 = \lim_{z\to 1} \frac{d}{dz}(1-z)^2 \bar{k}^g(z) = \lim_{z\to 1} \frac{d}{dz} \frac{{}_fp^g(z)}{{}^{>}p^g(z)} $$

$$ = \lim_{z\to 1} \frac{-{}^{>}p^g(z)\,{}_fp^{g\prime}(z) + {}_fp^g(z)\,{}^{>}p^{g\prime}(z)}{[{}^{>}p^g(z)]^2} = \frac{-{}^{>}p^g(1)\,{}_fp^{g\prime}(1) + {}_fp^g(1)\,{}^{>}p^{g\prime}(1)}{[{}^{>}p^g(1)]^2} $$

$$ = \frac{-\bar{l}\,{}_f\bar{l} + (1/2)(\overline{l^2} - \bar{l})}{\bar{l}^2} $$

$$ = \frac{1}{2\bar{l}}\left[-1 + \frac{\overline{l^2}}{\bar{l}} - 2\,{}_f\bar{l}\right]. \tag{7.2.67}$$

Here we have used ${}_f\bar{l}$ for the mean of the ${}_fp(\cdot)$ distribution. Then from Equation 7.2.64 we have

$$ {}_f\bar{k}(n) = a_1 n + a_1 + a_2 $$

$$ = \frac{1}{\bar{l}}\,n + \frac{1}{2\bar{l}}\left[1 + \frac{\overline{l^2}}{\bar{l}} - 2\,{}_f\bar{l}\right] \qquad \text{for large } n. \tag{7.2.68}$$

This equation shows how the expected number of renewals in the interval $(1, n)$ grows linearly with $n$ when $n$ is large. Further, it reveals that ${}_f\bar{k}(n)$ for large $n$ has a constant component that may be positive, negative, or zero depending on ${}_f\bar{l}$, the mean time to the first renewal. Clearly, if ${}_f\bar{l}$ is very large, the constant term will be negative and will represent the dearth of early renewals due to the exceptionally long-lived first unit. If a renewal is known to have occurred at time zero, ${}_f\bar{l} = \bar{l}$ and the constant term may be either positive or negative.

**Random starting.** We obtain an interesting check on Equation 7.2.68, from the case of random starting. For this case ${}_f\bar{l} = {}_r\bar{l}$, the mean of the ${}_rp(\cdot)$ distribution whose transform appears in Equation 7.1.54. Using Equation 7.2.56 we find

$$
\begin{aligned}
{}_r\bar{l} &= \frac{d}{dz}\,{}_rp^g(z)\Big|_{z=1} = \frac{d}{dz}\frac{z}{\bar{l}}\frac{1 - p^g(z)}{1 - z}\Big|_{z=1} \\
&= \frac{d}{dz}\frac{z}{\bar{l}}\,{}^{>}p^g(z)\Big|_{z=1} \\
&= \frac{1}{\bar{l}}\,[{}^{>}p^g(z) + z\,{}^{>}p^{g\prime}(z)]_{z=1} = \frac{1}{\bar{l}}\,[{}^{>}p^g(1) + {}^{>}p^{g\prime}(1)].
\end{aligned} \tag{7.2.69}
$$

Then from Equations 7.2.59 and 7.2.62,

$$
\begin{aligned}
{}_r\bar{l} &= \frac{1}{\bar{l}}\left[\bar{l} + \frac{1}{2}(\overline{l^2} - \bar{l})\right] \\
&= \frac{1}{2}\left[1 + \frac{\overline{l^2}}{\bar{l}}\right].
\end{aligned} \tag{7.2.70}
$$

When we substitute this value of ${}_r\bar{l}$ for ${}_f\bar{l}$ in Equation 7.2.68, we see that the constant term disappears—a result expected because Equation 7.2.50 shows that $(1/\bar{l})n$ is in fact an exact expression for ${}_r\bar{k}(n)$ for all $n$.

***Asymptotic variance of renewals***

Now we proceed to obtain an asymptotic expression for the variance of renewals. We begin by calculating the asymptotic second moment. First we introduce Equation 7.2.56 into Equation 7.2.44 to produce

$$
{}_f\overline{k^2}{}^g(z) = \frac{{}_fp^g(z)[1 + p^g(z)]}{(1 - z)^3[{}^{>}p^g(z)]^2}. \tag{7.2.71}
$$

The three nontransient components of $\overline{k^2}(n)$ when $n$ is large are those corresponding to the expansion of ${}_f\overline{k^2}{}^g(z)$ in the form

$$
{}_f\overline{k^2}{}^g(z) = \frac{b_1}{(1 - z)^3} + \frac{b_2}{(1 - z)^2} + \frac{b_3}{(1 - z)}, \tag{7.2.72}
$$

with inverse,

$$
\begin{aligned}
{}_f\overline{k^2}(n) &= b_1(1/2)(n + 1)(n + 2) + b_2(n + 1) + b_3 \\
&= (1/2)b_1n^2 + [(3/2)b_1 + b_2]n + \text{constant} \qquad \text{for large } n.
\end{aligned} \tag{7.2.73}
$$

We find $b_1$ and $b_2$ from

$$
b_1 = \lim_{z\to 1}(1 - z)^3\,{}_f\overline{k^2}{}^g(z) = \lim_{z\to 1}\frac{{}_fp^g(z)[1 + p^g(z)]}{[{}^{>}p^g(z)]^2} = \frac{{}_fp^g(1)[1 + p^g(1)]}{[{}^{>}p^g(1)]^2} = \frac{2}{\bar{l}^2} \tag{7.2.74}
$$

and

$$b_2 = \lim_{z\to 1} -\frac{d}{dz}(1-z)^3\, {}_f\overline{k^2}^g(z)$$

$$= \lim_{z\to 1} -\frac{d}{dz}\frac{{}_fp^g(z)[1+p^g(z)]}{[{}^{>}p^g(z)]^2} = \frac{2\overline{l^2} - 2\bar{l}\,{}_f\bar{l} - \bar{l}^2 - 2\bar{l}}{\bar{l}^3}. \qquad (7.2.75)$$

Now we combine the results of Equations 7.2.65 and 7.2.73 to write the asymptotic variance in the form,

$${}_f\check{k}(n) = {}_f\overline{k^2}(n) - [{}_f\bar{k}(n)]^2$$

$$= n^2\left(\frac{b_1}{2} - a_1^2\right) + n[(3/2)b_1 + b_2 - 2a_1^2 - 2a_1a_2] + \text{constant} \qquad \text{for large } n. \qquad (7.2.76)$$

From Equations 7.2.66 and 7.2.74 we see that the coefficient of the $n^2$ term is zero. The coefficient of the linear term turns out to be just $(\overline{l^2} - \bar{l}^2)/\bar{l}^3$ or

$${}_f\check{k}(n) = n\frac{\overline{l^2} - \bar{l}^2}{\bar{l}^3} + \text{constant} = n\frac{\check{l}}{\bar{l}^3} + \text{constant} \qquad \text{for large } n. \qquad (7.2.77)$$

Thus the variance of the number of renewals in the interval $(1, n)$ grows asymptotically linearly with $n$ when $n$ is large at a rate equal to the variance of the lifetime distribution divided by its mean cubed. Comparing this result with Equation 5.2.129 leads us to think of analyzing renewal processes by means of state occupancy statistics—we shall do just this.

***Renewal moment examples***

For the example with lifetime equally likely to be 1 or 2 units, we have $\bar{l} = 3/2$, $\overline{l^2} = 5/2$, and $\check{l} = 1/4$. Therefore, assuming that we start after a renewal so that ${}_f\bar{l} = \bar{l}$, the asymptotic moment equations 7.2.68 and 7.2.77 become

$$\bar{k}(n) = (2/3)n - 1/9 \qquad \text{for large } n, \qquad (7.2.78)$$

and

$$\check{k}(n) = (2/27)n + \text{constant} \qquad \text{for large } n. \qquad (7.2.79)$$

These asymptotic expressions show the behavior for $\bar{k}(n)$ and $\check{k}(n)$ that we would expect in Table 7.2.1 if $n$ were large. Actually the mean expression is quite close for $n = 5$, but the variance expression $(2/27)n$ at this point is still off by about 50%. The variance expression would naturally be considerably closer if the constant term were included in the approximation.

For the Bernoulli process with the geometrically distributed lifetime of Equation 7.1.34, we have from Section 4.2: $\bar{l} = 1/p$, $\overline{l^2} = (2 - p)/p^2$, and $\check{l} = (1 - p)/p^2$. If we start after a renewal so that ${}_f\bar{l} = \bar{l}$, Equations 7.2.68 and 7.2.77 take the form

$$\bar{k}(n) = pn \qquad \text{for large } n, \tag{7.2.80}$$

and

$$\check{k}(n) = p(1 - p)n + \text{constant} \qquad \text{for large } n. \tag{7.2.81}$$

We have, of course, previously developed exact expressions for these moments in Equations 7.2.26 and 7.2.29. These earlier findings are not only consistent with the asymptotic results—they also show that the residual terms are in fact zero for this case. Since the lifetime distribution is geometric, we obtain the same results for the case of starting at random.

## Renewals as State Occupancies

The similarity of Equations 5.2.129 and 7.2.70 suggests interpreting renewals as state occupancies. By analogy with our approach in Section 5.1, we could define $x(r)$ to be a random variable that is equal to one if a renewal occurs at time $r$ and is zero otherwise. Then ${}_fk(n)$, the number of renewals in $(1, n)$ would be given by

$${}_fk(n) = \sum_{r=1}^{n} x(r). \tag{7.2.82}$$

By finding the moments of ${}_fk(n)$ directly, we shall both review this basic analytic technique and observe some interesting similarities with the occupancy statistic section.

### *Mean*

First, we compute the mean of ${}_fk(n)$, ${}_f\bar{k}(n)$. We write

$${}_f\bar{k}(n) = \sum_{r=1}^{n} \bar{x}(r). \tag{7.2.83}$$

Since the probability that $x(r)$ equals 1 is just ${}_f\phi(r)$, the expectation of $x(r)$ is ${}_f\phi(r)$ and we obtain

$${}_f\bar{k}(n) = \sum_{r=1}^{n} {}_f\phi(r), \tag{7.2.84}$$

the analog of Equation 7.2.20 for $\bar{k}(n)$. When we transform this equation and use Equation 7.1.45, we find

$$_{\mathrm{f}}\bar{k}^{g}(z) = \frac{1}{1-z}\,{}_{\mathrm{f}}\phi^{g}(z) = \frac{1}{1-z}\cdot\frac{{}_{\mathrm{f}}p^{g}(z)}{1-p^{g}(z)}, \tag{7.2.85}$$

which is, of course, just Equation 7.2.42. Table 7.2.2 confirms the result of Equation 7.2.84 for the values of ${}_{\mathrm{f}}\phi(r)$ given in Equation 7.1.51.

***Second moment***

For the second moment of ${}_{\mathrm{f}}k(n)$, ${}_{\mathrm{f}}\overline{k^2}(n)$, we write

$$_{\mathrm{f}}\overline{k^2}(n) = \overline{\sum_{r=1}^{n} x(r) \sum_{s=1}^{n} x(s)} = \sum_{r=1}^{n}\sum_{s=1}^{n} \overline{x(r)x(s)}. \tag{7.2.86}$$

Now,

$$x(r)x(s) = 1 \text{ with probability } \begin{cases} {}_{\mathrm{f}}\phi(r)\phi(s-r) & \text{if } r < s \\ {}_{\mathrm{f}}\phi(r) & \text{if } r = s \\ {}_{\mathrm{f}}\phi(s)\phi(r-s) & \text{if } r > s. \end{cases} \tag{7.2.87}$$

Therefore,

$$\overline{x(r)x(s)} = \begin{cases} {}_{\mathrm{f}}\phi(r)\phi(s-r) & \text{if } r < s \\ {}_{\mathrm{f}}\phi(r) & \text{if } r = s \\ {}_{\mathrm{f}}\phi(s)\phi(r-s) & \text{if } r > s, \end{cases} \tag{7.2.88}$$

and

$$\begin{aligned} {}_{\mathrm{f}}\overline{k^2}(n) &= \sum_{r=1}^{n}\sum_{s=r+1}^{n} {}_{\mathrm{f}}\phi(r)\phi(s-r) + \sum_{s=1}^{n}\sum_{r=s+1}^{n} {}_{\mathrm{f}}\phi(s)\phi(r-s) + \sum_{r=1}^{n} {}_{\mathrm{f}}\phi(r) \\ &= 2\sum_{r=1}^{n}\sum_{s=r+1}^{n} {}_{\mathrm{f}}\phi(r)\phi(s-r) + \sum_{r=1}^{n} {}_{\mathrm{f}}\phi(r) \\ &= 2\sum_{r=1}^{n} {}_{\mathrm{f}}\phi(r) \sum_{m=1}^{n-r} \phi(m) + \sum_{r=1}^{n} {}_{\mathrm{f}}\phi(r). \end{aligned} \tag{7.2.89}$$

We simplify by using Equations 7.2.20 and 7.2.84,

$$_{\mathrm{f}}\overline{k^2}(n) = 2\sum_{r=1}^{n} {}_{\mathrm{f}}\phi(r)\bar{k}(n-r) + {}_{\mathrm{f}}\bar{k}(n). \tag{7.2.90}$$

When we transform this equation and use Equations 7.1.45, 7.2.13, and 7.2.85, we obtain

$$
\begin{aligned}
{}_{f}\overline{k^2}^g(z) &= 2\,{}_{f}\phi^g(z)\bar{k}^g(z) + {}_{f}\bar{k}^g(z) \\
&= 2\,\frac{{}_{f}p^g(z)}{1 - p^g(z)}\cdot\frac{p^g(z)}{(1 - z)[1 - p^g(z)]} + \frac{{}_{f}p^g(z)}{(1 - z)[1 - p^g(z)]} \\
&= \frac{{}_{f}p^g(z)[1 + p^g(z)]}{(1 - z)[1 - p^g(z)]^2},
\end{aligned}
\qquad (7.2.91)
$$

which is, of course, the value for the transform given in Equation 7.2.45.

***Relationship to state occupancy results***

With the proper interpretations, Equations 7.2.84 and 7.2.90 are identical in form to Equations 5.1.89 and 5.1.95 developed for state occupancy statistics in the case where the initial placement of the process in a state is not counted. First, we identify a renewal with occupancy of state $j$ in a Markov process and place the system initially in state $i$. The time required for the system to get from state $i$ to state $j$ for the first time, the first passage time from state $i$ to state $j$, is just the starting time for the renewal process whose distribution we have called ${}_{f}p(\cdot)$. The time between occupancies of state $j$, the recurrence time of state $j$, is the lifetime of the renewal process. Starting to observe the renewal process after a renewal corresponds to placing the Markov process in state $j$ initially. There is thus a direct correspondence between ${}_{f}\bar{k}(n)$ and ${}_{1}\overline{\nu_{ij}}(n)$, ${}_{f}\phi(r)$ and $\phi_{ij}(r)$, $\bar{k}(n)$ and ${}_{1}\overline{\nu_{jj}}(n)$. The equivalence of Equations 7.2.77 and 5.2.129 is a direct result of this interpretation.

## Importance of Renewal Processes

One might ask at this point why we have developed special relations for the renewal process if it is just a special case of the Markov process. The answer is, first, that renewal processes are of sufficient practical importance to deserve special treatment. Second, the renewal process point of view provides a somewhat deeper insight into the operation of the Markov process. Third, and overriding, we have not restricted our starting lifetime distributions in the renewal processes to be distributions that *could* have been first passage time and recurrence time distributions for some finite-state Markov process. All we require is that these distributions have a finite mean, and a finite variance if our variance expressions are to be valid; otherwise, they could be very peculiar indeed. Therefore, the results of this section stand independent of the Markov process theory we have developed —but they are a very useful supplement to it.

## 7.3 PATTERNS IN BERNOULLI TRIALS

The theory of renewal processes has an important application in the study of patterns of successes in Bernoulli trials. The importance of the application arises from the existence of physical systems whose operation (or lack of operation) depends on the particular sequence of random inputs they receive. In Section 4.2 we used the transient process formulation to study how many trials would be required to obtain a particular pattern for the first time. Now we shall generalize these results and in addition examine the repetition of patterns.

The models we discuss in this section are probably the simplest obtainable for this type of problem. We shall find that our work in this section is not tedious—solution of what used to be difficult probabilistic questions is reduced to a satisfying, but not difficult, exercise in flow graph analysis.

### Success Runs

We shall begin by analyzing success runs in Bernoulli trials. We can think of a success as a Head produced in tossing a coin and give it the symbol H. We shall assume that the probability of success or Head on each trial is $p$. A success run of length $r$ is just a succession of $r$ Heads in a row, HHH $\cdots$ H. Immediately, we face a problem in deciding how success runs are to be counted. For example, if we are looking for success runs of length 3 and find the pattern $\cdots$THHHH$\cdots$, is that one or two success runs of length 3? Different ways of taking the count may be appropriate in different circumstances. For the moment, we shall assume that whenever a success run of length $r$ occurs, it wipes the slate clean for counting the next success run of length $r$. Thus the pattern $\cdots$THHHH$\cdots$ would count as one success run of length three plus a start of one success toward whatever develops in the next tosses.

#### *Flow graph analysis*

Let us now find statistical properties of the occurrence of success runs of length $r$ in a series of Bernoulli trials. We can consider the generation of success runs of length $r$ as a renewal process whose lifetime distribution is just the distribution of the number of tosses required to produce such a run starting from scratch. The time to produce a run of length $r$ can be modeled as the time to leave a transient Markov process whose states represent the various stages of completion of a run

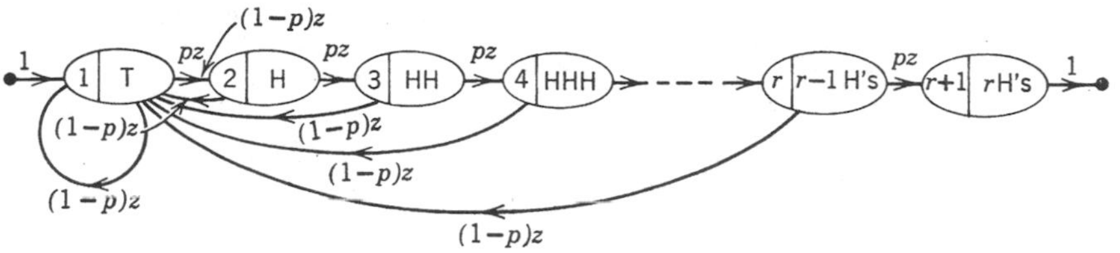


**Figure 7.3.1** Flow graph for success run of length $r$.

of length $r$. These stages are: 1) the last toss was a Tail; 2) the last toss was a Head preceded by a Tail; 3) the last toss was a Head preceded by a Head and a Tail $\cdots$; $r$) the last toss was a head preceded by $r - 1$ Heads then a Tail; $r + 1$) a run of length $r$ has been completed. Figure 7.3.1 shows the $r + 1$-node flow graph implied by these state descriptions. State $r + 1$ is, of course, the trapping state that terminates the transient process. The graph shows that the occurrence of a Head in any state advances the process to the state with next higher index; this advance happens with probability $p$. Regardless of state, the occurrence of a Tail returns the process to state 1: The steps toward building a success run must begin anew. The probability of a Tail on any toss is, of course, $1 - p$. Thus after adding $z$'s to all transition probabilities to create a flow graph we obtain the one shown in Figure 7.3.1. Note that this flow graph is just an extension of Figure 4.2.8 to the case where the sequence of Heads to be generated is of general length $r$ rather than 3.

Although the flow graph of Figure 7.3.1 is of general size, we find it easy to write its transmission because all loops and paths pass through node 1. The transmission $\vec{t}_{1,r+1}$ of this graph, the transform of the lifetime distribution, is therefore

$$\vec{t}_{1,r+1} = p^{\mathscr{g}}(z) = \frac{p^r z^r}{1 - (1-p)z - (1-p)pz^2 - (1-p)p^2z^3 - \cdots - (1-p)p^{r-1}z^r}$$

$$= \frac{p^r z^r}{1 - \dfrac{(1-p)z}{1-pz}[1 - p^r z^r]}$$

$$= \frac{(1-pz)p^r z^r}{1 - z + (1-p)p^r z^{r+1}}. \qquad (7.3.1)$$

We see immediately that $p_R = p^{\mathscr{g}}(1) = 1$. Occurrences of success runs of length $r$ are therefore recurrent events.

### *Moments*

We can find the mean $\bar{l}$ and variance $\check{l}$ of the number of tosses to obtain the first success run of length $r$, or equivalently of the number of tosses between completions of these runs, by differentiating Equation 7.3.1. We write

$$\bar{l} = p^{\mathscr{g}\prime}(1) = \frac{1 - p^r}{(1 - p)p^r} \tag{7.3.2}$$

$$\overline{l^2} = p^{\mathscr{g}\prime\prime}(1) + p^{\mathscr{g}\prime}(1) = \frac{2(1 - p^r)}{[(1 - p)p^r]^2} - \frac{2r + 1}{(1 - p)p^r} + \frac{1}{1 - p} \tag{7.3.3}$$

and

$$\check{l} = \overline{l^2} - \bar{l}^2 = \frac{1}{[(1 - p)p^r]^2} - \frac{2r + 1}{(1 - p)p^r} - \frac{p}{(1 - p)^2}. \tag{7.3.4}$$

The differentiation required to produce these results is made somewhat easier by multiplying Equation 7.3.1 by the denominator of the transmission expression before differentiating. When $r = 3$, Equations 7.3.2 and 7.3.4 specialize to the expressions found in Equations 4.2.18 and 4.2.23 for the moments of the time to produce a success run of length 3. Although we cannot see this result from casual inspection of the relations, we can easily establish it algebraically.

### *Moments from the transition diagram*

We have available as an alternative method for finding the moments of the lifetime distribution the inspection methods of Section 4.1 for determining the moments of the time spent in transient processes. The key to these methods is finding the matrix of transmissions of the transition diagram for the transient process. We can construct this transition diagram mentally by removing all the $z$'s from the flow graph of Figure 7.3.1. Then by inspection we find the matrix of transmissions among the $r$ transient states of the process to be,

$$\vec{T} = \begin{array}{c} (1) \\ (2) \\ (3) \\ \\ \vdots \\ (r) \end{array} \overset{\begin{array}{cccccc} (1) & (2) & (3) & & & (r) \end{array}}{\begin{bmatrix} 1 & p & p^2 & p^3 & \cdots & p^{r-1} \\ 1 - p^{r-1} & p & p^2 & p^3 & \cdots & p^{r-1} \\ 1 - p^{r-2} & p(1 - p^{r-2}) & p^2 & p^3 & \cdots & p^{r-1} \\ 1 - p^{r-3} & p(1 - p^{r-3}) & p^2(1 - p^{r-3}) & p^3 & \cdots & p^{r-1} \\ \vdots & \vdots & \vdots & \vdots & & \vdots \\ 1 - p & p(1 - p) & p^2(1 - p) & & \cdots & p^{r-1} \end{bmatrix}} \cdot \frac{1}{p^r} \tag{7.3.5}$$

When $r = 3$ this matrix of transmissions becomes the same as that shown in Equation 4.2.14 for the run of three Heads. Another check is that when $p = 1$ in Equation 7.3.5 all elements of $\vec{T}$ below the diagonal become zero while the rest become 1. This result is just what we expect from the flow graph and from our interpretation of $\vec{t}_{ij}$ for a transient process as the expected number of transitions made into state $j$ in an infinite number of tosses if the process is started in state $i$.

**Mean.** The row sums of $\vec{T}$ are the mean times spent in the process for each starting state. Thus from Equation 4.1.38,

$$\bar{\boldsymbol{\nu}} = \vec{T}\mathbf{s} = \begin{matrix} (1) \\ (2) \\ (3) \\ \vdots \\ (r-1) \\ (r) \end{matrix} \begin{bmatrix} \dfrac{1-p^r}{(1-p)p^r} = \dfrac{1+p+p^2+\cdots+p^{r-1}}{p^r} \\ \dfrac{1-p^{r-1}}{(1-p)p^r} = \dfrac{1+p+p^2+\cdots+p^{r-2}}{p^r} \\ \dfrac{1-p^{r-2}}{(1-p)p^r} = \dfrac{1+p+p^2+\cdots+p^{r-3}}{p^r} \\ \vdots \\ \dfrac{1-p^2}{(1-p)p^r} = \dfrac{1+p}{p^r} \\ \dfrac{1-p}{(1-p)p^r} = \dfrac{1}{p^r} \end{bmatrix}. \tag{7.3.6}$$

In particular, we see that the first component in this vector, $\bar{\nu}_1$, is just the mean lifetime $\bar{l}$ of Equation 7.3.2. In general, any component of $\bar{\boldsymbol{\nu}}$, say $\bar{\nu}_k$, represents the expected number of trials to obtain a run of length $r$ if a run of length $k - 1$ has just been completed. Therefore, we see from $\bar{\nu}_r$ that $1/p^r$ is the expected number of trials to complete a success run of length $r$ if we need just one more successful trial to complete the run.

**Second moment and variance.** We can also find the variance of the time spent in the transient process for any starting point, but we shall write it only for the case we are interested in—starting in state 1. We begin by calculating the second moment of time spent in the transient process for this starting state using Equation 4.1.69,

$$\overline{\nu_1^2} = 2\sum_{j=1}^{r} \vec{t}_{1j}\bar{\nu}_j - \bar{\nu}_1. \tag{7.3.7}$$

Thus we postmultiply the first row of the matrix $\vec{T}$ by the column vector $\bar{\boldsymbol{\nu}}$, double the result and subtract $\bar{\nu}_1$. The result is that $\overline{\nu_1^2} = \overline{l^2}$ in Equation 7.3.3 as it should. Therefore $\overset{\vee}{\nu}_1 = \overset{\vee}{l}$ in Equation 7.3.4, and we have established the important moments of the lifetime distribution by inspection of the transition diagram. The inspection alternative for the moments is a valuable check in problems where algebraic mistakes are likely.

**The coin-tossing example** Section 4.2 already illustrates our results for the case where $r = 3$ and the coin is fair, $p = 1/2$. For future reference we shall calculate the moments of the lifetime distribution for a run of 6 Heads with a fair coin. From Equations 7.3.2, 7.3.3, and 7.3.4 with $r = 6$ and $p = 1/2$ we find

$$\bar{l} = 126, \qquad \overline{l^2} = 30594, \qquad \overset{v}{l} = 14718. \tag{7.3.8}$$

Note that since the standard deviation $\overset{s}{l}$,

$$\overset{s}{l} = \sqrt{\overset{v}{l}} \approx 121 \tag{7.3.9}$$

is approximately equal to the mean $\bar{l}$, the time to complete a Head run of length 6 for a fair coin has a very broad distribution relative to its mean. We observed the same kind of behavior in Section 4.2 for the Head run of length 3.

***Dominant component analysis***

If we need the entire lifetime distribution rather than just its moments, then our task becomes considerably more difficult. One approach is to write the difference equation implied by the transform in Equation 7.3.1 and then solve it recursively as we did in Section 4.2. Direct, exact inversion of the transform is complicated because the degree of the denominator polynomial is $r + 1$—solution for all the roots of an $r$th order polynomial is easy only when $r$ is small. However, as we found in Section 4.2, we can use the method of dominant components to find approximate expressions for the lifetime distribution. The method involves finding the root of the denominator that governs the form of $p(n)$ when $n$ is large. We learned in Section 4.2 that this dominant component could be an excellent approximation to the exact solution.

To obtain the dominant component for success runs of length $r$ we must solve

$$z = 1 + (1 - p)p^r z^{r+1} \tag{7.3.10}$$

to find the root of the denominator of Equation 7.3.1 that is closest to one in magnitude. We have available many successful approximate methods for accomplishing this task, including Newton's method. The reciprocal of the dominant root is then the decay factor for the dominant geometric component of $p(n)$.

**The coin-tossing example.** Table 7.1 shows the dominant roots and dominant geometric component decay factors for Head runs of length $r$ for a fair coin. We see that the decay factor approaches one more and more closely as $r$ increases,

**Table 7.3.1** Dominant Component of Lifetime Distribution for Success Runs of Length $r$ When the Probability of Success on Each Trial is 1/2

| Run Length $r$ | $z_d$ = Dominant Root of Equation $z = 1 + [(1/2)z]^{r+1}$ | Decay Factor of Dominant Geometric Component = $1/z_d$ |
|---|---|---|
| 1 | 2 | 0.50000 |
| 2 | 1.23607 | 0.80902 |
| 3 | 1.08738 | 0.91964 |
| 4 | 1.03758 | 0.96378 |
| 5 | 1.01732 | 0.98297 |
| 6 | 1.00828 | 0.99179 |

thus indicating that the lifetime distribution $p(n)$ will not fall off so rapidly with $n$ as the length of the run increases. As we discussed in Section 4.2, as long as $n$ is greater than the run length $r$, modeling the run process as a Bernoulli process with a probability of success equal to one minus the dominant geometric decay factor provides a very useful approximation to the actual lifetime distribution.

***Probability of run completion on trial n***

The probability $\phi(n)$ of completing a run on the $n$th trial, not necessarily for the first time, has a transform $\phi^{g}(z)$ related to the transform of the lifetime distribution by Equation 7.1.16,

$$\phi^{g}(z) = \frac{1}{1 - p^{g}(z)}. \tag{7.3.11}$$

If we substitute $p^{g}(z)$ from Equation 7.3.1, we obtain

$$\phi^{g}(z) = \frac{1 - z + (1 - p)p^{r}z^{r+1}}{(1 - z)(1 - p^{r}z^{r})}. \tag{7.3.12}$$

This expression for the transform of the renewal probability is no easier to invert than the transform $p^{g}(z)$. However, it does allow us to confirm one of our results,

$$\phi = \phi(\infty) = \lim_{z \to 1} (1 - z)\phi^{g}(z) = \frac{(1 - p)p^{r}}{1 - p^{r}} = \frac{1}{\bar{l}}; \tag{7.3.13}$$

the probability of a renewal at any trial in the steady state is the reciprocal of the mean lifetime.

***Trial-dependent success probabilities***

We should note in passing that there is no real difficulty in having the probability of success on a trial depend on the number of successes in a row that have been produced before the trial. The flow graph of Figure 7.3.1 would merely have different probabilities for the various transitions. All the results we have obtained for the case of constant probability of success could be extended to this more general model at the expense of algebraic complexity.

## Success Runs or Failure Runs

Sometimes we are interested in the occurrence of one of a specified set of patterns on any trial. For example, suppose that we want to examine occurrences of either a success run of length $r$ or a failure run of length $s$. Whenever either of these occurs, we begin afresh to look for the next occurrence. We let $p$ be the probability of a success or Head and $q = 1 - p$ be the probability of a failure or Tail. We use $p(n)$ for the probability that one of these patterns will occur for the first time on the $n$th trial. The probability $p(n)$ is, of course, the lifetime distribution for the renewal process whose renewals are just completions of either success runs of length $r$ or failure runs of length $s$.

### *Flow graph analysis*

The transform of the lifetime distribution, $p^{g}(z)$, we find as the transmission of the transient process shown in Figure 7.3.2. This flow graph has a very similar structure to the flow graph for success runs in Figure 7.3.1. The differences arise because we must account for both success and failure runs and because the trial outcome that prevents a pattern of one type from coming to completion also starts a pattern of the other type. The states are named according to the number of Heads or Tails in succession that they represent. State 0 is the renewal state—it, of course, appears at both ends of the transient process.

We can find the transmission of the graph of Figure 7.3.2 directly by the inspection method. However, we shall prefer to place the flow graph in the form of Figure 7.3.3 by applying the residual node method to the original graph, and then to find the transmission of the flow graph of Figure 7.3.3 by the inspection method. We note that four paths contribute to the transmission and obtain

$$p^{g}(z) = \frac{p^r z^r(1 + qz + q^2z^2 + \cdots + q^{s-1}z^{s-1}) + q^s z^s(1 + pz + p^2z^2 + \cdots + p^{r-1}z^{r-1})}{1 - (pz + p^2z^2 + \cdots + p^{r-1}z^{r-1})(qz + q^2z^2 + \cdots + q^{s-1}z^{s-1})}$$

$$= \frac{p^r z^r \dfrac{1 - q^s z^s}{1 - qz} + q^s z^s \dfrac{1 - p^r z^r}{1 - pz}}{1 - \dfrac{pz - p^r z^r}{1 - pz} \cdot \dfrac{qz - q^s z^s}{1 - qz}}$$

$$= \frac{(1 - pz)p^r z^r(1 - q^s z^s) + (1 - qz)q^s z^s(1 - p^r z^r)}{1 - z + qp^r z^{r+1} + pq^s z^{s+1} - p^r q^s z^{r+s}}. \tag{7.3.14}$$

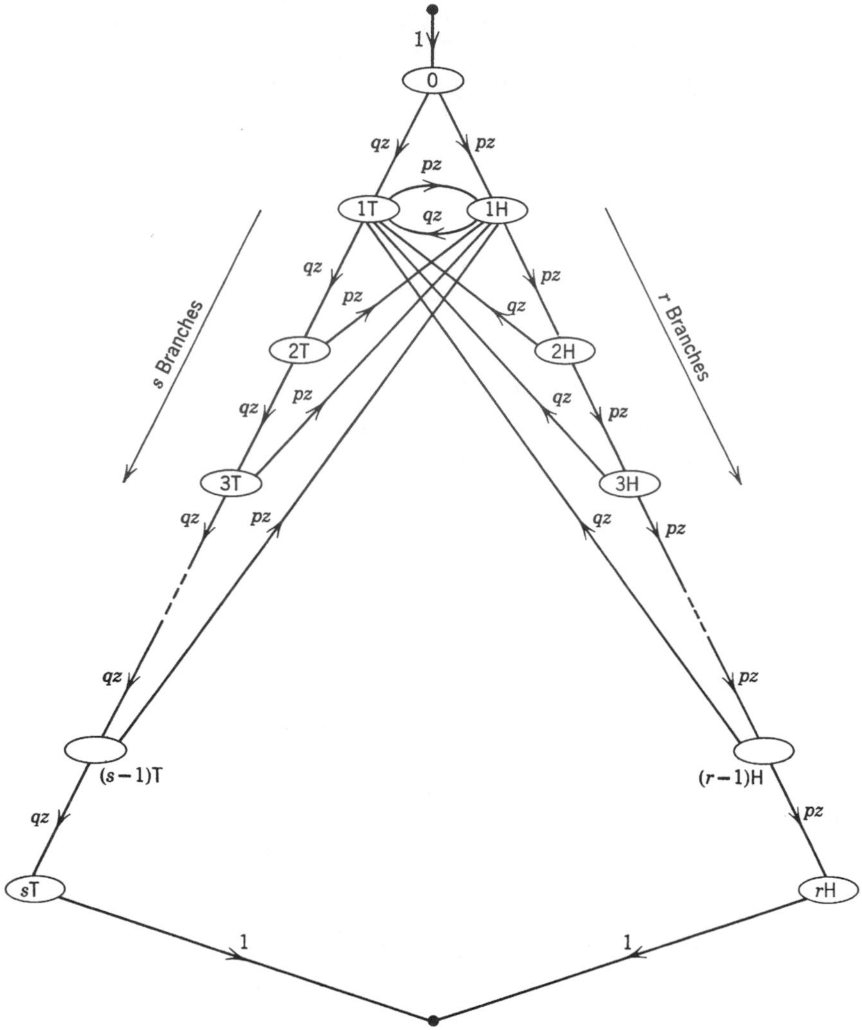


**Figure 7.3.2** Flow graph for success run of length $r$ or failure run of length $s$.

We can verify this result by checking $p^{\mathscr{g}}(1) = 1$ and by noting that as $s$ becomes very large $p^{\mathscr{g}}(z)$ approaches the transform of the lifetime distribution for success runs of length $r$ found in Equation 7.3.1.

***Mean time to first run***

We find the mean lifetime, the expected time to the first run of either type, by multiplying both sides of Equation 7.3.14 by the denominator, differentiating with

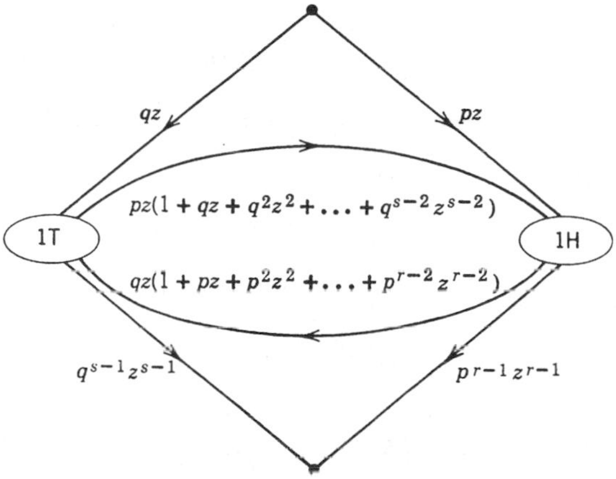


**Figure 7.3.3** Reduced flow graph for success run of length $r$ or failure run of length $s$.

respect to $z$, and evaluating the result at $z = 1$. We obtain

$$\bar{l} = p^{g\prime}(1) = \frac{(1 - p^r)(1 - q^s)}{qp^r + pq^s - p^rq^s}. \tag{7.3.15}$$

This expression reduces to Equation 7.3.2 as $s$ becomes very large.

***Renewal probability***

The transform of the renewal probability, $\phi^g(z)$, we write using Equation 7.1.16 and the result of Equation 7.3.14,

$$\phi^g(z) = \frac{1}{1 - p^g(z)} = \frac{1 - z + qp^rz^{r+1} + pq^sz^{s+1} - p^rq^sz^{r+s}}{(1 - z)(1 - p^rz^r)(1 - q^sz^s)}. \tag{7.3.16}$$

We note that this equation becomes Equation 7.3.12 as $s$ grows very large. We can use this transform as an alternate method for finding the mean of the lifetime distribution or as a check on it:

$$\phi = \phi(\infty) = \lim_{z\to 1}(1 - z)\phi^g(z) = \frac{1}{\bar{l}} = \frac{qp^r + pq^s - p^rq^s}{(1 - p^r)(1 - q^s)}. \tag{7.3.17}$$

The amount of manipulation involved in finding $\bar{l}$ by either method is often about the same.

We can also find $\phi^g(z)$ directly from the results on success runs alone. Let $\phi_S(n)$ be the renewal probability for a success run of length $r$, and $\phi_F(n)$ be the renewal probability for a failure run of length $s$. The transform $\phi_S{}^g(z)$ is shown in Equation

7.3.12. The transform $\phi_F{}^g(z)$ is the same except for an interchange of $p$ and $q$ and of $r$ and $s$. Since runs of the two types are mutually exclusive, we have

$$\phi(n) = \phi_S(n) + \phi_F(n) - \delta(n) \qquad n = 0, 1, 2, \ldots, \tag{7.3.18}$$

where the subtraction of $\delta(n)$ is necessary to meet the requirement that

$$\phi(0) = \phi_S(0) = \phi_F(0) = 1. \tag{7.3.19}$$

The transform of Equation 7.3.18 is

$$\begin{aligned} \phi^g(z) &= \phi_S{}^g(z) + \phi_F{}^g(n) - 1 \\ &= \frac{1 - z + qp^r z^{r+1}}{(1 - z)(1 - p^r z^r)} + \frac{1 - z + pq^s z^{s+1}}{(1 - z)(1 - q^s z^s)} - 1 \end{aligned} \tag{7.3.20}$$

which leads directly to Equation 7.3.16.

**Success Run before Failure Run**

An interesting aspect of the previous example is the calculation of the probability that a success run of length $r$ will be completed before a failure run of length $s$. We can visualize the problem in the flow graph of Figure 7.3.2 by considering states $r$H and $s$T to be trapping states. The probability that a success run of length $r$ will occur before a failure run of length $s$ is just the probability that the transient process will be trapped by state $r$H. Since we already know that the probability that a transient process will be trapped by any trapping state is the transmission of the transition diagram from the input to that state, all we have to do is draw the transition diagram corresponding to Figure 7.3.2 and find its transmission from state 0 to state $r$H with state $s$T a trapping state. We construct this transition diagram easily from the flow graph of Figure 7.3.3—it appears in Figure 7.3.4. The transmission of the transition diagram from state 0 to state $r$H is

$$\vec{t} = \frac{p^r[1 + q + q^2 + \cdots + q^{s-1}]}{1 - [p + p^2 + p^3 + \cdots + p^{r-1}][q + q^2 + q^3 + \cdots + q^{s-1}]}$$

$$= \frac{p^r \dfrac{1 - q^s}{1 - q}}{1 - \left[\dfrac{p - p^r}{1 - p}\right]\left[\dfrac{q - q^s}{1 - q}\right]} = \frac{qp^r(1 - q^s)}{qp^r + pq^s - p^r q^s} = \frac{p^{r-1}(1 - q^s)}{p^{r-1} + q^{s-1} - p^{r-1}q^{s-1}}. \tag{7.3.21}$$

We check by noting that if $r = s = 1$, then the probability is equal to $p$, and that as $s$ becomes very large the probability approaches 1.

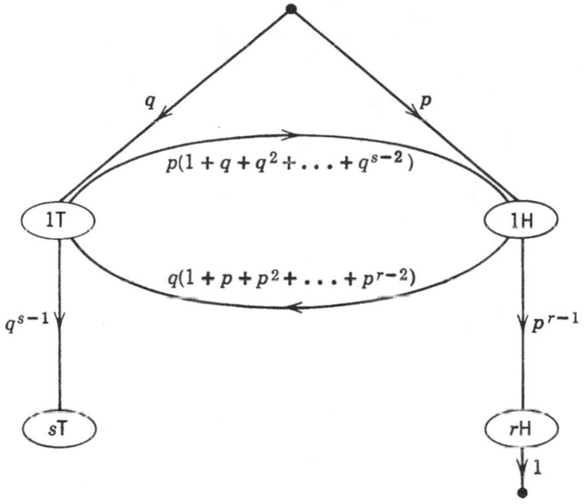


**Figure 7.3.4** Reduced transition diagram for success run of length $r$ before failure run of length $s$.

### The Pattern HHTTHH

Runs are not the only interesting patterns of successes and failures in Bernoulli trials. In analyzing complex equipment we are often faced with the necessity of treating rather arbitrary patterns of trial outcomes. The techniques are the same as those used for runs. We shall illustrate them by investigating the pattern HHTTHH. We assume for the moment that the occurrences of such a pattern form a renewal process: Every time a pattern is completed the next pattern starts from scratch.

#### *Flow graph analysis*

The transient process corresponding to this pattern has the flow graph shown in Figure 7.3.5. We need 7 states to represent the various stages of completion of the pattern. The transitions in the flow graph reflect the consequence of either a Head or a Tail in each state. For example, if the process is in state 4 by having completed HHT of the pattern, then it will proceed to state 5, HHTT, if the next trial is a Tail and fall back to state 2, H, if the trial is a Head. Since Tail and Head have probability $q$ and $p$ the branch transmissions for these two transitions are $qz$ and $pz$.

The process may be started in state 1, T, without prejudicing the statistics of the pattern HHTTHH. Therefore, the transform of the lifetime distribution is just the

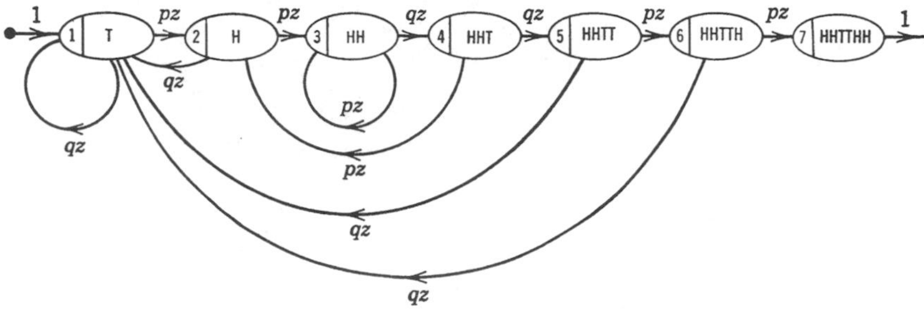


**Figure 7.3.5** Flow graph for pattern HHTTHH.

transmission of the flow graph of Figure 7.3.5 from state 1 to state 7. We find it by the inspection method,

$$\vec{t}_{17}(z) = p^{\mathscr{g}}(z) = \frac{p^4q^2z^6}{1 - z + p^2q^2z^4 - p^2q^3z^5 - p^3q^3z^6}. \tag{7.3.22}$$

We verify immediately that $p^{\mathscr{g}}(1) = 1$.

***Mean lifetime***

Differentiation of this transform and evaluation at $z = 1$ produces the mean of the lifetime distribution,

$$\bar{l} = p^{\mathscr{g}\prime}(1) = \frac{1}{p^4q^2}[1 + p^2q^2 + p^3q^2]. \tag{7.3.23}$$

We can also find the mean lifetime by inspection of the transition diagram for the process. The transition diagram is just the flow graph of Figure 7.3.5 with all $z$'s removed. We compute the mean lifetime by summing the transmissions of this transition diagram from node 1 to each of the nodes 1 through 6. By evaluating the denominator of $p^{\mathscr{g}}(z)$ in Equation 7.3.22 at $z = 1$, we find that the common denominator of all transmissions of the transition diagram is

$$p^2q^2 - p^2q^3 - p^3q^3 = p^2q^2(1 - q - pq) = p^4q^2.$$

Thus we can write

$$\begin{aligned}\bar{l} &= \vec{t}_{11}(1) + \vec{t}_{12}(1) + \vec{t}_{13}(1) + \vec{t}_{14}(1) + \vec{t}_{15}(1) + \vec{t}_{16}(1)\\ &= \frac{1}{p^4q^2}[(1 - p - p^2q) + p(1 - p) + p^2 + p^2q + p^2q^2 + p^3q^2]\\ &= \frac{1}{p^4q^2}[1 + p^2q^2 + p^3q^2]. \end{aligned} \tag{7.3.24}$$

***Renewal probability***

The probability $\phi(n)$ of completing a pattern of this type at trial $n$, not necessarily for the first time, has a transform $\phi^g(z)$ given by

$$\phi^g(z) = \frac{1}{1 - p^g(z)} = \frac{1 - z + p^2q^2z^4[1 - qz - pqz^2]}{(1 - z)[1 + p^2q^2z^4(1 + pz)]}. \tag{7.3.25}$$

Inversion of this transform is not easy, but we can use it to verify the relationship between $\phi(\infty)$ and $\bar{l}$,

$$\phi = \phi(\infty) = \lim_{z \to 1} (1 - z)\phi^g(z) = \frac{p^4q^2}{1 + p^2q^2 + p^3q^2} = \frac{1}{\bar{l}}. \tag{7.3.26}$$

If $p$ is assigned a numerical value then we can use Equations 7.3.22 and 7.3.25 to write difference equations for solving for $p(n)$ and $\phi(n)$ by recursion in our usual way.

***Effect of pattern on mean lifetime***

We can calculate the mean number of trials that will separate recurrences of the pattern HHTTHH from Equation 7.3.23. When $p = 1/2$, we find

$$\bar{l} = 70. \tag{7.3.27}$$

Recall from Equation 7.3.8 that the mean number of trials for success runs of length 6 with a fair coin was 126. The patterns HHHHHH and HHTTHH are of the same length and have elements that occur with the same probability when $p = 1/2$. However, as we have seen, the mean number of trials between recurrences is quite different for them. There is a fundamental difference between success runs and other patterns of the same length even if success and failure are equally probable. The difference is most clearly revealed by the variation in the topology of Figures 7.3.1 and 7.3.5.

## Overlapping Patterns

Up to this point we have assumed that the completion of a pattern wipes the slate clean for the building of another pattern of the same type. We shall now consider the case where parts of old patterns are allowed to count toward the next completion of the pattern. Thus in the last example we analyzed recurrences of the pattern HHTTHH under the assumption that every time the pattern was completed the process started anew—we shall call this the non-overlapping case. Now we shall treat the overlapping case of the same example, where completion of the pattern means that we already have two successes or Heads toward the next pattern. Thus the sequence THHTTHHHTTHHTH contains one pattern

HHTTHH under the non-overlapping assumption and two patterns HHTTHH under the overlapping assumption.

*Flow graph analysis*

The flow graph for this Markov process is the same as that of Figure 7.3.5 with the exception that now state 7 is really the same as state 3. We represent this relationship in the flow graph of Figure 7.3.6 by the dotted unity transmission between nodes 7 and 3. We recognize this flow graph as the flow graph of a general renewal process. The lifetime distribution $p(\cdot)$ for the process is the recurrence time distribution of state 3. The transform of this distribution, $p^g(z)$, is the transform of the flow graph from state 3 to 7 with the dotted branch removed. The probability distribution for time to the first occurrence, ${}_fp(\cdot)$, is the first passage time distribution from state 1 to state 7. The transform of this distribution is just the transmission from node 1 to node 7 of the flow graph with the dotted branch removed; of course, this is just the transmission found in Equation 7.3.22. Thus we write

$$p^g(z) = \vec{t}_{37}(z) = \frac{p^2q^2z^4[1 - qz - pqz^2]}{1 - z + p^2q^2z^4 - p^2q^3z^5 - p^3q^3z^6} \quad (7.3.28)$$

(dotted branch removed)

and

$${}_fp^g(z) = \vec{t}_{17}(z) = \frac{p^4q^2z^6}{1 - z + p^2q^2z^4 - p^2q^3z^5 - p^3q^3z^6}. \quad (7.3.29)$$

(dotted branch removed)

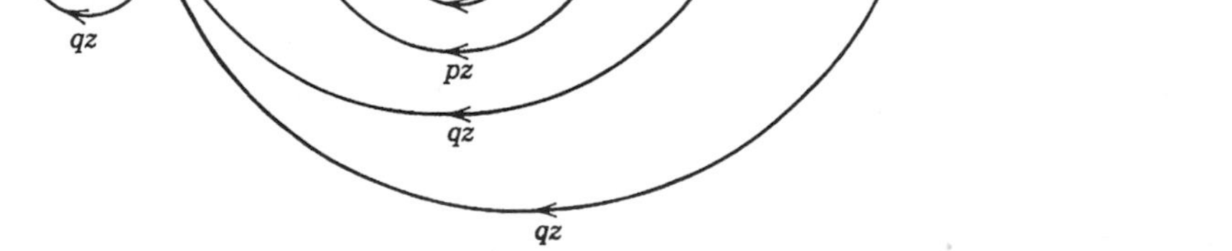

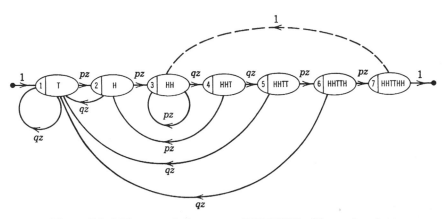

**Figure 7.3.6** Flow graph for pattern HHTTHH with overlapping.

***Mean time to first completion***

We have already found in Equation 7.3.23 the mean time to the first pattern HHTTHH, ${}_f\bar{l}$, by differentiating $\vec{t}_{17}(z) = {}_fp^g(z)$ and evaluating the result at $z = 1$,

$$ {}_f\bar{l} = \frac{1}{p^4q^2}[1 + p^2q^2 + p^3q^2]. \tag{7.3.30} $$

When we were working under the non-overlapping assumption, this expression also represented the mean of the lifetime distribution, the mean time between recurrences.

***Mean time between completions***

However, now that we are using the overlapping assumption we must find $\bar{l}$, the the mean time between occurrences of the pattern HHTTHH, by differentiating the transform of the lifetime distribution in Equation 7.3.28 and evaluating the result at $z = 1$. We find

$$ \bar{l} = \frac{1}{p^4q^2}. \tag{7.3.31} $$

If we are tossing a fair coin ($p = 1/2$), this equation shows that the mean number of trials between recurrences of the HHTTHH under the overlapping assumption is 64. Thus the overlapping assumption reduces the mean time between patterns from the value of 70 in Equation 7.3.27 to 64, a reduction of 6 trials. Equation 7.3.2 with $p = 1/2$, $r = 2$ reveals that the mean number of trials to produce a Head run of length 2 is 6. This is just the advantage toward the next pattern that the overlapping assumption provides.

***Renewal probability***

We can calculate the probability of completing a pattern HHTTHH on the $n$th trial, possibly not for the first time, from our previous results on the general renewal process. Thus, if we start the process in state 1, we call the renewal probability ${}_f\phi(n)$ and write its transform using Equation 7.1.45,

$$ {}_f\phi^g(z) = \frac{{}_fp^g(z)}{1 - p^g(z)}. \tag{7.3.32} $$

We substitute the results of Equations 7.3.28 and 7.3.29 to produce

$$ {}_f\phi^g(z) = \frac{p^4q^2z^6}{1 - z}, \tag{7.3.33} $$

then invert this transform to obtain

$$ {}_f\phi(n) = \begin{cases} 0 & n = 0, 1, 2, 3, 4, 5 \\ p^4q^2 & n = 6, 7, 8, \dots \end{cases}. \tag{7.3.34} $$

As we would expect, we have the same probability of producing the pattern HHTTHH on the $n$th trial if the number of trials is at least 6 and no chance of producing it if the number of trials is less than 6.

If we start to count trials after a pattern is completed we are entering the renewal process at node 3. In this case we call the renewal probability $\phi(n)$ and compute its transform from

$$\phi^{\mathscr{g}}(z) = \frac{1}{1 - p^{\mathscr{g}}(z)} = 1 + \frac{p^2q^2z^4(1 - qz - pqz^2)}{1 - z}. \qquad (7.3.35)$$

The inverse transform is

$$\phi(n) = \begin{cases} 1 & n = 0 \\ 0 & n = 1, 2, 3 \\ p^2q^2 & n = 4 \\ p^3q^2 & n = 5 \\ p^4q^2 & n \geq 6 \end{cases} \qquad (7.3.36)$$

This equation shows that there is a probability $p^2q^2$ of completing a pattern after only 4 additional trials. This probability is just the probability of obtaining two Tails then two Heads to finish the pattern from the headstart of two Heads. The quantity $\phi(5)$ has a similar interpretation. The transient effect of the headstart is over when the number of trials is 6 or more: $\phi(n) = {}_{\mathrm{f}}\phi(n) = p^4q^2$. When $n$ is at least 6, both $\phi(n)$ and ${}_{\mathrm{f}}\phi(n)$ are just the probability that the last 6 tosses form the pattern HHTTHH. In general when we consider any overlapping sequence, the value of $\phi(n)$ or ${}_{\mathrm{f}}\phi(n)$ for large $n$ is just the probability of the pattern of successes and failures in the sequence. The mean time between these sequences is therefore just the reciprocal of the probability of the pattern, as shown for this example in Equation 7.3.31.

**Separated Runs**

We have now completed an analysis of runs and more general patterns in Bernoulli trials. The issue of overlapping versus non-overlapping sequences showed the importance of defining how patterns are to be counted. For example, if a run of three Heads with a fair coin is counted according to the non-overlapping rule, we found in Equation 4.2.17 that the mean number of trials between occurrences of this run would be 14. If the run is counted according to the overlapping rule, we have just found that the mean time between occurrences of the run would be one over the probability of Heads cubed, or 8. However, we can make a strong case that in some problems neither of these methods of counting success runs is appropriate.

Suppose that we asked someone to count the number of times he saw three Heads in a row while observing the outcomes of coin tossing. The chances are probably good that he would count the pattern THHHHT not as the occurrence of 1 success run of length 3 or of 2 success runs of length 3, but rather as a success run of length 4. Thus he would not include such a pattern in his tally.

We might say that our observer is counting in a manner different from either of the overlapping or non-overlapping rules we have prescribed, but, in fact, we shall see that it is just a matter of proper interpretation. Let us call the type of runs he is counting "separated runs." A separated success run of length $r$ is a sequence of $r$ successes preceded by and followed by at least one failure. If the sequence of trials is just beginning, then having the first $r$ trials result in success and the next trial result in failure would also constitute a separated success run of length $r$.

***Flow graph analysis***

Therefore, our observer of separated success runs of length 3 is, in fact, counting overlapping patterns of the form THHHT. Figure 7.3.7 shows the flow graph that represents the process. The probability of a success or Head is $p$, that of a failure or Tail is $q = 1 - p$. Every time the system enters state 6, THHHT, it makes an instantaneous transition to state 2, T, and counting begins again. Therefore, the lifetime distribution of the renewal process is the transmission of the flow graph from state 2 to state 6 with the dotted branch removed. Note that starting the system in state 2 initially assumes a Tail on the zeroth trial and hence assures that the sequence HHHT on the first 4 trials will also be counted as a separated success run of length 3.

Now that we have modeled the separated success run of length 3 as the overlapping sequence THHHT, we can apply the results of our work on overlapping trials to find the mean time between recurrences. We know that with overlapping the probability of an occurrence of the pattern at any trial in the steady state is just the probability of the events that make up the pattern; therefore,

$$\phi = \phi(\infty) = q \cdot p \cdot p \cdot p \cdot q = p^3 q^2. \tag{7.3.37}$$

We know further that the mean time between recurrences of the pattern is then the reciprocal of this probability,

$$\bar{l} = \frac{1}{\phi} = \frac{1}{p^3 q^2}. \tag{7.3.38}$$

The same result is obtained, of course, by setting $z = 1$ in Figure 7.3.7 and then computing $\bar{l}$ from

$$\bar{l} = \vec{t}_{21} + \vec{t}_{22} + \vec{t}_{23} + \vec{t}_{24} + \vec{t}_{25} \qquad \text{(dotted branch removed)}. \tag{7.3.39}$$

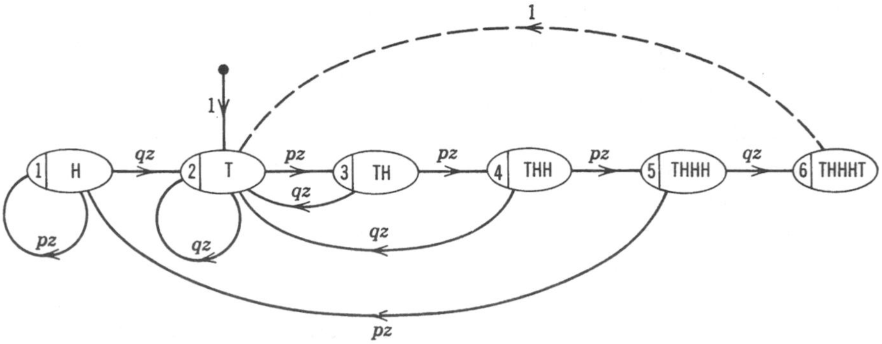


**Figure 7.3.7** Flow graph for separated success run HHH.

If we are dealing with a fair coin, then $p = q = 1/2$, and $\bar{l} = 32$. The expected number of tosses of a fair coin between separated runs of three Heads is therefore 32. This is considerably larger than the 14 tosses expected between success runs of length 3 without overlapping and the 8 tosses expected between such runs with overlapping. We see once again the importance of being clear on the type of event to be counted. However, once the method of counting is made definite, we shall usually find that one of the methods we have discussed will apply directly.

### *Separated runs of many lengths*

If we perform a sequence of Bernoulli trials, we shall observe a certain number of separated success runs of length 1, 2, 3, etc. The numbers of runs of each length are clearly dependent random variables because large numbers of runs of any one length in a given number of trials will necessarily imply smaller numbers of runs of other lengths. Let $\phi(k_1, k_2, \ldots, k_r|n)$ be the probability of $k_1$ separated success runs of length 1, $k_2$ of length 2, up to $k_r$ of length $r$ in a sequence of $n$ Bernoulli

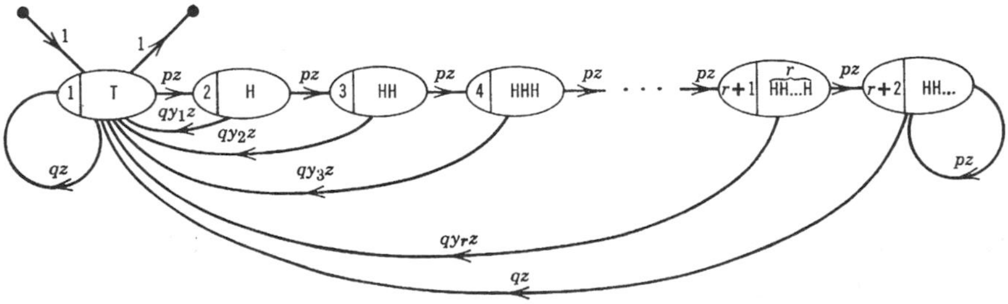


**Figure 7.3.8** Tagged flow graph for separated runs of many lengths.

trials. We choose $r$ as the length of the longest separated run we wish to count. Since in $n$ trials there can be no separated run longer than $n - 1$, there is never any point in choosing $r$ larger than $n - 1$.

The geometric transform of $\phi(k_1, k_2, \ldots, k_r|n)$ is $\phi^{gg\cdots}(y_1, y_2, \ldots, y_r|n)$. We can find this transform as the transmission $\vec{t}_{11}$ of the tagged flow graph of Figure 7.3.8. Here again $p$ is the probability of a success or Head, $q = 1 - p$ the probability of a failure or Tail. The flow graph shows that every time a series of Heads is followed by a Tail a separated run is completed and the system returns to the Tail state through a branch tagged by a tagging variable corresponding to the length of the separated runs.

The problem involved in inverting the transform of the joint probability of separated runs of various lengths is of the type we discussed in Section 6.1. The difference equations implied by the transform allow us to calculate the joint probability or any moments of the joint distribution.

## 7.4 RANDOM WALKS

A particularly interesting class of renewal processes are those called random walks. Implicit in the idea of a random walk is motion of a particle on some lattice of states according to a probabilistic mechanism that depends only on the state last occupied.

### Infinite Random Walks

The first random walk we shall analyze can be thought of as the amount of net winnings a player has at any time when he is tossing a coin for a unit payment. That is, if the coin turns up Heads, our player receives a dollar; if it comes up Tails, he loses one. We assume that the probability of Heads on any toss is $p$ and that the probability of Tails is $q = 1 - p$. Then we let the state of the system be the net amount that the player has won after a given number of tosses. We consider the player and his opponent to have infinite capital, and the game to continue indefinitely. Therefore, the net winnings at any time could be any integer in the range from $-\infty$ to $+\infty$.

#### *Flow graph representation*

If we want to represent the operation of this game as a Markov process, we require that the process have an infinite number of states. Although we have not considered processes of this type previously, they are in principle a direct extension of our earlier work. In particular, we can draw the flow graph of Figure 7.4.1 for the process. Although the graph is infinite in extent, we have no problem in interpreting its relationships. Note that the way the graph is drawn indicates that

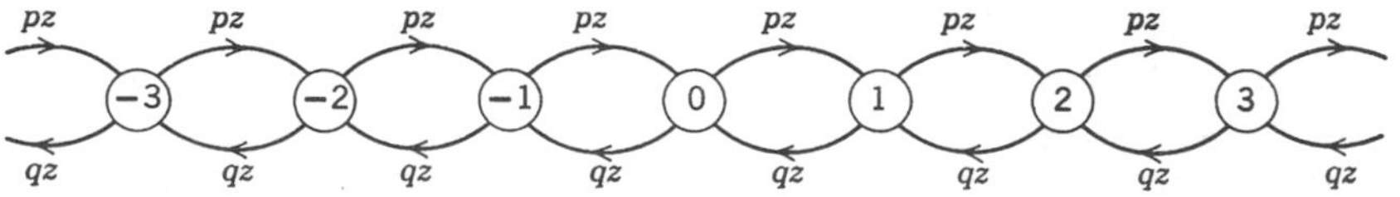


**Figure 7.4.1** Flow graph for a bi-infinite random walk.

the process will move one step to the right when a Head occurs and one step to th left on the occurrence of a Tail. Each state is numbered according to the ne amount the player has won since he started at a reference state numbered 0.

The problem we shall study concerns returns of the process to state 0. Ever: return represents a situation in which the player has won on as many tosses as he ha lost. Is a return certain for any value of $p$? What is the expected number of tosse before a return occurs? How many returns do we expect to occur in an infinit number of tosses? The key to answering these questions is the lifetime distributio of the renewal process. We can find the transform of this lifetime distribution b: constructing the transient process corresponding to the departure from, and firs return to, state 0. The flow graph for this transient process appears in Figur 7.4.2. If we could find the transmission of this flow graph, we would have th transform of the lifetime distribution.

### *Infinite-state flow graph analysis*

We are not yet equipped to find the transmissions of infinite flow graphs lik the one in Figure 7.4.2. Clearly, the inspection method will lead us nowhere becaus of the infinite loop structure. However, the regularity of the graph does let us fin its transmission very easily. We first consider the transmission of the genera repeated infinite flow graph of Figure 7.4.3. Although we do not know the valu of its transmission $\vec{t}_{\mathrm{I}}$, we do know from the infinite structure of the graph that thi transmission $\vec{t}_{\mathrm{I}}$ is also the loop transmission around the first node in the infinit

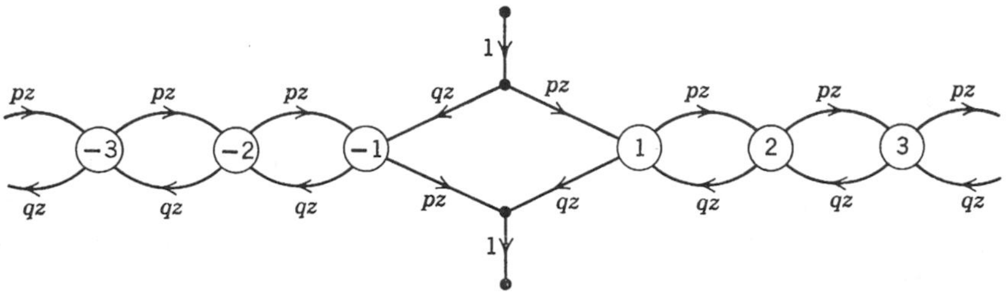


**Figure 7.4.2** Flow graph for the transform of the lifetime distribution for the bi-infinite random walk.

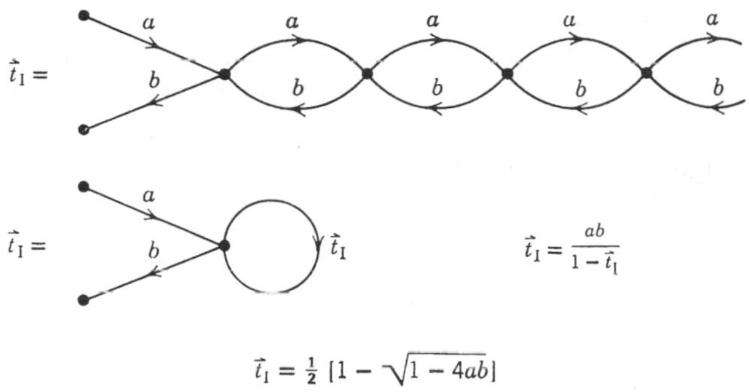


**Figure 7.4.3** Transmission of a general infinite-state flow graph.

graph. Therefore, we can draw the infinite graph as an equivalent one-loop graph and establish the transmission relation

$$\vec{t}_I = \frac{ab}{1 - \vec{t}_I}, \tag{7.4.1}$$

which implies the quadratic equation,

$$\vec{t}_I^{\,2} - \vec{t}_I + ab = 0, \tag{7.4.2}$$

with solutions,

$$\vec{t}_I = \frac{1}{2}[1 \pm \sqrt{1 - 4ab}]. \tag{7.4.3}$$

Since we require from the original graph in Figure 7.4.3 that $\vec{t}_I$ equal zero when $a$ or $b$ equals zero, the root corresponding to the minus sign is chosen and we have, finally,

$$\vec{t}_I = \frac{1}{2}[1 - \sqrt{1 - 4ab}]. \tag{7.4.4}$$

Thus we have obtained an expression for the transmission of an infinite graph. Note that it is expressed by a square root rather than by the ratio of polynomials that we have always found in the finite-state case. Note also that the transmission expression is symmetric in $a$ and $b$, a result that satisfies intuition only when we observe that every path in the flow graph of Figure 7.4.3 has a transmission that is a multiple of $ab$.

Now we can apply these results to find the transmission of the flow graph in Figure 7.4.2. The total transmission is just the sum of the transmissions of two infinite flow graphs of the type we analyzed in Figure 7.4.3. The transmission of the right part of the flow graph is $\vec{t}_{\mathrm{I}}$ with $a = pz$, $b = qz$; the transmission of the left part is $\vec{t}_{\mathrm{I}}$ with $a = qz$, $b = pz$. However, since we know that the transmission of the infinite graph in Figure 7.4.3 is symmetric in $a$ and $b$, the transmission of the flow graph in Figure 7.4.2 is just twice $\vec{t}_{\mathrm{I}}$ in Equation 7.4.4 with $a = pz$, $b = qz$. Therefore, we obtain for the transform $p^{\mathscr{g}}(z)$ of the lifetime distribution

$$p^{\mathscr{g}}(z) = 1 - \sqrt{1 - 4pqz^2}. \tag{7.4.5}$$

***Probability of eventual renewal***

We see immediately that this lifetime distribution is unlike any other geometric transform we have seen up to this time: it is an irrational function of $z$. However, we can use it in the usual way to find the probability of eventual renewal for this process, $p_R$,

$$\begin{aligned} p_R = p^{\mathscr{g}}(1) &= 1 - \sqrt{1 - 4pq} \\ &= 1 - \sqrt{1 - 4p + 4p^2} \\ &= 1 - \sqrt{(1 - 2p)^2}. \end{aligned} \tag{7.4.6}$$

Since

$$\sqrt{1 - 4pq} = \sqrt{(1 - 2p)^2} = \begin{cases} 1 - 2p & 0 \le p \le 1/2 \\ 2p - 1 & 1/2 \le p \le 1, \end{cases} \tag{7.4.7}$$

we can write $p_R$ as

$$p_R = \begin{Bmatrix} 2p & 0 \le p \le 1/2 \\ 2(1 - p) & 1/2 \le p \le 1 \end{Bmatrix} = 1 - |p - q|. \tag{7.4.8}$$

Figure 7.4.4 shows the probability of eventual renewal $p_R$ as a function of $p$, the probability of Heads on each toss. Note that $p_R = 1$ only when $p = 1/2$; the

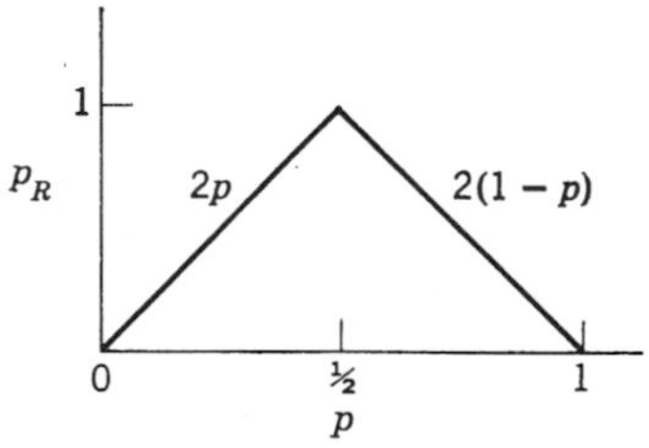


**Figure 7.4.4** Probability of eventual return to state 0 in the infinite coin-tossing game.

renewal process is recurrent only when the coin is fair. When $p \neq 1/2$, there is a finite non-zero probability that a return to state zero will never occur. For example, if $p = 3/4$ then there is probability 1/2 that the fortune of the player will never again be the same after he begins to toss the coins—the renewal process is transient.

***Renewal moments***

Equations 7.1.6 and 7.1.7 show how to compute the mean and variance of the geometric distribution of the number of renewals in an infinite time for a transient renewal process, a renewal process with probability of eventual renewal $p_R < 1$. The results for this example are:

$$\bar{r} = \frac{p_R}{1 - p_R} = \begin{cases} \dfrac{2p}{1 - 2p} & 0 \leq p \leq 1/2 \\ \dfrac{2(1 - p)}{2p - 1} & 1/2 \leq p \leq 1 \end{cases} \tag{7.4.9}$$

and

$$\overset{v}{r} = \frac{p_R}{(1 - p_R)^2} = \begin{cases} \dfrac{2p}{(1 - 2p)^2} & 0 \leq p \leq 1/2 \\ \dfrac{2(1 - p)}{(2p - 1)^2} & 1/2 \leq p \leq 1. \end{cases} \tag{7.4.10}$$

These quantities are plotted in Figure 7.4.5. As we expect, the mean and variance of the number of renewals become infinite when $p = 1/2$ because the renewal process is then recurrent. For other values of $p$ the renewal process is transient—we observe a finite mean and variance for the number of renewals in an infinite number of tosses.

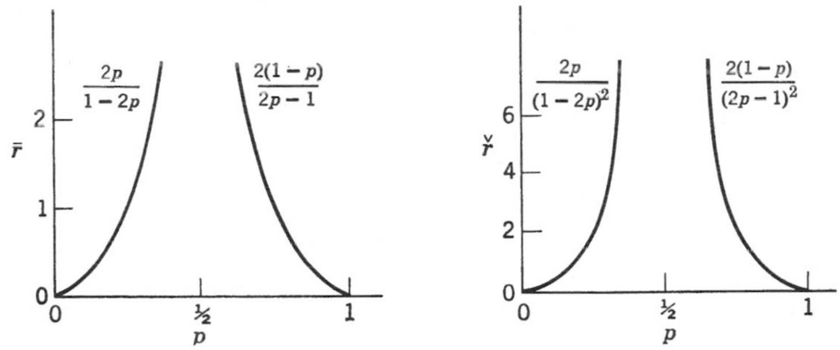


**Figure 7.4.5** Mean and variance of the number of returns to state 0 in the infinite coin-tossing game.

***Mean trials to renewal***

Since $p_R < 1$ when $p \neq 1/2$, the lifetime distribution is not a probability mass function unless the coin is fair. Therefore, the mean of the lifetime distribution will be infinite when $p \neq 1/2$ because of the non-zero probability that a renewal will never occur. However, we can find the mean lifetime conditional on the event that a renewal occurs, $\langle n|R\rangle$, by computing $p^{\mathscr{g}\prime}(1)$ and dividing by $p_R$,

$$\langle n|R\rangle = \frac{p^{\mathscr{g}\prime}(1)}{p_R}. \tag{7.4.11}$$

This equation will hold even when $p = 1/2$ because we are then conditioning on a certain event. From Equation 7.4.5 we find

$$p^{\mathscr{g}\prime}(z) = \frac{4pqz}{\sqrt{1 - 4pqz^2}}. \tag{7.4.12}$$

Then,

$$p^{\mathscr{g}\prime}(1) = \frac{4pq}{\sqrt{1 - 4pq}} = \frac{4p(1 - p)}{\sqrt{(1 - 2p)^2}}, \tag{7.4.13}$$

or, from Equation 7.4.7,

$$p^{\mathscr{g}\prime}(1) = \begin{cases} \dfrac{4p(1 - p)}{1 - 2p} & 0 \leq p \leq 1/2 \\ \dfrac{4p(1 - p)}{2p - 1} & 1/2 \leq p < 1. \end{cases} \tag{7.4.14}$$

Now we write Equation 7.4.11 using the results of Equations 7.4.8 and 7.4.14,

$$\langle n|R\rangle = \begin{cases} \dfrac{2(1 - p)}{1 - 2p} & 0 \leq p \leq 1/2 \\ \dfrac{2p}{2p - 1} & 1/2 \leq p \leq 1. \end{cases} \tag{7.4.15}$$

This equation states the expectation of the time to the first renewal given that a renewal occurs; it is plotted as a function of $p$ in Figure 7.4.6. The expectation is 2 as $p$ approaches zero or one because in these cases of great tendency to move to the left or right, the knowledge that a return has occurred makes it extremely probable that the return occurred after the minimum possible number of tosses, namely, two.

Note that the expectation is infinite when $p = 1/2$ even though a renewal is certain to occur. Results like this for infinite-state processes seem to defy intuition. A return to state zero is certain only when the coin is fair. For such a

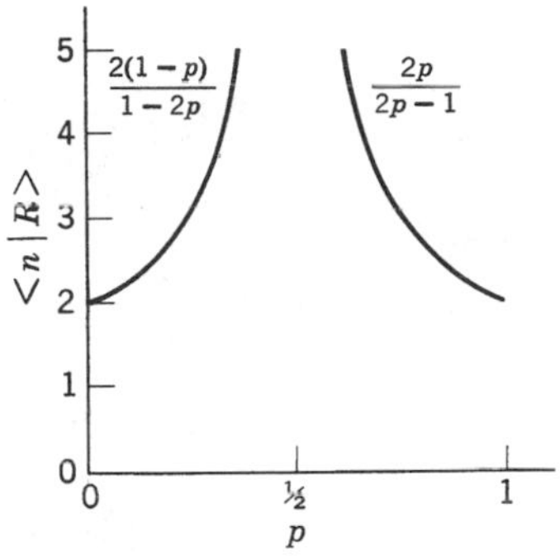


**Figure 7.4.6** The expectation of the number of tosses to produce a return to state 0 given that a return occurs.

fair coin, moreover, the expected number of returns in an infinite number of tosses is infinite. Yet the expected number of tosses to the first return is infinite. These statements are consistent, but puzzling. They indicate why in most problems we prefer the finite-state model to the infinite.

***The lifetime distribution***

The strange nature of infinite-state systems need not prevent us from obtaining explicit expressions for the lifetime distribution $p(\cdot)$ and the probability of a renewal at the $n$th toss, $\phi(n)$. We already have the transform of the lifetime distribution in Equation 7.4.5. The difficulty is that the transform does not appear in our transform table. However, we can easily develop the transform from a few elementary results. We begin by writing Newton's formula for the infinite binomial expansion,

$$(1 + x)^{\beta} = \sum_{n=0}^{\infty} \binom{\beta}{n} x^n \qquad \text{any real } \beta,\ -1 < x < 1, \tag{7.4.16}$$

where

$$\binom{\beta}{n} = \frac{\beta(\beta - 1)(\beta - 2)\cdots(\beta - n + 1)}{n!} \qquad n = 1, 2, 3, \ldots \tag{7.4.17}$$

and

$$\binom{\beta}{0} = 1. \tag{7.4.18}$$

We can now write Equation 7.4.5 as

$$\begin{aligned} p^{\mathscr{g}}(z) &= 1 - (1 - 4pqz^2)^{1/2} = 1 - \sum_{n=0}^{\infty} \binom{1/2}{n}(-4pqz^2)^n \\ &= \sum_{n=1}^{\infty} \binom{1/2}{n}(-1)^{n+1}(4pq)^n z^{2n}. \end{aligned} \tag{7.4.19}$$

By comparing this expression term by term with the transform equation

$$p^{\mathscr{g}}(z) = \sum_{n=0}^{\infty} p(n)z^n, \tag{7.4.20}$$

we see immediately that

$$\begin{aligned} p(0) &= 0 & & \\ p(n) &= 0 & & n \text{ odd} \\ p(2n) &= \binom{1/2}{n}(-1)^{n+1}(4pq)^n & & n = 1, 2, 3, \ldots . \end{aligned} \tag{7.4.21}$$

We must now evaluate the factor $\binom{1/2}{n}$ in a more convenient form using Equation 7.4.17. After some manipulation we find

$$\binom{1/2}{n} = \frac{(-1)^{n+1}(2n-2)!}{2^{2n-1}n!\,(n-1)!} \qquad n = 1, 2, 3, \ldots . \tag{7.4.22}$$

Finally, we substitute this result in Equation 7.4.21 to produce

$$\begin{aligned} p(0) &= 0 & & \\ p(n) &= 0 & & n \text{ odd} \\ p(2n) &= 2\frac{(2n-2)!}{n!\,(n-1)!}(pq)^n & & n = 1, 2, 3, \ldots . \end{aligned} \tag{7.4.23}$$

When $p = q = 1/2$, we compute the first few non-zero values of $p(n)$ to be

$$p(2) = 1/2 \qquad p(4) = 1/8 \qquad p(6) = 1/16 \qquad p(8) = 5/128. \tag{7.4.24}$$

We can easily interpret these results. The probability of a first return to state zero after two tosses is the probability of a Head followed by a Tail or vice versa on the first two tosses. The probability of a first return to state zero after four tosses is the probability of occurrence of one of the sequences HHTT or TTHH, each of which has probability 1/16. However, the probability of a first return after six tosses is the probability of occurrence of one of the four sequences HHHTTT, TTTHHH, HHTHTT, or TTHTHH, each having probability 1/64.

***Probability of renewal at time n***

The probability $\phi(n)$ that a return to state zero will occur on the $n$th toss after the beginning of the game, but not necessarily for the first time, has a transform $\phi^g(z)$ related to the transform of the lifetime distribution $p^g(z)$ by Equation 7.1.16,

$$\phi^g(z) = \frac{1}{1 - p^g(z)} = \frac{1}{\sqrt{1 - 4pqz^2}}. \tag{7.4.25}$$

Now we use the result of Equation 7.4.16 to write

$$\phi^g(z) = (1 - 4pqz^2)^{-1/2} = \sum_{n=0}^{\infty} \binom{-1/2}{n} (-4pqz^2)^n$$

$$= \sum_{n=0}^{\infty} \binom{-1/2}{n} (-1)^n (4pq)^n z^{2n}. \tag{7.4.26}$$

By direct comparison with the terms in the expansion of $\phi^g(z)$ we see that the inverse $\phi(n)$ is given by

$$\begin{aligned} \phi(n) &= 0 & & n \text{ odd} \\ \phi(2n) &= \binom{-1/2}{n} (-1)^n (4pq)^n & & n = 1, 2, 3, \ldots. \end{aligned} \tag{7.4.27}$$

Direct evaluation of $\binom{-1/2}{n}$ using Equation 7.4.17 produces

$$\binom{-1/2}{n} = \frac{(2n)!}{n!\,n!} (-1)^n 4^{-n}. \tag{7.4.28}$$

When we substitute this result into Equation 7.4.27, we obtain

$$\begin{aligned} \phi(n) &= 0 & & n \text{ odd} \\ \phi(2n) &= \frac{(2n)!}{n!\,n!} (pq)^n = \binom{2n}{n} (pq)^n & & n = 1, 2, 3, \ldots. \end{aligned} \tag{7.4.29}$$

The first few non-zero values of $\phi(n)$ for a fair coin are:

$$\phi(0) = 1 \quad \phi(2) = 1/2 \quad \phi(4) = 3/8 \quad \phi(6) = 5/16 \quad \phi(8) = 35/128. \tag{7.4.30}$$

Of course, the values of $p(n)$ and $\phi(n)$ in Equations 7.4.24 and 7.4.30 are related by the difference equation 7.1.13.

The result of Equation 7.4.29 can also be established by direct consideration of the problem. The probability of a return to state zero on the $2n$th toss after the beginning of the game is the binomial probability that $n$ out of the $2n$ tosses produced Heads—this is just the probability expressed by Equation 7.4.29.

## Semi-Infinite Random Walks

An interesting variant of the random walk problem we have analyzed is the semi-infinite random walk where the process occupies only states 0, 1, 2, .... For each of the states 1, 2, 3, ... the probability of a step to the right is $p$, and of a step to the left is $q = 1 - p$. When the system is in state zero, the process cannot make any further steps to the left; it remains in state zero with probability $1 - \alpha$ and moves to state 1 with probability $\alpha$.

### *Flow graph representation*

The flow graph for this process appears in Figure 7.4.7. Note that the special structure of the graph in state zero allows us to make state zero a trapping state by choosing $\alpha = 0$, or a state that must be vacated immediately by choosing $\alpha = 1$. In the terminology of random walks, state zero is said to be an absorbing barrier when $\alpha = 0$ and a reflecting barrier when $\alpha = 1$. The parameter $\alpha$ therefore permits us to make the barrier at state zero either of these extremes or partially reflecting to any degree when $0 < \alpha < 1$. Models with barriers are important in many physical process models, like models of heat flow or conduction in solid-state devices.

We find the transform of the lifetime distribution from the flow graph of Figure 7.4.8. Note that the infinite state part of the flow graph to the right of node 1 has a transmission given by Equation 7.4.4 with $a = pz$, $b = qz$. Then we easily find the transform of the lifetime distribution to be

$$p^{\mathscr{g}}(z) = (1 - \alpha)z + \frac{\alpha q z^2}{1 - 1/2[1 - \sqrt{1 - 4pqz^2}]}$$

$$= (1 - \alpha)z + \frac{2\alpha q z^2}{1 + \sqrt{1 - 4pqz^2}}. \qquad (7.4.31)$$

### *Probability of eventual renewal*

The probability of eventual renewal $p_R$ is given by

$$p_R = p^{\mathscr{g}}(1) = 1 - \alpha + \frac{2\alpha q}{1 + \sqrt{1 - 4pq}}. \qquad (7.4.32)$$

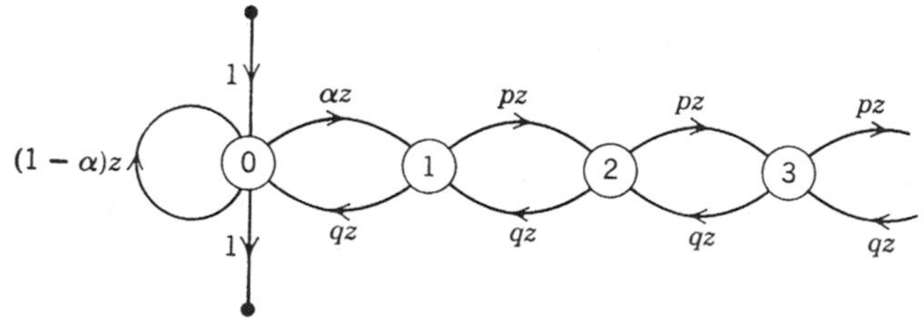


**Figure 7.4.7** Flow graph for a semi-infinite random walk.

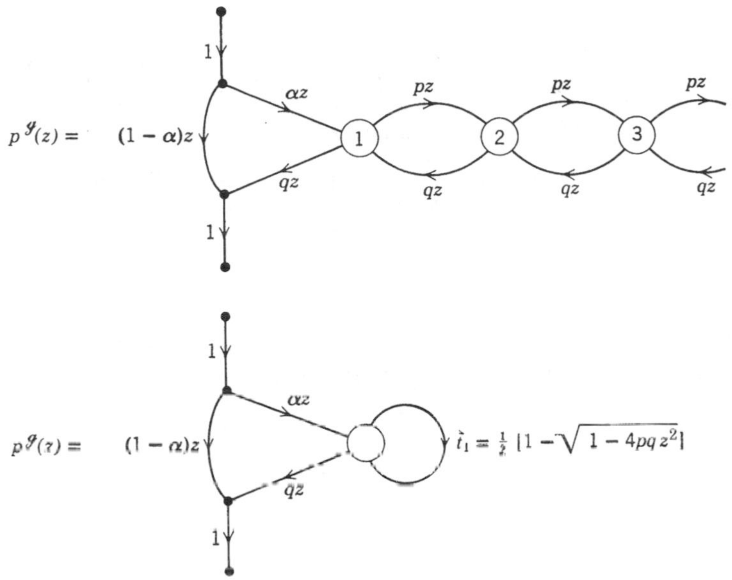


**Figure 7.4.8** Flow graph for the transform of the lifetime distribution for the semi-infinite random walk.

From Equation 7.4.7 we write

$$p_R = \begin{cases} 1 & 0 \le p \le 1/2 \\ 1 - \alpha\dfrac{2p-1}{p} & 1/2 \le p \le 1. \end{cases} \tag{7.4.33}$$

The renewal process is recurrent when $0 \le p \le 1/2$. Note that if $p > 1/2$ and $\alpha = p$, then $p_R = 2(1-p)$, the same result we obtained for the original infinite random walk in Equation 7.4.8.

***Moments of number of returns in an infinite number of trials***

Since returns to state zero are a recurrent renewal process when $p \le 1/2$, the number of returns in an infinite number of tosses is infinite in this case. However, when $1/2 \le p \le 1$, the renewal process is transient and the number of returns in an infinite number of tosses is geometrically distributed with parameter $p_R$. From Equations 7.1.6 and 7.1.7 we find that the mean and variance of the number of returns in an infinite number of tosses for $1/2 \le p \le 1$ are

$$\bar{r} = \frac{p_R}{1-p_R} = \frac{p}{\alpha(2p-1)} - 1 \qquad 1/2 \le p \le 1 \tag{7.4.34}$$

and

$$\check{r} = \frac{p_R}{(1-p_R)^2} = \frac{p^2}{\alpha^2(2p-1)^2} - \frac{p}{\alpha(2p-1)} \qquad 1/2 \le p \le 1. \tag{7.4.35}$$

***Mean trials to return given that return occurs***

The expected number of tosses required to produce the first return given that a return occurs, $\langle n|R\rangle$, is obtained from

$$\langle n|R\rangle = \frac{p^{\mathscr{g}\prime}(1)}{p_R}. \tag{7.4.36}$$

Since a return is certain when $0 \le p \le 1/2$, the lifetime distribution is a probability mass function in this region; consequently, $\langle n|R\rangle$ equals $p^{\mathscr{g}\prime}(1)$, the mean of the lifetime distribution. From Equation 7.4.31 we obtain

$$p^{\mathscr{g}\prime}(1) = 1 - \alpha + \frac{4\alpha q}{1+\sqrt{1-4pq}} + \frac{8\alpha p q^2}{\sqrt{1-4pq}[1+\sqrt{1-4pq}]^2}. \tag{7.4.37}$$

By using the results of Equation 7.4.7 we produce

$$p^{\mathscr{g}\prime}(1) = \begin{cases} 1 - \alpha + 2\alpha\dfrac{1-p}{1-2p} & 0 \le p \le 1/2 \\ 1 - \alpha + 2\alpha\dfrac{1-p}{2p-1} & 1/2 \le p \le 1. \end{cases} \tag{7.4.38}$$

When we substitute this result and Equation 7.4.33 into Equation 7.4.36, we obtain

$$\langle n|R\rangle = \begin{cases} 1 + \dfrac{\alpha}{1-2p} & 0 \le p \le 1/2 \\ \dfrac{1 - \alpha + 2\alpha\dfrac{1-p}{2p-1}}{1 - \alpha\dfrac{2p-1}{p}} & 1/2 \le p \le 1. \end{cases} \tag{7.4.39}$$

Note that when $p = 0$, $\langle n|R\rangle = 1 + \alpha$, and when $p = 1$, $\langle n|R\rangle = 1$. The results correspond directly to our understanding of the flow graph in Figure 7.4.7 in these special cases.

## 7.5 BIRTH AND DEATH PROCESSES

The random walks we discussed in the previous section can be generalized by allowing the transition probabilities to be different at different points on the lattice. The most frequent use of such models arises when we identify the state number with the number of elements in a system. For example, we might say that a population of experimental animals was in state $k$ if there were $k$ animals in the population. The state index changes as the number of elements in the system increases or decreases. A particularly useful model is one in which the state index can change by at most one unit at each transition. Thus, if the system is now in state $k$, its state after the next transition will be $k - 1$, $k$, or $k + 1$. Because of the analogy with populations, these models are called birth and death models. The state index in birth and death models usually takes on the values $0, 1, 2, \ldots$; the maximum state index may be finite or infinite.

### Process Description

To describe a birth and death process we must specify for each state the probability that the next transition will be toward a higher and toward a lower state index. The probability that the transition returns the system to the same state is, of course, one minus the sum of these two numbers. We can exclude the possibility of a return to the same state by simply requiring that the sum of these numbers be one. For convenience we shall call a transition to a higher state index a "birth" and a transition to a lower state index a "death." Thus associated with state $k$ there is a birth probability $b_k$ and a death probability $d_k$. Finally, we have for a time-invariant birth and death process

$$\begin{aligned} b_k &= \mathscr{P}\{s(n+1) = k+1 \mid s(n) = k\} \qquad k = 0, 1, 2, \ldots \\ d_k &= \mathscr{P}\{s(n+1) = k-1 \mid s(n) = k\} \qquad n = 0, 1, 2, \ldots. \end{aligned} \tag{7.5.1}$$

When $b_k + d_k = 1$, the system cannot return to state $k$ on its next transition.

If the system has $N$ as its largest possible state index, then it has $N + 1$ states. States 0 and $N$ require special treatment because a death is impossible in state 0 and a birth is impossible in state $N$: $d_0 = 0$, $b_N = 0$. Thus a birth and death process with $N + 1$ states is completely specified by $b_0$, $d_N$, and the $N - 1$ pairs $(b_k, d_k)$, $k = 1, 2, \ldots, N - 1$. If the number of states is infinite, then we require $b_0$, and all pairs $(b_k, d_k)$, $k = 1, 2, 3, \ldots$.

### Transition Diagram

We often find it convenient to use a transition diagram to specify the birth and death probabilities. Figure 7.5.1 shows such a diagram for a finite-state birth and death process. We have already encountered an infinite-state birth and death process in the semi-infinite random walk of Section 7.4. We see from the flow graph for this process in Figure 7.4.7 that the birth and death probabilities are:

$$\begin{aligned} b_0 = \alpha \qquad b_k &= p \qquad\qquad\quad k = 1, 2, 3, \ldots \\ d_k &= q = 1 - p \qquad k = 1, 2, 3, \ldots. \end{aligned} \tag{7.5.2}$$

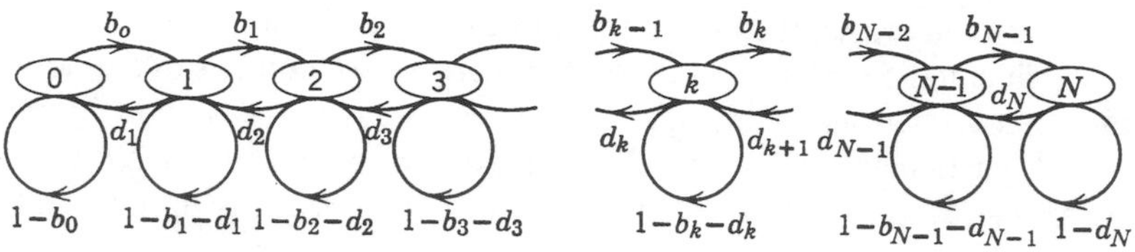


**Figure 7.5.1** Transition diagram for birth and death process.

Since the birth and death process is just a special case of the Markov process, all of the developments that we have learned are pertinent. However, the special structure of the birth and death process allows us to establish more general results than we can obtain for less well-ordered Markov processes. Since the transient behavior of the birth and death process is not very different from that of the general Markov process, we shall find it advantageous to restrict our attention to the limiting state probabilities for the birth and death process.

### Limiting State Probabilities

We begin by writing the transition probability matrix $P$ for the birth and death process. We can write it directly from the transition diagram of Figure 7.5.1 for a finite-state process,

$$P=\begin{bmatrix} 1-b_0 & b_0 & 0 & 0 & 0 & & & & & & & 0 \\ d_1 & 1-b_1-d_1 & b_1 & 0 & 0 & & & & & & & 0 \\ 0 & d_2 & 1-b_2-d_2 & b_2 & 0 & & & & & & & 0 \\ 0 & & & & & & & & & & & 0 \\ & & & 0 & d_k & 1-b_k-d_k & b_k & 0 & & & & \\ 0 & & & & & & & & & & & 0 \\ 0 & & & & & & & 0 & d_{N-1} & 1-b_{N-1}-d_{N-1} & & b_{N-1} \\ 0 & & & & & & & 0 & 0 & d_N & & 1-d_N \end{bmatrix}. \tag{7.5.3}$$

We shall develop all our relations for the finite-state case; the extension to infinite-state processes is immediate. Our equations for the limiting state probability vector $\boldsymbol{\pi}$, which now has $N+1$ elements, require

$$\boldsymbol{\pi} = \boldsymbol{\pi} P \tag{7.5.4}$$

and

$$\sum_{k=0}^{N} \pi_k = 1. \tag{7.5.5}$$

We can write Equation 7.5.4 in the form

$$\pi(P - I) = 0, \tag{7.5.6}$$

and use Equation 7.5.3 to construct the matrix $P - I$,

$$P - I = \begin{bmatrix} -b_0 & b_0 & 0 & 0 & 0 & & & & & 0 \\ d_1 & -b_1-d_1 & b_1 & 0 & 0 & & & & & 0 \\ 0 & d_2 & -b_2-d_2 & b_2 & 0 & & & & & 0 \\ 0 & & & & & & & & & 0 \\ & & 0 & d_k & -b_k-d_k & b_k & 0 & & & \\ 0 & & & & & & & & & 0 \\ 0 & & & & & & 0 & d_{N-1} & -b_{N-1}-d_{N-1} & b_{N-1} \\ 0 & & & & & & 0 & 0 & d_N & -d_N \end{bmatrix}. \tag{7.5.7}$$

Note that the matrix $P - I$ is a differential matrix since its rows sum to zero. We return to Equation 7.5.6 and write the relations for the limiting state probabilities using the matrix 7.5.7,

$$\begin{aligned} -b_0\pi_0 + d_1\pi_1 &= 0 \\ b_0\pi_0 - (b_1 + d_1)\pi_1 + d_2\pi_2 &= 0 \\ b_1\pi_1 - (b_2 + d_2)\pi_2 + d_3\pi_3 &= 0 \\ &\vdots \\ b_{k-1}\pi_{k-1} - (b_k + d_k)\pi_k + d_{k+1}\pi_{k+1} &= 0 \\ &\vdots \\ b_{N-1}\pi_{N-1} - d_N\pi_N &= 0. \end{aligned} \tag{7.5.8}$$

***Equilibrium of probabilistic flows interpretations***

These equations form what we have called in Chapter 1 an "equilibrium of probabilistic flows" relationship among the limiting state probabilities, as in Equation 1.3.29. The partition is considered to exist between each state and the remaining states of the process. For example, the first equation states that after the system has been operating for some time, the chance of seeing a transition out of state 0 must be the same as the chance of seeing a transition into state 0,

$$b_0\pi_0 = d_1\pi_1. \tag{7.5.9}$$

The second equation states that the chance of seeing a transition out of state 1 must be the same as the chance of seeing a transition into state 1,

$$(b_1 + d_1)\pi_1 = b_0\pi_0 + d_2\pi_2. \tag{7.5.10}$$

In fact, all of the Equations 7.5.8 can be written from the requirement that in the steady state the chance of seeing a transition out of a state must be the same as the chance of seeing a transition into one. Therefore, it is usually easier to write Equations 7.5.8 directly from the transition diagram than it is to use the differential matrix $P - I$ of Equation 7.5.7.

There is one further observation we can make. If we add together the first one, first two, first three equations of 7.5.8, we obtain successively

$$\begin{aligned} b_0\pi_0 &= d_1\pi_1 \\ b_1\pi_1 &= d_2\pi_2 \\ b_2\pi_2 &= d_3\pi_3 \\ &\vdots \\ b_k\pi_k &= d_{k+1}\pi_{k+1} \\ &\vdots \\ b_{N-1}\pi_{N-1} &= d_N\pi_N. \end{aligned} \tag{7.5.11}$$

Equations 7.5.11 form another "equilibrium of probabilistic flows" relationship. Here, however, we imagine the partition constructed between adjacent states. The first shows that the probability of observing a transition from state 0 to state 1 in the steady state is the same as the probability of observing a transition from state 1 to state 0. The successive equations make the same statement about the successively higher pairs of states, ending with the last equation which shows that the chance of observing a transition from state $N - 1$ to state $N$ is the same as the chance of observing a transition from state $N$ to state $N - 1$. Of course, these relations could also have been written directly from the transition diagram.

Although we have written two types of equilibrium equations for birth and death processes, we know that a more general relation holds for any monodesmic Markov process; if we divide the transition diagram of the process into two parts by a partition that cuts transition branches, the steady-state probability of a transition in one direction through the partition must be the same as the steady-state probability of a transition in the other direction.

*General solution*

Equations 7.5.11 provide more than physical insight; they also show the solution for the remaining state probabilities. They reveal successively

$$\pi_1 = \frac{b_0}{d_1}\pi_0$$

$$\pi_2 = \frac{b_1}{d_2}\pi_1 = \frac{b_1}{d_2}\frac{b_0}{d_1}\pi_0 \tag{7.5.12}$$

$$\pi_3 = \frac{b_2}{d_3}\pi_2 = \frac{b_2}{d_3}\frac{b_1}{d_2}\frac{b_0}{d_1}\pi_0,$$

or, in general,

$$\pi_k = \frac{b_{k-1}}{d_k}\pi_{k-1} = \pi_0\prod_{i=1}^{k}\frac{b_{i-1}}{d_i} \qquad k = 1, 2, \ldots, N. \tag{7.5.13}$$

Equation 7.5.13 shows that the limiting state probability of any state from 1 through $N$ is equal to the limiting state probability of state 0 multiplied by successive ratios of birth to death probabilities. The problem that remains, then, is how to find $\pi_0$. But this is no problem at all because we still have at our disposal the normalizing equation 7.5.5; it serves to evaluate $\pi_0$.

To summarize, we find the steady-state probabilities for a birth and death process by first using Equation 7.5.13 to relate them to $\pi_0$, i.e., the limiting state probability of state 0. Then we use Equation 7.5.5 to find $\pi_0$, and our solution is complete. Of course, our selection of $\pi_0$ as the probability to be evaluated using Equation 7.5.5 was arbitrary—the probability of any other state would have served as well if with somewhat less convenience.

Perhaps the best way of visualizing the solution given by Equation 7.5.13 is to return to the transition diagram of Figure 7.5.1. We see that the limiting state probability of any state is equal to the limiting state probability of its left-hand neighbor multiplied by the ratio of the birth to death probabilities on the branches joining them. This observation permits us to write the result of Equation 7.5.13 directly from the transition diagram in any problem.

## Some Applications

*The general two-state process*

The general two-state process with transition probability matrix

$$P = \begin{bmatrix} 1-a & a \\ b & 1-b \end{bmatrix} \tag{7.5.14}$$

is, of course, also a birth and death process. If we renumber the states as 0 and 1 to correspond to our birth and death notation, then the birth and death probabilities that must be specified are

$$b_0 = a, \qquad d_1 = b. \tag{7.5.15}$$

From Equation 7.5.13 we have

$$\pi_1 = \frac{b_0}{d_1}\pi_0 = \frac{a}{b}\pi_0. \tag{7.5.16}$$

We normalize using Equations 7.5.5,

$$\pi_0 + \pi_1 = 1 = \pi_0 + \frac{a}{b}\pi_0,$$

to obtain

$$\pi_0 = \frac{b}{a+b} \tag{7.5.17}$$

and

$$\pi_1 = \frac{a}{a+b}. \tag{7.5.18}$$

Equations 7.5.17 and 7.5.18 are the limiting state probabilities for the general two-state process that we have observed on a number of occasions.

***The semi-infinite random walk***

A more interesting example of a birth and death process is the semi-infinite random walk of Section 7.4. We have already written the birth and death probabilities of this infinite-state process in Equation 7.5.2. As usual, we must be careful in applying our results to an infinite-state process because of the possibility of encountering the peculiar effects we observed in Section 7.4. You will recall that for this problem a return to state 0 was certain only when $p < q$, or equivalently, $p < 1/2$. If $p > 1/2$, we see intuitively that the process is likely to occupy higher and higher numbered states as it makes more and more transitions; it has a "drift" toward higher state indices. Since such a behavior would preclude the possibility of a limiting state probability distribution, we would expect our result of Equation 7.5.13 to be inappropriate when $p \geq 1/2$ in this problem. However, when $p < 1/2$ we should obtain a valid limiting state probability distribution.

We substitute the birth and death probabilities of Equation 7.5.2 into Equation 7.5.13 or use the transition diagram implied by Figure 7.4.7 to obtain

$$\pi_k = \pi_0 \frac{\alpha p^{k-1}}{q^k} \qquad k = 1, 2, 3, \ldots. \tag{7.5.19}$$

Now we use Equation 7.5.5 for normalization,

$$\sum_{k=0}^{\infty} \pi_k = 1 = \pi_0\left[1 + \frac{\alpha}{q} + \frac{\alpha p}{q^2} + \frac{\alpha p^2}{q^3} + \cdots\right]$$

$$= \pi_0\left[1 + \frac{\alpha}{q}\left(1 + \frac{p}{q} + \frac{p^2}{q^2} + \cdots\right)\right]. \quad (7.5.20)$$

The series $1 + p/q + p^2/q^2 + \cdots$ will have a finite sum only if $p/q < 1$ or equivalently, if $p < 1/2$. Therefore, even if we had not stopped to think about the implications of the infinite number of states but instead had used our method blindly, we would have been forced to the same conclusion about the values of $p$ for which a limiting state probability distribution will exist. When we restrict $p$ so that $p < 1/2$, the series has the sum $\{1/[1 - (p/q)]\} = q/(q - p)$ and we obtain

$$1 = \pi_0\left[1 + \frac{\alpha}{q}\cdot\frac{q}{q-p}\right] \quad (7.5.21)$$

or

$$\pi_0 = \frac{q - p}{q - p + \alpha} = \frac{1 - 2p}{1 - 2p + \alpha} \qquad p < \frac{1}{2}. \quad (7.5.22)$$

We then find the limiting state probabilities of the other states from Equation 7.5.19,

$$\pi_k = \frac{1 - 2p}{1 - 2p + \alpha}\left(\frac{\alpha p^{k-1}}{(1 - p)^k}\right) \qquad p < \frac{1}{2};\ k = 1, 2, 3, \ldots. \quad (7.5.23)$$

We observe that the limiting state probability distribution is a geometric distribution in the state index.

Equation 7.5.22 for $\pi_0$ provides an interesting verification of an earlier result we have obtained. Since returns to state 0 form a recurrent renewal process when $p < 1/2$, we expect the steady-state probability of a renewal at any time to be the reciprocal of the mean of the lifetime distribution. The steady-state probability of renewal is just the $\pi_0$ found in Equation 7.5.22. It should be the reciprocal of the mean lifetime for this process found in Equation 7.4.39, and it is.

### A Finite-State Random Walk

We can use the birth and death structure to provide a model for a random walk on a finite number of states. This problem is like the semi-infinite random walk except that at some state $N$ we create a partially reflecting barrier rather than extending the number of states indefinitely. The transition diagram for the process appears in Figure 7.5.2. When the process is in state 0 it has a probability $1 - \alpha$

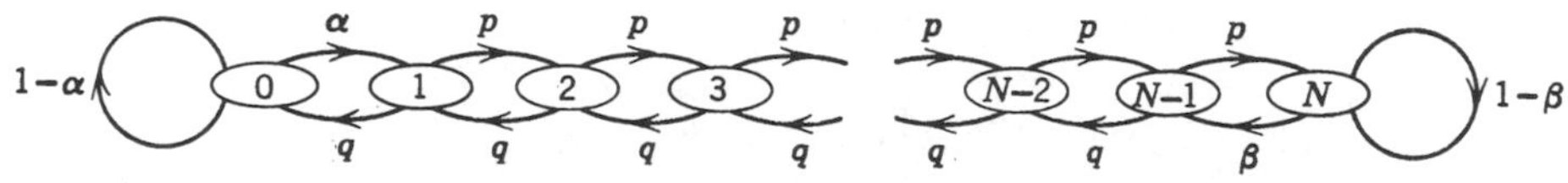


**Figure 7.5.2** A finite-state random walk.

of staying there and a probability $\alpha$ of moving to state 1. Similarly, when the process is in state $N$ it has a probability $1 - \beta$ of remaining in that state and a probability $\beta$ of moving to state $N - 1$. In all other states the process moves one state to the right with a probability $p$ and one state to the left with a probability $q = 1 - p$.

The birth and death probabilities for this walk are

$$\begin{aligned} b_0 &= \alpha \neq 0; \quad b_k = p, \quad k = 1, 2, \ldots, N - 1 \\ d_k &= q = 1 - p, \quad k = 1, 2, \ldots, N - 1; \quad d_N = \beta \neq 0. \end{aligned} \tag{7.5.24}$$

Then, either from Equation 7.5.13 or directly from the flow graph we write

$$\begin{aligned} \pi_k &= \pi_0 \frac{\alpha}{q}\left(\frac{p}{q}\right)^{k-1} \qquad k = 1, 2, \ldots, N - 1 \\ \pi_N &= \pi_0 \frac{\alpha}{\beta}\left(\frac{p}{q}\right)^{N-1}. \end{aligned} \tag{7.5.25}$$

We now normalize by setting the summed limiting state probabilities equal to one,

$$\begin{aligned} \sum_{k=0}^{N} \pi_k = 1 &= \pi_0\left\{1 + \frac{\alpha}{q}\left[1 + \frac{p}{q} + \left(\frac{p}{q}\right)^2 + \cdots + \left(\frac{p}{q}\right)^{N-2}\right] + \frac{\alpha}{\beta}\left(\frac{p}{q}\right)^{N-1}\right\} \\ &= \pi_0\left[1 + \frac{\alpha}{q}\cdot\frac{1 - \left(\dfrac{p}{q}\right)^{N-1}}{1 - \dfrac{p}{q}} + \frac{\alpha}{\beta}\left(\frac{p}{q}\right)^{N-1}\right] \end{aligned} \tag{7.5.26}$$

or

$$\pi_0 = \frac{1}{1 + \alpha\dfrac{1 - \left(\dfrac{p}{q}\right)^{N-1}}{q - p} + \dfrac{\alpha}{\beta}\left(\dfrac{p}{q}\right)^{N-1}}. \tag{7.5.27}$$

When we substitute this value for $\pi_0$ into Equation 7.5.25, we have the complete set of limiting state probabilities for the process.

Note that because of the finite state structure we did not have to assume $p < 1/2$ to obtain convergence of the series in Equation 7.5.26. If we do assume $p < 1/2$, then Equation 7.5.27 approaches Equation 7.5.22 as $N$ grows larger and larger.

However, the expression we have written in Equation 7.5.27 is not appropriate when $p = q$. We can easily write the correct expression by returning to the series in Equation 7.5.26 for this case. We find

$$\pi_0 = \frac{1}{1 + 2\alpha(N-1) + \dfrac{\alpha}{\beta}} \cdot \qquad p = q = 1/2. \tag{7.5.28}$$

The remainder of the solution is then

$$\pi_k = \frac{2\alpha}{1 + 2\alpha(N-1) + \dfrac{\alpha}{\beta}} \qquad p = q = 1/2;\ k = 1, 2, \ldots, N-1$$

$$\pi_N = \frac{\dfrac{\alpha}{\beta}}{1 + 2\alpha(N-1) + \dfrac{\alpha}{\beta}} \qquad p = q = 1/2. \tag{7.5.29}$$

Equations 7.5.28 and 7.5.29 show that when $p = q = 1/2$ all states with the exception of the end states have the same limiting state probability.

Since returns to state 0 are recurrent renewal processes in this finite-state system, we can find the mean time between such renewals in the steady state by taking the reciprocal of $\pi_0$ in Equation 7.5.27 or Equation 7.5.28. Choosing the case $p = q = 1/2$, we find from Equation 7.5.28 that the mean of the lifetime distribution is

$$\frac{1}{\pi_0} = 1 + 2\alpha(N-1) + \frac{\alpha}{\beta} \cdot \tag{7.5.30}$$

Thus the mean lifetime increases with increased $N$ and $\alpha$, and decreases with increased $\beta$ in correspondence with our intuition based on the transition diagram of Figure 7.5.2.

## 7.6 RUIN PROBLEMS

We can interpret the finite-state random walk of Figure 7.5.2 as a gambling game in which the total capital of the two players is $N$ and the state index $i$ represents the capital of one of them, whom we shall call our player. Thus when $i$ equals 0 our player is without funds and when $i$ equals $N$ he has won all the money available. Plays of the game are decided by flipping a coin with probability of Heads $p$. If it comes up Heads our player wins 1 unit from his opponent. The outcome Tails means that he loses one unit, a result that occurs with probability $q = 1 - p$. As the game appears in Figure 7.5.2, whenever our player wins all the capital in the game, he flips another coin with probability of Heads $\beta$ and generously gives one

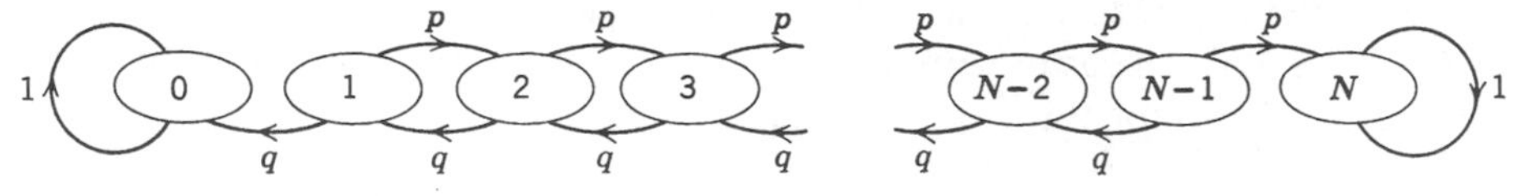


**Figure 7.6.1** Transition diagram for ruin problem.

unit to his opponent the first time the coin produces a Head. Then the game continues as before. Similarly, if our player loses all his capital, his opponent carries out the same procedure, but with a coin whose probability of Heads is $\alpha$. Thus the game will never terminate—the fortunes of the players will fluctuate indefinitely.

However, human nature being what it is, the most frequent form of the game is the one in which $\alpha = \beta = 0$. That is, once either player wins all the capital, the game ceases. Such games are called games of ruin. The transition diagram corresponding to this game appears in Figure 7.6.1. Unlike the finite-state random walk of Figure 7.5.2, the ruin problem is not a monodesmic Markov process. The system may be trapped by either of the two states 0 or $N$. Consequently, rather than being interested in limiting state probabilities, we are interested in the probability of being trapped by each state and in the duration of the game—typical considerations in a transient process. In particular, we would like to know how our player's initial capital affects his probability of winning, his expected winnings, and the expected duration of the game.

## Probability of Winning

We begin with the probability that our player will win if his initial capital is $i$. We call this probability $\phi_i$. It is, of course, the probability that the system will reach state $N$ before state 0 if it is started in state $i$. We could find this quantity as the transmission of the transition diagram of Figure 7.6.1 from node $i$ to node $N$ with the self-loop around nodes 0 and $N$ removed. However, it is easier and more convenient to find an expression for $\phi_i$ by writing a difference equation.

Suppose that our player's capital is $i$ at some intermediate point in the game. With probability $p$ the next trial will produce a Head and his capital will move to $i + 1$; he will then have a probability $\phi_{i+1}$ of winning the game ultimately. However, with probability $q = 1 - p$ the next toss will produce a Tail, his capital will decrease to $i - 1$, and his ultimate probability of winning will become $\phi_{i-1}$. Thus we can relate the probability of his winning with capital $i$ to the probability of his winning with capital $i - 1$ and capital $i + 1$ by the equation

$$\phi_i = p\phi_{i+1} + q\phi_{i-1} \qquad i = 1, 2, \ldots, N - 1. \tag{7.6.1}$$

Note that this equation holds only for the internal states of the game, states 1 through $N - 1$. At the two boundary states 0 and $N$ we can easily write the probability of our player's winning since he has no chance of winning when he is in state 0, and he has already won when he is in state $N$; thus,

$$\phi_0 = 0, \qquad \phi_N = 1. \tag{7.6.2}$$

***Geometric transformation on the state index***

Equation 7.6.1 and the boundary conditions 7.6.2 completely determine the probabilities $\phi_i$. Although there are several methods available for solving such equations, we shall find it most instructive to use geometric transforms. Note that this use of transforms is a departure from our past practice, because we have previously used geometric transforms only to transform time sequences whereas now we shall transform on $i$, the index of the starting state. First we adjust the index in Equation 7.6.1 so that it starts at $i = 0$,

$$\phi_{i+1} = p\phi_{i+2} + q\phi_i \qquad i = 0, 1, 2, \ldots, N - 2. \tag{7.6.3}$$

Next we write the geometric transform of $\phi_i$ as $\phi_z^{g}$,

$$\phi_z^{g} = \sum_{i=0}^{\infty} \phi_i z^i. \tag{7.6.4}$$

Note that the transform variable $z$ appears as a subscript because we are transforming the subscript $i$. From first principles or from our transform table we can now write the transform of Equation 7.6.3 as

$$z^{-1}[\phi_z^{g} - \phi_0] - pz^{-2}[\phi_z^{g} - \phi_0 - z\phi_1] + q\phi_z^{g}. \tag{7.6.5}$$

Since $\phi_0 = 0$ from Equation 7.6.2, we can place Equation 7.6.5 in the form

$$\phi_z^{g}(z - p \quad qz^2) = -pz\phi_1 \tag{7.6.6}$$

or

$$\phi_z^{g} - \frac{z\phi_1}{1 - \dfrac{1}{p}z + \dfrac{q}{p}z^2}. \tag{7.6.7}$$

***Transform inversion***

This equation expresses the transform $\phi_z^{g}$ in terms of the probability of winning starting in state 1, $\phi_1$. Of course, this probability is not yet known, but we can determine it by writing the inverse transform $\phi_i$ in terms of $\phi_1$ and then using the condition $\phi_N = 1$ from Equation 7.6.2 to evaluate $\phi_1$.

**Case $p \neq q$.** To find the inverse transform we first separate the two cases $p \neq q$ and $p = q$. If $p \neq q$ then we factor Equation 7.6.7 in the form

$$\phi_z^{\mathscr{g}} = \frac{z\phi_1}{(1 - z)\left(1 - \dfrac{q}{p}z\right)} \tag{7.6.8}$$

where now both denominator factors are distinct. By partial fraction expansion we obtain

$$\phi_z^{\mathscr{g}} = \frac{\dfrac{p}{p-q}\phi_1}{1 - z} + \frac{\dfrac{-p}{p-q}\phi_1}{1 - \dfrac{q}{p}z} \qquad p \neq q. \tag{7.6.9}$$

The inverse geometric transform is then

$$\phi_i = \frac{p}{p-q}\phi_1\left[1 - \left(\frac{q}{p}\right)^i\right] \qquad i = 0, 1, 2, \ldots, N. \tag{7.6.10}$$

Finally, we substitute the boundary condition $\phi_N = 1$ from Equation 7.6.2,

$$\phi_N = 1 = \frac{p}{p-q}\phi_1\left[1 - \left(\frac{q}{p}\right)^N\right]. \tag{7.6.11}$$

When we divide Equation 7.6.10 by Equation 7.6.11, we produce the desired result

$$\phi_i = \frac{1 - \left(\dfrac{q}{p}\right)^i}{1 - \left(\dfrac{q}{p}\right)^N} \qquad p \neq q,\ i = 0, 1, 2, \ldots, N. \tag{7.6.12}$$

This equation shows that the probability of winning starting with capital $i$ is very sensitive to the ratio of $q$ to $p$. There is nothing like an edge on the odds of each trial to produce a high probability of winning. It is easy to see from Equation 7.6.12 that if the game is unfavorable to our player ($p < 1/2$), he will maximize his chance of winning for fixed capital of both players by making the number of units staked at each trial as large as possible.

We can also use Equation 7.6.12 to investigate the case of playing against an infinitely rich opponent by letting $N$ approach infinity while maintaining $i$ finite. If $q > p$, the game favors the opponent, and we see that $\phi_i = 0$. If $q < p$, then the game favors our player and he has a probability $1 - (q/p)^i$ of winning against even an infinitely rich adversary. This observation is directly related to the result in Section 7.4 that returns to the origin in semi-infinite random walks are not recurrent renewal processes when $q < p$.

**Case $p = q$.** We still must consider the case of the fair game, where $p = q = 1/2$. For this case the transform of Equation 7.6.7 becomes

$$\phi_z^{\mathcal{G}} = \frac{z\phi_1}{1 - 2z + z^2} = \frac{z\phi_1}{(1 - z)^2} \qquad p = q = 1/2. \tag{7.6.13}$$

We write the inverse transform immediately as

$$\phi_i = i\phi_1 \qquad i = 0, 1, 2, \ldots, N. \tag{7.6.14}$$

Using the boundary condition $\phi_N = 1$ from Equation 7.6.2 we obtain

$$\phi_N = 1 = N\phi_1. \tag{7.6.15}$$

Finally, we divide Equation 7.6.14 by Equation 7.6.15 to produce

$$\phi_i = \frac{i}{N} \qquad p = q = 1/2. \tag{7.6.16}$$

Thus when the game is fair, the probability of our player's winning is directly proportional to the ratio of his capital to the total capital. We see that if our player were to play a fair game against an infinitely rich opponent he would have a zero probability of winning.

## Expected Winnings

We can obtain more insight into the result of Equation 7.6.16 by considering the expected winnings for our player in an entire play of the game. The expected winnings starting with initial capital $i$ are computed by multiplying the probability of winning, $\phi_i$, by the profit from winning, $N - i$, and then adding the probability of not winning, $1 - \phi_i$, times the loss from not winning, $-i$, or

$$\phi_i(N - i) + (1 - \phi_i)(-i) = \phi_i N - i. \tag{7.6.17}$$

If the game is to be fair, the expected winnings should be 0. Setting them equal to zero establishes immediately the value for $\phi_i$ given by Equation 7.6.16 for the fair game. Of course, if $p \neq q$ so that the game is not fair, then we use Equations 7.6.12 and 7.6.17 to compute the non-zero expected winnings from the game.

## Games with Draws on Each Trial

We can also consider ruin games where each trial can result in a draw. Then the state of our player's capital need not change on each trial. If $b$ is the probability of his winning on a trial and $d$ is the probability of his losing, then $1 - b - d$ is the probability of a draw. With this notation, we construct the transition diagram of

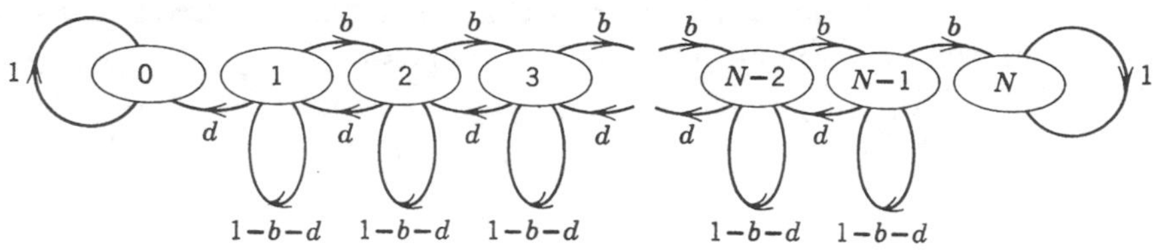


**Figure 7.6.2** Transition diagram for ruin problem with draws.

Figure 7.6.2 to represent the problem. By converting each of the self-loops about the internal nodes to an equivalent transmission $1/(b + d)$ on all the branches leaving those nodes we change the transition diagram to the form shown in Figure 7.6.3. Note that this transition diagram is the same as that of Figure 7.6.1 with $p = b/(b + d)$, $q = d/(b + d)$. Thus the expressions we have already developed for the probability of winning in the case of no draws apply directly. However, we would expect the expected duration of the equivalent no-draw game to be $b + d$ times the expected duration of the game with draws, because in the draw case there is only a probability $b + d$ that each trial will produce a change of state.

### Expected Duration

Turning now to the question of the expected duration of the game without draws, we define $\bar{\nu}_i$ to be the expected number of trials necessary to terminate the game if our player starts with capital $i$, regardless of whether he wins or loses. The difference equation we write for $\bar{\nu}_i$ is very similar to the one we wrote for $\phi_i$. The expected number of trials $\bar{\nu}_i$ is composed of the one trial required for the next transition plus the expected number of trials after that transition. The expected number of trials after the next trial is $\bar{\nu}_{i+1}$ with probability $p$ and $\bar{\nu}_{i-1}$ with probability $q = 1 - p$. Therefore, we can write

$$\bar{\nu}_i = 1 + p\bar{\nu}_{i+1} + q\bar{\nu}_{i-1} \qquad i = 1, 2, \ldots, N - 1. \tag{7.6.18}$$

Like Equation 7.6.1, this equation applies only for the internal states of the system,

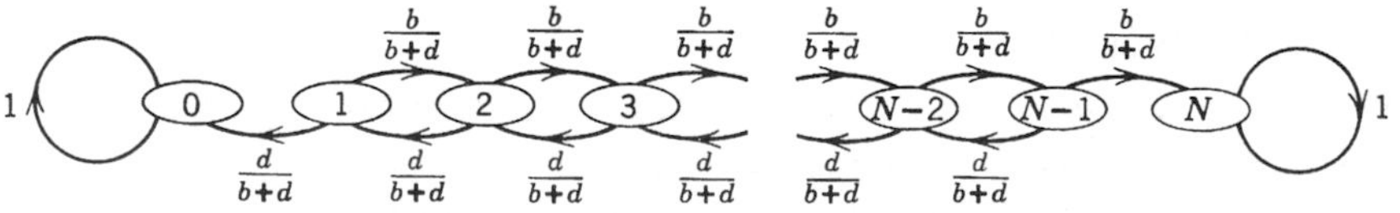


**Figure 7.6.3** Equivalent no-draw transition diagram for ruin problem with draws.

1 through $N - 1$. We write the boundary condition directly from the requirement that the expected duration of the game be zero in either state 0 or state $N$,

$$\bar{\nu}_0 = \bar{\nu}_N = 0. \tag{7.6.19}$$

***Geometric transformation on the state index***

Once more we shall use geometric transforms to solve Equations 7.6.18. We begin by changing the index so that it starts at $i = 0$,

$$\bar{\nu}_{i+1} = 1 + p\bar{\nu}_{i+2} + q\bar{\nu}_i \qquad i = 0, 1, 2, \ldots, N - 2. \tag{7.6.20}$$

We write $\bar{\nu}_z^{\mathscr{g}}$ as the geometric transform of $\bar{\nu}_i$ and transform Equation 7.6.20 to produce

$$z^{-1}(\bar{\nu}_z^{\mathscr{g}} - \bar{\nu}_0) = \frac{1}{1 - z} + pz^{-2}(\bar{\nu}_z^{\mathscr{g}} - \bar{\nu}_0 - z\bar{\nu}_1) + q\bar{\nu}_z^{\mathscr{g}}. \tag{7.6.21}$$

After noting $\bar{\nu}_0 = 0$ from Equation 7.6.19, we place Equation 7.6.21 in the form

$$\bar{\nu}_z^{\mathscr{g}}\left(1 - \frac{1}{p}z + \frac{q}{p}z^2\right) = z\bar{\nu}_1 - \frac{\frac{1}{p}z^2}{(1 - z)} \tag{7.6.22}$$

or

$$\bar{\nu}_z^{\mathscr{g}} = \frac{z\bar{\nu}_1}{(1 - z)\left(1 - \frac{q}{p}z\right)} - \frac{\frac{1}{p}z^2}{(1 - z)^2\left(1 - \frac{q}{p}z\right)}$$

$$= \frac{(1 - z)z\bar{\nu}_1 - \frac{1}{p}z^2}{(1 - z)^2\left(1 - \frac{q}{p}z\right)}. \tag{7.6.23}$$

***Inverse transformation***

Once more we must distinguish between the cases $p \neq q$ and $p = q$.

**Case $p \neq q$.** When $p \neq q$ we expand $\bar{\nu}_z^{\mathscr{g}}$ in the partial fraction form

$$\bar{\nu}_z^{\mathscr{g}} = \frac{-\frac{1}{p - q}}{(1 - z)^2} + \frac{\frac{1}{p - q} + \frac{p}{p - q}\bar{\nu}_1 + \frac{p}{(p - q)^2}}{1 - z} + \frac{\frac{-p}{p - q}\bar{\nu}_1 + \frac{-p}{(p - q)^2}}{1 - \frac{q}{p}z}. \tag{7.6.24}$$

The inverse transform is

$$\bar{\nu}_i = -\left(\frac{1}{p-q}\right)(i+1) + \frac{1}{p-q} + \frac{p}{p-q}\bar{\nu}_1 + \frac{p}{(p-q)^2} - \left[\frac{p}{p-q}\bar{\nu}_1 + \frac{p}{(p-q)^2}\right]\left(\frac{q}{p}\right)^i$$

$$= \frac{-1}{p-q}i + \left[\frac{p}{p-q}\bar{\nu}_1 + \frac{p}{(p-q)^2}\right]\left[1 - \left(\frac{q}{p}\right)^i\right]$$

$$p \neq q \quad i = 0, 1, 2, \ldots, N. \qquad (7.6.25)$$

Now we write the requirement $\bar{\nu}_N = 0$ from Equation 7.6.19,

$$\bar{\nu}_N = 0 = -\frac{N}{p-q} + \left[\frac{p}{p-q}\bar{\nu}_1 + \frac{p}{(p-q)^2}\right]\left[1 - \left(\frac{q}{p}\right)^N\right]. \qquad (7.6.26)$$

When we solve for the bracketed expression containing $\bar{\nu}_1$ in Equation 7.6.26 and substitute it for the same expression in Equation 7.6.25, we obtain

$$\bar{\nu}_i = \frac{1}{p-q}\left[N\frac{1-\left(\frac{q}{p}\right)^i}{1-\left(\frac{q}{p}\right)^N} - i\right] \qquad p \neq q,\ i = 0, 1, 2, \ldots, N. \qquad (7.6.27)$$

Equation 7.6.12 allows us to write this result in the simple form

$$\bar{\nu}_i = \frac{1}{p-q}(N\phi_i - i) \qquad p \neq q,\ i = 0, 1, 2, \ldots, N. \qquad (7.6.28)$$

The expected duration of the game is thus directly related to the probability of our player's winning.

**Case $p = q$.** We find the transform of the expected duration for the case $p = q = 1/2$ from Equation 7.6.23,

$$\bar{\nu}_z^{\mathscr{G}} = \frac{z\bar{\nu}_1}{(1-z)^2} - \frac{2z^2}{(1-z)^3} \qquad p = q = 1/2\,. \qquad (7.6.29)$$

The inverse geometric transform is

$$\bar{\nu}_i = i\bar{\nu}_1 - i(i-1) \qquad p = q = 1/2,\ i = 0, 1, 2, \ldots, N. \qquad (7.6.30)$$

The condition $\bar{\nu}_N = 0$ produces

$$\bar{\nu}_N = 0 = N\bar{\nu}_1 - N(N-1), \qquad (7.6.31)$$

which implies

$$\bar{\nu}_1 = N - 1. \qquad (7.6.32)$$

Then Equation 7.6.30 yields

$$\bar{\nu}_i = i(N-i) \qquad p = q = 1/2,\ i = 0, 1, 2, \ldots, N. \qquad (7.6.33)$$

This equation shows that for the fair game $p = q = 1/2$, the maximum expected duration occurs when the total capital in the game is approximately equally divided between the two players. If $N$ is even and each player has initial capital $N/2$, then the expected duration of the game is

$$\bar{\nu}_{N/2} = (1/4)N^2. \tag{7.6.34}$$

Thus the expected duration of the symmetric game increases as the square of the total capital.

***Expected winnings per trial***

If we divide the expected profit from each play of the game as given in Equation 7.6.17 by the expected number of trials per game from Equation 7.6.28, we obtain $p - q$, the expected winnings per trial, a result that checks with our intuition and that will be formally shown to be a consequence of compound distribution theory in Section 8.2. Therefore, by using the fact that the expected winnings per trial must be $p - q$ we could have established the result of Equation 7.6.28 directly from the simple calculation of Equation 7.6.17. However, we could not use this method to compute the expected duration of a fair game shown in Equation 7.6.33 because in this case $p - q = 0$.

## An Example

Let us verify the results we have obtained for the ruin problem by considering an example we can solve easily using flow graph methods. We shall investigate the ruin problem with total capital $N = 4$ where our player has probability $p = 2/3$ of winning on each trial. From Equation 7.6.12 we find

$$\phi_1 = 8/15 \qquad \phi_2 = 4/5 \qquad \phi_3 = 14/15. \tag{7.6.35}$$

Equation 7.6.28 then produces

$$\bar{\nu}_1 = 17/5 \qquad \bar{\nu}_2 = 18/5 \qquad \bar{\nu}_3 = 11/5. \tag{7.6.36}$$

The highest expected duration occurs when our player starts with initial capital 2. We know that the expected winnings per game starting in state $i$ are just the expected winnings per trial, $p - q = 1/3$, times the expected number of trials in a game starting in state $i$, or 1/3 of the values given by Equation 7.6.36.

***Transient process analysis***

The modified transition diagram for this example appears in Figure 7.6.4. The sum of the loop products for this graph is

$$\Delta = 1 - 2/9 - 2/9 = 5/9.$$

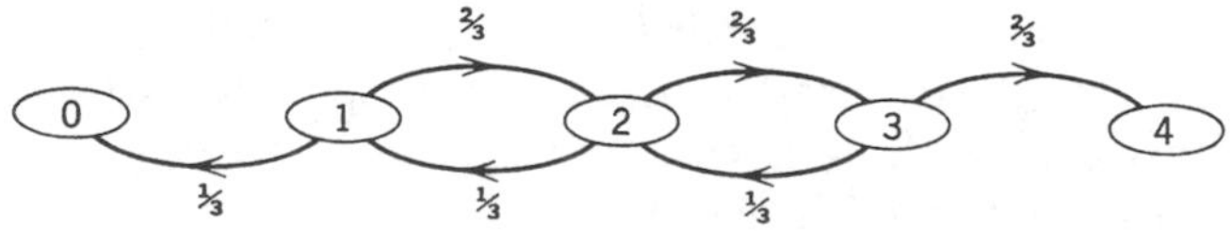


**Figure 7.6.4** Modified transition diagram for ruin example.

From our relations for transient processes we know that the probability that the process will be trapped by state 4 if it starts in state $i$ is $\vec{t}_{i4}$. Thus,

$$\begin{aligned} \phi_1 &= \vec{t}_{14} = \frac{1}{\Delta}\cdot 8/27 = 8/15 \\ \phi_2 &= \vec{t}_{24} = \frac{1}{\Delta}\cdot 4/9 = 4/5 \\ \phi_3 &= \vec{t}_{34} = \frac{1}{\Delta}\cdot (2/3)[1 - (2/9)] = 14/15, \end{aligned} \tag{7.6.37}$$

and we have obtained the same results given by Equation 7.6.35.

Similarly, the total expected time spent in the transient process is the sum of the expected time spent in all of the transient states. The expected time spent in a transient state $j$ if the system starts in state $i$ is, as we know, given by $\vec{t}_{ij}$. Therefore, we can write

$$\begin{aligned} \bar{\nu}_1 &= \vec{t}_{11} + \vec{t}_{12} + \vec{t}_{13} = \frac{1}{\Delta}[[1 - (2/9)] + 2/3 + 4/9] = 17/5 \\ \bar{\nu}_2 &= \vec{t}_{21} + \vec{t}_{22} + \vec{t}_{23} = \frac{1}{\Delta}[1/3 + 1 + 2/3] = 18/5 \\ \bar{\nu}_3 &= \vec{t}_{31} + \vec{t}_{32} + \vec{t}_{33} = \frac{1}{\Delta}[1/9 + 1/3 + [1 - (2/9)]] = 11/5, \end{aligned} \tag{7.6.38}$$

and we have produced the results of Equation 7.6.36.

We must always bear in mind what this example illustrates; namely, that ruin problems are special cases of Markov processes and therefore are susceptible to the general methods of analysis we have developed.

## PROBLEMS

**1.** A coin with probability $p$ of Head (success) is flipped repeatedly.

a) Find the probability $b(k|n)$ of achieving $k$ successes in $n$ trials (the binomial distribution).

b) Find $p(n)$, the probability that the first success occurs on the $n$th trial; find its geometric transform, $p^{g}(z)$, and both its mean and variance.

c) Find $q(k, n)$, the probability that the $k$th success occurs on the $n$th trial; find its geometric transform, $q^{\mathscr{G}}(k, z)$, and both its mean and variance. What is the relationship of this distribution to the one found in a)?

**2.** A coin with probability $p$ of Head is tossed repeatedly.

a) What is the average number of tosses necessary to produce the sequence HTTH if no overlapping of sequences is allowed?

b) What is the average number of tosses necessary to produce the sequence if overlapping is allowed, i.e., if the last H of a sequence is allowed to be the first H of the next sequence?

c) Find the transforms of the mean and variance of the number of HTTH sequences completed in $n$ tosses of the coin. Do this for the overlapping and non-overlapping cases.

d) Find the asymptotic form of the mean and variance in part c).

e) Let $p = 1/2$. Can you prove for all $n > 0$ that the mean number of occurrences of HTTH is greater with overlap than without overlap. Do not invert the transforms.

**3.** A fair coin is being tossed.

a) Determine the geometric transform of the probability that the sequence HHTT occurs for the first time on trial $n$. Find the expected number of trials to obtain this sequence.

b) Determine the expected number of trials to obtain the sequence HHHH for the first time.

c) Why are the answers to a) and b) different?

**4.** Mr. Jones has $N$ lightbulbs in his house and once a year on January 1 he replaces all of the bulbs that have burned out. He has found that on the average one-half of the bulbs will burn out during their first year of operation, one-fourth of the bulbs will burn out during their second year of operation, and one-fourth of the bulbs will burn out during their third year of operation.

a) If he has been doing this for many years, what will be the steady-state values of $\bar{u}_1$, $\bar{u}_2$, and $\bar{u}_3$ where $\bar{u}_k$ is the average number of bulbs in the house that are $k$ years old on December 31?

b) Find the transforms of the mean and variance of the number of lightbulbs Mr. Jones will have replaced by December 31, $(19xy + n)$ if he started off on January 1, $19xy$ with all new bulbs.

c) Find the asymptotic values of the mean and variance in part b).

**5.** Mr. George Roy has a systematic method of replacing his $R$ handkerchiefs. Once a year, on January 2nd when he can take advantage of special sales, he replaces all handkerchiefs that are torn or frayed. He always buys handkerchiefs having the following properties:

$$\mathscr{P}\{\text{handkerchief will tear or fray during 1st year of use}\} = 3/8$$
$$\mathscr{P}\{\text{handkerchief will tear or fray during 2nd year of use}\} = 1/2$$
$$\mathscr{P}\{\text{handkerchief will tear or fray during 3rd year of use}\} = 1/8.$$

Mr. Roy's rich uncle has died and his will states that on January 2, $19xy$ part of his money will be used to replace all $R$ of his nephew's handkerchiefs regardless of whether they are worn or not. Assume that Mr. Roy will then revert to his old replacement policy described above.

a) What will be the expected numbers, $\bar{u}_{11}(k)$, $\bar{u}_{12}(k)$, $\bar{u}_{13}(k)$, of 1, 2, and 3 year old handkerchiefs, respectively, on January 1, $(19xy + k)$, i.e., $k$ years later (just before the replacements for the year are made)?

b) What are the steady-state values of the quantities found in part a)?

*Hint:* Although this is a three-state Markov process the calculations are somewhat involved. You may wish to express your answers in part a) in terms of real numbers (sines and cosines). They will then be in convenient form for rapid evaluation of the $\bar{u}$'s for any specific $k$ values.

**6.** Consider a coin whose probability of Heads is $p$. We are interested in the quantity $q(k)$, the probability of no TT sequence in the first $k$ tosses.

a) Draw the appropriate flow graph and find the geometric transform of $q(k)$.

b) For the special case of $p = 1/3$, determine $q(k)$.

c) Consider an event that is certain to happen. Let $q(k) = \mathscr{P}\{$the event has not occurred by the end of trial $k\}$ $k \geq 0$. Let $n$ = number of trial on which the event occurs. Then prove that the expected value of $n$ is

$$\bar{n} = \sum_{k=0}^{\infty} q(k).$$

d) Use the results of parts a) and c) to determine the expected number of tosses required to obtain the sequence TT for the first time with a coin whose probability of Heads is $p$. Check your result by calculating the same quantity some other way.

**7.** a) Determine the expected number of occurrences of the sequence (4, 3) in $n$ rolls of a fair die. What is the variance of the number of such sequences?

b) In a large number of throws what fraction of them will result in the completion of the sequence (4, 3)?

c) Let $q(n)$ be the probability that the sequence (4, 3) has not occurred in $n$ rolls. Determine the geometric transform of $q(n)$ (do not attempt to invert it).

d) Show by a *simple* calculation that $q(n) \to 0$ as $n \to \infty$.

**8.** Let $q^{gg}(y, z)$ be the double geometric transform of the quantity $q(r, n)$ defined in Section 7.1. Draw a flow graph whose transmission is this transform. Interpret the properties of $q(r, n)$ discussed in the section in terms of the flow graph representation.

**9.** Sam has one dollar and needs three dollars to pay his rent. In an attempt to acquire the remaining amount he has decided to enter a "friendly" game of chance in which the chances of winning a dollar on every play of the game are $p$. Unfortunately, the chances of losing a dollar are $q = 1 - p$. Sam plays until he has the \$3 to pay his rent. Every time Sam goes broke he borrows one dollar to get back into the game. (Assume that Sam does not repay any borrowed money until next month.)

a) What is the probability that Sam pays his rent without having borrowed any money?

b) If Sam has to borrow money at least twice, what is the expected number of plays of the game between the times when he borrows the \$1?

c) What is Sam's expected debt when he pays his rent?

**10.** Two people are playing a coin game in which $A$ collects a dollar from $B$ every time a Head is turned up and $B$ collects a dollar from $A$ every time a Tail is turned up. They are playing with two distinguishable coins, one with known probability of Heads $p \geq 1/2$ and the other with known probability of Heads equal to $(1 - p)$. To start the game they toss a *fair* coin to see who gets to select the coins to be used on the first toss. Thereafter, the winner of the last toss gets to select the coin for the next toss. In an infinite number of tosses (and assuming each player plays to maximize his expected winnings), what will be the expected number of times the players will be even? If the players are even (i.e., the total winnings of each are zero), how many tosses will it take on the average for them to be even again?

**11.** Big Jack has an infinite number of silver dollars at his disposal. He decides to gamble on two slot machines as follows: He uses machine $A$ on every odd trial and machine $B$ on every even trial. The machines have the following characteristics: Each trial requires a deposit of 1 dollar into the machine. Machine $A$ returns 2 dollars (including the 1 dollar deposit) with probability $p$. Machine $A$ returns nothing with probability $q = 1 - p$. The corresponding probabilities for machine $B$ are $r$ and $s = 1 - r$.

a) What is the probability that he will ever return to the break-even point (i.e., the same financial status at which he starts)? What relation (involving $p$ and $r$) must hold for this probability to be one?

b) Express the expected number of returns to the break-even point as a function of $p$ and $r$. Sketch this function.

c) Suppose that he starts with only 2 dollars (instead of an infinite amount) and stops when he first becomes broke or when he has a total of 5 dollars. What is the probability as a function of $p$ and $r$ that he quits as a winner? Evaluate your answer for the case $p = 3/4$, $r = 1/2$.

**12.** Two very hardy and evenly matched teams begin a volleyball game. They volley for service, and each team has half a chance of winning this opportunity. When the initial server is decided, however, the game proceeds as follows. The server has probability $p$ of winning a point, in which case he continues to serve. If he does not win, with probability $q = 1 - p$, then the ball goes to the other team with no loss of point. Now the other team is serving, and also has probability $p$ of winning the point and probability $q$ of losing the ball with no loss of point. As the number of services grows very large, find:

a) The geometric transform of the probability that the game is tied for the first time on the $n$th service. (Do *not* find the inverse transform.)

b) The expected number of times the game will be tied in an infinite number of services.

c) The expected time between tied scores.

d) In volleyball a tied game is won when one team or the other scores two consecutive points. Given that the game is tied, find the expected number of services necessary to end the game. Plot your answer as a function of $p$.

**13.** The following model has been proposed for the fluctuations in price of a market item. If the item is priced at $\$k$ today, then with probability $\alpha^{k+1}$ the price tomorrow will be $\$(k + 1)$, while with probability $k\beta^k$ it will be $\$(k - 1)$. If neither of these two events materializes, the price tomorrow will remain at $\$k$. If the item started off at a price of \$1 and has been on the market for a long time, what will be the probability mass function for its price? What will be the mean and variance of its price?

**14.** A certain social club is concerned about the number of members that it has on its rolls. Through long experience it has found that if there are $k$ members at the end of the $n$th month, then the probability of increasing the membership to $k + 1$ by the end of the next month is $p/(k + 1)$ for all $k \geq 0$. The decrease in the probability is due to crowded conditions in the club house, decrease in potential members, etc. On the other hand, the probability of a net decrease to $k - 1$ is equal to $q$ for all $k \geq 1$. The third alternative is that the membership remains equal to $k$.

a) If the club has been operating for a long time and is willing to have any number of members, what is the probability that there is no one in the club?

b) What is the expected membership in the club?

# 8 MARKOVIAN POPULATION MODELS

In most of the previous work our attention has been focused on how a single unit behaves in a system. However, in many applications of Markov processes there exists not one unit of special interest, but rather an entire population of units whose distribution over the states of the system changes with time. In the marketing problem, for example, where the state of the system was specified by the last brand purchased by a customer, the number of customers purchasing each brand will change from time to time. As we shall see, when units behave independently of each other, total system response can be obtained from superposition of the Markovian trajectories of each individual unit in the population. Often the statistics of this total population distribution are more relevant to analysis than the behavior of any particular unit.

Furthermore, there often arises the additional complication that the total number of units in the system may change in the course of time. In the marketing example, these changes would be caused by variations in the number of customers for the product in question arising from such influences as the vagaries of taste and population growth and mobility. In this chapter we shall discuss Markovian population models where the change in the size of the population can be due to either external or internal forces.

## 8.1 THE VECTOR MARKOV PROCESS

We shall consider a system in which the behavior of each unit is governed by the same Markov process, and in which all units are acted upon independently. Each state or node of the system receives new units as specified by a given time function; the time function could be deterministic or random. The fundamental description of this process is the vector that represents the populations of each state at any time instant. For this reason we call the process a vector Markov process. The quantity of most general interest is the joint probability mass function governing the random behavior of this state population vector as a function of time. However, in many applications the marginal mass functions of each element of the vector will adequately characterize the random process.

We shall now present an approach to the problem of obtaining transient and steady-state population statistics for a vector Markov process, and illustrate its application to a general class of replacement problems.

## Defining Relations

We begin by defining $f_i(m)$ to be the number of units introduced at time $m$ to node $i$ of a Markov process. Initially, we shall prefer the word "node" to "state" in this development because the true state of the system is prescribed by the state population vector rather than by a single state index. The Markov process that acts on each of the units in the system is, of course, completely characterized by its transition probability matrix $P$.

Figure 8.1.1 illustrates a vector Markov process diagrammatically. Here $f_1(\cdot)$ and $f_2(\cdot)$ are time functions describing the introduction of new units to nodes 1 and 2. Typical forms for $f_1(\cdot)$ and $f_2(\cdot)$ are shown in Figure 8.1.2. Both functions can take on only non-negative integral values at integral units of time. The time origin is taken at some arbitrary time 0 for convenience. For example, the function $f_1(\cdot)$ specifies that 3 units are introduced to node 1 at time 0; 2 units are introduced at time 1; 1 at time 2; and 4 at time 3.

As each unit is introduced it is acted upon by the system in a manner described by the transition probabilities. We assume that each unit is treated independently—that the presence or absence of other units in the system does not affect the transitions of the unit in question. The vector Markov process can be interpreted as our usual lily pond with the feature that new frogs are continually being added to various pads. As the amphibian population grows and each frog is subjected to the whims of the $P$ matrix, we may focus our attention on a particular node (pad) and inquire about the probability that we shall find $k$ units (frogs) in that node

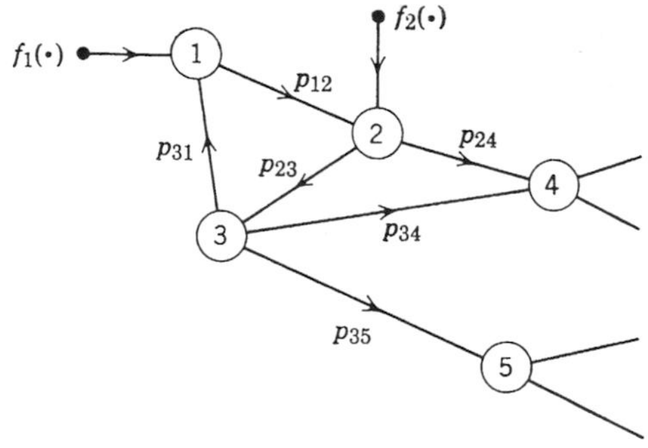


**Figure 8.1.1** A vector Markov process.

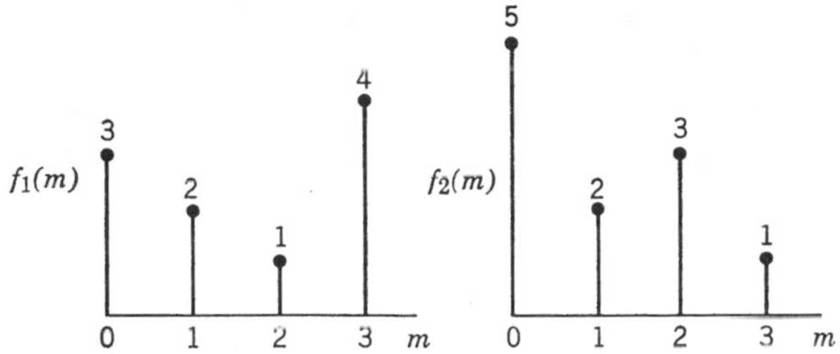


**Figure 8.1.2** Typical input time functions.

after $n$ time periods have passed since the process was started. Let $h_j(k|n)$ be the probability that $k$ units are in node $j$ at time $n$ if the input functions to each node are known.

We find this quantity as follows. Let $h_{ij}(k|m, n)$ be the probability that $k$ units are in node $j$ at time $n$ if $f_i(m)$ units are introduced to node $i$ at some specific time $m \leq n$ and if there is no other input to the system. Each of the $f_i(m)$ units introduced to the system at time $m$ has the multistep transition probability $\phi_{ij}(n - m)$ of reaching node $j$ at time $n$, and hence a probability $1 - \phi_{ij}(n - m)$ of reaching some other node at that time. Since each unit is treated independently, the probability that $k$ units will be situated at node $j$ at time $n$ is just the binomial probability of $k$ successes in $f_i(m)$ Bernoulli trials with probability of success $\phi_{ij}(n - m)$. Therefore, we have

$$h_{ij}(k|m, n) = \binom{f_i(m)}{k}[\phi_{ij}(n - m)]^k[1 - \phi_{ij}(n - m)]^{f_i(m)-k} \quad 0 \leq k \leq f_i(m). \quad (8.1.1)$$

If the input to node $i$ is distributed over time according to a time function $f_i(\cdot)$, but node $i$ remains the only input node in the process, then the number of units at node $j$ at time $n$ will be the number of units that arrive at node $j$ at time $n$ from each input batch summed over all times at which input to the process occurs. Since all inputs to the process are treated independently, the probability $h_{ij}(k|n)$ that there will be $k$ units in node $j$ at time $n$ for a given input time pattern $f_i(\cdot)$ is given by the value at $k$ of the convolution of each $h_{ij}(\cdot|m, n)$ over all input times. Thus we obtain

$$h_{ij}(\cdot|n) = h_{ij}(\cdot|0, n) * h_{ij}(\cdot|1, n) * \cdots * h_{ij}(\cdot|n, n)$$

or

$$h_{ij}(k|n) = \left[\mathop{\circledast}_{m=0}^{n} h_{ij}(\cdot|m, n)\right](k) \qquad 0 \leq k \leq \sum_{m=0}^{n} f_i(m), \qquad (8.1.2)$$

where $\circledast$ represents manifold convolution and the symbol $(k)$ indicates evaluation of the result at the argument $k$.

Finally, if there are several input nodes $i$, each makes an independent contribution to the number of units in $j$ at time $n$. Consequently, to obtain $h_j(k|n)$, the probability that there are $k$ units in node $j$ at time $n$ given the input functions $f_1(\cdot)$, $f_2(\cdot)$, etc., for each node, we must convolve $h_{ij}(\cdot|n)$ over all input nodes. Our result is

$$h_j(k|n) = \left[\mathop{\ast}_{i=1}^{N} h_{ij}(\cdot|n)\right](k), \tag{8.1.3}$$

or finally,

$$h_j(k|n) = \left[\mathop{\ast}_{i=1}^{N} \mathop{\ast}_{m=0}^{n} h_{ij}(\cdot|m, n)\right](k) \qquad 0 \le k \le \sum_{i=1}^{N} \sum_{m=0}^{n} f_i(m). \tag{8.1.4}$$

This equation enables us to compute the probability distribution of the number of units at node $j$ at time $n$ for a general set of node input functions. However, it is important to note that this is only the *marginal* distribution of the number of units in node $j$ at time $n$, or the marginal distribution of the elements of the state population vector. If we are interested in the joint probability distribution of the elements of the vector at time $n$, then we must consider each new input unit to generate a multinomial rather than a binomial process. The various kinds of success for the multinomial correspond to the different states in which the input unit may be found at time $n$. This multinomial would have to be convolved over all input times and nodes to yield the joint probability distribution of the number of units in each node at time $n$. However, for our purposes the marginal distribution derived from the binomial will suffice, and even in this case we shall concentrate on only the simplest questions.

### Population Moments

Let $r_j(n)$ be the number of units in node $j$ of the system at time $n$, the $j$th element of the state population vector $\mathbf{r}(n)$ at time $n$. This random variable has the marginal probability mass function given by Equation 8.1.4. We shall denote its mean and variance by $\bar{r}_j(n)$ and $\check{r}_j(n)$. The quantity $r_j(n)$ is the sum of independent random variables governed by the probability mass function of Equation 8.1.1, where the sum is taken over all values of state $i$ and time $m$ for which input occurs. Consequently, the mean and variance of $r_j(n)$ will be just the mean and variance of $h_{ij}(k|m, n)$ summed over the same values. Since the mean of the binomial expressed by Equation 8.1.1 is $f_i(m)\phi_{ij}(n - m)$ and since its variance is $f_i(m)\phi_{ij}(n - m) \times [1 - \phi_{ij}(n - m)]$, we obtain

$$\bar{r}_j(n) = \sum_{i=1}^{N} \sum_{m=0}^{n} f_i(m)\phi_{ij}(n - m) \qquad \begin{array}{l} j = 1, 2, \ldots, N \\ n = 0, 1, 2, \ldots \end{array} \tag{8.1.5}$$

and

$$\overset{\vee}{r}_j(n) = \sum_{i=1}^{N} \sum_{m=0}^{n} f_i(m)\phi_{ij}(n-m)[1-\phi_{ij}(n-m)] \qquad \begin{array}{l} j = 1, 2, \ldots, N \\ n = 0, 1, 2, \ldots . \end{array} \tag{8.1.6}$$

Thus the mean and variance of the number of units in each node at each time can be calculated from the multistep transition probabilities for the process and the various input time functions.

We find it useful to define a variance impulse response $v_{ij}(n)$ by

$$v_{ij}(n) = \phi_{ij}(n)[1-\phi_{ij}(n)] \qquad \begin{array}{l} i = 1, 2, \ldots, N \\ j = 1, 2, \ldots, N \\ n = 0, 1, 2, \ldots . \end{array} \tag{8.1.7}$$

This is just the variance of the variable $x_{ij}(n)$ defined in Equation 4.1.26. Then we can write Equation 8.1.6 in the form

$$\overset{\vee}{r}_j(n) = \sum_{i=1}^{N} \sum_{m=0}^{n} f_{ij}(m)v_{ij}(n-m) \qquad \begin{array}{l} j = 1, 2, \ldots, N \\ n = 0, 1, 2, \ldots . \end{array} \tag{8.1.8}$$

***Transform analysis***

We place these expressions in the transform domain in our usual notation by writing Equation 8.1.5 as

$$\bar{r}_j^g(z) = \sum_{i=1}^{N} f_i^g(z)\phi_{ij}^g(z) \qquad j = 1, 2, \ldots, N. \tag{8.1.9}$$

In this equation it is natural to regard $f_i^g(z)$ as the transform of an input $i$ that is multiplied by $\phi_{ij}^g(z)$, the transfer function of the system between nodes $i$ and $j$, to obtain a contribution to the transformed output at node $j$, $r_j^g(z)$. The total output at $j$ is the sum of the effects due to the input at each node.

In the transform domain Equation 8.1.8 becomes

$$\overset{\vee}{r}_j^g(z) = \sum_{i=1}^{N} f_i^g(z)v_{ij}^g(z). \tag{8.1.10}$$

In this case we interpret $v_{ij}^g(z)$ as the variance transfer function that must be used to multiply a transformed input at node $i$ to evaluate its contribution to the variance of the transformed output at node $j$.

Equations 8.1.9 and 8.1.10 provide the transform relationships between the population moments, the transformed input patterns, and the ordinary and variance transfer functions of the process. Although these results have theoretical uses, we shall generally find it far easier to work with the convolution expressions for the moments given by Equations 8.1.5 and 8.1.8. We now proceed to use this analysis of vector Markov processes as the background for applications in the theory of replacement processes.

### The Replacement Process

We define the replacement process to be a vector Markov process in which each unit is governed by a renewal process of the type we discussed in Section 7.1. Thus we consider a system composed of identical units. New units are added to the system according to a program of installations. The life of each unit is a random variable with a known probability distribution that we shall call, as before, the lifetime distribution. Each unit lasts at least one period, but when it fails it is immediately replaced by a new unit. The total demand for new units is then composed of the sum of new installation demand and replacement demand.

Let us define the following quantities:

$$f(n) = \text{number of new units installed at time } n$$

$$r(n) = \text{number of units in system that fail at time } n$$

$$p(j) = \mathscr{P}\{\text{a new unit lasts exactly } j \text{ periods}\}(\text{the lifetime distribution})$$

$$^{>}p(k) = \mathscr{P}\{\text{a new unit lasts more than } k \text{ periods}\} = \sum_{j=k+1}^{\infty} p(j).$$

Each of these quantities has an associated geometric transform in our usual transform notation.

***Flow graph representation***

We have already analyzed in Section 7.1 the properties of the renewal process. In Figure 7.1.2, we showed how to construct the transient process that corresponds to a given lifetime distribution. We could also have drawn this figure in the form of Figure 8.1.3. The alternate form has the advantage that we can identify the states of the process with the age of a unit, where by age we mean the number of time periods since the unit was last replaced. Thus state 0 represents new units, state 1 represents one-period-old units, state 2, two-period-old units, etc. We have also

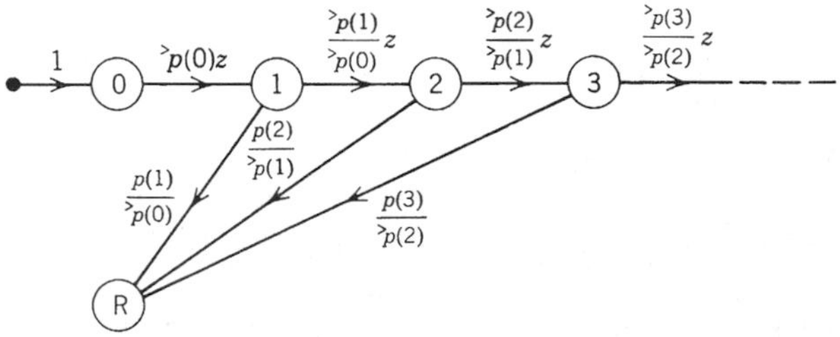


**Figure 8.1.3** A flow graph for the transient replacement process.

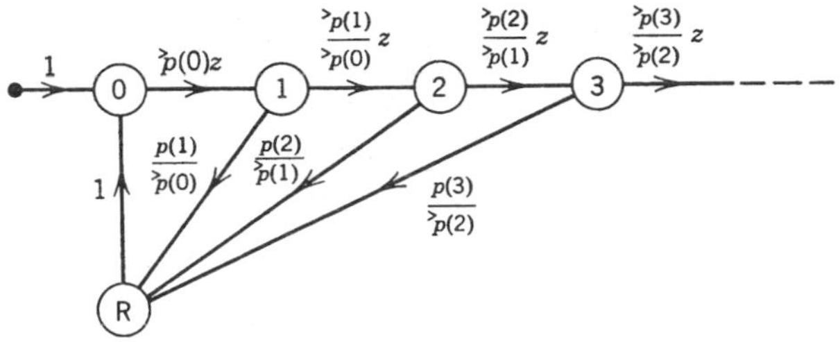


**Figure 8.1.4** A flow graph for the replacement process.

designated a state R that corresponds to units that have just failed and need replacement. Note that the flow graph for the transient process does not assign any delay to transitions to the failed state—the units that fail pass instantly from their age state to the failure state R. The transmission of this flow graph from state 0 to state R, $\vec{t}_{0R}$, is, of course, the transform of the lifetime distribution,

$$\vec{t}_{0R} = p^g(z). \tag{8.1.11}$$

The representations of Figures 7.1.2 and 8.1.3 are therefore externally equivalent.

Since replacements of failed elements are instantaneous, every unit that enters state R immediately moves to state 0. We can show this property in the flow graph by adding a unit transmission branch from state R to state 0; we obtain Figure 8.1.4. The transmissions of this flow graph provide the basic probabilistic information about the replacement process. For example, the transmission from state 0 to state $k$, $k = 0, 1, 2, \ldots$ is the transform of the probability $\phi_{0k}(n)$ that a unit introduced at the time 0 will be represented by a unit of age $k$ at time $n$,

$$\vec{t}_{0k} = \phi_{0k}{}^g(z) \qquad k = 0, 1, 2, \ldots. \tag{8.1.12}$$

Similarly, the transmission from state 0 to state R is the transform of the replacement probability $\phi_{0R}(n)$ that a unit of age 0 at time 0 or one of its replacements will fail at time $n$,

$$\vec{t}_{0R} = \phi_{0R}{}^g(z). \tag{8.1.13}$$

If we are primarily interested in the number of replacements made in the process, we can take advantage of Equation 8.1.11 to draw the flow graph of Figure 8.1.4 in the form of Figure 8.1.5. Then from Equation 8.1.13, we compute $\phi_{0R}{}^g(z)$ as the transmission of this flow graph from state 0 to state R,

$$\phi_{0R}{}^g(z) = \vec{t}_{0R} = \frac{p^g(z)}{1 - p^g(z)}. \tag{8.1.14}$$

Comparing Figure 8.1.5 for the replacement process with Figure 7.1.1 for the basic renewal process shows the slight difference in process definition: the replacement process does not count the original installation of the unit. Thus $\phi_{0R}{}^g(z) = \phi^g(z) - 1$.

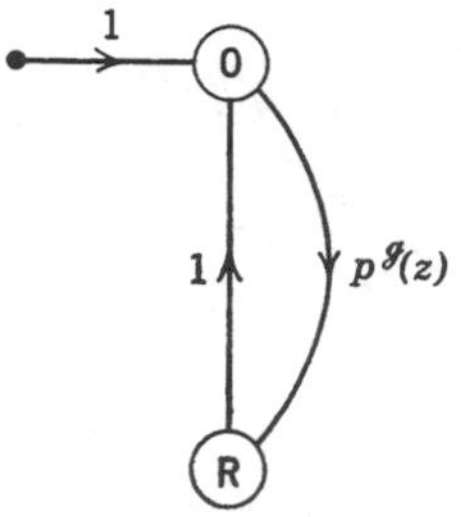


**Figure 8.1.5** A simplified flow graph for the replacement process.

***Replacement moments***

The inverse $\phi_{0R}(n)$ of the transform $\phi_{0R}{}^{\mathscr{g}}(z)$ is the impulse response of the Markovian system between states 0 and R. Therefore, we can compute the transformed mean $\bar{r}^{\mathscr{g}}(z)$ and variance $\check{r}^{\mathscr{g}}(z)$ of the number of replacements at time $n$ for a given installation pattern $f(\cdot)$ by direct application of Equations 8.1.9, 8.1.7, and 8.1.10. We obtain

$$\bar{r}^{\mathscr{g}}(z) = f^{\mathscr{g}}(z)\phi_{0R}{}^{\mathscr{g}}(z) = f^{\mathscr{g}}(z)\frac{p^{\mathscr{g}}(z)}{1 - p^{\mathscr{g}}(z)}, \tag{8.1.15}$$

$$v_{0R}{}^{\mathscr{g}}(z) = \sum_{n=0}^{\infty} \phi_{0R}(n)[1 - \phi_{0R}(n)]z^n, \tag{8.1.16}$$

$$\check{r}^{\mathscr{g}}(z) = f^{\mathscr{g}}(z)v_{0R}{}^{\mathscr{g}}(z). \tag{8.1.17}$$

These equations provide a useful transform technique for computing the mean and variance of the number of replacements as a function of time when the installation pattern and the lifetime distribution are known. However, for large problems the transform operations become prohibitive and we revert to the underlying equations to perform the computations:

$$\phi_{0R}(n) = p(n) + \sum_{m=0}^{n} p(m)\phi_{0R}(n - m) \qquad n \geq 0;\ \phi_{0R}(0) = p(0) = 0 \tag{8.1.18}$$

$$\bar{r}(n) = \sum_{m=0}^{n} f(m)\phi_{0R}(n - m) \qquad n \geq 0 \tag{8.1.19}$$

$$v_{0R} = \phi_{0R}(n)[1 - \phi_{0R}(n)] \qquad n \geq 0 \tag{8.1.20}$$

$$\check{r}(n) = \sum_{m=0}^{n} f(m)v_{0R}(n - m) \qquad n \geq 0. \tag{8.1.21}$$

Equations 8.1.19 and 8.1.21 show that the mean and variance of the number of replacements can be calculated simply by convolution of the installation pattern with the replacement probability $\phi_{OR}(\cdot)$ and the variance impulse response $v_{OR}(\cdot)$, respectively. Thus, all we have to do to find the expected number of replacements as a function of time is to multiply the number of units introduced to the system at each time $m$, that is, $f(m)$, by the replacement probability function delayed by $m$ periods and then sum over all input times. The variance computation is the same except that we delay the variance impulse response rather than the replacement probability function.

### *A replacement example*

Suppose that new units in the system are equally likely to last one or two periods and that three units are installed initially, two units the next period, then one unit, and none thereafter. What are the mean and variance of the number of replacements necessary as a function of time?

**Probability of replacement.** Figure 8.1.6 shows the lifetime distribution, installation pattern, and flow graphs for the example. Note that this lifetime distribution is the same one we investigated in Section 7.1. We find

$$\phi_{OR}{}^{g}(z) = \frac{p^{g}(z)}{1 - p^{g}(z)} = \frac{(1/2)z + (1/2)z^2}{1 - (1/2)z - (1/2)z^2} = \frac{(1/2)z + (1/2)z^2}{(1 - z)[1 + (1/2)z]}$$

$$= \frac{(2/3)z}{1 - z} + \frac{-(1/6)z}{1 + (1/2)z}. \tag{8.1.22}$$

The inverse transform is

$$\phi_{OR}(n) = \begin{cases} 0 & n = 0 \\ (2/3) - (1/6)(-1/2)^{n-1} = (2/3) + (1/3)(-1/2)^n & n \geq 1. \end{cases} \tag{8.1.23}$$

Note that this equation differs from the result of Equation 7.1.25 only when $n = 0$. The difference arises because in defining $\phi_{OR}(n)$ we could not count the original installation as a replacement. We observe that as $n$ increases, $\phi_{OR}(n)$ approaches the value 2/3, the reciprocal of the mean lifetime $\bar{l} = 1.5$. In the steady state each unit has a probability 2/3 of being replaced at any time.

**Mean Replacements.** We could compute $\bar{r}^{g}(z)$ from the transform equation 8.1.15,

$$\bar{r}^{g}(z) = f^{g}(z)\phi_{OR}{}^{g}(z)$$

$$= (3 + 2z + z^2)\left[\frac{(2/3)z}{1 - z} + \frac{(-1/6)z}{1 + (1/2)z}\right], \tag{8.1.24}$$

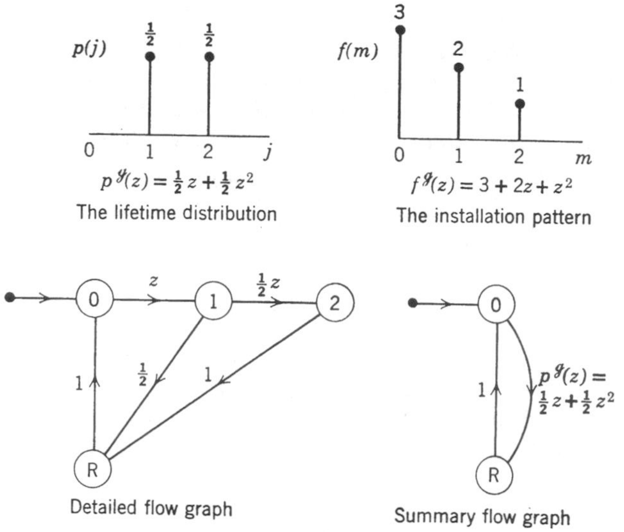


**Figure 8.1.6** A replacement example.

and then invert the transform to obtain $\bar{r}(n)$. However, it is easier to use the result of Equation 8.1.23 in Equation 8.1.19. We find

$$\begin{aligned}\bar{r}(n) &= \sum_{m=0}^{n} f(m)\phi_{\text{OR}}(n-m) \\ &= 3\phi_{\text{OR}}(n) + 2\phi_{\text{OR}}(n-1) + \phi_{\text{OR}}(n-2),\end{aligned} \tag{8.1.25}$$

or

$$\bar{r}(n) = \begin{cases} 0 & n = 0 \\ 3/2 & n = 1 \\ 13/4 & n = 2 \\ 4 + (-1/2)^n & n \geq 3\,. \end{cases} \tag{8.1.26}$$

This equation gives the expected number of units that will be replaced at each time.

Note that $\bar{r}(n)$ approaches the value 4 asymptotically for large $n$. We explain this result by reflecting that in the steady state 6 units will be operating, each with a replacement probability of 2/3. We therefore expect to replace $6 \times 2/3 = 4$ units per period.

**Variance of replacements.** To obtain the variance of the number of replacements, we first calculate

$$\begin{aligned}v_{\text{OR}}(n) &= \phi_{\text{OR}}(n)[1 - \phi_{\text{OR}}(n)] = [(2/3) + (1/3)(-1/2)^n][(1/3) - (1/3)(-1/2)^n] \\ &= (2/9) - (1/9)(-1/2)^n - (1/9)(1/4)^n \qquad n \geq 0.\end{aligned} \tag{8.1.27}$$

We could use Equation 8.1.17 to find the transform of $\overset{v}{r}(n)$ by first transforming $v_{OR}(n)$,

$$v_{OR}{}^{g}(z) = \frac{2/9}{1 - z} + \frac{(-1/9)}{1 + (1/2)z} + \frac{(-1/9)}{1 - (1/4)z}. \qquad (8.1.28)$$

However, once more we prefer to use Equation 8.1.21 and the variance impulse response of Equation 8.1.27. We find

$$\begin{aligned}\overset{v}{r}(n) &= \sum_{m=0}^{n} f(m)v_{OR}(n-m)\\ &= 3v_{OR}(n) + 2v_{OR}(n-1) + v_{OR}(n-2) \qquad n \geq 0,\end{aligned} \qquad (8.1.29)$$

or

$$\overset{v}{r}(n) = \begin{cases} 0 & n = 0 \\ 3/4 & n = 1 \\ (4/3) - (1/3)(-1/2)^n - 3(1/4)^n & n \geq 2\,. \end{cases} \qquad (8.1.30)$$

We observe from this equation that the variance of the number of replacements approaches a value of 4/3 as $n$ increases. Equation 8.1.21 shows that when $n$ is large and $f(m)$ becomes zero after a few periods, the limiting variance $\overset{v}{r}(\infty)$ is just the total number installed times the limiting variance impulse response, $v_{OR}(\infty)$. Since $v_{OR}(\infty)$ is given by

$$\begin{aligned}v_{OR}(\infty) &= \phi_{OR}(\infty)[1 - \phi_{OR}(\infty)] = \frac{1}{\bar{l}}\left[1 - \frac{1}{\bar{l}}\right]\\ &= (2/3)(1/3)\\ &= 2/9,\end{aligned} \qquad (8.1.31)$$

the limiting variance is just six times 2/9 or 4/3. Notice that the limiting mean and variance of the number of replacements both depend only on the mean of the lifetime distribution and the total number of installations.

Table 8.1.1 shows the mean, the variance, and the standard deviation of the number of replacements at any time, $\overset{s}{r}(n) = \sqrt{\overset{v}{r}(n)}$. The mean is plotted in Figure 8.1.7, with one standard deviation limits indicated. We see that the system approaches its asymptotic behavior rather quickly. Of course, we must remember that the actual number of replacements at any time must be integral.

### *The population age distribution moments*

Sometimes we are interested not only in the number of units replaced in a replacement process, but also in the age distribution of the units in the system. Let $u_k(n)$ be the number of units of age $k$ in the system at time $n$. Of special interest are

**Table 8.1.1** Moments of the Number of Replacements at Any Time

| $n =$ | 0 | 1 | 2 | 3 | 4 | 5 | $\infty$ |
|---|---|---|---|---|---|---|---|
| $\bar{r}(n) =$ | 0 | 3/2 | 13/4 | 31/8 | 65/16 | 127/32 | 4 |
| | 0 | 1.500 | 3.250 | 3.875 | 4.063 | 3.969 | 4.000 |
| $\overset{v}{r}(n) =$ | 0 | 3/4 | 17/16 | 85/64 | 333/256 | 1373/1024 | 4/3 |
| | 0 | 0.750 | 1.063 | 1.328 | 1.301 | 1.341 | 1.333 |
| $\overset{s}{r}(n) =$ | 0 | $1/2\sqrt{3}$ | $1/4\sqrt{17}$ | $1/8\sqrt{85}$ | $1/16\sqrt{333}$ | $1/32\sqrt{1373}$ | $2/3\sqrt{3}$ |
| | 0 | 0.866 | 1.031 | 1.153 | 1.141 | 1.158 | 1.155 |

the mean and variance of this random variable, $\bar{u}_k(n)$ and $\overset{v}{u}_k(n)$, with geometric transforms $\bar{u}_k{}^g(z)$ and $\overset{v}{u}_k{}^g(z)$. Fortunately, these transforms may be obtained from a simple application of our earlier results.

The quantity $\phi_{0k}(n)$ is the probability that a unit introduced at time zero or one of its replacements will be of age $k$ at time $n$. The unit may have been replaced many times, but its age is the time since its last replacement. Since all units last at least one time period, if we observe the process just before the replacements occur the ages we shall see are the integers 1, 2, 3, . . . . Equation 8.1.12 shows that we can find the transform of $\phi_{0k}(n)$ as the transmission of the flow graph of Figure 8.1.4 from node 0 to node $k$,

$$\phi_{0k}{}^g(z) = \vec{t}_{0k} = \frac{{}^{>}p(k-1)z^k}{1 - p^g(z)} \qquad k = 1, 2, 3, \ldots . \tag{8.1.32}$$

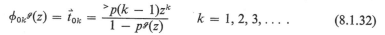

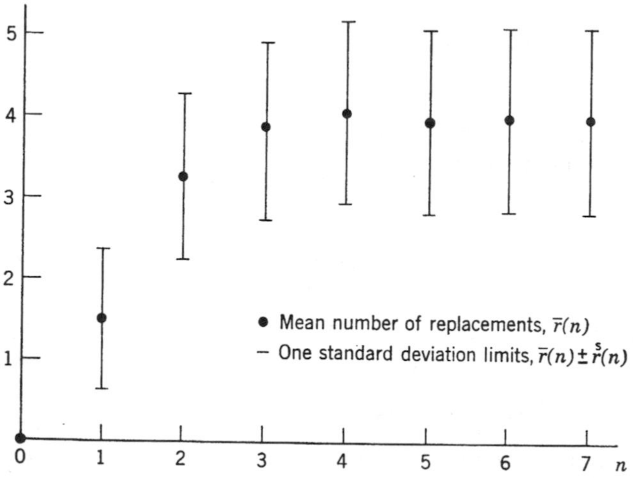


**Figure 8.1.7** Moments of the number of replacements at time $n$.

We now use Equations 8.1.9, 8.1.7, and 8.1.10 to write

$$\bar{u}_k{}^g(z) = f^g(z)\phi_{0k}{}^g(z) \qquad k = 1, 2, 3, \ldots \tag{8.1.33}$$

$$v_{0k}{}^g(z) = \sum_{n=0}^{\infty} \phi_{0k}(n)[1 - \phi_{0k}(n)]z^n \qquad k = 1, 2, 3, \ldots \tag{8.1.34}$$

$$\check{u}_k{}^g(z) = f^g(z)v_k{}^g(z) \qquad k = 1, 2, 3, \ldots. \tag{8.1.35}$$

We thus have a method for finding the transforms of the age distribution moments directly from the transforms of the installation pattern and the lifetime distribution. For most computational purposes we want these equations in their explicit forms,

$$\phi_{0k}(n) = {}^{>}p(k-1)\,\delta(n-k) + \sum_{m=0}^{n} p(m)\phi_{0k}(n-m) \qquad n \geq 0, k = 1, 2, 3, \ldots \tag{8.1.36}$$

$$\bar{u}_k(n) = \sum_{m=0}^{n} f(m)\phi_{0k}(n-m) \qquad n \geq 0, k = 1, 2, 3, \ldots \tag{8.1.37}$$

$$v_{0k}(n) = \phi_{0k}(n)[1 - \phi_{0k}(n)] \qquad n \geq 0, k = 1, 2, 3, \ldots \tag{8.1.38}$$

$$\check{u}_k(n) = \sum_{m=0}^{n} f(m)v_{0k}(n-m) \qquad n \geq 0, k = 1, 2, 3, \ldots. \tag{8.1.39}$$

Thus the calculation of the mean and variance of the number of units of each age in the population is exactly analogous to the calculation of replacement statistics.

***Age distribution moments for replacement example***

For the replacement example of Figure 8.1.6, all units in the system are either one or two years old. How does the expected number of each age change with time? Using the results of the previous section, we first find either from Equation 8.1.32 or by inspection of the detailed flow graph in Figure 8.1.6 that

$$\begin{aligned} \phi_{01}{}^g(z) &= \frac{z}{1 - (1/2)z - (1/2)z^2} = \frac{2/3}{1 - z} + \frac{-(2/3)}{1 + (1/2)z} \\ \phi_{02}{}^g(z) &= \frac{(1/2)z^2}{1 - (1/2)z - (1/2)z^2} = \frac{1/3z}{1 - z} + \frac{-(1/3)z}{1 + (1/2)z}. \end{aligned} \tag{8.1.40}$$

Inversion of the transforms then produces

$$\begin{aligned} \phi_{01}(n) &= (2/3) - (2/3)(-1/2)^n \qquad n \geq 0 \\ \phi_{02}(n) &= \begin{cases} 0 & n = 0 \\ (1/3) - (1/3)(-1/2)^{n-1} & n \geq 1. \end{cases} \end{aligned} \tag{8.1.41}$$

Notice that $\phi_{01}(n)$ and $\phi_{02}(n)$ sum to one for $n \geq 1$.

**Means.** Then the expected number of units of each age are computed as

$$\begin{aligned} \bar{u}_1(n) &= \sum_{m=0}^{n} f(m)\phi_{01}(n-m) = 3\phi_{01}(n) + 2\phi_{01}(n-1) + \phi_{01}(n-2) \\ \bar{u}_2(n) &= \sum_{m=0}^{n} f(m)\phi_{02}(n-m) = 3\phi_{02}(n) + 2\phi_{02}(n-1) + \phi_{02}(n-2), \end{aligned} \tag{8.1.42}$$

or

$$\begin{aligned} \bar{u}_1(n) &= \left\{\begin{matrix} 0 & n = 0 \\ 3 & n = 1 \\ 4 + (-1/2)^{n-1} & n \geq 2 \end{matrix}\right\} \\ \bar{u}_2(n) &= \left\{\begin{matrix} 0 & n = 0, 1 \\ 3/2 & n = 2 \\ 2 - (-1/2)^{n-1} & n \geq 3 \end{matrix}\right\}. \end{aligned} \tag{8.1.43}$$

We have thus found expressions for the expected number of units of each age at any time. We observe that $\bar{u}_1(n)$ plus $\bar{u}_2(n)$ is always equal to the number of units already installed in the system (but not including units installed at the time of observation).

Note that in the steady state we expect to have 4 units of age 1 and 2 units of age 2 in the system. This is a direct result of the fact that the limiting state probability of state 1 is 2/3, ($\phi_{01}(\infty) = 2/3$), while the limiting state probability of state 2 is 1/3, ($\phi_{02}(\infty) = 1/3$). We encountered this property in another form in Equation 7.1.59, where we found for the same lifetime distribution we are using now that if we entered the system in the steady state there was a 2/3 probability of observing the first renewal after one period and a 1/3 probability of observing it after two periods.

**Variances.** The variances of the age distribution follow from Equations 8.1.38 and 8.1.39. First we find the variance impulse responses,

$$\begin{aligned} v_{01}(n) &= \phi_{01}(n)[1 - \phi_{01}(n)] \\ &= (2/9) + (2/9)(-1/2)^n - (4/9)(1/4)^n \qquad n \geq 0, \\ v_{01}(n) &= \phi_{02}(n)[1 - \phi_{02}(n)] \\ &= (2/9) + (2/9)(-1/2)^n - (4/9)(1/4)^n \qquad n \geq 0. \end{aligned} \tag{8.1.44}$$

We note that $v_{01}(n) = v_{02}(n)$ because every unit in the system must either be one or two periods old at the time of observation. Therefore, when we compute the variance of the age distribution from Equation 8.1.39, we obtain:

$$\overset{v}{u}_1(n) = \sum_{m=0}^{n} f(m)v_{01}(n - m) = \overset{v}{u}_2(n) = \sum_{m=0}^{\infty} f(m)v_{02}(n - m) \tag{8.1.45}$$

$$\overset{v}{u}_1(n) = \overset{v}{u}_2(n) = 3v_{01}(n) + 2v_{01}(n - 1) + v_{01}(n - 2) \tag{8.1.46}$$

$$\overset{v}{u}_1(n) = \overset{v}{u}_2(n) = \begin{cases} 0 & n = 0, 1 \\ (4/3) - (1/3)(-1/2)^{n-1} - 3(1/4)^{n-1} & n \geq 2 \end{cases} \tag{8.1.47}$$

The limiting variances of the age distributions are 4/3 by the same type of reasoning that we used in finding the limiting variances of the number of replacements at any time.

Table 8.1.2 shows the mean, variance, and standard deviation of the number of units of each age as a function of time, $\overset{s}{u}_k(n) = \sqrt{\overset{v}{u}_k(n)}$; the means are plotted in Figure 8.1.8, with one standard deviation limits indicated. Once again the system approaches its steady-state behavior rapidly.

**Table 8.1.2** Moments of the Population Age Distributions at Time $n$

| $n=$ | 0 | 1 | 2 | 3 | 4 | 5 | 6 | $\infty$ |
|---|---|---|---|---|---|---|---|---|
| $\bar{u}_1(n)=$ | 0 | 3 | 7/2 | 17/4 | 31/8 | 65/16 | 127/32 | 4 |
| | 0 | 3.000 | 3.500 | 4.250 | 3.875 | 4.063 | 3.969 | 4.000 |
| $\bar{u}_2(n)=$ | 0 | 0 | 3/2 | 7/4 | 17/8 | 31/16 | 65/32 | 2 |
| | 0 | 0 | 1.500 | 1.750 | 2.125 | 1.938 | 2.031 | 2.000 |
| $\overset{v}{u}_1(n)=\overset{v}{u}_2(n)=$ | 0 | 0 | 3/4 | 17/16 | 85/64 | 333/256 | 1373/1024 | 4/3 |
| | 0 | 0 | 0.750 | 1.063 | 1.328 | 1.301 | 1.341 | 1.333 |
| $\overset{s}{u}_1(n)=\overset{s}{u}_2(n)=$ | 0 | 0 | $1/2\sqrt{3}$ | $1/4\sqrt{17}$ | $1/8\sqrt{85}$ | $1/16\sqrt{333}$ | $1/32\sqrt{1373}$ | $2/3\sqrt{3}$ |
| | 0 | 0 | 0.866 | 1.031 | 1.153 | 1.141 | 1.158 | 1.155 |

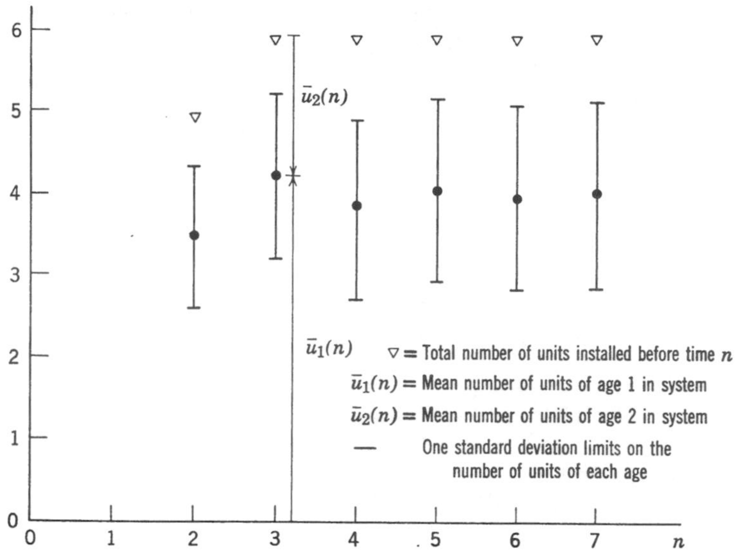


**Figure 8.1.8** Moments of the population age distribution at time $n$.

## Random Input Functions

Up to this point we have assumed that the input functions to the vector Markov process are deterministic. Let us now suppose that the input function to any node $i$ can be governed by any joint probability distribution over the non-negative integral values it takes on at successive time points. Thus, all input functions to the process can be random sequences. We already know how to compute the probability distribution of the state population vector if all input functions are specified. If we are faced with random time functions, what we must do is this: calculate the probability distribution of the state population vector for every conceivable input pattern at all nodes; multiply each distribution by the probability that that pattern will occur as given by the joint probability distribution of input patterns; and then sum the result for all possible input patterns. Such a procedure is complicated beyond our interest in all but the simplest problems.

### *State population vector moments given input patterns*

Therefore, we shall focus our attention on the moments of the state population vector, just as we did before. If we denote by $\mathbf{f}(\cdot)$ the input function at all nodes for all times, then Equations 8.1.5 and 8.1.6 show us how to compute the moments of the state population vector when $\mathbf{f}(\cdot)$ is known. We find for the mean $\langle r_j(n)|\mathbf{f}(\cdot)\rangle$,

$$\langle r_j(n)|\mathbf{f}(\cdot)\rangle = \bar{r}_j(n) = \sum_{i=1}^{N} \sum_{m=0}^{n} f_i(m)\phi_{ij}(n - m), \tag{8.1.48}$$

and for the variance ${}^{v}\langle r_j(n)|\mathbf{f}(\cdot)\rangle$,

$$\begin{aligned} {}^{v}\langle r_j(n)|\mathbf{f}(\cdot)\rangle &= \langle (r_j(n))^2|\mathbf{f}(\cdot)\rangle - \langle r_j(n)|\mathbf{f}(\cdot)\rangle^2 \\ &= \sum_{i=1}^{N} \sum_{m=0}^{n} f_i(m)\phi_{ij}(n - m)[1 - \phi_{ij}(n - m)]. \end{aligned} \tag{8.1.49}$$

***Mean state population vector***

If we now use the joint probability distribution $p(\mathbf{f}(\cdot))$ that governs all input functions, we obtain the mean of the state population vector unconditional on $\mathbf{f}(\cdot)$, i.e., $\langle r_j(n)\rangle$, from

$$\begin{aligned} \langle r_j(n)\rangle &= \sum_{\mathbf{f}(\cdot)} \langle r_j(n)|\mathbf{f}(\cdot)\rangle p(\mathbf{f}(\cdot)) \\ &= \sum_{\mathbf{f}(\cdot)} \sum_{i=1}^{N} \sum_{m=0}^{n} f_i(m)\phi_{ij}(n - m)p(\mathbf{f}(\cdot)) \\ &= \sum_{i=1}^{N} \sum_{m=0}^{n} \bar{f}_i(m)\phi_{ij}(n - m). \end{aligned} \tag{8.1.50}$$

Here we have used $\bar{f}_i(m)$ to denote the expected number of units introduced to node $i$ at time $m$. We see by comparing this equation with Equation 8.1.5 that the mean of the state population vector can be obtained by considering the inputs to be deterministic time functions equal to the means of the random time functions.

***Variance of the state population vector***

We cannot use the same procedure directly to find the variance of the state population vector unconditional on $\mathbf{f}(\cdot)$, i.e., ${}^{v}\langle r_j(n)\rangle$. However, we can evaluate this quantity using the conditional and unconditional second moments, $\langle (r_j(n))^2|\mathbf{f}(\cdot)\rangle$ and $\langle (r_j(n))^2\rangle$. We first write

$${}^{v}\langle r_j(n)\rangle = \langle (r_j(n))^2\rangle - \langle r_j(n)\rangle^2. \tag{8.1.51}$$

We know $\langle r_j(n)\rangle$ and we compute $\langle (r_j(n))^2\rangle$ from

$$\langle (r_j(n))^2\rangle = \sum_{\mathbf{f}(\cdot)} \langle (r_j(n))^2|\mathbf{f}(\cdot)\rangle p(\mathbf{f}(\cdot)). \tag{8.1.52}$$

We evaluate $\langle (r_j(n))^2 | \mathbf{f}(\cdot)\rangle$ from Equations 8.1.49 and 8.1.48,

$$\begin{aligned}
\langle (r_j(n))^2 | \mathbf{f}(\cdot)\rangle &= {}^{v}\langle r_j(n) | \mathbf{f}(\cdot)\rangle + \langle r_j(n) | \mathbf{f}(\cdot)\rangle^2 \\
&= \sum_{i=1}^{N} \sum_{m=0}^{n} f_i(m)\phi_{ij}(n-m)[1-\phi_{ij}(n-m)] \\
&\quad + \left[\sum_{i=1}^{N} \sum_{m=0}^{n} f_i(m)\phi_{ij}(n-m)\right]^2 \\
&= \sum_{i=1}^{N} \sum_{m=0}^{n} f_i(m)\phi_{ij}(n-m)[1-\phi_{ij}(n-m)] \\
&\quad + \sum_{i=1}^{N} \sum_{m=0}^{n} f_i(m)\phi_{ij}(n-m) \sum_{k=1}^{N} \sum_{r=0}^{n} f_k(r)\phi_{kj}(n-r) \\
&= \sum_{i=1}^{N} \sum_{m=0}^{n} f_i(m)\phi_{ij}(n-m)[1-\phi_{ij}(n-m)] \\
&\quad + \sum_{i=1}^{N} \sum_{k=1}^{N} \sum_{m=0}^{n} \sum_{r=0}^{n} f_i(m) f_k(r)\phi_{ij}(n-m)\phi_{kj}(n-r). \qquad (8.1.53)
\end{aligned}$$

When we take the expectation of this equation as indicated in Equation 8.1.52, we find

$$\begin{aligned}
\langle (r_j(n))^2 \rangle &= \sum_{i=1}^{N} \sum_{m=0}^{n} \bar{f}_i(m)\phi_{ij}(n-m)[1-\phi_{ij}(n-m)] \\
&\quad + \sum_{i=1}^{N} \sum_{k=1}^{N} \sum_{m=0}^{n} \sum_{r=0}^{n} \overline{f_i(m) f_k(r)}\phi_{ij}(n-m)\phi_{kj}(n-r). \qquad (8.1.54)
\end{aligned}$$

Here we have used $\overline{f_i(m) f_k(r)}$ to represent the expected value of the product of the input to node $i$ at time $m$ and the input to node $k$ at time $r$. Now we substitute this result into Equation 8.1.51 and use Equation 8.1.50 to obtain the final expression for the unconditional variance of the state population vector,

$$\begin{aligned}
{}^{v}\langle r_j(n)\rangle &= \langle (r_j(n))^2\rangle - \langle r_j(n)\rangle^2 \\
&= \sum_{i=1}^{N} \sum_{m=0}^{n} \bar{f}_i(m)\phi_{ij}(n-m)[1-\phi_{ij}(n-m)] \\
&\quad + \sum_{i=1}^{N} \sum_{k=1}^{N} \sum_{m=0}^{n} \sum_{r=0}^{n} \overline{f_i(m) f_k(r)}\phi_{ij}(n-m)\phi_{kj}(n-r) \\
&\quad - \left(\sum_{i=1}^{N} \sum_{m=0}^{n} \bar{f}_i(m)\phi_{ij}(n-m)\right)\left(\sum_{k=1}^{N} \sum_{r=0}^{n} \bar{f}_k(r)\phi_{kj}(n-r)\right), \qquad (8.1.55)
\end{aligned}$$

or

$$
{}^{v}\langle r_j(n)\rangle = \sum_{i=1}^{N}\sum_{m=0}^{n} \bar{f}_i(m)\phi_{ij}(n-m)[1-\phi_{ij}(n-m)]
$$

$$
+\sum_{i=1}^{N}\sum_{k=1}^{N}\sum_{m=0}^{n}\sum_{r=0}^{n} (\overline{f_i(m)f_k(r)} - \bar{f}_i(m)\bar{f}_k(r))\phi_{ij}(n-m)\phi_{kj}(n-r). \tag{8.1.56}
$$

***Covariance***

The quantity $\overline{f_i(m)f_k(r)} - \bar{f}_i(m)\bar{f}_k(r)$ is called the covariance of $f_i(m)$ and $f_k(r)$. Therefore, we have found that when the inputs are random functions, the marginal variance of the state population vector at any time depends only on the multistep transition probabilities, the expectations of the inputs, and the covariance function for all possible times and nodes.

A particularly important case of this result occurs when all inputs to the process have independent values at successive times and from one node to another. In this case,

$$
\overline{f_i(m)f_k(r)} = (1-\delta_{ik}\,\delta_{mr})\bar{f}_i(m)\bar{f}_k(r) + \delta_{ik}\,\delta_{mr}\overline{(f_i(m))^2}, \tag{8.1.57}
$$

where the $\delta$ factors are required to account for the special situation when $i = k$ and $m = r$. Then Equation 8.1.56 reduces to

$$
{}^{v}\langle r_j(n)\rangle = \sum_{i=1}^{N}\sum_{m=0}^{n} \bar{f}_i(m)\phi_{ij}(n-m)[1-\phi_{ij}(n-m)] + \sum_{i=1}^{N} \overset{v}{f}_i(m)[\phi_{ij}(n-m)]^2, \tag{8.1.58}
$$

where we have written $\overset{v}{f}_i(m) = \overline{(f_i(m))^2} - (\bar{f}_i(m))^2$ for the variance of the number of units introduced to node $i$ at time $m$. Note that this equation is consistent with Equation 8.1.6 for the case of deterministic inputs.

To summarize, we can compute the mean of the state population vector for dependent or independent random inputs by replacing the random input functions by their expectations and then solving the problem by the methods we use for deterministic inputs. To find the variance of the state population vector, we must compute the covariance functions for the inputs and then use Equation 8.1.56. When the random inputs are independent, the variance of the state population vector depends on the means and variances of the random input functions as shown in Equation 8.1.58.

The implication of these observations for the replacement example is that if the installation pattern we used in the example is the *expected value* of an independent random pattern, then all our results for the mean of the number of replacements at any time are unchanged. Furthermore, our results for the means of the population age distribution are exactly the same as the ones we computed for the deterministic pattern.

### Cumulative Population Moments

On some occasions we may want to examine not the state population vector at a particular time, but the sums of the state population vectors over some interval. We shall consider the problem for the case of deterministic inputs, and restrict our attention to the moments. Let ${}^{c}r_j(n)$ be the total number of units that have occupied node $j$ in the period 0 through $n$,

$$ {}^{c}r_j(n) = \sum_{s=0}^{n} r_j(s). \tag{8.1.59}$$

We can find the mean ${}^{c}\bar{r}_j(n)$ and variance ${}^{c}\overset{\vee}{r}_j(n)$ by remembering that each unit introduced into the system is acted upon independently by the Markov process. If a unit is introduced to node $i$ at time $m$, the mean and variance of the number of times it will occupy node $j$ through time $n$, $n \geq m$, are given by the mean $\bar{\nu}_{ij}(n - m)$ and variance $\overset{\vee}{\nu}_{ij}(n - m)$ of the state occupancy statistics developed in Section 5.1. Because of the independence of each unit introduced, we can find the mean and variance of the total number of units that will occupy node $j$ through time $n$ by multiplying $\bar{\nu}_{ij}(n - m)$ and $\overset{\vee}{\nu}_{ij}(n - m)$ by the number of units introduced to node $i$ at time $m$, $f_i(m)$, and summing over all input nodes and times,

$$ {}^{c}\bar{r}_j(n) = \sum_{i=0}^{N} \sum_{m=0}^{n} f_i(m)\bar{\nu}_{ij}(n - m) \tag{8.1.60}$$

and

$$ {}^{c}\overset{\vee}{r}_j(n) = \sum_{i=1}^{N} \sum_{m=0}^{n} f_i(m)\overset{\vee}{\nu}_{ij}(n - m). \tag{8.1.61}$$

Thus the cumulative population moments follow directly from the state occupancy statistics.

#### *The replacement problem*

We can apply these results to the replacement problem by noting that the occupancy statistics for renewal processes were examined at length in Section 7.2. In particular, we developed expressions for $\bar{k}(n)$ and $\overset{\vee}{k}(n)$, the mean and variance of the number of renewals in the interval (0, $n$). We can therefore write the statistics for the cumulative number of replacements ${}^c r(n)$ in the replacement problem as

$$ {}^c\bar{r}(n) = \sum_{m=0}^{n} f(m)\bar{k}(n-m) \tag{8.1.62} $$

and

$$ {}^c\check{r}(n) = \sum_{m=0}^{n} f(m)\check{k}(n-m). \tag{8.1.63} $$

Table 7.2.1 shows some values of $\bar{k}(n)$ and $\check{k}(n)$ for the same lifetime distribution we are using in this section. By following Equations 8.1.62 and 8.1.63, we can use the values of Table 7.2.1 to produce values of ${}^c\bar{r}(n)$ and ${}^c\check{r}(n)$ that appear in Table 8.1.3.

***Alternate expression for the mean***

By taking the expectation of Equation 8.1.59, we write an alternate expression for the mean cumulative number of replacements. In general,

$$ {}^c\bar{r}_j(n) = \sum_{s=0}^{n} \bar{r}_j(s), \tag{8.1.64} $$

which is easily shown to be consistent with Equation 8.1.60. For the replacement problem,

$$ {}^c\bar{r}(n) = \sum_{s=0}^{n} \bar{r}(s). \tag{8.1.65} $$

Therefore, we can find the mean cumulative number of replacements in an interval by summing the expected number of replacements at each time within the interval. We check this result by applying Equation 8.1.65 to the results of Table 8.1.1 and producing the values for ${}^c\bar{r}(n)$ already found in Table 8.1.3.

**Table 8.1.3** Moments of Cumulative Replacements through Time $n$

| $n =$ | 0 | 1 | 2 | 3 | 4 | 5 |
|---|---|---|---|---|---|---|
| ${}^c\bar{r}(n) =$ | 0 | 3/2 | 19/4 | 69/8 | 203/16 | 533/32 |
| | 0 | 1.500 | 4.750 | 8.625 | 12.688 | 16.656 |
| ${}^c\check{r}(n) =$ | 0 | 3/4 | 17/16 | 109/64 | 517/256 | 2613/1024 |
| | 0 | 0.750 | 1.063 | 1.703 | 2.020 | 2.552 |

Note, however, that because successive values of $r_j(n)$ are not independent, it is not true in general that

$$ {}^{c}\overset{v}{r}_j(n) = \sum_{s=0}^{n} \overset{v}{r}_j(s), \tag{8.1.66}$$

nor is it true for the replacement example that

$$ {}^{c}\overset{v}{r}(n) = \sum_{s=0}^{n} \overset{v}{r}(s), \tag{8.1.67}$$

as Tables 8.1.1 and 8.1.3 show by direct evaluation.

*Asymptotic cumulative population moments*

We can gain some insight into the asymptotic forms of the cumulative population moments for this example from the results of Section 7.2. Equations 7.2.78 and 7.2.79 show that the mean and variance of the number of renewals in the interval (0, $n$) grow with $n$ at the rates

$$\bar{k}(n) = \frac{1}{\bar{l}}n = (2/3)n \qquad \text{for large } n \tag{8.1.68}$$

$$\overset{v}{k}(n) = \frac{\overset{v}{l}}{\bar{l}^3}n = (2/27)n \qquad \text{for large } n. \tag{8.1.69}$$

Therefore, when all six units have been installed and the system has been allowed to approach the steady state, the cumulative replacement moments should have the asymptotic growth rates,

$${}^{c}\bar{r}(n) = 6(2/3)n = 4n \qquad \text{for large } n \tag{8.1.70}$$

$${}^{c}\overset{v}{r}(n) = 6(2/27)n = (4/9)n \qquad \text{for large } n. \tag{8.1.71}$$

The results of Table 8.1.3 begin to exhibit these asymptotic tendencies.

## 8.2 BRANCHING PROCESSES

In many systems the size of the population in the system changes not as the result of external activity, but because of the action of the population itself. The birth and death processes we studied in Chapter 7 are an important special case of this type of system. However, here we want to study a much more general model, called a branching process. We consider a population composed initially of some number of members. Each member may produce a number 0, 1, 2, . . . of new population members according to a specified probability distribution. In correspondence with the terminology applied to populations of living entities, we call the new members produced by the initial population the "first generation." Each member of the first generation can then produce new members which form the second generation, and so on. We call the members of a succeeding generation produced by the same member of the last generation his "direct descendants"; the probability distribution of direct descendants in each generation therefore specifies the branching process.

Figure 8.2.1 illustrates the first two generations of a branching process. The zeroth generation consisted of one unit. It produced three direct descendants to constitute the first generation. The first of these three produced a single direct descendant for the second generation; the second, two direct descendants; and the third, no direct descendants at all. Therefore, the total size of the second generation was three.

For convenience we shall assume that all members of a population vanish before the next generation arrives. Then the size of each generation is the size of the total population. We could include the case where the members instead live forever and breed at each generation by shifting the probability distribution of direct descendants by one unit toward the higher integers to allow for re-creating the members of the preceding generation.

For simplicity, if not for the pleasure of the entities involved, we assume that all population members reproduce independently.

We can easily think of several applications for this type of process. Family names are passed on by the males in our society. By establishing a probability distribution for the number of male children each male will produce, we ought to

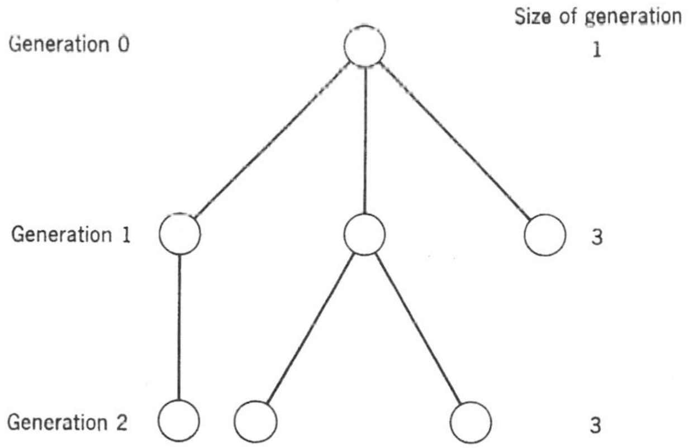


**Figure 8.2.1** A branching process.

be able to use the theory of branching processes to predict the probability distribution for the prevalence of the name in future generations. The same model is applicable to the spread of mutations due to gene changes.

Perhaps the most dramatic application is to nuclear chain reactions. Each neutron generated either will miss all fissionable atoms with some probability or will hit a fissionable atom and produce several neutrons with one minus that probability. These neutrons in turn have the same opportunities for future reproduction. If the probability of missing is low enough, the number of descendants per hit is high enough, and the process continues long enough, man produces the world's largest mushroom.

### The General Branching Process

We shall now make specific our model for the general branching process. By the general branching process we mean one in which the probability distribution of direct descendants may be different for successive generations. We shall let $n$ be the index of generations, $n = 0, 1, 2, \ldots$, and let $k_n$ be the number of direct descendants in the $n$th generation produced by a member of the $n - 1$ generation. We shall use $p_n(\cdot)$ for the probability distribution of $k_n$,

$$p_n(k) = \mathscr{P}\{k_n = k\}. \tag{8.2.1}$$

Our notation for the total size of the $n$th generation will be $m_n$: it is the sum of $m_{n-1}$ independent samples from the probability distribution $p_n(\cdot)$. We shall let $q_n(\cdot)$ be the probability distribution of $m_n$,

$$q_n(m) = \mathscr{P}\{m_n = m\}. \tag{8.2.2}$$

Since $m_n$ is the sum of $m_{n-1}$ independent samples from $p_n(\cdot)$, the distribution of $m_n$ given $m_{n-1}$ must be just the $m_{n-1}$-fold convolution of $p_n(\cdot)$. If we let $q_{n|n-1}(\cdot|\cdot)$ be the probability distribution of $m_n$ given $m_{n-1}$,

$$q_{n|n-1}(s|r) = \mathscr{P}\{m_n = s \,|\, m_{n-1} = r\}, \tag{8.2.3}$$

then we write

$$q_{n|n-1}(\cdot|r) = \underset{r}{\ast}\, p_n(\cdot). \tag{8.2.4}$$

#### *Transform analysis*

If we denote the geometric transforms of $p_n(\cdot)$ and $q_{n|n-1}(\cdot|r)$ by $p_n{}^{\mathscr{g}}(z)$ and $q^{\mathscr{g}}_{n|n-1}(z|r)$,

$$p_n{}^{\mathscr{g}}(z) = \sum_{k=0}^{\infty} p_n(k) z^k \tag{8.2.5}$$

and

$$q^{g}_{n|n-1}(z|r) = \sum_{m=0}^{\infty} q_{n|n-1}(m|r)z^m, \tag{8.2.6}$$

then in the transform domain Equation 8.2.4 becomes

$$q^{g}_{n|n-1}(z|r) = [p_n{}^g(z)]^r. \tag{8.2.7}$$

Of course, we really want the transform of $q_n(\cdot)$ defined by

$$q_n{}^g(z) = \sum_{m=0}^{\infty} q_n(m)z^m. \tag{8.2.8}$$

Therefore, we write

$$\begin{aligned} q_n{}^g(z) &= \sum_{m=0}^{\infty} q_n(m)z^m = \sum_{m=0}^{\infty} z^m \sum_{r=0}^{\infty} q_{n|n-1}(m|r)q_{n-1}(r) \\ &= \sum_{r=0}^{\infty} q_{n-1}(r) \sum_{m=0}^{\infty} q_{n|n-1}(m|r)z^m. \end{aligned} \tag{8.2.9}$$

By using Equations 8.2.6 and 8.2.7, we produce

$$\begin{aligned} q_n{}^g(z) &= \sum_{r=0}^{\infty} q_{n-1}(r)q^{g}_{n|n-1}(z|r) \\ &= \sum_{r=0}^{\infty} q_{n-1}(r)[p_n{}^g(z)]^r. \end{aligned} \tag{8.2.10}$$

The only difference between the right-hand side of Equation 8.2.10 and the transform $q^{g}_{n-1}(z)$ implied by Equation 8.2.8 is that $p_n{}^g(z)$ has been substituted for $z$. Therefore, we obtain

$$q_n{}^g(z) = q^{g}_{n-1}(p_n{}^g(z)) \qquad n = 1, 2, 3, \ldots. \tag{8.2.11}$$

Equation 8.2.11 shows how to relate the transform of the probability distribution of generation size at any generation to the probability distribution of generation size at the preceding generation and to the transform of the probability distribution of direct descendants that governed the process between the two generations. All we need to complete our analysis is the boundary condition that describes the size of the zeroth generation. We shall assume for simplicity that the zeroth generation had exactly one member:

$$q_0(m) = \delta(m-1), \qquad q_0{}^g(z) = z. \tag{8.2.12}$$

By repeated application of Equation 8.2.11 we therefore develop the transforms of successive generation size distributions

$$\begin{aligned}
q_1^{\mathcal{g}}(z) &= p_1^{\mathcal{g}}(z) \\
q_2^{\mathcal{g}}(z) &= q_1^{\mathcal{g}}(p_2^{\mathcal{g}}(z)) = p_1^{\mathcal{g}}(p_2^{\mathcal{g}}(z)) \\
q_3^{\mathcal{g}}(z) &= q_2^{\mathcal{g}}(p_3^{\mathcal{g}}(z)) = p_1^{\mathcal{g}}(p_2^{\mathcal{g}}(p_3^{\mathcal{g}}(z))) \\
q_n^{\mathcal{g}}(z) &= p_1^{\mathcal{g}}(p_2^{\mathcal{g}}(\cdots(p_n^{\mathcal{g}}(z))\cdots)).
\end{aligned} \tag{8.2.13}$$

***An example***

To illustrate the behavior of the general branching process, we shall now consider an example where the probability distribution of direct descendants is uniform with an upper limit that increases with the number of the generation,

$$p_n(k) = \frac{1}{n+1} \qquad k = 0, 1, 2, \ldots, n;\ n = 1, 2, 3, \ldots. \tag{8.2.14}$$

Thus, for the first three generations the probability distributions of direct descendants are

$$\begin{aligned}
p_1(k) &= (1/2)\,\delta(k) + (1/2)\,\delta(k-1) \\
p_2(k) &= (1/3)\,\delta(k) + (1/3)\,\delta(k-1) + (1/3)\,\delta(k-2) \\
p_3(k) &= (1/4)\,\delta(k) + (1/4)\,\delta(k-1) + (1/4)\,\delta(k-2) + (1/4)\,\delta(k-3).
\end{aligned} \tag{8.2.15}$$

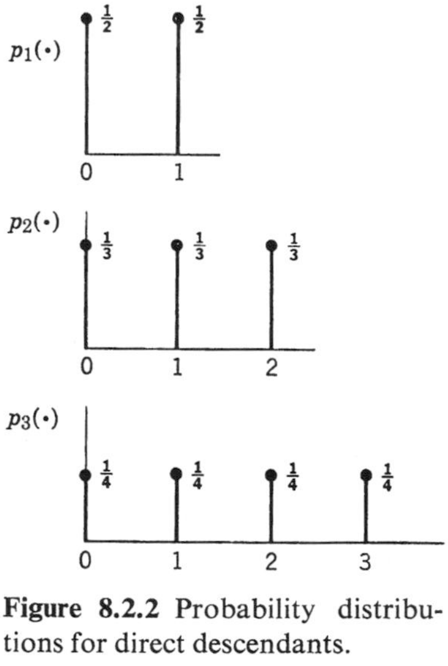


**Figure 8.2.2** Probability distributions for direct descendants.

These distributions are shown in Figure 8.2.2. We see that this example corresponds to a system in which the population members are likely to be more prolific on successive generations.

These first few direct descendant distributions have geometric transforms,

$$\begin{aligned} p_1^g(z) &= (1/2) + (1/2)z \\ p_2^g(z) &= (1/3) + (1/3)z + (1/3)z^2 \\ p_3^g(z) &= (1/4) + (1/4)z + (1/4)z^2 + (1/4)z^3. \end{aligned} \tag{8.2.16}$$

The geometric transform of the general direct descendant distribution is

$$p_n^g(z) = \frac{1}{n+1}\sum_{k=0}^{n} z^k \qquad n = 1, 2, 3, \ldots. \tag{8.2.17}$$

Either by direct calculation or by differentiation of the transform Equation 8.2.17, we find that the mean $\bar{k}_n$ and variance $\check{k}_n$ of the number of direct descendants for the $n$th generation are

$$\bar{k}_n = \sum_{k=0}^{n} k p_n(k) = (1/2)n \qquad n = 1, 2, 3, \ldots \tag{8.2.18}$$

and

$$\check{k}_n = \sum_{k=0}^{n} k^2 p_n(k) - \bar{k}_n^{\,2} = (1/12)n(n+2) \qquad n = 1, 2, 3, \ldots. \tag{8.2.19}$$

We can readily verify these relations for the distributions of Equation 8.2.15. As we expect, both the mean and the variance of the probability distribution of direct descendants increase with $n$: the mean increases linearly, and the variance approximately as the square.

**Probability distributions for size of successive generations.** Now we can proceed to find the probability distributions for the size of the first few generations. From Equations 8.2.13 and 8.2.16, we find

$$\begin{aligned} q_1^g(z) &= p_1^g(z) = 1/2 + (1/2)z \\ q_2^g(z) &= q_1^g(p_2^g(z)) = (1/2) + (1/2)p_2^g(z) = 2/3 + (1/6)z + (1/6)z^2 \\ q_3^g(z) &= q_2^g(p_3^g(z)) = 2/3 + (1/6)p_3^g(z) + (1/6)[p_3^g(z)]^2 \\ &= 23/32 + (1/16)z + (7/96)z^2 + (1/12)z^3 + (1/32)z^4 + (1/48)z^5 + (1/96)z^6. \end{aligned} \tag{8.2.20}$$

By putting $z = 1$ in these results we verify that each of the probability distributions $q_1(\cdot)$, $q_2(\cdot)$, and $q_3(\cdot)$ sums to one. Figure 8.2.3 shows these three probability distributions of generation size as found by direct identification of the transforms from Equation 8.2.20. We see that the range of the $q_n(\cdot)$ distribution is constantly

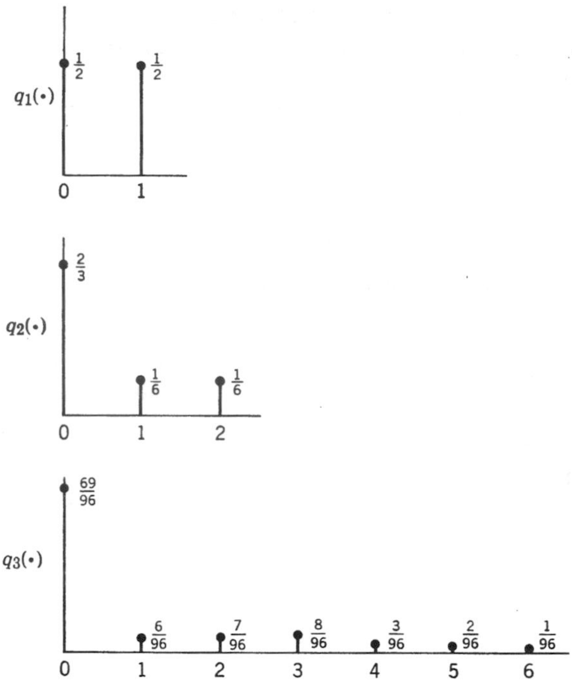


**Figure 8.2.3** Probability distributions of generation size.

increasing; at the $n$th generation it is $n!$. A moment's reflection shows that the probability of achieving this maximum value at the $n$th generation is

$$q_n(n!) = (1/2)(1/3)^1(1/4)^2(1/5)^6 \cdots \left(\frac{1}{n+1}\right)^{(n-1)!} = \prod_{m=1}^{n} \left(\frac{1}{m+1}\right)^{(m-1)!}$$

$$n = 1, 2, 3, \ldots . \tag{8.2.21}$$

**Moments of size of successive generations.** By direct calculation from the distributions of Figure 8.2.3 or by differentiation of the transforms in Equation 8.2.20, we can find the mean $\overline{m}_n$ and variance $\check{m}_n$ of the size of the $n$th generation for the first three generations,

$$\overline{m}_1 = 1/2 \qquad \overline{m}_2 = 1/2 \qquad \overline{m}_3 = 3/4 \tag{8.2.22}$$

and

$$\check{m}_1 = 1/4 \qquad \check{m}_2 = 7/12 \qquad \check{m}_3 = 31/16. \tag{8.2.23}$$

Although it is always possible, at least theoretically, to obtain the moments of the size of each generation from the distribution, we find it more convenient to work with the expressions that give the moments directly.

*Moments of generation size*

**Mean.** By differentiating Equation 8.2.11 with respect to $z$ and evaluating the result at $z = 1$, we can find a recursive expression for $\bar{m}_n$, the mean size of the $n$th generation. We obtain

$$q_n^{\mathcal{g}\prime}(z) = q_{n-1}^{\mathcal{g}\prime}(p_n^{\mathcal{g}}(z)) \cdot p_n^{\mathcal{g}\prime}(z), \tag{8.2.24}$$

then

$$\begin{aligned} q_n^{\mathcal{g}\prime}(1) &= q_{n-1}^{\mathcal{g}\prime}(p_n^{\mathcal{g}}(1)) \cdot p_n^{\mathcal{g}\prime}(1) \\ &= q_{n-1}^{\mathcal{g}\prime}(1) \cdot p_n^{\mathcal{g}\prime}(1). \end{aligned} \tag{8.2.25}$$

Since the derivative of the geometric transform at $z = 1$ is the mean of the corresponding probability distribution, we immediately write

$$\bar{m}_n = \bar{m}_{n-1}\bar{k}_n \qquad n = 1, 2, 3, \ldots. \tag{8.2.26}$$

The mean size of the $n$th generation is the mean size of the $n - 1$st generation multiplied by the mean of the direct descendant distribution that produces the $n$th generation. Since $\bar{m}_0 = 1$, Equation 8.2.26 implies

$$\bar{m}_n = \bar{k}_1 \cdot \bar{k}_2 \cdot \bar{k}_3 \cdots \bar{k}_n \qquad n = 1, 2, 3, \ldots. \tag{8.2.27}$$

The expected size of the $n$th generation is the product of the means of all direct descendant generations through the $n$th generation.

**Mean by probabilistic argument.** We can alternately derive Equation 8.2.26 from basic probabilistic arguments. If we knew that the size of the $n - 1$st generation was $m_{n-1}$, then, because each population member reproduces independently, we would say that the expected value of the size of the $n$th generation was $\bar{k}_n m_{n-1}$. Symbolically, using $\langle x|y\rangle$ for the expected value of $x$ given $y$, we have

$$\begin{aligned} \langle m_n\rangle = \bar{m}_n &= \sum_{m_{n-1}} \langle m_n|m_{n-1}\rangle \mathscr{P}\{m_{n-1}\} = \sum_{m_{n-1}} \bar{k}_m m_{n-1}\mathscr{P}\{m_{n-1}\} \\ &= \bar{k}_n \sum_{m_{n-1}} m_{n-1}\mathscr{P}\{m_{n-1}\} \\ &= \bar{k}_n \bar{m}_{n-1} \qquad n = 1, 2, 3, \ldots, \end{aligned} \tag{8.2.28}$$

which is the result of Equation 8.2.26.

**Variance.** We can proceed to find the variance $\overset{\vee}{m}_n$ by both these methods. Starting again with transforms, we know from Chapter 4 that we can find the second moment of any random variable by adding together the first and second derivatives of the geometric transform of its probability distribution evaluated at $z = 1$. In particular,

$$\overline{m_n^2} = q_n^{g''}(1) + q_n^{g'}(1). \tag{8.2.29}$$

Using Equation 8.2.24, we compute

$$q_n^{g''}(z) = q_{n-1}^{g''}(p_n^g(z))[p_n^{g'}(z)]^2 + q_{n-1}^{g'}(p_n^g(z))p_n^{g''}(z). \tag{8.2.30}$$

Then we evaluate this expression at $z = 1$,

$$\begin{aligned} q_n^{g''}(1) &= q_{n-1}^{g''}(p_n^g(1))[p_n^{g'}(1)]^2 + q_{n-1}^{g'}(p_n^g(1))p_n^{g''}(1) \\ &= q_{n-1}^{g''}(1)\bar{k}_n^{\,2} + q_{n-1}^{g'}(1)[\overline{k_n^2} - \bar{k}_n] \\ &= [\overline{m_{n-1}^2} - \bar{m}_{n-1}]\bar{k}_n^{\,2} + \bar{m}_{n-1}[\overline{k_n^2} - \bar{k}_n]. \end{aligned} \tag{8.2.31}$$

Finally, we can write

$$\check{m}_n = \overline{m_n^2} - \bar{m}_n^{\,2} = [\overline{m_{n-1}^2} - \bar{m}_{n-1}]\bar{k}_n^{\,2} + \bar{m}_{n-1}[\bar{k}_n^{\,2} - \bar{k}_n] + \bar{m}_n - \bar{m}_n^{\,2} \tag{8.2.32}$$

or, by using Equation 8.2.28,

$$\begin{aligned} \check{m}_n &= [\overline{m_{n-1}^2} - \bar{m}_{n-1}]\bar{k}_n^{\,2} + \bar{m}_{n-1}[\overline{k_n^2} - \bar{k}_n] + \bar{k}_n\bar{m}_{n-1} - \bar{k}_n^{\,2}\bar{m}_{n-1}^2 \\ &= [\overline{m_{n-1}^2} - \bar{m}_{n-1}^2]\bar{k}_n^{\,2} + \bar{m}_{n-1}[\overline{k_n^2} - \bar{k}_n^{\,2}] \\ &= \check{m}_{n-1}\bar{k}_n^{\,2} + \bar{m}_{n-1}\check{k}_n \qquad n = 1, 2, 3, \ldots. \end{aligned} \tag{8.2.33}$$

Thus the variance of the size of the $n$th generation is equal to the variance of the size of the $n - 1$st generation multiplied by the square of the mean of the $n$th direct descendant distribution plus the mean of the size of the $n - 1$st generation multiplied by the variance of the $n$th direct descendant distribution.

We can see the general explicit form of $\check{m}_n$ by writing it for the first few generations,

$$\begin{aligned} \check{m}_1 &= \check{k}_1 \\ \check{m}_2 &= \check{m}_1\bar{k}_2^{\,2} + \bar{m}_1\check{k}_2 = \check{k}_1\bar{k}_2^{\,2} + \bar{k}_1\check{k}_2 \\ \check{m}_3 &= \check{m}_2\bar{k}_3^{\,2} + \bar{m}_2\check{k}_3 = \check{k}_1\bar{k}_2^{\,2}\bar{k}_3^{\,2} + \bar{k}_1\check{k}_2\bar{k}_3^{\,2} + \bar{k}_1\bar{k}_2\check{k}_3 \\ \check{m}_4 &= \check{m}_3\bar{k}_4^{\,2} + \bar{m}_3\check{k}_4 = \check{k}_1\bar{k}_2^{\,2}\bar{k}_3^{\,2}\bar{k}_4^{\,2} + \bar{k}_1\check{k}_2\bar{k}_3^{\,2}\bar{k}_4^{\,2} \\ &\quad + \bar{k}_1\bar{k}_2\check{k}_3\bar{k}_4^{\,2} + \bar{k}_1\bar{k}_2\bar{k}_3\check{k}_4. \end{aligned} \tag{8.2.34}$$

Although the general form for $\check{m}_n$ is awkward to put into words or to write as an equation, it is clearly indicated by $\check{m}_4$ in Equation 8.2.34. An interesting observation is that if all direct descendant distributions have a mean equal to one, then the variance of the size of the $n$th generation is $\sum_{m=1}^{n} \check{k}_m$, the sum of the variances of the direct descendant distributions.

**Variance by probabilistic argument.** To find the variance by a probabilistic argument we note that if we knew there were $m_{n-1}$ members in the $n-1$st generation, the variance of the size of the $n$th generation would be $\check{k}_n m_{n-1}$. If we denote the variance of $m_n$ given $m_{n-1}$ by ${}^{v}\langle m_n | m_{n-1}\rangle$, then

$$ {}^{v}\langle m_n | m_{n-1}\rangle = \check{k}_n m_{n-1}. \tag{8.2.35} $$

This result requires only that the members of a generation act independently in producing the next generation. However, to use this result we begin with the second moment of $m_n$ given $m_{n-1}$, $\langle m_n^2 | m_{n-1}\rangle$. We write

$$ \begin{aligned} \langle m_n^2 | m_{n-1}\rangle &= {}^{v}\langle m_n | m_{n-1}\rangle + \langle m_n | m_{n-1}\rangle^2 \\ &= \check{k}_n m_{n-1} + \bar{k}_n^2 m_{n-1}^2. \end{aligned} \tag{8.2.36} $$

Next, we find the unconditional second moment of $m_n^2$, $\langle m_n^2\rangle$,

$$ \begin{aligned} \langle m_n^2\rangle = \overline{m_n^2} &= \sum_{m_{n-1}} \langle m_n^2 | m_{n-1}\rangle \mathscr{P}\{m_{n-1}\} \\ &= \sum_{m_{n-1}} (\check{k}_n m_{n-1} + \bar{k}_n^2 m_{n-1}^2)\mathscr{P}\{m_{n-1}\} \\ &= \check{k}_n \sum_{m_{n-1}} m_{n-1}\mathscr{P}\{m_{n-1}\} + \bar{k}_n^2 \sum_{m_{n-1}} m_{n-1}^2 \mathscr{P}\{m_{n-1}\} \\ &= \check{k}_n \bar{m}_{n-1} + \bar{k}_n^2 \overline{m_{n-1}^2}. \end{aligned} \tag{8.2.37} $$

Finally, we compute the unconditional variance of $m_n$, ${}^{v}\langle m_n\rangle$,

$$ \begin{aligned} {}^{v}\langle m_n\rangle = \check{m}_n &= \overline{m_n^2} - \bar{m}_n^2 \\ &= \check{k}_n \bar{m}_{n-1} + \bar{k}_n^2 \overline{m_{n-1}^2} - \bar{k}_n^2 \bar{m}_{n-1}^2 \\ &= \check{k}_n \bar{m}_{n-1} + \bar{k}_n^2 \check{m}_{n-1}, \end{aligned} \tag{8.2.38} $$

in correspondence with Equation 8.2.33.

**Compound distributions.** Equations 8.2.26 and 8.2.33 are illustrations of what is called the theory of compound distributions. Suppose that a random variable $m_n$ is the sum of $m_{n-1}$ independent samples of a random variable $k_n$, and that $m_{n-1}$ is itself a random variable. Equation 8.2.26 shows that the mean of $m_n$ is the mean of $m_{n-1}$ times the mean of $k_n$. Equation 8.2.33 shows that the variance of $m_n$ is the mean of $m_{n-1}$ times the variance of $k_n$ plus the variance of $m_{n-1}$ times the square of the mean of $k_n$. These results apply to any random variables $m_n$, $m_{n-1}$, and $k_n$; consequently, they have implications beyond branching process theory.

**Moments of generation size in the example.** We can use the results of Equations 8.2.26 and 8.2.33 to find the mean and variance of the size of several successive generations for the problem whose direct descendant distribution appears in Equation 8.2.14. We found the mean and variance of these direct descendant distributions in Equations 8.2.18 and 8.2.19. Table 8.2.1 summarizes the computation. We have also included two columns showing the maximum size of each generation and the standard deviation of generation size, $\overset{s}{m}_n = \sqrt{\overset{v}{m}_n}$. The results of the Table are, of course, consistent with Equations 8.2.22 and 8.2.2.23.

Each generation $n$ has an expected size equal to that of the preceding generation multiplied by the expected number of direct descendants per individual, $\bar{k}_n = (1/2)n$. Furthermore, for this example and for large $n$, the standard deviation of the size of generation $n$ is approximately equal to that of the preceding generation again multiplied by $\bar{k}_n = (1/2)n$. The reason is that when $n$ is large, the term $\overset{v}{m}_{n-1}\bar{k}_n^2$ contributes much more to $\overset{v}{m}_n$ than does the term $\bar{m}_{n-1}\overset{v}{k}_n$.

## The Homogeneous Branching Process

A very important special case of the general branching process occurs when all the direct descendant probability distributions are the same,

$$p_n(k) = p(k) \qquad n = 1, 2, 3, \ldots. \tag{8.2.39}$$

We shall call this special case the homogeneous branching process. We can now specialize our earlier results by expressing them in terms of the common direct descendant distribution $p(\cdot)$ and its geometric transform $p^g(z)$. For example, Equation 8.2.11 for the geometric transform of generation size becomes

$$q_n^g(z) = q_{n-1}^g(p^g(z)) \qquad n = 1, 2, 3, \ldots, \tag{8.2.40}$$

and Equation 8.2.13 is, in general,

$$q_n^g(z) = \overbrace{p^g(p^g(\cdots(p^g(z))\cdots))}^{n} \qquad n = 1, 2, 3, \ldots. \tag{8.2.41}$$

### *Mean generation size*

By introducing the mean $\bar{k}$ of the common direct descendant distribution, we specialize Equation 8.2.26 to

$$\bar{m}_n = \bar{m}_{n-1}\bar{k} \qquad n = 1, 2, 3, \ldots. \tag{8.2.42}$$

**Table 8.2.1** Moments of Generation Size for Example

| | Generation | Mean of direct descendant distribution | Variance of direct descendant distribution | Maximum generation size | Mean of generation size | Variance of generation size | Standard deviation of generation size |
|---|---|---|---|---|---|---|---|
| | $n$ | $\bar{k}_n = (1/2)n$ | $\overset{v}{k}_n = (1/12)n(n+2)$ | $n!$ | $\bar{m}_n = \bar{m}_{n-1}\bar{k}_n$ | $\overset{v}{m}_n = \overset{v}{m}_{n-1}\bar{k}_n^2 + \bar{m}_{n-1}\overset{v}{k}_n$ | $\overset{s}{m}_n = \sqrt{\overset{v}{m}_n}$ |
| Boundary condition | 0 | 1 | 0 | 0 | 1 = 1.00 | 0 = 0 | 0 |
| | 1 | 1/2 = 0.50 | 1/4 = 0.25 | 1 | 1/2 = 0.50 | 1/4 = 0.25 | 0.50 |
| | 2 | 1 = 1.00 | 2/3 = 0.67 | 2 | 1/2 = 0.50 | 7/12 = 0.58 | 0.76 |
| | 3 | 3/2 = 1.50 | 5/4 = 1.25 | 6 | 3/4 = 0.75 | 31/16 = 1.94 | 1.39 |
| | 4 | 2 = 2.00 | 2 = 2.00 | 24 | 3/2 = 1.50 | 37/4 = 9.25 | 3.04 |
| | 5 | 5/2 = 2.50 | 35/12 = 2.92 | 120 | 15/4 = 3.75 | 995/16 = 62.19 | 7.89 |
| | 6 | 3 = 3.00 | 4 = 4.00 | 720 | 45/4 = 11.25 | 9195/16 = 574.69 | 24.0 |
| | 7 | 7/2 = 3.50 | 21/4 = 5.25 | 5040 | 315/8 = 39.38 | 454,335/64 = 7099 | 84.3 |

Then Equation 8.2.27 for the mean size of the $n$th generation becomes simply

$$\bar{m}_n = \bar{k}^n \qquad n = 1, 2, 3, \ldots. \tag{8.2.43}$$

The expected size of the $n$th generation is just the $n$th power of the mean of the common direct descendant distribution.

### *Variance of generation size*

Next, by introducing the variance $\check{k}$ of the common direct descendant distribution we specialize Equation 8.2.33 to

$$\check{m}_n = \check{m}_{n-1}\bar{k}^2 + \bar{m}_{n-1}\check{k} \qquad n = 1, 2, 3, \ldots. \tag{8.2.44}$$

We can also write a general expression for $\check{m}_n$ from the relations of Equation 8.2.34,

$$\begin{aligned} \check{m}_n &= \check{k}(\bar{k}^{n-1} + \bar{k}^n + \cdots + \bar{k}^{2n-3} + \bar{k}^{2n-2}) \\ &= \check{k}\bar{k}^{n-1}(1 + \bar{k} + \cdots + \bar{k}^{n-2} + \bar{k}^{n-1}) \end{aligned} \qquad n = 1, 2, 3, \ldots \tag{8.2.45}$$

or

$$\check{m}_n = \left\{ \begin{matrix} \check{k}\bar{k}^{n-1} \cdot \dfrac{1 - \bar{k}^n}{1 - \bar{k}} & \bar{k} \neq 1 \\ \check{k}n & \bar{k} = 1 \end{matrix} \right\} \qquad n = 1, 2, 3, \ldots. \tag{8.2.46}$$

This equation expresses the variance of the generation size distributions explicitly in terms of the mean and variance of the common direct descendant distribution.

### *Homogeneous branching process examples*

Let us consider concurrently three homogeneous branching processes with common direct descendant distributions ${}^1p(\cdot)$, ${}^2p(\cdot)$, and ${}^3p(\cdot)$ given by

$$\begin{aligned} {}^1p(k) &= (1/2)\,\delta(k) + (1/4)\,\delta(k-1) + (1/4)\,\delta(k-2) \\ {}^2p(k) &= (1/4)\,\delta(k) + (1/2)\,\delta(k-1) + (1/4)\,\delta(k-2) \\ {}^3p(k) &= (1/4)\,\delta(k) + (1/4)\,\delta(k-1) + (1/2)\,\delta(k-2) \qquad 0 \le k. \end{aligned} \tag{8.2.47}$$

These distributions are plotted in Figure 8.2.4. Their geometric transforms are:

$$\begin{aligned} {}^1p^{g}(z) &= 1/2 + (1/4)z + (1/4)z^2 \\ {}^2p^{g}(z) &= 1/4 + (1/2)z + (1/4)z^2 \\ {}^3p^{g}(z) &= 1/4 + (1/4)z + (1/2)z^2 \end{aligned} \tag{8.2.48}$$

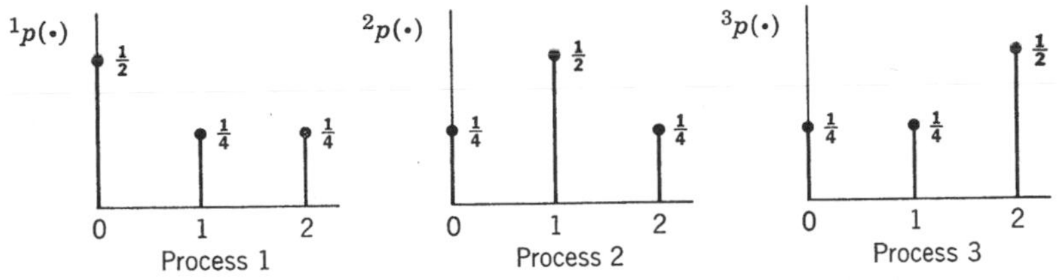


**Figure 8.2.4** Three homogeneous branching processes.

By direct calculation or differentiation of the transform, we find the means and variances of these distributions to be

$$ {}^1\bar{k} = 3/4 \qquad {}^2\bar{k} = 1 \qquad {}^3\bar{k} = 5/4 \tag{8.2.49} $$

$$ {}^1\check{k} = 11/16 \qquad {}^2\check{k} = 1/2 \qquad {}^3\check{k} = 11/16. \tag{8.2.50} $$

**Moments of generation size.** We could use Equation 8.2.41 to develop probability distributions of generation size for the examples, but we shall find it most interesting to compute simply the means and variances of these distributions using Equations 8.2.43 and 8.2.46. We find

$$ {}^1\bar{m}_n = (3/4)^n \qquad {}^2\bar{m}_n = 1 \qquad {}^3\bar{m}_n = (5/4)^n \qquad n = 1, 2, 3, \ldots \tag{8.2.51} $$

and

$$ {}^1\check{m}_n = (11/16)(3/4)^{n-1}\frac{1-(3/4)^n}{1-(3/4)} \qquad {}^2\check{m}_n = (1/2)n \qquad {}^0\check{m}n = (11/16)(5/4)^{n-1}\frac{1-(5/4)^n}{1-(5/4)} \qquad n = 1, 2, 3, \ldots $$

$$ = (11/4)(3/4)^{n-1}[1-(3/4)^n] \qquad\qquad = (11/4)(5/4)^{n-1}[(5/4)^n - 1]. \tag{8.2.52} $$

**Extinction.** We observe that the mean and variance of the size of the $n$th generation for the first process both decrease to zero as $n$ increases. This implies that ultimately the population will become extinct. The mean of the size of the $n$th generation for the second process remains fixed at one, while the variance increases linearly with $n$. This situation is more difficult to interpret, but here again, as we shall see, the process is certain to become extinct. The third process has a mean and variance of the size of the $n$th generation that grow larger and larger as $n$ increases—we expect the third process to grow perpetually.

The behavior of the process therefore depends critically on the mean of the common direct descendant distribution. We shall now explore the nature of this dependence.

## Extinction Probabilities

Often our primary interest in branching processes is whether or not they tend to become extinct as the number of generations increases. This is certainly an important issue in the examples of name transmission and nuclear fission we discussed earlier. The probability that a branching process will be extinct by its $n$th generation is the probability $q_n(0)$ that the $n$th generation has 0 members. By the initial value theorem for geometric transforms, this probability is just the transform of $q_n(\cdot)$ evaluated at $z = 0$,

$$q_n(0) = q_n^{g}(0). \tag{8.2.53}$$

Equations 8.2.11 and 8.2.13 then allow us to write

$$q_n(0) = q_n^{g}(0) = q_{n-1}^{g}(p_n^{g}(0)) = q_{n-1}^{g}(p_n(0)) \qquad n = 1, 2, 3, \ldots, \tag{8.2.54}$$

and

$$q_n(0) = q_n^{g}(0) = p_1^{g}(p_2^{g}(\cdots(p_{n-1}^{g}(p_n(0)))\cdots)). \tag{8.2.55}$$

We can use the transforms of Equations 8.2.20 directly to find the extinction probability for the first few generations of the general branching process example we considered previously. We obtain

$$q_1(0) = q_1^{g}(0) = 1/2 \qquad q_2(0) = q_2^{g}(0) = 2/3 \qquad q_3(0) = q_3^{g}(0) = 23/32. \tag{8.2.56}$$

Since $q_n(0)$ is the probability of extinction <u>by</u> the $n$th generation, the $q_n(0)$ can never decrease with increasing $n$.

### *Extinction in homogeneous branching processes*

Homogeneous branching processes provide a more interesting examination of the extinction question. For the case of a common direct descendant distribution, Equation 8.2.55 becomes

$$q_n(0) = \overbrace{p^{g}(p^{g}(\cdots(p^{g}}^{n-1}(p(0)))\cdots)). \tag{8.2.57}$$

This equation states that the probability of extinction by the $n$th generation is the result of evaluating the function $p^{g}(\cdot)$ recursively $n - 1$ times starting with an initial value of argument equal to the probability of no direct descendant, $p(0)$. We see immediately that the probability of extinction at all generations is zero if $p(0) = 0$, and one if $p(0) = 1$. Therefore, we shall restrict our future considerations to the interesting case $0 < p(0) < 1$.

Equation 8.2.57 suggests that we can compute $q_n(0)$ by plotting $p^{g}(z)$ as a function of $z$ between 0 and 1, reading the ordinate of the function for the abscissa $p(0)$, then substituting this ordinate as the next abscissa, reading a new ordinate and continuing until the substitution has been done $n - 1$ times. The final ordinate is then the probability of extinction by the $n$th generation.

***Extinction in the homogeneous branching process examples***

Figure 8.2.5 shows the plots of the transforms of the direct descendant distributions for the three branching processes of Figure 8.2.4. Note that each transform is convex in $z$ in the interval (0, 1) because each is a power series in $z$ having only positive coefficients. Each transform intercepts the $z = 0$ axis at $p(0)$, since $p^{\mathscr{g}}(0) = p(0)$; obviously, $p(0)$ is a lower bound on the probability of ultimate extinction. All transforms pass through one when $z = 1$ because $p^{\mathscr{g}}(1) = 1$.

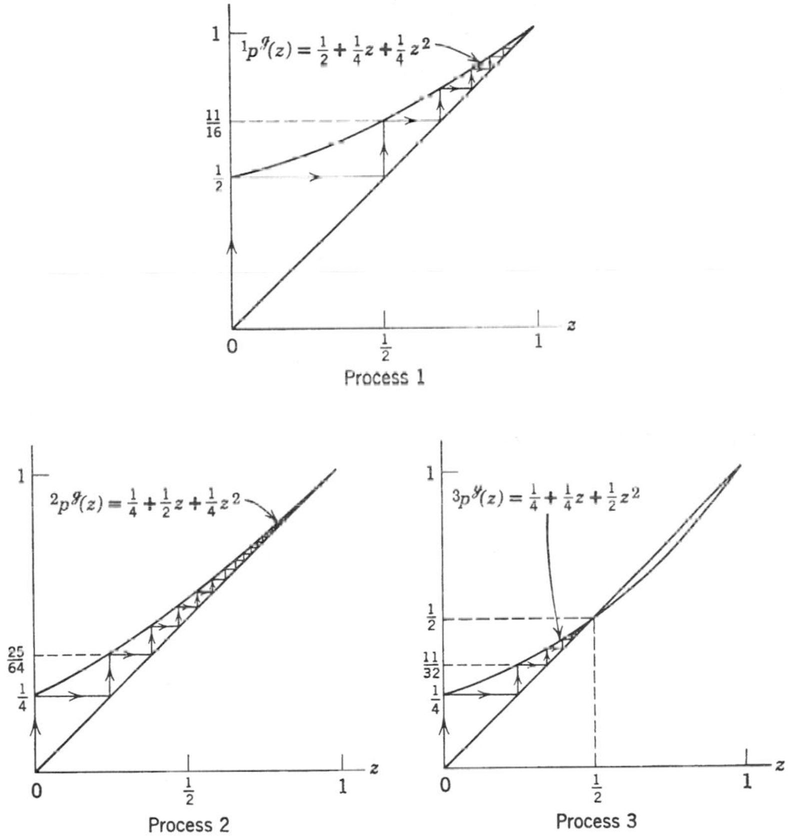


**Figure 8.2.5** Calculation of extinction probabilities for three homogeneous branching processes.

Rather than carry out the procedure of switching ordinates and abscissas that we described, we can, to the same end, simply reflect successive values of the transform off the straight line passing through (0, 0) and (1, 1). The arrows in Figure 8.2.5 show the result of this process. Each time the arrow trajectory touches the curve representing the transform of the direct descendant distribution, the ordinate at that point is the next extinction probability. Thus for the first process we obtain

$$ {}^1q_1(0) = 1/2 = 0.500 \qquad {}^1q_2(0) = 11/16 = 0.688 \qquad {}^1q_3(0) = 809/1024 = 0.790. \tag{8.2.58} $$

Notice that as this process continues the extinction probability ${}^1q_n(0)$ will become larger and larger and ultimately approach 1. The process is certain to become extinct ultimately, as we predicted earlier on the basis of its moments alone.

We see from the figure that the extinction probability for the second process also approaches one as the generation number is increased. Because the two curves $z$ and ${}^2p^g(z)$ approach each other asymptotically, the rate of convergence to one is very slow. Yet, the process is certain to become extinct sooner or later in support of the statement made to this effect when considering the moments of the process.

The third process has a different type of asymptotic behavior. As the number of generations is increased, the extinction probability approaches not one, but the value 1/2. We observe that 1/2 is a root of the equation ${}^3p^g(z) = z$. Thus, even if we allow the process to continue for many generations, the probability that the process will become extinct never rises above 1/2. If the process survives its first few critical generations, it is assured a long life.

***Conditions for a chance at survival***

The question naturally arises, when will the transform of the direct descendant distribution $p^g(z)$ and the line $z$ have a crossing? The answer is when the equation $p^g(z) = z$ has a root less than one. Because of the convexity of $p^g(z)$, there can be at most one crossing or root in the (0, 1) interval. Because of the continuity of $p^g(z)$ and the fact that it passes through $(0, p(0))$ and (1, 1), we see immediately that such a crossing will occur if the derivative of $p^g(z)$ at $z = 1$ is greater than 1. However, the derivative of $p^g(z)$ at $z = 1$ is just the mean of the direct descendant distribution. Therefore, the condition for the existence of an ultimate extinction probability less than one is that the mean number of direct descendants per population member be greater than one. When this condition is met, we can find the probability of ultimate extinction as the root of the equation

$$ p^g(z) = z \tag{8.2.59} $$

that lies between 0 and 1.

To confirm our findings, let us solve Equation 8.2.59 for all three processes. We obtain

$$\begin{array}{ccc}
{}^{1}p^{g}(z) = z & {}^{2}p^{g}(z) = z & {}^{3}p^{g}(z) = z \\
1/2 + (1/4)z + (1/4)z^2 = z & 1/4 + (1/2)z + (1/4)z^2 = z & 1/4 + (1/4)z + (1/2)z^2 = z \\
z^2 - 3z + 2 = 0 & z^2 - 2z + 1 = 0 & 2z^2 - 3z + 1 = 0 \\
z = 1, 2 & z = 1, 1 & z = 1/2, 1.
\end{array} \tag{8.2.60}$$

Thus neither the first nor the second process has any root within the (0, 1) interval, whereas the third process produces the root 1/2 in this interval. Therefore, although the first two processes are certain ultimately to become extinct, the third process has an ultimate extinction probability of only 1/2.

## PROBLEMS

**1.** The general 2-state discrete-time Markov process is subjected to deterministic inputs $f_1(n)$ and $f_2(n)$, where $f_i(n)$ is the number of arrivals at state $i$ immediately after the $n$th transition. Let $r_j(k, n)$ be the number of occupants of state $j$ at the conclusion of the $n$th transition *who have made exactly k transitions from 2 to 2.*

Draw a flow graph whose outputs are $\bar{r}_1^{gg}(y, z)$ and $\bar{r}_2^{gg}(y, z)$.

**2.** The Plutocratic Club has found that the wealth of any one of its members can be modeled by a discrete-time Markov process in which each state corresponds to a certain level of wealth. The time interval for transitions is one year, and all the parameters of the process are known.

The club only allows members to join the club on the first day of each year and furthermore stipulates that all entering members must enter in the same state: State 1. The mean number of members entering at the beginning of year $n$ is $\bar{f}(n)$.

State $m$ corresponds to a wealth of a million dollars and a tradition has grown up in which a member is referred to as a "$k$-year man" if it took him $k$ years to reach state $m$ *for the first time.*

If $\nu(n)$ is the number of "$k$-year men" in state $m$ at the end of year $n$, find an expression for the mean value of $\nu(n)$ in terms of any of the common quantities of the Markov process (e.g., first passage time mass functions, interval transition probabilities, etc.). Assume that the club is started at the beginning of the zeroth year.

**3.** Consider two armed forces, I and II. Assume that whenever there are $m$ and $n$ units on the two sides, respectively, the probability that the next casualty is on the second side is $m/(m + n)$. Let $p(m, n)$ denote the probability that the first side wins if there are $m$ and $n$ units in I and II, respectively. Write the difference equation that governs $p(m, n)$. Can you find a closed form expression for $p(m, n)$? Is the battle a Markov process? Determine $p(4, 3)$. If each casualty takes a unit time, what can be said about the duration of the battle? Illustrate with the (4, 3) case.

**4.** After treatment by irradiation, a certain insect is equally likely to produce zero or three descendants, which continue to reproduce with the same results. What is the ultimate extinction probability of each insect's family-line?

**5.** An experimental study once found that the geometric transform of the number of male offspring for a male in the population was

$$p^{g}(z) = \frac{0.482 - 0.041z}{1 - 0.559z}.$$

Assuming that this transform is the transform of the common direct descendant distribution for a homogeneous branching process, find

a) the mean and variance of the size of the $n$th generation of males.

b) the extinction probability, the probability that a surname unique to one male will ultimately become extinct.

**6.** The Job Shop is set up to perform any combination of $N$ discrete operations on items while they are in the shop. The process of performing operations on each item can be described by a Markov process in which $p_{ij}$ is the probability that an item having the $i$th operation performed on it today will have the $j$th operation tomorrow. Assume that each operation takes exactly one day. For each operation $i$ there is also a probability $p_{i,N+1}$ that the item will be finished at the completion of this operation and will leave the shop. On each day $n$, there is a probability $\{[\lambda(n)]^k/k!\}e^{-\lambda(n)}$, $k = 0, 1, 2, \ldots$, (where $\lambda(n) > 0$ is a parameter particular to day $n$) that there will be $k$ items arriving at the shop. The number of arrivals on any day is independent of the number of arrivals on previous days. There is a probability $p_{0i}$ that the first operation for each item is the $i$th one. Each item is operated upon independently of the others and the initial operation for each item is performed on the day of its arrival. Let $r_i(n)$ = the number of items in the $i$th operation on day $n$. Determine whether or not $r_i(n)$ and $r_j(n)$ are dependent or independent random variables for $i \neq j$.

(You may find it useful to use the result that a necessary and sufficient condition for two random variables $\alpha$ and $\beta$ to be independent is for $\overline{y^{\alpha}z^{\beta}} = \overline{y^{\alpha}}\,\overline{z^{\beta}}$.

**7.** Consider the Job Shop problem above for $N = 3$ and

$$P = \begin{array}{c} \\ 0 \\ 1 \\ 2 \\ 3 \\ 4 \end{array}\begin{array}{c} \begin{array}{ccccc} 0 & 1 & 2 & 3 & 4 \end{array} \\ \begin{bmatrix} 0 & 0.2 & 0.5 & 0.3 & 0 \\ 0 & 0 & 0.5 & 0 & 0.5 \\ 0 & 0.4 & 0 & 0.2 & 0.4 \\ 0 & 0 & 0.5 & 0.2 & 0.3 \\ 0 & 0 & 0 & 0 & 1 \end{bmatrix} \end{array}.$$

The mean number of arrivals per day is the same for all days and equal to 5.

a) If the Job Shop begins operations on day 0, then on day $n$ what will be the mean and variance of $r_1(n)$, the number of items in operation number one?

b) Explain anything unusual about your answers in part a). What is the necessary and sufficient condition on the arrival statistics for this to occur?

c) Evaluate the mean in part a) for large $n$ and compare your answer to $5\bar{\nu}_{\sim 1}$ for the process.

**8.** Consider a transient Markov process subjected to random inputs at each node such that the mean number of arrivals at node $i$ at time $n$, $\bar{f}_i(n)$, is the same for all $n$, i.e., $\bar{f}_i(n) = f_i$. If state $j$ is a transient state, find an expression for the asymptotic mean population at node $j$ in terms of easily calculated quantities of the process. What can you say about the asymptotic variance of the population at node $j$ if $\overset{\vee}{f}_i(n) = \bar{f}_i(m) = f_i$ for all $i$?

**9.** The Old Rumor Mill has found from experience that each person who knows the rumor will pass it on with probability $p$ to someone else during each day. They never pass the rumor on to more than one person during any one day; a person hearing of the rumor begins his discussions on the following day.

a) Calculate the mean and variance of the number of persons knowing a rumor at the end of day $n$ if only one person knows it at the beginning of day 1.

b) Calculate the asymptotic ratio of the mean to the standard deviation.

# 9 TIME-VARYING MARKOV PROCESSES

The Markov processes we have discussed up to this time have all had the same transition probabilities to govern successive transitions. Now we shall expand the Markov model to allow transition probabilities that can change from one transition to the next. Such expanded models are useful in analyzing processes affected by seasonality or growth.

## 9.1 THE CHAPMAN-KOLMOGOROV† EQUATIONS

Let us return to the basic relations that a Markov process must satisfy. Figure 9.1.1 will aid in visualizing the type of system trajectory we shall consider. We shall use the notation of Section 6.2: $S_{nj}$ represents the event that the state of the system at time $n$ is $j$. The probability $\mathscr{P}\{S_{nj}|S_{mi}\}$ that the system will be in state $j$ at time $n$ given that it was in state $i$ at time $m$ must be related to the joint probability $\mathscr{P}\{S_{nj}S_{rk}|S_{mi}\}$ of being in state $j$ at time $n$ and in state $k$ at time $r$ for the same starting conditions by the equation

$$\mathscr{P}\{S_{nj}|S_{mi}\} = \sum_{k=1}^{N} \mathscr{P}\{S_{nj}S_{rk}|S_{mi}\} \qquad m \leq r \leq n. \tag{9.1.1}$$

This equation holds for any process because the states occupied at time $r$ form a mutually exclusive and collectively exhaustive set. By the definition of conditional probability we can then write,

$$\mathscr{P}\{S_{nj}|S_{mi}\} = \sum_{k=1}^{N} \mathscr{P}\{S_{rk}|S_{mi}\}\mathscr{P}\{S_{nj}|S_{rk}S_{mi}\}. \tag{9.1.2}$$

Since the process we are discussing is a Markov process, the probability $\mathscr{P}\{S_{nj}|S_{rk}S_{mi}\}$ that the system will occupy state $j$ at time $n$ given that it occupied

† It has come to my attention that the Chapman-Kolmogorov equation was apparently first written by Louis Bachelier in "Theory of Speculation" *Ann. Sci. Ecole Norm Sup.* (*3*) No. 1018 (Paris, Gauther-Villars, 1900).

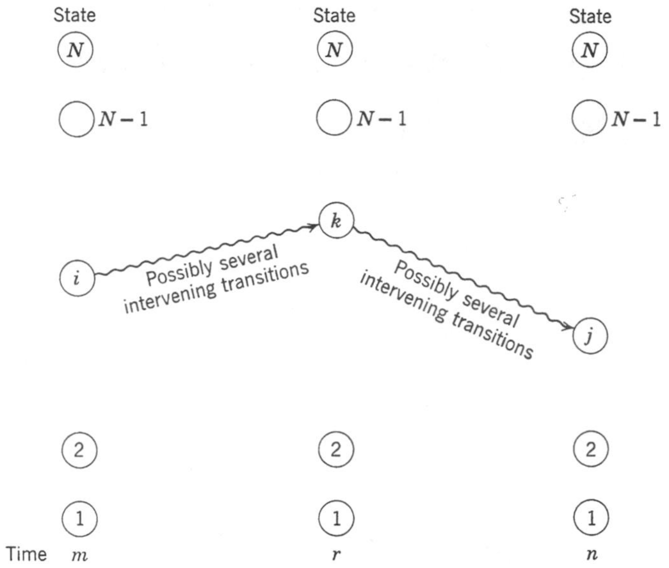


**Figure 9.1.1** A system trajectory.

state $k$ at time $r$ and state $i$ at time $m$ must depend only on the state it occupied most recently, namely, state $k$ at time $r$,

$$\mathscr{P}\{S_{nj}|S_{rk}S_{mi}\} = \mathscr{P}\{S_{nj}|S_{rk}\} \qquad m \le r \le n. \tag{9.1.3}$$

Therefore, Equation 9.1.2 becomes

$$\mathscr{P}\{S_{nj}|S_{mi}\} = \sum_{k=1}^{N} \mathscr{P}\{S_{rk}|S_{mi}\}\mathscr{P}\{S_{nj}|S_{rk}\} \qquad m \le r \le n. \tag{9.1.4}$$

We now define a multistep transition probability from time $m$ to time $n$, $\phi_{ij}(m, n)$, by

$$\phi_{ij}(m, n) = \mathscr{P}\{S_{nj}|S_{mi}\}. \tag{9.1.5}$$

This multistep transition probability contains the starting time explicitly; it is related to the multistep transition probability we have previously used by

$$\phi_{ij}(n) = \phi_{ij}(0, n) = \mathscr{P}\{S_{nj}|S_{0i}\}. \tag{9.1.6}$$

With this notation we place Equation 9.1.4 in the form

$$\phi_{ij}(m, n) = \sum_{k=1}^{N} \phi_{ik}(m, r)\phi_{kj}(r, n) \qquad \begin{array}{l} i = 1, 2, \ldots, N \\ j = 1, 2, \ldots, N \\ m \le r \le n \end{array}. \tag{9.1.7}$$

This equation states that we can decompose any multistep transition probability over an interval into the sums of products of multistep transition probabilities to and from the states that might have been occupied at some time within the interval. We can perform this decomposition for any time within the interval that we choose. Equation 9.1.7 is called the Chapman-Kolmogorov equation for discrete-time Markov processes.

### Matrix Form

If we define $\Phi(m, n)$ to be an $N$ by $N$ multistep transition probability matrix from time $m$ to time $n$ with elements $\phi_{ij}(m, n)$, then we can write the Chapman-Kolmogorov equation in matrix form as

$$\Phi(m, n) = \Phi(m, r)\Phi(r, n) \qquad m \le r \le n. \tag{9.1.8}$$

For consistency we must require,

$$\Phi(n, n) = I. \tag{9.1.9}$$

Nothing we have done thus far implies that the transition probabilities are the same at all transition times. We now define $p_{ij}(n)$ to be the transition probability from state $i$ to state $j$ at time $n$,

$$p_{ij}(n) = \mathscr{P}\{s(n + 1) = j \mid s(n) = i\} = \mathscr{P}\{S_{n+1,j} \mid S_{ni}\} = \phi_{ij}(n, n + 1). \tag{9.1.10}$$

The transition probabilities $p_{ij}(n)$ form the elements of an $N$ by $N$ transition probability matrix $P(n)$ at time $n$,

$$P(n) = \Phi(n, n + 1). \tag{9.1.11}$$

To specify a time-varying Markov process over some interval $(m, n)$, we must therefore know the transition probability matrices $P(m), P(m + 1), \ldots, P(n - 1)$ that govern the process throughout this interval. Therefore, unless these matrices can be expressed analytically, the data requirements of the model are very large.

### The Forward Chapman-Kolmogorov Equations

Since we can choose any value of $r$ in Equation 9.1.8 that lies between $m$ and $n$, inclusive, let us choose $r = n - 1$. The equation then becomes,

$$\Phi(m, n) = \Phi(m, n - 1)\Phi(n - 1, n) \qquad m \le n - 1 \tag{9.1.12}$$

or in view of Equation 9.1.11,

$$\Phi(m, n) = \Phi(m, n - 1)P(n - 1) \qquad m \le n - 1. \tag{9.1.13}$$

Equation 9.1.13 is called the *forward* Chapman-Kolmogorov equation for discrete-time Markov processes because it is written at the forward or front end of the $(m, n)$ interval.

Let us define a difference operator $\Delta_n$ by

$$\Delta_n f(n) = f(n + 1) - f(n). \tag{9.1.14}$$

Then we can write,

$$\Delta_n \Phi(m, n - 1) = \Phi(m, n) - \Phi(m, n - 1). \tag{9.1.15}$$

By using Equation 9.1.13 we obtain

$$\Delta_n \Phi(m, n - 1) = \Phi(m, n - 1)P(n - 1) - \Phi(m, n - 1) = \Phi(m, n - 1)[P(n - 1) - I]. \tag{9.1.16}$$

After defining the differential matrix $D(n)$ by

$$D(n) = P(n) - I, \tag{9.1.17}$$

we place Equation 9.1.16 in the form

$$\Delta_n \Phi(m, n - 1) = \Phi(m, n - 1)D(n - 1) \qquad m \leq n - 1. \tag{9.1.18}$$

This equation is the forward Chapman-Kolmogorov equation for discrete-time Markov processes in difference equation form.

### The Backward Chapman-Kolmogorov Equations

Another choice for $r$ in Equation 9.1.8 is to make $r$ equal to $m + 1$; then

$$\Phi(m, n) = \Phi(m, m + 1)\Phi(m + 1, n) \qquad m \leq n - 1 \tag{9.1.19}$$

or

$$\Phi(m, n) = P(m)\Phi(m + 1, n) \qquad m \leq n - 1. \tag{9.1.20}$$

This equation is called the backward Chapman-Kolmogorov equation for discrete-time Markov processes because it is written at the back end of the $(m, n)$ interval.

The difference operator $\Delta_m$,

$$\Delta_m f(m) = f(m + 1) - f(m), \tag{9.1.21}$$

allows us to write

$$\Delta_m \Phi(m, n) = \Phi(m + 1, n) - \Phi(m, n). \tag{9.1.22}$$

When we substitute the result of Equation 9.1.20 we have

$$\begin{aligned} \Delta_m \Phi(m, n) &= \Phi(m + 1, n) - P(m)\Phi(m + 1, n) \\ &= [I - P(m)]\Phi(m + 1, n) \\ &= -D(m)\Phi(m + 1, n) \qquad m \leq n - 1. \end{aligned} \tag{9.1.23}$$

Equation 9.1.23 is the backward Chapman-Kolmogorov equation in difference equation form. We write the Chapman-Kolmogorov equations in difference equation form primarily because we can use them to develop differential equations for the continuous-time processes we shall consider in Volume II.

### The Common Solution

The forward and backward Chapman-Kolmogorov equations have the same solution, as we would expect. We can find it by writing Equation 9.1.13 for successive values of $n$:

$$\begin{aligned} \Phi(m, m+1) &= \Phi(m, m)P(m) = P(m) \\ \Phi(m, m+2) &= \Phi(m, m+1)P(m+1) = P(m)P(m+1) \\ \Phi(m, m+3) &= \Phi(m, m+2)P(m+2) = P(m)P(m+1)P(m+2) \\ &\vdots \end{aligned} \tag{9.1.24}$$

The general form is

$$\Phi(m, n) = P(m)P(m+1)\cdots P(n-1) \qquad m \le n - 1. \tag{9.1.25}$$

We must always remember in using this result that the order of multiplication of the transition probability matrices is crucial. We see readily that this solution satisfies the backward equation 9.1.20 and also the basic equation 9.1.8.

Of course, if the system is not time-varying, then $P(n) = P$, and Equation 9.1.25 becomes

$$\Phi(m, n) = P^{n-m} \qquad m \le n. \tag{9.1.26}$$

This relation corresponds to the equation for the multistep transition probability matrix $\Phi(n)$ we developed in Chapter 1,

$$\Phi(0, n) = \Phi(n) = P^n. \tag{9.1.27}$$

### State Probabilities

We can use the multistep transition probabilities $\phi_{ij}(m, n)$ to relate the state probabilities of the time-varying process at two times $m$ and $n$. We write

$$\mathscr{P}\{S_{nj}\} = \sum_{i=1}^{N} \mathscr{P}\{S_{mi}\}\mathscr{P}\{S_{nj}|S_{mi}\} \qquad m \le n, \tag{9.1.28}$$

or in our usual notation for state probabilities,

$$\pi_j(n) = \sum^{N} \pi_i(m)\phi_{ij}(m, n) \qquad m \le n. \tag{9.1.29}$$

In matrix form this equation becomes,

$$\boldsymbol{\pi}(n) = \boldsymbol{\pi}(m)\Phi(m, n) \qquad m \leq n. \tag{9.1.30}$$

If we substitute $m = 0$ then we obtain a relationship from Chapter 1,

$$\boldsymbol{\pi}(n) = \boldsymbol{\pi}(0)\Phi(0, n) = \boldsymbol{\pi}(0)\Phi(n). \tag{9.1.31}$$

Equation 9.1.30 also shows

$$\boldsymbol{\pi}(n + 1) = \boldsymbol{\pi}(n)\Phi(n, n + 1) = \boldsymbol{\pi}(n)P(n) \qquad n = 0, 1, 2, \ldots, \tag{9.1.32}$$

which is the generalization of Equation 1.3.8 to the time-varying case.

### Computation and Practice

The computational requirements of the time-varying process are not greatly different from those of the constant transition probability Markov process if we are interested only in the multistep transition probabilities. We simply use the result of Equation 9.1.25 to find the multistep transition probability matrix for any interval by multiplying together, in order, the transition probability matrices for each of the transition times that make up that interval.

However, there is little we can say in general about the behavior of a time-varying Markov process. Since the transition probabilities can change at each transition, there is usually little meaning to the idea of limiting state probabilities. First passage times become difficult to consider because their statistics will be nonstationary—every first passage time calculation would have to contain the time at which observation of the process was begun. We would have to ask questions such as, "What is the probability that $n$ more transitions will be required to reach state $j$ for the first time if the system is now in state $i$ and is about to make its $m$th transition?" The situation is much the same with respect to state occupancy statistics, except that Equation 5.1.3 still applies, as its derivation readily shows. Because of the difficulties in specification and in computation, time-varying Markov models are not often used.

## 9.2 TIME-VARYING MARKOV PROCESS EXAMPLES

### The Periodically-Varying Process

One type of time-varying Markov process where we do not encounter difficult problems in specification is the one with transition probability matrices that vary periodically with successive transitions. The simplest case, and the one we shall study in detail, arises when two different transition probability matrices are used for alternate transitions. Thus, suppose the transition probability matrix $P(n)$ is given by one stochastic matrix ${}^{e}P$ when $n$ is even and by another stochastic matrix ${}^{o}P$ when $n$ is odd,

$$P(n) = \begin{cases} {}^{e}P & n = 0, 2, 4, 6, 8, \ldots \\ {}^{o}P & n = 1, 3, 5, 7, 9, \ldots . \end{cases} \tag{9.2.1}$$

If we now use Equation 9.1.32 to write the state probability vector $\pi(n)$ for successive $n$, we find

$$\begin{aligned} \pi(1) &= \pi(0)P(0) = \pi(0)\,{}^{e}P \\ \pi(2) &= \pi(1)P(1) = \pi(1)\,{}^{o}P = \pi(0)\,{}^{e}P\,{}^{o}P \\ \pi(3) &= \pi(2)P(2) = \pi(2)\,{}^{e}P = \pi(0)({}^{e}P\,{}^{o}P)\,{}^{e}P \\ \pi(4) &= \pi(3)P(3) = \pi(3)\,{}^{o}P = \pi(0)({}^{e}P\,{}^{o}P)({}^{e}P\,{}^{o}P) \\ \pi(5) &= \pi(4)P(4) = \pi(4)\,{}^{e}P = \pi(0)({}^{e}P\,{}^{o}P)({}^{e}P\,{}^{o}P)\,{}^{e}P. \end{aligned} \tag{9.2.2}$$

***State probabilities for n even***

We see from these equations that we can obtain $\pi(n)$ for $n$ even from

$$\pi(n) = \pi(0)({}^{e}P\,{}^{o}P)^{n/2} \qquad n = 0, 2, 4, \ldots . \tag{9.2.3}$$

By defining

$${}^{eo}P = {}^{e}P\,{}^{o}P, \tag{9.2.4}$$

Equation 9.2.3 becomes

$$\pi(n) = \pi(0)\,{}^{eo}P^{n/2} \qquad n = 0, 2, 4, \ldots . \tag{9.2.5}$$

Since the product of two stochastic matrices is stochastic, ${}^{eo}P$ is a stochastic matrix. If ${}^{eo}P$ corresponds to a monodesmic Markov process, then as it is raised to higher and higher powers, it converges to a stochastic matrix with all rows equal. Consequently $\pi(n)$ in Equation 9.2.5 tends to a limit ${}^{e}\pi$ when $n$ is large and even. The vector ${}^{e}\pi$ must satisfy the equations

$${}^{e}\pi = {}^{e}\pi\,{}^{eo}P = {}^{e}\pi\,{}^{e}P\,{}^{o}P \tag{9.2.6}$$

and

$$\sum_{i=1}^{N} {}^{e}\pi_i = 1. \tag{9.2.7}$$

***State probabilities for n odd***

Now that we have established a limiting behavior for even-transition state probabilities, we can show that the state probabilities on odd transitions also have a limiting behavior. We note from Equation 9.2.2 that $\pi(n)$ for $n$ odd is given by

$$\pi(n) = \pi(n-1)\,{}^{e}P \qquad n = 1, 3, 5, \ldots . \tag{9.2.8}$$

Thus the state probabilities for odd transitions are equal to the state probabilities at the preceding even transitions multiplied by the stochastic matrix ${}^{e}P$. Therefore, if the even-transition state probability vector converges to a limit ${}^{e}\pi$, the odd-transition state probability vector will also converge to a limit ${}^{o}\pi$. The vector ${}^{o}\pi$ will be related to the vector ${}^{e}\pi$ by

$$ {}^{o}\pi = {}^{e}\pi \, {}^{e}P. \tag{9.2.9} $$

***Summary of limiting state probability behavior***

Thus for a monodesmic process the state probabilities at even and odd transitions each approach a unique limit. We can find the limiting state probability vector for even transitions from Equations 9.2.6 and 9.2.7 and then use Equation 9.2.9 to find the limiting state probability vector for odd transitions. Alternatively, we could find the limiting state probability vector for odd transitions from

$$ {}^{o}\pi = {}^{o}\pi \, {}^{oe}P = {}^{o}\pi \, {}^{o}P \, {}^{e}P \tag{9.2.10} $$

and

$$ \sum_{i=1}^{N} {}^{o}\pi_i = 1, \tag{9.2.11} $$

by analogy with Equations 9.2.6 and 9.2.7. Then

$$ {}^{e}\pi = {}^{o}\pi \, {}^{o}P \tag{9.2.12} $$

would produce the limiting state probability vector for even transitions.

***An example***

A numerical example will be helpful in understanding these results. We consider a two-state process with even and odd transition matrices,

$$ {}^{e}P = \begin{bmatrix} 3/4 & 1/4 \\ 1/2 & 1/2 \end{bmatrix} \qquad {}^{o}P = \begin{bmatrix} 2/3 & 1/3 \\ 3/4 & 1/4 \end{bmatrix}. \tag{9.2.13} $$

From Equation 9.2.4 we compute ${}^{eo}P$,

$$ {}^{eo}P = {}^{e}P \, {}^{o}P = \begin{bmatrix} 11/16 & 5/16 \\ 17/24 & 7/24 \end{bmatrix}. \tag{9.2.14} $$

**Limiting state probabilities.** Now from Equations 9.2.6 and 9.2.7, or by inspection for this two-state problem, we find the limiting state probability vector at even transition times to be

$$ {}^{e}\pi = [34/49 \quad 15/49] = [0.694 \quad 0.306]. \tag{9.2.15} $$

Then we use Equation 9.2.9 to produce

$$^o\pi = {^e\pi}\,{^eP} = [33/49 \quad 16/49] = [0.673 \quad 0.327]. \tag{9.2.16}$$

Of course, these values for the $^e\pi$ and $^o\pi$ vectors also satisfy Equations 9.2.10 and 9.2.12.

**Comparison with limiting state probabilities of related processes.** It is interesting to compare the two limiting state probability vectors of Equations 9.2.15 and 9.2.16 with the state probability vectors that would result if the process were governed by $^eP$ or $^oP$ alone. We easily find that the limiting state probability vector for a process with transition matrix $^eP$ is

$$\pi = [2/3 \quad 1/3] = [0.667 \quad 0.333]. \tag{9.2.17}$$

Similarly, for a process governed by $^oP$ we obtain

$$\pi = [9/13 \quad 4/13] = [0.692 \quad 0.308]. \tag{9.2.18}$$

Notice that the probability of being in state 1 at even transition times in the steady state, 0.694, is higher than the probability of being in state 1 in the steady state under either $^eP$ or $^oP$ acting alone. This type of resonance phenomenon is not unusual in Markov processes with periodically changing transition matrices.

### *Cycles*

In the most general case of periodically changing transition probability matrices, the Markov process is governed in succession by transition probability matrices $^1P, {^2P}, \ldots, {^kP}$—and then the cycle is repeated over and over in that order. This process is readily interpreted as a $kN$-state time-invariant Markov process partioned into $k$ periodic superstates linked by the successive transition probability matrices.

To solve for the limiting state probabilities in such a system, we first form a composite stochastic matrix $^{1,2,\ldots,k}P$ according to

$$^{1,2,\ldots,k}P = {^1P}\,{^2P}\,{^3P}\cdots{^kP}. \tag{9.2.19}$$

Then we find the limiting state probability vector $^0\pi$ that holds at the end of each cycle of $k$ transitions from

$$^0\pi = {^0\pi}\,{^{1,2,\ldots,k}P} \tag{9.2.20}$$

and

$$\sum_{i=1}^{N} {^0\pi_i} = 1. \tag{9.2.21}$$

Next we find the limiting state probability vectors ${}^1\boldsymbol{\pi}, {}^2\boldsymbol{\pi}, \ldots, {}^{k-1}\boldsymbol{\pi}$ from ${}^0\boldsymbol{\pi}$ by using the equations,

$$\begin{aligned} {}^1\boldsymbol{\pi} &= {}^0\boldsymbol{\pi}\, {}^1P \\ {}^2\boldsymbol{\pi} &= {}^1\boldsymbol{\pi}\, {}^2P \\ {}^3\boldsymbol{\pi} &= {}^2\boldsymbol{\pi}\, {}^3P \\ &\vdots \\ {}^{k-1}\boldsymbol{\pi} &= {}^{k-2}\boldsymbol{\pi}\, {}^{k-1}P. \end{aligned} \tag{9.2.22}$$

The cyclical nature of the process assures that ${}^0\boldsymbol{\pi}$ also satisfies

$${}^0\boldsymbol{\pi} = {}^{k-1}\boldsymbol{\pi}\, {}^kP. \tag{9.2.23}$$

A Markov process with periodically-changing transition probability matrices is an important model for systems whose behavior is periodic with respect to a year. Thus in a hydroelectric problem we might let the head of water in a dam be represented in discrete increments by the state of a Markov process. Since rainfall will vary in the course of a year we might consider either the four seasons or the twelve months as appropriate transition times, each with its own transition probability matrix. Such a model would form the basis for deciding what policy of consuming the head of water as opposed to using auxiliary generating facilities would be most profitable.

## A Search Problem

As a second example of time-varying Markov processes we shall consider a search problem. Suppose that someone is looking for a document in a file in which the document is certain to be present. We assume initially that the searcher has probability $p$, $0 < p < 1$, of finding the document on each search, and that he continues to search until he finds it.

### *Transition diagram*

We can represent the search by the Markov process whose transition diagram appears in Figure 9.2.1. The searcher starts in state 1, has a probability $p$ of passing to state 2 on each transition, and remains in state 2 when he arrives there. Thus state 1 represents the situation where the searcher has not yet found the document, and state 2 represents the situation where he has already found it. The transition probability matrix for the process is

$$P = \begin{bmatrix} 1-p & p \\ 0 & 1 \end{bmatrix}. \tag{9.2.24}$$

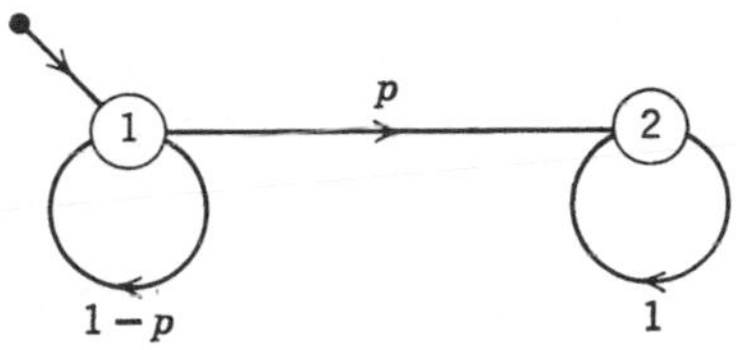


**Figure 9.2.1** Transition diagram for a search problem.

***State probabilities***

Since the initial state probability vector for the process is $\pi(0) = [1 \quad 0]$, we can compute successive state probabilities as:

$$\begin{aligned}
\pi(1) &= \pi(0)P = [1 - p \quad p] \\
\pi(2) &= \pi(1)P = [(1 - p)^2 \quad 1 - (1 - p)^2] \\
\pi(3) &= \pi(2)P = [(1 - p)^3 \quad 1 - (1 - p)^3] \\
&\vdots
\end{aligned} \tag{9.2.25}$$

In general, we see that the probability of not having found the document after $n$ searches is given by

$$\pi_1(n) = (1 - p)^n \qquad n = 0, 1, 2, \ldots . \tag{9.2.26}$$

Of course, we could easily have written this result by fundamental consideration of the process.

***A tiring search***

Now we turn to the process we really want to analyze, one where the searcher tires as the search progresses. In particular, we shall assume that the probability of the searcher finding the document on the search conducted at time $n$ is $p^{n+1}$, $0 < p < 1$. The same process might serve as a model for a student trying to learn, but becoming discouraged.

***Transition diagram***

We can draw the transition diagram of Figure 9.2.2 to represent the problem. The corresponding time-varying transition probability matrix at transition time $n$ is

$$P(n) = \begin{bmatrix} 1 - p^{n+1} & p^{n+1} \\ 0 & 1 \end{bmatrix} \qquad n = 0, 1, 2, \ldots . \tag{9.2.27}$$

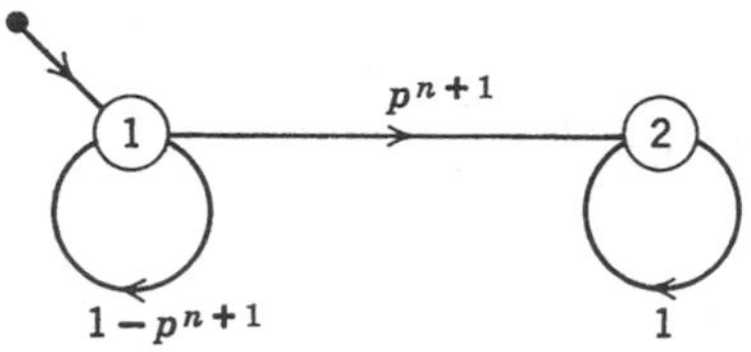


**Figure 9.2.2** Transition diagram for search by a tiring searcher.

The first few transition probability matrices are thus,

$$P(0) = \begin{bmatrix} 1-p & p \\ 0 & 1 \end{bmatrix} \qquad P(1) = \begin{bmatrix} 1-p^2 & p^2 \\ 0 & 1 \end{bmatrix} \qquad P(2) = \begin{bmatrix} 1-p^3 & p^3 \\ 0 & 1 \end{bmatrix}. \tag{9.2.28}$$

***State probabilities***

Since the initial state probability vector is still $\boldsymbol{\pi}(0) = [1 \quad 0]$, we use Equation 9.1.32 to compute successive state probability vectors:

$$\begin{aligned} \boldsymbol{\pi}(1) &= \boldsymbol{\pi}(0)P(0) = [1-p \quad p] \\ \boldsymbol{\pi}(2) &= \boldsymbol{\pi}(1)P(1) = [(1-p)(1-p^2) \quad 1-(1-p)(1-p^2)] \\ \boldsymbol{\pi}(3) &= \boldsymbol{\pi}(2)P(2) = [(1-p)(1-p^2)(1-p^3) \quad 1-(1-p)(1-p^2)(1-p^3)] \\ &\vdots \end{aligned} \tag{9.2.29}$$

The general expression for $\pi_1(n)$, the probability that the document will not have been found after $n$ searches by the tiring searcher, is

$$\pi_1(n) = \prod_{i=1}^{n} (1-p^i) \qquad n = 1, 2, 3, \ldots. \tag{9.2.30}$$

***The effect of tiring***

By comparing the results of Equations 9.2.26 and 9.2.30 we determine the effect of tiring on the probability that the document will not be found after any number of searches. Note that both the tiring and nontiring searchers have the same probability of finding the document on the first search. However, as the searches continue the nontiring searcher will always have a lower probability of the document's remaining unfound than will the tiring searcher, since $(1-p)$ is less than $(1-p^i)$ for $i$ greater than or equal to 2.

If the searches continue indefinitely, the nontiring searcher is sure to find the document, since

$$\pi_1(\infty) = \lim_{n\to\infty} \pi_1(n) = \lim_{n\to\infty} (1-p)^n = 0 \qquad 0 < p < 1. \tag{9.2.31}$$

Is the tiring searcher also assured of success? For him,

$$\pi_1(\infty) = \lim_{n\to\infty} \pi_1(n) = \lim_{n\to\infty} \prod_{i=1}^{n} (1 - p^i) = \prod_{i=1}^{\infty} (1 - p^i) \qquad 0 < p < 1. \qquad (9.2.32)$$

The value of the limit in this equation is not clear. For example, if $p = 1/2$, we have

$$\pi_1(\infty) = (1/2)\cdot(3/4)\cdot(7/8)\cdot(15/16)\ldots. \qquad (9.2.33)$$

The product is composed of successive factors that are less than one, but increasingly close to it.

***Bounds on the probability of futility***

Although the limit in Equation 9.2.32 is difficult to evaluate, we can place bounds upon it. We start with

$$\pi_1(\infty) = \prod_{i=1}^{\infty} (1 - p^i). \qquad (9.2.34)$$

If we take the logarithm of both sides, we obtain

$$\ln \pi_1(\infty) = \sum_{i=1}^{\infty} \ln (1 - p^i). \qquad (9.2.35)$$

We can express the summation in an alternate form by using the following development. We write

$$\frac{1}{1-x} = 1 + x + x^2 + \cdots \qquad |x| < 1 \qquad (9.2.36)$$

and then integrate

$$\int_0^y \frac{1}{1-x}\,dx = \int_0^y (1 + x + x^2 + \cdots)\,dx \qquad |y| < 1 \qquad (9.2.37)$$

to obtain

$$-\ln (1 - y) = y + \frac{y^2}{2} + \frac{y^3}{3} + \cdots = \sum_{j=1}^{\infty} \frac{y^j}{j} \qquad |y| < 1. \qquad (9.2.38)$$

Since $|p^i| < 1$, we can use this result to write

$$-\ln (1 - p^i) = p^i + \frac{p^{2i}}{2} + \frac{p^{3i}}{3} + \cdots. \qquad (9.2.39)$$

When we substitute this expression into Equation 9.2.35, we find

$$
\begin{aligned}
-\ln \pi_1(\infty) &= \sum_{i=1}^{\infty} -\ln(1 - p^i) \\
&= \sum_{i=1}^{\infty}\left(p^i + \frac{p^{2i}}{2} + \frac{p^{3i}}{3} + \cdots\right) \\
&= \frac{p}{1-p} + \frac{p^2}{2(1-p^2)} + \frac{p^3}{3(1-p^3)} + \cdots \\
&= \sum_{j=1}^{\infty} \frac{p^j}{j(1-p^j)} \qquad 0 < p < 1.
\end{aligned} \tag{9.2.40}
$$

From this equation we produce $\pi_1(\infty)$ in the form

$$
\pi_1(\infty) = \exp\left(-\sum_{j=1}^{\infty} \frac{p^j}{j(1-p^j)}\right). \tag{9.2.41}
$$

The series in the exponent is not itself easy to evaluate, but the expression serves to place bounds on $\pi_1(\infty)$.

**An upper bound.** To establish an upper bound on $\pi_1(\infty)$, we truncate the series at some finite number of terms $k$. Since all of the terms in the summation are positive,

$$
\pi_1(\infty) < \exp\left(-\sum_{j=1}^{k} \frac{p^j}{j(1-p^j)}\right). \tag{9.2.42}
$$

If we choose $k = 2$, we obtain

$$
\pi_1(\infty) < \exp\left(-\frac{p}{1-p} - \frac{p^2}{2(1-p^2)}\right) = \exp\left(-\frac{p\left(1 + \frac{3}{2}p\right)}{1-p^2}\right). \tag{9.2.43}
$$

**A lower bound.** To establish a lower bound we recall

$$
1 - p^j > 1 - p \qquad j = 2, 3, 4, \ldots. \tag{9.2.44}
$$

Then if we substitute $1 - p$ for $1 - p^j$ in Equation 9.2.41 we have

$$
\pi_1(\infty) > \exp\left(-\sum_{j=1}^{\infty} \frac{p^j}{j(1-p)}\right) = \exp\left(-\frac{1}{1-p}\sum_{j=1}^{\infty} \frac{p^j}{j}\right). \tag{9.2.45}
$$

**Table 9.2.1** Lower and Upper Bounds on $\pi_1(\infty)$, the Probability that the Document Will Never Be Found by a Tiring Searcher

| $p$ | Lower Bound $(1-p)^{1/(1-p)}$ | Upper Bound $e^{\{-p[1+(3/2)p]/(1-p^2)\}}$ |
|---|---|---|
| 0.1 | 0.890 | 0.890 |
| 0.2 | 0.757 | 0.763 |
| 0.3 | 0.601 | 0.620 |
| 0.4 | 0.427 | 0.466 |
| 0.5 | 0.250 | 0.311 |
| 0.6 | 0.101 | 0.168 |
| 0.7 | 0.018 | 0.060 |
| 0.8 | 0.00032 | 0.0075 |
| 0.9 | $10^{-10}$ | 0.00069 |

By using Equation 9.2.38 we form

$$\pi_1(\infty) > \exp\left(-\frac{1}{1-p}[-\ln(1-p)]\right) \tag{9.2.46}$$

or

$$\pi_1(\infty) > (1-p)^{1/(1-p)}. \tag{9.2.47}$$

**Computational results.** Table 9.2.1 shows the values of the upper bound of Equation 9.2.43 and the lower bound of Equation 9.2.47 for several values of $p$. The lower bound is most interesting because it shows that for any $p$ in the range $0 > p > 1$ there is a finite nonzero probability $\pi_1(\infty)$ that the tiring searcher will never find the document, even in an infinite number of searches. Results like this should be expected when transition probabilities are allowed to change in time.

## 9.3 INVERSE MARKOV PROCESSES

An interesting question that can be asked of a Markov model is: What is the probability that the system was in each state of the process at some time in the past if we know its present state? This is just the inverse to the problem we usually pose, namely, given the present state of the process find the probability that it will occupy each state after a given number of transitions. The inverse problem is really an inference problem, as we shall see. We assign a probability to the system having occupied a certain state at some time in the past and then modify this probability using Bayes' theorem to take into account the information that the present state is known. We consider this inference problem in this chapter because it has an intriguing interpretation in terms of a time-varying Markov process.

**The Inverse Transition Probability Structure**

We want to find the probability $\mathscr{P}\{S_{mi}|S_{nj}\}$ that the system was in state $i$ at time $m$ given that it is in state $j$ at time $n$, $m \leq n$. By using the definition of conditional probability, or Bayes' theorem, we can write this probability as

$$\mathscr{P}\{S_{mi}|S_{nj}\} = \frac{\mathscr{P}\{S_{mi}S_{nj}\}}{\mathscr{P}\{S_{nj}\}} = \frac{\mathscr{P}\{S_{mi}\}\mathscr{P}\{S_{nj}|S_{mi}\}}{\mathscr{P}\{S_{nj}\}} \qquad m \leq n. \tag{9.3.1}$$

The denominator $\mathscr{P}\{S_{nj}\}$ in this equation is just the sum of the numerator over all values of $i$,

$$\mathscr{P}\{S_{nj}\} = \sum_{i=1}^{N} \mathscr{P}\{S_{mi}S_{nj}\} = \sum_{i=1}^{N} \mathscr{P}\{S_{mi}\}\mathscr{P}\{S_{nj}|S_{mi}\}. \tag{9.3.2}$$

We shall represent $\mathscr{P}\{S_{mi}|S_{nj}\}$ by $\phi_{ji}(n, m)$ in accordance with Equation 9.1.5,

$$\phi_{ji}(n, m) = \mathscr{P}\{S_{mi}|S_{nj}\} \qquad m \leq n. \tag{9.3.3}$$

Note, however, that in this equation the second argument $m$ will generally be less than the first argument $n$. This is just opposite in sense to the usage in Equation 9.1.5, but no confusion will result. We are simply empowering ourselves to use the $\phi$ notation to ask questions about the past as well as the future.

We can use our usual notation $\pi_i(n) = \mathscr{P}\{S_{ni}\}$ for state probabilities to write Equations 9.3.1 and 9.3.2 in the forms,

$$\phi_{ji}(n, m) = \frac{\pi_i(m)\phi_{ij}(m, n)}{\pi_j(n)} \qquad m \leq n \tag{9.3.4}$$

and

$$\pi_j(n) = \sum_{i=1}^{N} \pi_i(m)\phi_{ij}(m, n) \qquad m \leq n. \tag{9.3.5}$$

When interpreted, these equations show how to solve our problem. We assume that in the absence of any other information we have assigned a prior probability $\pi_i(m)$ to the event that the system was in state $i$ at time $m$. Then we find that at some later time $n$ the system is for a fact in state $j$. This information causes us to change our assignment of the probability that the system was in state $i$ at time $m$ from $\pi_i(m)$ to $\phi_{ji}(n, m)$.

Equations 9.3.4 and 9.3.5 show exactly how to make the change. Since the Markov process we wish to study may be time-varying, we first compute $\phi_{ij}(m, n)$ for the process using the methods of Section 9.1. Then we compute $\pi_j(n)$ from $\pi_i(m)$ using Equation 9.3.5. Finally we introduce all these results into Equation 9.3.4 to obtain $\phi_{ji}(n, m)$.

On some occasions our prior probability distribution over the states of the process may be given, not at the time $m$ about which we wish to make the inference, but at some earlier time, perhaps 0. All we do in this case is multiply the probability vector $\boldsymbol{\pi}(0)$ by $\Phi(0, m)$ to produce the needed prior $\boldsymbol{\pi}(m)$ at time $m$.

When we have computed $\phi_{ji}(n, m)$ for all values of $i$ and $j$, we can represent the result as an $N$ by $N$ matrix $\Phi(n, m)$ with elements $\phi_{ji}(n, m)$. We shall call this matrix the inverse multistep transition probability matrix. Again, its second argument $m$ will never exceed its first argument $n$. The inverse multistep transition probability matrix is the key to analyzing the inverse Markov process just as the multistep transition probability matrix was the key to analyzing the future trajectory of the ordinary Markov processes.

## The Inverse Markov Process

We shall now show that the inverse multistep transition probability matrix $\Phi(n, m)$, $m \leq n$, satisfies the Chapman-Kolmogorov equation. The important consequence of this result is that the inverse process is thus shown to be itself a Markov process. Happily, therefore, all of our earlier findings on Markov processes also apply in this case.

In our development of the Chapman-Kolmogorov equation for the inverse process we shall use once more the diagram of Figure 9.1.1. We draw on basic probability theory to write

$$\mathscr{P}\{S_{mi}|S_{nj}\} = \sum_{k=1}^{N} \mathscr{P}\{S_{rk}|S_{nj}\}\mathscr{P}\{S_{mi}|S_{rk}S_{nj}\} \qquad m \leq r \leq n. \tag{9.3.6}$$

The probability $\mathscr{P}\{S_{mi}|S_{rk}S_{nj}\}$ that the system was in state $i$ at time $m$ given that it was in state $k$ at time $r$ and in state $j$ at time $n$ can be written as

$$\mathscr{P}\{S_{mi}|S_{rk}S_{nj}\} = \frac{\mathscr{P}\{S_{mi}|S_{rk}\}\mathscr{P}\{S_{nj}|S_{rk}S_{mi}\}}{\mathscr{P}\{S_{nj}|S_{rk}\}}$$

$$= \mathscr{P}\{S_{mi}|S_{rk}\} \qquad m \leq r \leq n \tag{9.3.7}$$

since, by the Markovian property, $\mathscr{P}\{S_{nj}|S_{rk}S_{mi}\} = \mathscr{P}\{S_{nj}|S_{rk}\}$. Therefore, Equation 9.3.6 becomes

$$\mathscr{P}\{S_{mi}|S_{nj}\} = \sum_{k=1}^{N} \mathscr{P}\{S_{rk}|S_{nj}\}\mathscr{P}\{S_{mi}|S_{rk}\} \qquad m \leq r \leq n. \tag{9.3.8}$$

We can write this equation in the notation of Equation 9.3.3 as

$$\phi_{ji}(n, m) = \sum_{k=1}^{N} \phi_{jk}(n, r)\phi_{ki}(r, m) \qquad m \leq r \leq n. \tag{9.3.9}$$

This is, of course, just the Chapman-Kolmogorov equation of Equation 9.1.7 with time running in the reverse direction. Equation 9.3.9 in matrix form is

$$\Phi(n, m) = \Phi(n, r)\Phi(r, m) \qquad m \leq r \leq n. \tag{9.3.10}$$

As in Equation 9.1.9, we require that $\Phi(n, m)$ be the identity matrix when its arguments are identical.

### Forward and Backward Chapman-Kolmogorov Equations for the Inverse Process

Just as we developed forward and backward forms for the Chapman-Kolmogorov equation for the ordinary process, so can we do the same for the inverse process. To develop the forward equation we choose $r = n - 1$ in Equation 9.3.10,

$$\Phi(n, m) = \Phi(n, n-1)\Phi(n-1, m) \qquad m \leq n - 1. \tag{9.3.11}$$

We now define an inverse transition probability matrix ${}^{I}P(n)$ at time $n$ to be

$${}^{I}P(n) = \Phi(n, n-1). \tag{9.3.12}$$

The element ${}^{I}p_{ij}(n)$ represents the probability that given the system is in state $i$ at time $n$ it was in state $j$ at the time before. With this definition Equation 9.3.11 becomes

$$\Phi(n, m) = {}^{I}P(n)\Phi(n-1, m) \qquad m \leq n - 1. \tag{9.3.13}$$

This equation is the forward Chapman-Kolmogorov equation for the inverse process. (Confusing, isn't it?)

We can find the solution $\Phi(n, m)$ to the forward equation by writing Equation 9.3.13 in explicit form,

$$\begin{aligned}\Phi(n, m) &= {}^{I}P(n)\Phi(n-1, m)\\ &= {}^{I}P(n)\,{}^{I}P(n-1)\Phi(n-2, m)\\ &= {}^{I}P(n)\,{}^{I}P(n-1)\,{}^{I}P(n-2)\Phi(n-3, m)\\ &\;\vdots\\ &= {}^{I}P(n)\,{}^{I}P(n-1)\,{}^{I}P(n-2)\cdots{}^{I}P(m+1) \qquad m \leq n - 1.\end{aligned} \tag{9.3.14}$$

If we write the backward equation by choosing $r = m + 1$ in Equation 9.3.10,

$$\begin{aligned}\Phi(n, m) &= \Phi(n, m+1)\Phi(m+1, m)\\ &= \Phi(n, m+1)\,{}^{I}P(m+1) \qquad m \leq n - 1,\end{aligned} \tag{9.3.15}$$

we find that it, too, is satisfied by the solution of Equation 9.3.14. Thus we see that the inverse multistep transition probability matrix is computed by a procedure that is completely analogous to the one used in Equation 9.1.25 for predicting the future trajectory.

### The Inverse Transition Probability Matrices

The problem that remains now is finding the inverse transition probability matrices defined by Equation 9.3.12. When this is done we have completely specified the inverse process and its solution. We proceed by writing an element of the transition probability matrix at some time $r$, $m < r \leq n$, using Equation 9.3.4,

$$^{I}p_{ji}(r) = \phi_{ji}(r, r-1) = \frac{\pi_i(r-1)\phi_{ij}(r-1, r)}{\pi_j(r)} \qquad m < r \leq n \tag{9.3.16}$$

or

$$^{I}p_{ji}(r) = \frac{\pi_i(r-1)}{\pi_j(r)} p_{ij}(r-1) \qquad m < r \leq n. \tag{9.3.17}$$

Thus the inverse transition probabilities are directly related to the state probabilities and the time-varying transition probabilities of the process. The state probabilities are computed recursively from Equation 9.3.5,

$$\begin{aligned} \pi_j(r) &= \sum_{i=1}^{N} \pi_i(r-1)\phi_{ij}(r-1, r) \\ &= \sum_{i=1}^{N} \pi_i(r-1)p_{ij}(r-1) \qquad r = m+1, m+2, \ldots, n, \end{aligned} \tag{9.3.18}$$

starting with the prior probabilities $\pi_i(m)$.

Therefore, we have found that the inverse Markov process corresponding to a Markov process with time-varying transition probabilities is also a Markov process with time-varying transition probabilities. Once we have solved the process to make future predictions, we can easily adapt the results to make the proper probability assignments to previous states that the process might have occupied.

### The Inverse Markov Process for Time-Invariant Markov Processes

Let us now restrict our discussion to finding the inverse Markov process corresponding to a Markov process with a constant transition probability matrix $P$. Although we might at first think that the inverse Markov process would also have a constant transition probability matrix, this is not the case. Equation 9.3.17 shows that the elements of the inverse transition probability matrices would be

$$^{I}p_{ji}(r) = \phi_{ji}(r, r-1) = \frac{\pi_i(r-1)}{\pi_j(r)} p_{ij} \qquad m < r \leq n \tag{9.3.19}$$

and therefore would depend on the time index $r$.

Consequently, the main simplification that a constant transition probability matrix $P$ introduces lies in the calculation of $\Phi(m, n)$, $n \geq m$, since

$$\Phi(m, n) = P^{n-m}\dagger \qquad m \leq n. \tag{9.3.20}$$

Then, too, the state probability vectors at the two times $m$, $n$ are related simply by

$$\pi(n) = \pi(m)\Phi(m, n) = \pi(m)P^{n-m} \qquad m \leq n. \tag{9.3.21}$$

## The Inverse Process for the Marketing Example

Let us now apply these results to find the inverse multistep transition probability matrix for the two-state marketing problem. The transition probability matrix for this problem is

$$P = \begin{bmatrix} 0.8 & 0.2 \\ 0.3 & 0.7 \end{bmatrix}. \tag{9.3.22}$$

We found in Equation 1.6.14 that the multistep transition probability matrix for this example was

$$\begin{aligned} \Phi(m, n) &= \Phi(n - m) = P^{n-m} \\ &= \begin{bmatrix} 0.6 & 0.4 \\ 0.6 & 0.4 \end{bmatrix} + (1/2)^{n-m}\begin{bmatrix} 0.4 & -0.4 \\ -0.6 & 0.6 \end{bmatrix} \qquad m \leq n. \end{aligned} \tag{9.3.23}$$

We now must specify a prior distribution. We shall assume that before knowing anything about the future of the process we had reason to believe that both states were equally likely to be occupied at time zero,

$$\pi(0) = [1/2 \quad 1/2]. \tag{9.3.24}$$

Then we compute the state probability vectors at time $m$ and $n$ from

$$\begin{aligned} \pi(m) &= \pi(0)\Phi(0, m) = [0.6 \quad 0.4] + (1/2)^m[-0.1 \quad 0.1] \\ \pi(n) &= \pi(0)\Phi(0, n) = [0.6 \quad 0.4] + (1/2)^n[-0.1 \quad 0.1]. \end{aligned} \tag{9.3.25}$$

† I might mention at this point a proposal that always seems to arise from new students encountering the inverse process problem. They note that we predict the location of the process $n$ transitions into the future by raising $P$ to the $n$th power. Then they ask why could we not assign the proper probabilities to previous states by raising $P^{-1}$ to the $n$th power. Remarking that there is no reason why $P$ should have an inverse seems to trigger the right kind of thinking about the problem.

***The inverse multistep transition probability matrix***

We now have all the information we need to compute the elements of the inverse multistep transition probability matrix from Equation 9.3.4:

$$
\begin{aligned}
\phi_{11}(n, m) &= \frac{\pi_1(m)\phi_{11}(m, n)}{\pi_1(n)} = \frac{[0.6 - 0.1(1/2)^m][0.6 + 0.4(1/2)^{n-m}]}{[0.6 - 0.1(1/2)^n]} \\
\phi_{12}(n, m) &= \frac{\pi_2(m)\phi_{21}(m, n)}{\pi_1(n)} = \frac{[0.4 + 0.1(1/2)^m][0.6 - 0.6(1/2)^{n-m}]}{[0.6 - 0.1(1/2)^n]} \\
\phi_{21}(n, m) &= \frac{\pi_1(m)\phi_{12}(m, n)}{\pi_2(n)} = \frac{[0.6 - 0.1(1/2)^m][0.4 - 0.4(1/2)^{n-m}]}{[0.4 + 0.1(1/2)^n]} \\
\phi_{22}(n, m) &= \frac{\pi_2(m)\phi_{22}(m, n)}{\pi_2(n)} = \frac{[0.4 + 0.1(1/2)^m][0.4 + 0.6(1/2)^{n-m}]}{[0.4 + 0.1(1/2)^n]}
\end{aligned}
\qquad (9.3.26)
$$

These four elements then form the inverse multistep transition probability matrix,

$$\Phi(n, m) = \begin{bmatrix} \phi_{11}(n, m) & \phi_{12}(n, m) \\ \phi_{21}(n, m) & \phi_{22}(n, m) \end{bmatrix} \qquad m \leq n. \qquad (9.3.27)$$

The complicated algebraic form of Equation 9.3.26 should be compared with the simpler form of Equation 9.3.23. The transients in the inverse multistep transition probability matrix are not geometric sequences.

We note first that $\Phi(n, m)$, $m \leq n$, is a stochastic matrix because its rows have positive elements that sum to one. We also observe $\Phi(n, n) = I$, as we would expect. If we choose $m = 0$, then we obtain *a posteriori* estimates of the starting probabilities of the process. For this case,

$$\Phi(n, 0) = \begin{bmatrix} \dfrac{0.3 + 0.2(1/2)^n}{0.6 - 0.1(1/2)^n} & \dfrac{0.3 - 0.3(1/2)^n}{0.6 - 0.1(1/2)^n} \\ \dfrac{0.2 - 0.2(1/2)^n}{0.4 + 0.1(1/2)^n} & \dfrac{0.2 + 0.3(1/2)^n}{0.4 + 0.1(1/2)^n} \end{bmatrix} \qquad 0 \leq n. \qquad (9.3.28)$$

In particular,

$$\Phi(\infty, 0) = \begin{bmatrix} 0.5 & 0.5 \\ 0.5 & 0.5 \end{bmatrix} \qquad (9.3.29)$$

showing that information about the state of the process after a great number of transitions has been made has in this case no bearing on the values of initial probabilities. This fact is further borne out by the calculation of

$$\Phi(\infty, m) = \begin{bmatrix} \pi_1(m) & \pi_2(m) \\ \pi_1(m) & \pi_2(m) \end{bmatrix} \qquad (9.3.30)$$

from Equation 9.3.4.

Figure 9.3.1 shows how the diagonal elements of $\Phi(n, 0)$ approach their limiting values.

Thus, although the calculation of the inverse multistep transition probability matrix is algebraically complicated, it follows directly from the calculation of the multistep transition probability matrix.

***The inverse transition probability matrices***

Let us turn now to computation of the inverse Markov process. By using Equations 9.3.19 and 9.3.26 we obtain:

$$\begin{aligned}
{}^{I}p_{11}(r) &= \phi_{11}(r, r-1) = \frac{0.48 - 0.08(1/2)^{r-1}}{0.6 - 0.1(1/2)^r} = \frac{0.48 - 0.16(1/2)^r}{0.6 - 0.1(1/2)^r} \\
{}^{I}p_{12}(r) &= \phi_{12}(r, r-1) = \frac{0.12 + 0.03(1/2)^{r-1}}{0.6 - 0.1(1/2)^r} = \frac{0.12 + 0.06(1/2)^r}{0.6 - 0.1(1/2)^r} \\
{}^{I}p_{21}(r) &= \phi_{21}(r, r-1) = \frac{0.12 - 0.02(1/2)^{r-1}}{0.4 + 0.1(1/2)^r} = \frac{0.12 - 0.04(1/2)^r}{0.4 + 0.1(1/2)^r} \\
{}^{I}p_{22}(r) &= \phi_{22}(r, r-1) = \frac{0.28 + 0.07(1/2)^{r-1}}{0.4 + 0.1(1/2)^r} = \frac{0.28 + 0.14(1/2)^r}{0.4 + 0.1(1/2)^r}
\end{aligned} \tag{9.3.31}$$

$$r = 1, 2, 3, \ldots$$

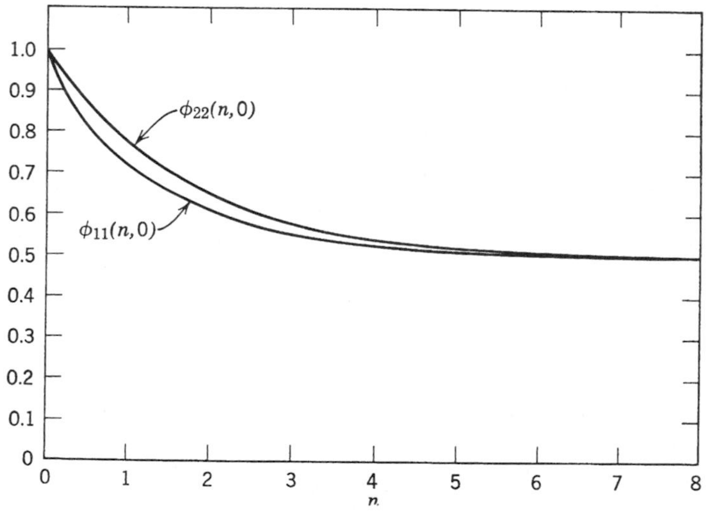


**Figure 9.3.1** Diagonal elements of the inverse multistep transition probability matrix for the marketing example with prior state probability vector $\pi(0) = [0.5\ 0.5]$.

These equations provide the elements for all the time-varying inverse transition probability matrices,

$$^{I}P(r) = \begin{bmatrix} ^{I}p_{11}(r) & ^{I}p_{12}(r) \\ ^{I}p_{21}(r) & ^{I}p_{22}(r) \end{bmatrix} \qquad r = 1, 2, 3, \ldots . \qquad (9.3.32)$$

The first few inverse matrices are

$$^{I}P(1) = \begin{bmatrix} 8/11 & 3/11 \\ 2/9 & 7/9 \end{bmatrix} \quad ^{I}P(2) = \begin{bmatrix} 88/115 & 27/115 \\ 22/85 & 63/85 \end{bmatrix} \quad ^{I}P(3) = \begin{bmatrix} 184/235 & 51/235 \\ 46/165 & 119/165 \end{bmatrix} \qquad (9.3.33)$$

or

$$^{I}P(1) = \begin{bmatrix} 0.727 & 0.273 \\ 0.222 & 0.778 \end{bmatrix} \quad ^{I}P(2) = \begin{bmatrix} 0.765 & 0.235 \\ 0.259 & 0.741 \end{bmatrix} \quad ^{I}P(3) = \begin{bmatrix} 0.783 & 0.217 \\ 0.279 & 0.721 \end{bmatrix}. \qquad (9.3.34)$$

As $r$ grows larger and larger, $^{I}P(r)$ approaches

$$^{I}P(\infty) = \begin{bmatrix} 0.8 & 0.2 \\ 0.3 & 0.7 \end{bmatrix} \qquad (9.3.35)$$

which is, of course, just the transition probability matrix $P$. Equation 9.3.35 states that if this two-state process has been running for a very long time after the point where our prior distribution was established, the inverse transition probability matrices are the same as the transition probability matrix $P$. We shall have more to say of this phenomenon.

### Summary of Inverse Process Calculations

Let us now summarize what we have found. If we observe the state of a Markov process at time $n$, and if we have a prior probability distribution on the state of the process at some earlier time, say 0, then we can assign a probability distribution over the states of the process at any intermediate time $0 < m < n$. One way to do this is to use the basic equations, 9.3.4 and 9.3.5. Another way is to construct the equivalent time-varying inverse Markov process as we did in Equations 9.3.31 and 9.3.32 for the example. Then we compute the inverse multistep transition probability matrix from Equation 9.3.14. The first method is more efficient, but the second method provides a valuable insight into the structure of the process.

### Inverse Processes in the Steady State

Suppose that a monodesmic Markov process with a constant transition probability matrix $P$ has been operating for some time. If we are going to observe the state of the process at some time point $n$, we might inquire about the probability distribution we should assign to the states of the process at a time $m \leq n$ before this information becomes available. The proper *a priori* probability distribution to assign to the states at the time $m$ is the limiting state probability distribution $\boldsymbol{\pi}$. The results that we have developed then become considerably simpler because we have conducted our inference when our knowledge about the process is in the steady state. Of course, we can eliminate any transient states from the process because we know that they cannot be occupied in these circumstances. When this is done all states will have a non-zero limiting state probability.

First, Equation 9.3.21 becomes

$$\boldsymbol{\pi}(n) = \boldsymbol{\pi}(m)P^{n-m} = \boldsymbol{\pi}P^{n-m} = \boldsymbol{\pi}. \tag{9.3.36}$$

This result allows us to write Equations 9.3.4 and 9.3.19 in the forms

$$\phi_{ji}(n, m) = \frac{\pi_i}{\pi_j}\,\phi_{ij}(m, n) \qquad m \leq n \tag{9.3.37}$$

and

$${}^{\mathrm{I}}p_{ji}(r) = \phi_{ji}(r, r-1) = \frac{\pi_i}{\pi_j}\,p_{ij}. \tag{9.3.38}$$

Note that ${}^{\mathrm{I}}p_{ji}$ is no longer a function of $r$. The inverse transition probability matrices are not time-varying—they have elements given simply by

$${}^{\mathrm{I}}p_{ji} = \frac{\pi_i}{\pi_j}\,p_{ij}. \tag{9.3.39}$$

The implication of this result is that processes whose observation is begun when they are in the steady state have inverse Markov processes that have constant inverse transition probability matrices. Computations in the inverse process thus become very simple.

***Matrix form***

We shall find it helpful to state the result of Equation 9.3.39 in matrix form. We therefore define an $N$ by $N$ matrix $B$ with elements

$$b_{ij} = \frac{\pi_i}{\pi_j} \qquad 1 \leq i, j \leq N. \tag{9.3.40}$$

Note that the diagonal elements of the $B$ matrix will always be unity. Using our box notation we can now write Equation 9.3.39 as

$${}^{\mathrm{I}}P = [B \,\square\, P]^t \tag{9.3.41}$$

and Equation 9.3.37 as

$$\Phi(n, m) = [B \square \Phi(m, n)]^t \qquad m \leq n, \tag{9.3.42}$$

where the superscript $t$ indicates that the matrix has been transposed. Since the diagonal elements of $B$ are one, the diagonal elements of ${}^{I}P$ will be the same as those of $P$ and the diagonal elements of $\Phi(n, m)$ will be identical to those of $\Phi(m, n)$.

***The marketing example***

Now let us apply this theory to the two-state marketing example of Equation 9.3.22. The multistep transition probability matrix $\Phi(m, n)$ already appears in Equation 9.3.23. It reveals what we have seen on many past occasions in this example: The limiting state probability vector of the process is

$$\pi = [0.6 \quad 0.4]. \tag{9.3.43}$$

We use these values for $\pi_i$ to compute the $B$ matrix defined in Equation 9.3.40,

$$B = \begin{bmatrix} 1 & 3/2 \\ 2/3 & 1 \end{bmatrix}. \tag{9.3.44}$$

Then we substitute this $B$ and $\Phi(m, n)$ from Equation 9.3.23 into Equation 9.3.42 to produce the inverse multistep transition probability matrix,

$$\Phi(n, m) = \begin{bmatrix} 0.6 & 0.4 \\ 0.6 & 0.4 \end{bmatrix} + (1/2)^{n-m}\begin{bmatrix} 0.4 & -0.4 \\ -0.6 & 0.6 \end{bmatrix} \qquad m \leq n. \tag{9.3.45}$$

Equation 9.3.45 shows that the inverse multistep transition probability matrix is the same as the multistep transition probability matrix of Equation 9.3.23. This means that probability assignments into the past in the steady state are computed in the same way as probability assignments into the future. If this is so, the inverse transition probability matrix should be the same as the transition probability matrix. Equation 9.3.41 confirms this implication,

$${}^{I}P = [B \square P]^t = \begin{bmatrix} 0.8 & 0.2 \\ 0.3 & 0.7 \end{bmatrix} = P. \tag{9.3.46}$$

**Reversibility**

When the inverse Markov process corresponding to steady-state inference in a monodesmic Markov process has the same transition probability matrix as the Markov process,

$${}^{I}P = P, \tag{9.3.47}$$

we say that the Markov process is reversible. The process describing the marketing problem is therefore reversible. This means, for example, that if we observe that a customer has just purchased a certain brand, then in the absence of any other information, we would assign the same probability to his purchasing that brand on, say, his third purchase into the past that we would assign to his purchasing that brand on his third purchase into the future. The multistep transition probability matrices for the inverse and ordinary Markov processes are identical.

### *The taxicab example*

Are all constant transition probability monodesmic Markov processes reversible? Let us investigate this question by considering the three-state taxicab problem of Chapter 1. The transition probability matrix for this problem was given in Equation 1.4.9:

$$P = \begin{bmatrix} 0.3 & 0.2 & 0.5 \\ 0.1 & 0.8 & 0.1 \\ 0.4 & 0.4 & 0.2 \end{bmatrix} \tag{9.3.48}$$

The multistep transition probability matrix for this example appears in Equation 1.6.44,

$$\Phi(m, n) = \Phi(n - m) = \begin{bmatrix} 1/5 & 3/5 & 1/5 \\ 1/5 & 3/5 & 1/5 \\ 1/5 & 3/5 & 1/5 \end{bmatrix} + (1/2)^{n-m} \begin{bmatrix} 13/35 & -26/35 & 13/35 \\ -1/5 & 2/5 & -1/5 \\ 8/35 & -16/35 & 8/35 \end{bmatrix}$$
$$+ (-1/5)^{n-m} \begin{bmatrix} 3/7 & 1/7 & -4/7 \\ 0 & 0 & 0 \\ -3/7 & -1/7 & 4/7 \end{bmatrix} \quad m \le n. \tag{9.3.49}$$

We can therefore see that the limiting state probability vector is

$$\pi = [1/5 \quad 3/5 \quad 1/5]. \tag{9.3.50}$$

We proceed now to find the inverse multistep transition probability matrix for this example. We compute $B$ from Equation 9.3.40,

$$B = \begin{bmatrix} 1 & 1/3 & 1 \\ 3 & 1 & 3 \\ 1 & 1/3 & 1 \end{bmatrix}. \tag{9.3.51}$$

Finally, we evaluate $\Phi(n, m)$ from Equations 9.3.42, 9.3.49, and 9.3.51,

$$\Phi(n, m) = \begin{bmatrix} 1/5 & 3/5 & 1/5 \\ 1/5 & 3/5 & 1/5 \\ 1/5 & 3/5 & 1/5 \end{bmatrix} + (1/2)^{n-m} \begin{bmatrix} 13/35 & -3/5 & 8/35 \\ -26/105 & 2/5 & -16/105 \\ 13/35 & -3/5 & 8/35 \end{bmatrix}$$

$$+ (-1/5)^{n-m} \begin{bmatrix} 3/7 & 0 & -3/7 \\ 1/21 & 0 & -1/21 \\ -4/7 & 0 & 4/7 \end{bmatrix} \qquad m \leq n. \tag{9.3.52}$$

Note that

$$\lim_{n-m\to\infty} \Phi(n, m) = \lim_{n-m\to\infty} \Phi(m, n) = \Phi, \tag{9.3.53}$$

the inverse and ordinary multistep transition probabilities have the same limiting matrix. We can establish this result in general from Equation 9.3.37,

$$\lim_{n-m\to\infty} \phi_{ji}(n, m) = \frac{\pi_i}{\pi_j} \lim_{n-m\to\infty} \phi_{ij}(m, n) = \frac{\pi_i}{\pi_j} \pi_j = \pi_i. \tag{9.3.54}$$

However, Equation 9.3.53 does not in itself imply that the process is reversible. In fact, comparing Equations 9.3.49 and 9.3.52 shows that their transient components are different in magnitude if not in kind. We can see by the nature of the operation we perform in Equation 9.3.42 that the same geometric decay rates will appear in the inverse and ordinary multistep transition probability matrices. Since their limiting behavior is the same, the only way they can differ is in the entries of the differential matrices that multiply the geometric decay factors. This is just the type of difference we observe in the three-state example.

Because the inverse and ordinary multistep transition probability matrices are different for this example, the transition probability matrix for the inverse process must be different from the transition probability matrix for the ordinary process. We compute it from Equations 9.3.41, 9.3.48, and 9.3.51,

$${}^{I}P = [B \,\square\, P]^t = \begin{bmatrix} 0.3 & 0.3 & 0.4 \\ 1/15 & 0.8 & 2/15 \\ 0.5 & 0.3 & 0.2 \end{bmatrix} = \begin{bmatrix} 0.300 & 0.300 & 0.400 \\ 0.067 & 0.800 & 0.133 \\ 0.500 & 0.300 & 0200 \end{bmatrix}. \tag{9.3.55}$$

Clearly this is different from the transition matrix of Equation 9.3.48 except for the diagonal elements. These will always be identical in the two matrices as a direct consequence of Equations 9.3.39. Of course, we can interpret Equation 9.3.52 as an expression for the $n - m$th power of ${}^{I}P$,

$$\Phi(n, m) = {}^{I}P^{n-m} \qquad m \leq n. \tag{9.3.56}$$

We have thus answered our question: All constant transition probability monodesmic Markov processes are not reversible in the steady state. We shall now find the condition under which reversibility will occur.

### The Condition for Reversibility

As we have said, a constant transition probability matrix monodesmic Markov process is reversible if ${}^{\mathrm{I}}P = P$, or, from Equation 9.3.39, if

$$ {}^{\mathrm{I}}p_{ji} = \frac{\pi_i}{\pi_j} p_{ij} = p_{ji}, \tag{9.3.57} $$

and therefore if,

$$ \pi_i p_{ij} = \pi_j p_{ji} \qquad \begin{array}{l} i = 1, 2, \ldots, N \\ j = 1, 2, \ldots, N \end{array}. \tag{9.3.58} $$

The process is reversible if and only if this equation holds for all pairs of states; it is therefore a necessary and sufficient condition for reversibility. We can state the result in matrix form by saying that for reversibility to exist the matrix $(\Phi \,\square\, I)P$ must be symmetric,

$$ (\Phi \,\square\, I)P = [(\Phi \,\square\, I)P]^t. \tag{9.3.59} $$

The condition expressed in Equation 9.3.58 has a simple physical interpretation. It states that a process is reversible if, in the absence of any information on its trajectory, one would have to assign the same probability to a transition from state $i$ to state $j$ that is assigned to a transition from state $j$ to state $i$. In view of Equation 7.5.11, all birth and death processes must be reversible.

#### *Examples*

We can perform this test for reversibility on the two examples. For the two-state marketing example we find

$$ (\Phi \,\square\, I)P = \begin{bmatrix} 0.6 & 0 \\ 0 & 0.4 \end{bmatrix} \begin{bmatrix} 0.8 & 0.2 \\ 0.3 & 0.7 \end{bmatrix} = \begin{bmatrix} 0.48 & 0.12 \\ 0.12 & 0.28 \end{bmatrix} \tag{9.3.60} $$

which is a symmetric matrix. Therefore, the two-state example should be reversible, as we found it was.

In fact, since the diagonal elements of ${}^{\mathrm{I}}P$ and $P$ must be equal in general, it follows from the fact that a 2 by 2 stochastic matrix is specified by its diagonal elements that all two-state Markov processes are reversible.

For the three-state taxicab example we find

$$(\Phi \,\Box\, I)P = \begin{bmatrix} 0.2 & 0 & 0 \\ 0 & 0.6 & 0 \\ 0 & 0 & 0.2 \end{bmatrix} \begin{bmatrix} 0.3 & 0.2 & 0.5 \\ 0.1 & 0.8 & 0.1 \\ 0.4 & 0.4 & 0.2 \end{bmatrix} = \begin{bmatrix} 0.06 & 0.04 & 0.10 \\ 0.06 & 0.48 & 0.06 \\ 0.08 & 0.08 & 0.04 \end{bmatrix} \tag{9.3.61}$$

which is not a symmetric matrix: the three-state example is not reversible.

***Further comments on reversibility***

Let us pursue the question of reversibility a little further. As we observed in Chapter 1, a Markov process with a doubly stochastic transition probability matrix has the same limiting state probabilities for all of its states. We therefore see from Equation 9.3.40 that when $P$ is doubly stochastic, the matrix $B$ is a unity matrix—all of its elements equal 1,

$$B = U. \tag{9.3.62}$$

Then Equations 9.3.41 and 9.3.42 show that

$${}^{I}P = P^t \tag{9.3.63}$$

and

$$\Phi(n, m) = [\Phi(m, n)]^t \qquad m \le n. \tag{9.3.64}$$

Thus a Markov process with a doubly stochastic transition probability matrix has an inverse process whose inverse transition probability matrix and inverse multistep transition probability matrix are the transposes of the corresponding quantities for the ordinary process. Such a process is not necessarily reversible according to our definition, but its inverse process has a very simple form.

If a Markov process has a symmetric transition probability matrix, then this matrix is also doubly stochastic and the process has all its limiting state probabilities equal. Therefore, Equation 9.3.63 holds for such a process; moreover, because of the symmetry the transition probability matrix is unaffected by the transposition. Consequently,

$${}^{I}P = P \tag{9.3.65}$$

and, of course, as for any reversible process,

$$\Phi(n, m) = \Phi(m, n) \qquad m \le n. \tag{9.3.66}$$

Monodesmic Markov processes with symmetric transition probability matrices are reversible. However, although symmetry is a sufficient condition for reversibility, it is not a necessary condition. The necessary and sufficient condition is symmetry of the $(\Phi \,\Box\, I)P$ matrix.

## PROBLEMS

**1.** a) Consider a random variable $x$ that can take on only the values 1, 2, 3, .... Let

$$\mathscr{P}\{x = k\} = f(k) \qquad k = 1, 2, 3, \ldots$$

and

$$f^g(z) = \sum_{k=1}^{\infty} f(k)z^k.$$

Find the geometric transform of $f(k - 1)/k$ in terms of $f^g(z)$.

b) Joe Smith is attempting to pass a final. He intends to keep trying each year until he finally passes. However, Joe is having his troubles; every time he fails he gets more discouraged so that the probability that he passes on his $n$th trial is $p/n$. Therefore,

$$P(n) = \begin{bmatrix} 1 - (p/n) & p/n \\ 0 & 1 \end{bmatrix}$$

where state 1 is the non-passed state, and state 2 is the passed state.

i) Verify that the probability he eventually passes is 1.

ii) Let $h(k) = \mathscr{P}\{\text{he passes on trial } k\}$ $k = 1, 2, \ldots$. Find the geometric transform $h^g(z)$.

iii) What is the mean number of trials required for Joe to pass?

**2.** Consider the 2-state, time-varying, discrete-time Markov process whose transition probability matrix is given by:

$$P(n) = \begin{bmatrix} p_{11}(n) & p_{12}(n) \\ p_{21}(n) & p_{22}(n) \end{bmatrix} \qquad n \geq 0$$

where $p_{ij}(n) = \mathscr{P}\{s(n + 1) = j | s(n) = i\}$.

a) Write a difference equation for $\pi_1(n + 1)$ in terms of $\pi_1(n)$ and the transition probabilities.

b) If for a particular process we have

$$p_{11}(n) - p_{21}(n) = c \quad \text{for all } n \geq 0$$

where $0 \leq c \leq 1$, find an expression for $\pi_1{}^g(z)$ in terms of $c$, $\pi_1(0)$, $\pi_2(0)$, and the elements of $P^g(z)$.

c) Find $\pi_1(n)$ if

$$P(n) = \begin{bmatrix} 1/4 + (1/2)^{n+1} & 3/4 - (1/2)^{n+1} \\ 1/8 + (1/2)^{n+1} & 7/8 - (1/2)^{n+1} \end{bmatrix} \qquad n \geq 0$$

and $\pi_1(0) = 0$.

**3.** Consider a periodically time-varying Markov process for which

$$P(n) = \begin{cases} \begin{bmatrix} 0.6 & 0.4 \\ 0.2 & 0.8 \end{bmatrix} = {}^{e}P & \text{for } n \text{ even} \\ \begin{bmatrix} 0.5 & 0.5 \\ 0.25 & 0.75 \end{bmatrix} = {}^{o}P & \text{for } n \text{ odd.} \end{cases}$$

a) Write an expression for $\Phi(0, n)$ in terms of $({}^{e}P\,{}^{o}P)$. Do this for $n$ even and $n$ odd.

b) Take the transform of the relations in a) and draw a flow graph whose transmission from the $i$th input to the $j$th output is the transform of $\phi_{ij}(0, n)$ for $n$ even. Do the same for $n$ odd.

c) Combine the two flow graphs in part b) into a single flow graph whose transmission from the $i$th input to the $j$th output is the complete transform of $\phi_{ij}(0, n)$.

d) Write an expression for $\Phi(0, n)$ in terms of $({}^{o}P\,{}^{e}P)$. Do this for $n$ even and $n$ odd. Now repeat parts b) and c) to obtain a different flow graph for the process. Verify that this flow graph yields the same results as the one in c).

e) What are the limiting state probabilities for $n$ even? For $n$ odd?

**4.** A frustrated meteorologist has decided to model the behavior of the weather with an $N$-state Markov process.

a) It is the end of the $n$th day. The meteorologist has a prior distribution on what the state of the process was at the end of the zeroth day: $\pi(0) = [\pi_1(0), \pi_2(0), \ldots]$. If he also knows that the state of the process at the end of the $n$th day is $j$, find an expression for $\gamma_{ji}(n, k)$, the probability that the state of the process at the end of the $(n - k)$th day was $i$, where $k \geq 0$. Of course, $\gamma_{ji}(n, k)$ is conditional upon all the information given above. Do this in terms of $\pi(0)$ and the multistep transition probabilities.

b) Assume that $n$ is fixed. Write an expression for $\gamma_{ji}(n, k + 1)$ in terms of all possible states that the process might have been in at time $n - k$. Your result should be in terms of the $\gamma_{j\Sigma}(n, k)$'s and $\gamma_{\Sigma i}(n - k, 1)$'s, where $\Sigma$ indicates summation over the corresponding index.

c) Let us consider the equations in Part b) as the Chapman-Kolmogorov equations for a time varying Markov process. What are the transition probabilities for this time-varying Markov process in terms of the elements of $\pi(0)$ and $\Phi(n)$? What are they equal to if

$$\pi(0) = \pi,$$

the limiting state probability vector of the process?

d) Is this time-varying Markov process time-invariant for

$$\pi(0) = \pi?$$

Why?

**5.** In an $N$-state Markov process, what is the probability that a system in state $j$ after $n$ transitions *was* in state $i$ after $n - 3$ transitions?

**6.** Is the multinomial process reversible?

# NOTATION

The following system of notation is first defined in general terms and then made specific for each chapter. Interpretation of a symbol in the text typically will require an understanding of both the general and the specific features.

For example, Equation 5.1.43 contains the symbol $\overline{\mathrm{N}^{2}}{}^{\mathscr{g}}(z)$. By referring to the notation specific to Chapter 5, we see that $\nu_{ij}(n)$ is defined as the number of times state $j$ is occupied through time $n$ given that the process started in state $i$ at time 0. Furthermore, we find that $\mathrm{N}(n)$ is the matrix of these quantities. Upon examining the general notation, we see that if $x$ is a variable, $x(n)$ is a discrete function, $x^{\mathscr{g}}(z)$ is the geometric transform of the discrete function, $\overline{x^2}$ is the second moment, and $X$ is a matrix. Consequently, $\overline{\mathrm{N}^{2}}{}^{\mathscr{g}}(z)$ must be the matrix of geometric transforms of second moments of $\nu_{ij}(n)$.

Fortunately most symbols are not so complicated. However complicated they are, this notation summary should make them clear.

## NOTATION (General)

$x$ = a variable (perhaps a random variable)

$\mathbf{x}$ = a vector (represented by an underline at the blackboard)

$x_i$ = the $i$th element of $\mathbf{x}$

$X$ = a matrix

$x_{ij}$ = the $ij$th element of $X$

$x(\cdot)$ = a discrete function

$x(n)$ = the value of a discrete function at argument $n$

$x^{\mathscr{g}}(z)$ = the geometric transform of $x(\cdot) = \sum_{n=0}^{\infty} z^n x(n)$

$x^{\mathscr{g}\prime}(z) = \dfrac{d}{dz} x^{\mathscr{g}}(z)$, the derivative of $x^{\mathscr{g}}(z)$

$x^{\mathscr{g}\prime\prime}(z) = \dfrac{d^2}{dz^2} x^{\mathscr{g}}(z)$, the second derivative of $x^{\mathscr{g}}(z)$

$x(\cdot, \cdot)$ = a discrete function with two arguments

$x(k, n)$ = the value of a discrete function at $(k, n)$

$x^{gg}(y, z)$ = the double geometric transform of $x(\cdot, \cdot) = \sum_{k=0}^{\infty} y^k \sum_{n=0}^{\infty} z^n x(k, n)$

$x^{\cdot g}(k, z)$ = the geometric transform of $x(k, \cdot)$

$x^{g\cdot}(y, n)$ = the geometric transform of $x(\cdot, n)$

$\bar{x} = \langle x \rangle$ = the expectation, or mean, of $x$

$\overline{x^m} = \langle x^m \rangle$ = the $m$th moment of $x$

$\overset{v}{x} = {}^{v}\langle x \rangle$ = the variance of $x$

$I$ = the identity matrix with elements $\delta_{ij} = \begin{cases} 1 & \text{if } i = j \\ 0 & \text{if } i \neq j \end{cases}$

$\mathbf{s}$ = a column vector with all components equal to 1

$U$ = the unity matrix, a square matrix with all elements = 1

## NOTATION (Specific)

### Chapter 1

| Matrix | Symbol | |
|---|---|---|
| | $N$ | = the number of states in the process |
| | $n$ | = transition time index |
| | $s(n)$ | = the state of the process at time $n$ |
| $P$ | $p_{ij}$ | = the transition probability from state $i$ to state $j$ = $\mathscr{P}\{s(n+1) = j \mid s(n) = i\}$ $n = 0, 1, 2, \ldots$ |
| $\Phi(n)$ | $\phi_{ij}(n)$ | = the $n$-step (or multistep) transition probability from state $i$ to state $j$ = $\mathscr{P}\{s(n) = j \mid s(0) = i\}$ $n = 0, 1, 2, \ldots$ |
| $\Phi$ | $\phi_{ij}$ | = $\lim_{n\to\infty} \phi_{ij}(n)$, the limiting multistep transition probability from state $i$ to state $j$ |
| $\boldsymbol{\pi}(n)$† | $\pi_i(n)$ | = the state probability at time $n = \mathscr{P}\{s(n) = i\}$ |
| $\boldsymbol{\pi}$ | $\pi_i$ | = $\lim_{n\to\infty} \pi_i(n)$ = the limiting state probability of state $i$ |
| | sf | = shrinkage factor for the possible region |
| $T(n)$ | $t_{ij}(n)$ | = the transient component of $\phi_{ij}(n)$ where $\phi_{ij}(n) = \phi_{ij} + t_{ij}(n)$ |

### Chapter 2

| Matrix | Symbol | |
|---|---|---|
| | $\vec{t}$ | = a flow graph transmission |
| $\vec{T}$ | $\vec{t}_{ij}$ | = the transmission of a flow graph from node $i$ to node $j$ |

† $\boldsymbol{\pi}(n)$, the state probability vector, is a row vector.

## CHAPTER 4

| Matrix | Symbol | |
|---|---|---|
| | $\nu$ = | the number of transitions (delay) until trapped for a transient process |
| | $p(n)$ = | $\mathscr{P}\{\nu = n\}$ |
| | $\phi(n)$ = | $\mathscr{P}\{\nu \leq n\}$ |
| | $x_{ij}(n)$ = | an indicator variable $\left\{\begin{matrix} = 1 & \text{if } s(n) = j \\ = 0 & \text{otherwise} \end{matrix} \middle\vert \text{given } s(0) = 1 \right\}$ |
| N | $\nu_{ij}$ = | the number of times state $j$ of a transient process is occupied in an infinite number of transitions given that the process started in state $i$ |
| | $\mathscr{T}$ = | the set of all transient states |
| | $\nu_i$ = | $\sum_{j \in \mathscr{T}} \nu_{ij}$ = the number of times transient states are occupied in an infinite number of transitions given that the process started in state $i$ |
| | $\phi_{ijk}(m, r)$ = | the joint transition probability $= \mathscr{P}\{s(m) = j, s(n) = k \mid s(0) = i\}$ |
| | $\nu_{\sim}$ = | total time in a transient process when the starting state is described by the state probability vector $\pi(0)$ |
| | $\nu_{\sim j}$ = | total time in state $j$ of a transient process when the starting state is described by $\pi(0)$ |
| | $\nu_{ij\mid k}$, $\nu_{i\mid k}$, etc. = | $\nu_{ij}$, $\nu_i$, etc. with proviso that the process is ultimately captured by trapping state $k$ |
| | $\tau_i$ = | the holding time for state $i$, the number of transitions after the process enters state $i$ until it enters a state different from $i$ |
| | $'$ = | superscript pertaining to a real-transition process |

## CHAPTER 5

| Matrix | Symbol | |
|---|---|---|
| $N(n)$ | $\nu_{ij}(n)$ = | the number of times state $j$ is occupied through time $n$ given that the process started in state $i$ at time 0 |
| | ${}_1\nu_{ij}(n)$ = | same, except occupancies at $n = 0$ are not counted |

| Matrix | Symbol |
|---|---|
| | $\zeta_{ij}(n) = \frac{1}{n+1}\nu_{ij}(n)$, the fractional number of occupancies |
| $\Theta$ | $\theta_{ij} =$ the number of transitions required from the time state $i$ is entered until state $j$ is entered for the first time, called the first passage time; when $i = j$, also called the recurrence time |
| $F(n)$ | $f_{ij}(n) = \mathscr{P}\{\theta_{ij} = n\}$, first passage-time mass function |

CHAPTER 6

| Matrix | Symbol |
|---|---|
| | $\phi_{ij}(k_1, k_2, \ldots, k_m \mid n) =$ the joint probability that a process started in state $i$ at time 0 would be in state $j$ at time $n$ and have made $k_1$ transitions of type 1, $k_2$ transitions of type 2, etc. |
| $\Phi(k \mid n)$ | $\phi_{ij}(k \mid n) =$ the joint probability that a process started in state $i$ at time 0 will be in state $j$ at time $n$ and will have made $k$ special (tagged) transitions |
| | $k_{ij}{}^m(n) = \sum_{n=0}^{\infty} k^m \phi_{ij}(k \mid n)$ |
| | $\overline{k_{ij}{}^m}(n) = \frac{k_{ij}{}^m(n)}{\phi_{ij}(n)} =$ the $m$th moment of the number of special transitions made by a process through time $n$ given that it started in state $i$ and ended in state $j$ |
| | $S_{mj} = \{s(m) = j\}$, the event that the state of the process at time $m$ is $j$ |
| | $S_j =$ the event that the process makes its last transition into state $j$ |
| | $C_n =$ the event that $n$ transitions have occurred |
| | $T_k =$ the event that $k$ special (tagged) transitions have occurred |
| ${}_sP$ | ${}_sp_{ij} =$ the portion of $p_{ij}$ pertinent to special transitions |
| | $Q_k =$ the event that the process is stopped after $k$ special transitions |

CHAPTER 7

| Matrix | Symbol |
|---|---|
| | $l =$ lifetime, time between renewals |
| | $p(n) =$ lifetime mass function, $\mathscr{P}\{l = n\}$ |
| | $p_R =$ probability of eventual renewal |

$q(r, n)$ = probability of $r$th renewal at time $n$
$r$ = number of renewals in infinite time
$g(r)$ = probability of $r$ renewals in infinite time
$\phi(n)$ = probability of renewal (not necessarily the first) at time $n$ given renewal at time 0
$\phi = \lim_{n\to\infty} \phi(n)$ = limiting probability of renewal
${}^{>}p(m) = \sum_{l=m+1}^{\infty} p(l) = \mathscr{P}\{l > m\}$, complementary cumulative probability distribution on lifetime
${}_{f}p(\cdot)$ = probability mass function for time to first renewal
${}_{f}\phi(n)$ = probability of renewal at time $n$ given time to first renewal governed by ${}_{f}p(\cdot)$; similarly for other quantities with presubscript f; presubscript r implies random selection of time zero
$k(n)$ = number of renewals at times 1 through $n$ given that renewal occurred at time zero
$\phi(k|n) = \mathscr{P}\{k(n) = k\}$
$b_k$ = birth probability $= \mathscr{P}\{s(n + 1) = k + 1|s(n) = k\}$
$d_k$ = death probability $= \mathscr{P}\{s(n + 1) = k - 1|s(n) = k\}$

CHAPTER 8

| Matrix | Symbol |
|---|---|
| | $f_i(m)$ = number of units added to node $i$ at time $m$ |
| | $h_j(k\|n)$ = probability of $k$ units at node $j$ at time $n$ |
| $\mathbf{r}(n)$ | $r_j(n)$ = number of units at node $j$ at time $n$ |
| $\mathbf{u}(n)$ | $u_k(n)$ = number of units of age $k$ at time $n$ |
| | $k_n$ = number of direct descendants in $n$th generation |
| | $p_n(\cdot)$ = probability mass function of $k_n$ |
| | $m_n$ = size of $n$th generation |
| | $q_n(\cdot)$ = probability mass function of $m_n$ |

CHAPTER 9

| Matrix | Symbol |
|---|---|
| $\Phi(m, n)$ | $\phi_{ij}(m, n)$ = multistep transition probability from time $m$ to time $n = \mathscr{P}\{s(n) = j\|s(m) = i\}$ |
| $P(n)$ | $p_{ij}(n) = \mathscr{P}\{s(n + 1) = j\|s(n) = i\}$ |
| ${}^{\mathrm{I}}P(n)$ | ${}^{\mathrm{I}}p_{ij}(n)$ = inverse transition probability $= \phi_{ij}(n, n - 1)$ |

APPENDIX

# A PROPERTIES OF CONGRUENT MATRIX MULTIPLICATION

We find it convenient to define a matrix operation called congruent matrix multiplication. If two matrices $A$ and $B$ have the same dimensions, we say they are commensurate. The congruent product $C$ of two commensurate matrices $A$ and $B$ is a matrix of the same dimensions as $A$ and $B$ and whose elements are related to those of $A$ and $B$ by

$$c_{ij} = a_{ij}b_{ij} \qquad \text{all } i, j. \tag{A.1}$$

Thus congruent matrix multiplication is multiplication of corresponding elements. We denote such multiplication by a "box" operator $\square$ and write Equation A.1 in the matrix form

$$C = A \square B. \tag{A.2}$$

Congruent multiplication is useful not only for manipulation of matrices but also for designation of portions of matrices. Recalling that the identity matrix $I$ has elements $\delta_{ij}$ and the unity matrix $U$ has elements 1, we see that if $A$ is a square matrix then $A \square I$ is a matrix composed only of the diagonal elements of $A$, with others set to zero, while $A \square (U - I)$ is the same as $A$ but with zero diagonal elements.

It is easy to show that congruent matrix multiplication has useful algebraic properties. For example, if $A$, $B$, and $C$ are commensurate matrices it is easy to prove from the defining relation (A.1) that congruent matrix multiplication is:

commutative, $\quad A \square B = B \square A, \qquad$ (A.3)

distributive, $\quad A \square (B + C) = A \square B + A \square C, \qquad$ (A.4)

and associative, $\quad (A \square B) \square C = A \square (B \square C). \qquad$ (A.5)

We can even define the congruent inverse $A^*$ of matrix $A$ from

$$A \square A^* = A^* \square A = U. \tag{A.6}$$

The elements of $A^*$ are the reciprocals of corresponding elements of $A$.

Generally, however, congruent matrix multiplication and ordinary matrix multiplication are not commutative, and mixtures of the two operations are best investigated from the defining relations. For example, consider the expression $D = [A(B \square I)] \square I$ where $A$ and $B$ are commensurate square matrices. The definitions of ordinary and congruent matrix multiplication yield the component form

$$d_{ij} = \left[\sum_k a_{ik}(b_{kj}\delta_{kj})\right] \cdot \delta_{ij} = \sum_k (a_{ik}\delta_{ik})(b_{kj}\delta_{kj}) \tag{A.7}$$

and thus show that

$$\begin{aligned} [A(B \square I)] \square I &= (A \square I)(B \square I) = (B \square I)(A \square I) \\ &= [B(A \square I)] \square I. \end{aligned} \tag{A.8}$$

We can use this result to solve a matrix equation that is of interest to us in Chapter 10. Consider the equation

$$A - B(A \square I) = C \tag{A.9}$$

where $A$, $B$, and $C$ are commensurate square matrices and we desire to find $A$ given $B$ and $C$. We first find the congruent product of this equation with $I$,

$$A \square I - [B(A \square I)] \square I = C \square I. \tag{A.10}$$

In view of Equation A.8, we have

$$A \square I - (B \square I)(A \square I) = C \square I \tag{A.11}$$

or

$$A \square I = [I - (B \square I)]^{-1}(C \square I) = (C \square I)[I - (B \square I)]^{-1}. \tag{A.12}$$

When we substitute this result into Equation A.9 we obtain the solution

$$A = C + B[I - (B \square I)]^{-1}(C \square I) = C + B(C \square I)[I - (B \square I)]^{-1}. \tag{A.13}$$

# APPENDIX

# B A PAPER BY A. A. MARKOV

It is fitting that we should include in this volume an example of the work of Andrei Andreivich Markov (1856–1922). Markov was a distinguished Russian mathematician of broad interests, but his greatest achievement was the devisal of a model for dependent trials that was the simplest generalization of independent trial processes.

Actually, Markov approached this distinction rather indirectly by first wondering whether the law of large numbers applied to dependent as well as independent random variables. He used what we now call the Markovian model of dependence and established that this was, in fact, the case. Then he examined the question of whether the sums of such dependent variables would satisfy the central limit theorem; that is, become, in the limit, normally distributed. In his paper "Investigation of an Important Case of Dependent Trials" he showed that this was true for variables that could take on only two possible values. However, the very special nature of this case and the cumbersome combinational form of the proof must have left him unsatisfied, for he then wrote the paper "Extension of the Limit Theorems of Probability Theory to a Sum of Variables Connected in a Chain," whose translation appears in this appendix.

This paper is of particular interest to us because it is one of the first to show the scope of Markov's thinking about dependent processes. Moreover, the results that he obtains could well have been included in the main text. He demonstrates, in our terms, that the state occupancies of a Markov process are asymptotically normally distributed. Furthermore, since his quantities $x_k$ could be interpreted as the rewards of a Markov process, he shows a new way to compute the gain of the process, the quantity $a$ in his notation.

However, the form of the paper is as interesting as the results. We have attempted to preserve the style and notation of the original to convey a feeling for how the subject was initially developed. There are coincidences—Markov uses the variable $z$ in writing his transform expressions just as we have used it in writing ours.

As you will see, Markov's careful yet creative exposition combines concern for the reader with logical elegance.

THE NOTES OF THE IMPERIAL ACADEMY OF SCIENCES
of
ST. PETERSBURG

VIII Series

Physio-Mathematical College
Volume XXII. No. 9.

Classe Physico-Mathématique
Volume XXII. No. 9.

# EXTENSION OF THE LIMIT THEOREMS OF PROBABILITY THEORY TO A SUM OF VARIABLES CONNECTED IN A CHAIN

by A. Markov

December 5, 1907

In the abstract, "Extension of the Law of Large Numbers to the Dependent Case," of Volume XV of the second series of notices of the Physio-Mathematical Society of the University of Kazan, I have shown that certain laws of large numbers established by Chebyshev for independent sequences apply also to many cases of dependent sequences.

Among these cases, in my opinion, are numbers connected into a chain in such a manner that, when the value of one becomes known, subsequent numbers become independent of the preceding ones.

Investigating one rather important dependent case and, it could be said, the simplest one,† I was able to prove the theorem that the mathematical expectation has the limiting probability distribution represented by the well-known Laplace's Integral (the central-limit theorem).

This result suggests that the central-limit theorem must also apply in other cases of numbers connected into a chain.

My proof was based on one of the peculiarities of the given case: symmetry of expressions referred to as $P$ and $Q$, with certain small changes to assure avoidance of special cases and to show the possibility of extending the reasoning and conclusions to other cases.

Rather than returning to the investigated case, we shall examine another more general one to achieve a proof of greater generality.

† "Investigation of an Important Case of Dependent Trials." *Izvestia Acad. Nauk SPB*, VI, ser. I, 61 (1907).

§1. Consider the sequence of numbers

$$x_1, x_2, \ldots, x_k, x_{k+1}, \ldots .$$

Assume that each of the numbers can have the three values:

$$-1, 0, +1.$$

Consider the system of numbers

$$p + q + r$$
$$p' + q' + r'$$
$$p'' + q'' + r''.$$

The first line represents, respectively, the probability of the events

$$x_{k+1} = -1, \qquad x_{k+1} = 0, \qquad x_{k+1} = +1$$

when $x_k = -1$. The second line represents the probability of the same events when $x_k = 0$, and, finally, the third line when $x_k = +1$.

Such probabilities could not, of course, be negative numbers and must satisfy the conditions

$$\begin{aligned} p + q + r &= 1, \\ p' + q' + r' &= 1, \\ p'' + q'' + r'' &= 1. \end{aligned} \tag{1}$$

In addition, we assume that none of them is equal to 1. This assumption is especially important when we refer to

$$p, q', r'',$$

since the sequence

$$x_1, x_2, x_3, \ldots$$

would become simple repetition of one number.

Finally we must introduce three additional quantities

$$P, Q, R,$$

to represent the probabilities that $x_1$ will have the following values:

$$-1, 0, +1.$$

With these assumptions, we undertake investigation of the probability distribution of the sum

$$x_1 + x_2 + \cdots + x_n,$$

for successively larger values of $n$.

With $n = 1$, our sum is the one number $x_1$ and, correspondingly, could have only the three values

$$-1, 0, +1,$$

the probabilities of which, as stipulated, must be equal to

$$P, Q, R.$$

With $n = 2$, we have the sum of two numbers

$$x_1 + x_2,$$

for which it is possible to have five values

$$-2, -1, 0, +1, +2,$$

the probabilities of which are, respectively,

$$Pp, Pq + Qp', Pr + Qq' + Rp'', Qr' + Rq'', Rr''.$$

For the sum

$$x_1 + x_2 + x_3$$

seven interpretations are possible,

$$-3, -2, -1, 0, +1, +2, +3,$$

the probabilities of which, in order, would be

$$Ppp, P(pq + qp') + Qp'p, \qquad P(pr + qq' + rp'') + Q(p'q + q'p') + Rp''p,$$
$$P(qr' + rq'') + Q(p'r + q'q' + r'p'') + R(p''q + q''p'),$$
$$Prr'' + Q(q'r' + r'q'') + R(p''r + q''q' + r''p''), \qquad Qr'r'' + R(q''r' + r''q''),$$
$$Rr''r''.$$

Proceeding according to the paper "Investigation of an Important Case of Dependent Trials," for each subsequent increase of $n$ by 1, we represent the probability of the event

$$x_1 + x_2 + \cdots + x_n = m,$$

for any given $n$ and $m$ as the sum

$$\overset{-}{P}_{m,n} + P_{m,n} + \overset{+}{P}_{m,n},$$

where the symbols

$$\overset{-}{P}_{m,n}, P_{m,n}, \overset{+}{P}_{m,n},$$

represent, respectively, the joint probability of the same event and the possibilities,

$$x_n = -1, \qquad x_n = 0, \qquad x_n = +1.$$

With these definitions it is not hard to establish the following equations:

$$
\begin{aligned}
\bar{P}_{m,n+1} &= p\bar{P}_{m+1,n} + p'P_{m+1,n} + p''\overset{+}{P}_{m+1,n}, \\
P_{m,n+1} &= q\bar{P}_{m,n} + q'P_{m,n} + q''\overset{+}{P}_{m,n}, \\
\overset{+}{P}_{m,n+1} &= r\bar{P}_{m-1,n} + r'P_{m-1,n} + r''\overset{+}{P}_{m-1,n},
\end{aligned}
\tag{2}
$$

for

$$n = 1, 2, 3, \ldots .$$

Using these equations it is possible to calculate the successive probabilities

$$\bar{P}_{m,n}, P_{m,n}, \overset{+}{P}_{m,n}$$

for

$$n = 2, 3, 4, \ldots;$$

by taking into consideration the basic equalities

$$\bar{P}_{-1,1} = P, \qquad P_{0,1} = Q, \qquad \overset{+}{P}_{1,1} = R \tag{3}$$

and noticing that the other expressions

$$\bar{P}_{m,1}, P_{m,1}, \overset{+}{P}_{m,1}$$

must be equal to 0, since they are probabilities of impossible events. We find therefore

$$
\begin{aligned}
\bar{P}_{-2,2} &= pP, & \bar{P}_{-1,2} &= p'Q, & \bar{P}_{0,2} &= p''R, \\
P_{-1,2} &= qP, & P_{0,2} &= q'Q, & P_{1,2} &= q''R, \\
\overset{+}{P}_{0,2} &= rP, & \overset{+}{P}_{1,2} &= r'Q, & \overset{+}{P}_{2,2} &= r''R,
\end{aligned}
$$

and that the other expressions

$$\bar{P}_{m,2}, P_{m,2}, \overset{+}{P}_{m,2}$$

are equal to 0, and thus we obtain

$$\bar{P}_{-3,3} = ppP, \qquad \bar{P}_{-2,3} = pp'Q + p'qP, \qquad P_{-2,3} = qpP, \qquad \overset{+}{P}_{-2,3} = 0,$$

$$\bar{P}_{-2,3} + P_{-2,3} + \overset{+}{P}_{-2,3} = P(pq + qp') + Qp'p, \qquad \text{etc.}$$

Eliminating from our investigation the case when the determinant

$$
\begin{vmatrix}
p, & p', & p'' \\
q, & q', & q'' \\
r, & r', & r''
\end{vmatrix}
$$

is equal to 0, we can introduce three more quantities

$$\bar{P}_{0,0},\ P_{0,0},\ \overset{+}{P}_{0,0},$$

determined from the equations

$$P = p\bar{P}_{0,0} + p'P_{0,0} + p''\overset{+}{P}_{0,0},$$
$$Q = q\bar{P}_{0,0} + q'P_{0,0} + q''\overset{+}{P}_{0,0},$$
$$R = r\bar{P}_{0,0} + r'P_{0,0} + r''\overset{+}{P}_{0,0},$$

which imply

$$\bar{P}_{0,0} + P_{0,0} + \overset{+}{P}_{0,0} = 1.$$

If we interpret the symbols

$$\bar{P}_{m,0},\ P_{m,0},\ \overset{+}{P}_{m,0}$$

to be 0 when $m$ is other than 0, we can extend our equations to the case where $n = 0$.

§2. By introducing the auxiliary variable $t$ and the functions

$$\bar{\phi}_n = \sum \bar{P}_{m,n}t^m, \qquad \phi_n = \sum P_{m,n}t^m, \qquad \overset{+}{\phi}_n = \sum \overset{+}{P}_{m,n}t^m,$$
$$\Phi_n = \bar{\phi}_n + \phi_n + \overset{+}{\phi}_n, \tag{4}$$

we can first ascertain the probability of the event

$$x_1 + x_2 + \cdots + x_n = m,$$

as the coefficient of $t^m$ in the expansion of $\Phi_n$ in the powers of $t$ and, second, from the equations (2) derive

$$t\bar{\phi}_{n+1} = p\bar{\phi}_n + p'\phi_n + p''\overset{+}{\phi}_n,$$
$$\phi_{n+1} = q\bar{\phi}_n + q'\phi_n + q''\overset{+}{\phi}_n, \tag{5}$$
$$\left(\frac{1}{t}\right)\overset{+}{\phi}_{n+1} = r\bar{\phi}_n + r'\phi_n + r''\overset{+}{\phi}_n.$$

These equations show that all the functions

$$\bar{\phi}_n,\ \phi_n,\ \overset{+}{\phi}_n,\ \Phi_n$$

satisfy the same linear equation representable in a simple symbolic expression for the function $\Phi_n$,

$$\begin{bmatrix} p - t\Phi, & p', & p'' \\ q, & q' - \Phi, & q'' \\ r, & r', & r'' - \frac{1}{t}\Phi \end{bmatrix} \Phi^n = 0, \tag{6}$$

where upon the completion of calculations, we must replace

$$\Phi^{n+3},\ \Phi^{n+2},\ \Phi^{n+1},\ \Phi^{n}$$

by

$$\Phi_{n+3},\ \Phi_{n+2},\ \Phi_{n+1},\ \Phi_{n},$$

for

$$n = 0, 1, 2, \ldots .$$

We emphasize the symbolic equation because it shows clearly how we can extend our conclusions to the more general case where $x_k$ could have more than three values.

In explicit form, equation (6) will be:

$$\Phi_{n+3} - A\Phi_{n+2} + B\Phi_{n+1} - D\Phi_n = 0,$$

where

$$A = \frac{p}{t} + q' + r''t, \qquad B = \frac{pq' - p'q}{t} + pr'' - p''r + (q'r'' - q''r')t, \tag{7}$$
$$D = pq'r'' - pq''r' + p'q''r - p'qr'' + p''qr' - p''q'r.$$

On the other hand, in accordance with our definition we have

$$\Phi_0 = 1$$

and, with direct evaluation at $n = 1$ and $n = 2$, we obtain

$$\Phi_1 = P\left(\frac{1}{t}\right) + Q + Rt$$

$$\Phi_2 = Pp\left(\frac{1}{t^2}\right) + (Pq + Qp')\left(\frac{1}{t}\right) + Pr + Qq' + Rp'' + (Qr' + Rq'')t + Rr''t^2.$$

With these results and equation (6) we can obtain the general $\Phi_n$ as the coefficient of $z^n$ in the power series in the auxiliary variable $z$ of the rational function

$$\frac{f(t, z)}{F(t, z)}.$$

The denominator is given by the formula

$$F(t, z) = -\begin{vmatrix} pz - t, & p'z, & p''z \\ qz, & q'z - 1, & q''z \\ rz, & r'z, & r''z - \dfrac{1}{t} \end{vmatrix} = 1 - Az + Bz^2 - Dz^3. \tag{8}$$

The numerator represents a polynomial of second degree in $z$.

The function $f(t, z)$ is determined by the first three terms of the right-hand side of the equation

$$\frac{f(t, z)}{F(t, z)} = \Phi_0 + \Phi_1 z + \Phi_2 z^2 + \cdots + \Phi_n z^n + \cdots. \tag{9}$$

After multiplication of the right-hand side of equation (9) by $F(t, z)$, we obtain

$$f(t, z) = \Phi_0 + (\Phi_1 - A\Phi_0)z + (\Phi_2 - A\Phi_1 + B\Phi_0)z^2.$$

It is important to note for our purpose that with $t = 1$, all functions $\Phi_n$ must be established as 1 and thus

$$\frac{f(1, z)}{F(1, z)} = \frac{1}{1 - z}. \tag{10}$$

§3. The equations we have found are applied in calculations of the mathematical expectation of various powers of the sum

$$x_1 + x_2 + \cdots + x_n - na,$$

with the number $a$ selected in such manner that the mathematical expectation of the first power of this sum remains finite as $n$ becomes infinite.

Just as the coefficient of any given power of $t$ in the expansion of the function $\Phi_n$ represented the probability that the sum

$$x_1 + x_2 + \cdots + x_n$$

is equal to the exponent of that power, so the probability that the sum

$$x_1 + x_2 + \cdots + x_n - na$$

is equal to a given number is the coefficient of the term in $t$ with that exponent in the expansion of

$$t^{-na}\Phi_n.$$

Thus it is not difficult to conclude that the mathematical expectation

$$(x_1 + x_2 + \cdots + x_n - na)^i,$$

where $i$ is any positive integer, can be obtained with the aid of the generating function

$$t^{-na}\Phi_n$$

in the following fashion.

Assume

$$t = e^u,$$

and then form the derivative

$$\frac{d^i(t^{-na}\Phi_n)}{du^i}$$

in which $u$ is set to 0. The result

$$\left\{\frac{d^i t^{-na}\Phi_n}{du^i}\right\}_{u=0}$$

gives us the mathematical expectation in question. The transformation from $t$ to $e^u$ is needed for the purpose of not changing the exponent while differentiating.

It is not difficult to see that

$$\Phi_0 + t^{-a}\Phi_1 z + t^{-2a}\Phi_2 z^2 + \cdots + t^{-na}\Phi_n z^n + \cdots$$

is obtained from

$$\Phi_0 + \Phi_1 z + \Phi_2 z^2 + \cdots + \Phi_n z^n + \cdots$$

by substituting for $z$ the product

$$t^{-a}z.$$

Thus in accordance with equation (9) we have

$$\frac{f(t, zt^{-a})}{F(t, zt^{-a})} = \Phi_0 + t^{-a}\Phi_1 z + t^{-2a}\Phi_2 z^2 + \cdots. \tag{11}$$

The mathematical expectation of

$$(x_1 + x_2 + \cdots + x_n - na)^i$$

is given by the coefficient of $z^n$ in the expansion of the expression

$$\frac{d^i}{du^i}\left(\frac{f(e^u, ze^{-au})}{F(e^u, ze^{-au})}\right)_{u=0} \tag{12}$$

as a power series in $z$.

§4. Let us pause at the mathematical expectation of first degree

$$x_1 + x_2 + \cdots + x_n - na,$$

in order to obtain $a$.

In accordance with our conclusions, this mathematical expectation is expressed by the coefficient of $z^n$ in the expansion as a power series in $z$ of the following function:

$$\frac{f'_{u=0}(e^u, ze^{-au})}{F(1, z)} - \frac{f(1, z)}{F(1, z)}\frac{F'_{u=0}(e^u, ze^{-au})}{F(1, z)}, \tag{13}$$

where the symbols

$$f'_{u=0}(e^u, ze^{-au}) \quad \text{and} \quad F'_{u=0}(e^u, ze^{-au})$$

are interpreted as the derivatives

$$\frac{df(e^u, ze^{-au})}{du} \quad \text{and} \quad \frac{dF(e^u, ze^{-au})}{du}$$

with $u = 0$.

For the investigation of the expansion of this function as a power series in $z$, we write the expansion in simple fractions and, for our purpose, it is important to consider only the fraction with denominator $(1 - z)^2$.

The denominators of the desired simple fractions could be only the simple factors of the polynomial $F(1, z)$.

One of the factors of $F(1, z)$ is

$$1 - z,$$

because the value of the function with $z = 1$ is the determinant

$$-\begin{vmatrix} p - 1, & p', & p'' \\ q, & q' - 1, & q'' \\ r, & r', & r'' - 1 \end{vmatrix},$$

which as a result of the basic requirements

$$p + q + r = p' + q' + r' = p'' + q'' + r'' = 1$$

is equal to 0.

Turning to other simple factors of $F(1, z)$, assume

$$F(1, z) = (1 - z)(1 - y_1 z)(1 - y_2 z), \tag{14}$$

where the numbers

$$1, y_1, y_2$$

represent three roots of the equation

$$\begin{vmatrix} p - y, & p', & p'' \\ q, & q' - y, & q'' \\ r, & r', & r'' - y \end{vmatrix} = 0. \tag{15}$$

With regard to this equation we will prove that one simple root is equal to 1 and that the magnitudes of the rest of its roots are less than 1.

For this purpose we compute the derivative with respect to $y$ of the left side of equation (15) and assume $y = 1$. The result can be represented as a sum of three differences

$$\{q''r' - (1 - q')(1 - r'')\} + \{p''r - (1 - p)(1 - r'')\} + \{p'q - (1 - p)(1 - q')\},$$

none of which could be positive, and which could all simultaneously equal 0 only in cases excluded by us when at least one of the numbers

$$p, q', r''$$

is equal to 1.

Being convinced that unity is a simple and not a multiple root of equation (15) we proceed to other roots

$$y_1, y_2,$$

of the equations that are different from 1.

Let $y$ be one of the two numbers

$$y_1, y_2.$$

We could select a system of numbers

$$\alpha, \beta, \gamma,$$

not consisting only of zeros satisfying the equations

$$\alpha y = p\alpha + q\beta + r\gamma,$$

$$\beta y = p'\alpha + q'\beta + r'\gamma,$$

$$\gamma y = p''\alpha + q''\beta + r''\gamma.$$

The differences

$$\alpha - \beta, \qquad \alpha - \gamma, \qquad \beta - \gamma$$

could not all equal 0 because with

$$\alpha = \beta = \gamma$$

the defining equations for $\alpha, \beta, \gamma$ show that

$$y = 1.$$

Let us investigate the largest in magnitude of the three differences,

$$\alpha - \beta, \qquad \alpha - \gamma, \qquad \beta - \gamma.$$

Let $\alpha - \beta$ correspond to this difference; subtract $\beta y$ from $\alpha y$, and obtain

$$(\alpha - \beta)y = (p - p')\alpha + (q - q')\beta + (r - r')\gamma.$$

Since the factors

$$p - p', \qquad q - q', \qquad r - r',$$

are less than 1 in magnitude and sum to 0, one of them has a sign opposite to the sign of the other two and has the same magnitude as their sum.

Assume that $p - p'$ and $q - q'$ have one sign, while $r - r'$ has the other.

With this assumption, $r - r'$ can be replaced by

$$p' - p + q' - q$$

to obtain

$$y = (p - p')\frac{\alpha - \gamma}{\alpha - \beta} + (q - q')\frac{\beta - \gamma}{\alpha - \beta},$$

and thus we have

$$|y| \leq |p - p'| + |q - q'| = |r - r'| < 1.$$

Thus the magnitudes of the coefficients $y_1$ and $y_2$ in the factored representation of the function $F(1, z)$ in equation (14) are less than 1.

In power series expansions in $z$ of fractions of the type

$$\frac{1}{(1 - y_1 z)^l} \quad \text{and} \quad \frac{1}{(1 - y_2 z)^l},$$

the coefficients of $z^n$ must approach 0 as $1/n$ approaches 0.

Considering power series expansions in $z$ of fractions of the type

$$\frac{1}{(1 - z)^l},$$

in accordance with the equation

$$E_n = \frac{(n + 1)(n + 2)\cdots(n + l - 1)}{1\cdot 2\cdots(l - 1)},$$

defining the coefficients of the transformation

$$\frac{1}{(1 - z)^l} = 1 + E_1 z + E_2 z^2 + \cdots + E_n z^n + \cdots,$$

we have

$$\lim_{n\to\infty} \frac{E_n}{n^{l-1}} = \frac{1}{1\cdot 2\cdots(l - 1)} \quad \text{and} \quad \lim_{n\to\infty} \frac{E_n}{n^{l-1+\varepsilon}} = 0,$$

where $\varepsilon$ is any given positive number.

Applying our deductions to the function (13), we conclude that the coefficient of $z^n$ in the power series expansion in $z$ is equal to the mathematical expectation of the sum

$$x_1 + x_2 + \cdots + x_n - na.$$

This expectation increases without bound with $n$ if and only if

$$\frac{f(1, z)}{F(1, z)} \cdot \frac{F'_{u=0}(e^u, ze^{-au})}{F(1, z)}$$

does not contain $1 - z$ as a denominator factor.

Desiring that this mathematical expectation should not grow without bound in $n$ and realizing that in accordance with the previously established relation

$$\frac{f(1, z)}{F(1, z)} = \frac{1}{1 - z}$$

the polynomial $f(1, z)$ could not contain a factor $1 - z$, we must assign to $a$ such a value that the factor $1 - z$ would be contained in the function

$$F'_{u=0}(e^u, ze^{-au}).$$

Thus we arrive at the equation

$$F'_{u=0}(e^u, e^{-au}) = 0,$$

which is not difficult to transform to the following:

$$-a\left\{\frac{dF(1, z)}{dz}\right\}_{z=1} + \left\{\frac{dF(t, 1)}{dt}\right\}_{t=1} = 0. \tag{16}$$

This equation yields one and only one solution which could be determined from the relations

$$\begin{aligned}\left\{\frac{dF(1, z)}{dz}\right\}_{z=1} &= q''r' - (1 - q')(1 - r'') + p''r - (1 - p)(1 - r'') \\ &\quad + p'q - (1 - p)(1 - q'), \\ \left\{\frac{dF(t, 1)}{dt}\right\}_{t=1} &= (1 - q')(1 - r'') - q''r' - (1 - p)(1 - q') + p'q.\end{aligned} \tag{17}$$

§5. Let us investigate higher powers of the sum

$$x_1 + x_2 + \cdots + x_n - na$$

using the value of $a$ that we have found.

As we have shown, the mathematical expectation of

$$(x_1 + x_2 + \cdots + x_n - na)^i$$

could be computed as the coefficient of $z^n$ in the expansion of

$$\frac{d^i}{du^i}\left(\frac{f(e^u, ze^{-au})}{F(e^u, ze^{-au})}\right)_{u=0}$$

as a power series in $z$.

To investigate this function, for the sake of brevity, assume that:

$$f(e^u, ze^{-au}) = U, \qquad F(e^u, ze^{-au}) = V,$$
$$\frac{d^iV}{du^i} = V^{(i)}, \qquad \frac{d^iU}{du^i} = U^{(i)}. \tag{18}$$

Using these definitions, we obtain, in accordance with the formula for differentiation of a product,

$$\frac{d^i}{du^i}\left(\frac{U}{V}\right) = U\frac{d^i}{du^i}\left(\frac{1}{V}\right) + \frac{i}{1}U'\frac{d^{i-1}}{du^{i-1}}\left(\frac{1}{V}\right) + \cdots. \tag{19}$$

Employing the formula for implicit differentiation, we obtain

$$\frac{d^i}{du^i}\left(\frac{1}{V}\right) = \sum \frac{i!\,j!}{V^{j+1}}\frac{(-V')^\lambda}{\lambda!}\cdot\frac{\left(-\frac{1}{2}V''\right)^\mu}{\mu!}\cdot\frac{\left(-\frac{1}{2.3}V'''\right)^\nu}{\nu!}\cdots. \tag{20}$$

where the summation extends over all possible combinations of the integers

$$j, \lambda, \mu, \nu, \ldots,$$

satisfying the equations

$$\lambda + \mu + \nu + \cdots = j,$$
$$\lambda + 2\mu + 3\nu + \cdots = i \tag{21}$$

and the inequalities

$$0 < j \leq i, \qquad \lambda \geq 0, \qquad \mu \geq 0, \qquad \nu \geq 0, \ldots.$$

We would now like to prove that as the number $n$ increases without limit the ratio of the mathematical expectation of

$$(x_1 + x_2 + \cdots + x_n - na)^i$$

to $n^{1/2}$ rapidly approaches the limit

$$\frac{1}{\sqrt{\pi}}C^{i/2}\int_{-\infty}^{+\infty} t^i e^{-t^2}\,dt,$$

where $C$ is some constant.

For this purpose, in view of previous deductions, it is necessary to demonstrate that the denominator of the rational function in $z$, represented by the derivative

$$\frac{d^i}{du^i}\left(\frac{U}{V}\right)$$

at $u = 0$, could with pertinent cancellations have a factor $1 - z$ to a power not exceeding $(i + 1)/2$ when $i$ is odd and not exceeding $i/2 + 1$ when $i$ is even.

Examining at $u = 0$ one of the terms

$$U^{(l)} \frac{d^{i-l}}{du^{i-l}}\left(\frac{1}{V}\right)$$

included in equation (19), we note that the factor $1 - z$ could not be contained in the denominator to a higher power than in the denominator of the second factor

$$\frac{d^{i-l}}{du^{i-l}}\left(\frac{1}{V}\right)_{u=0}.$$

We can deduce rather important conclusions regarding this second factor on the basis of Equation (20) by substituting for $i$ the difference $i - l$.

Indeed it is not difficult to see that with the selection of $a$ in the term

$$\frac{i!\,j!}{V^{j+1}} \frac{(-V')^{\lambda}}{\lambda!} \cdot \frac{\left(-\frac{1}{2} V''\right)^{\mu}}{\mu!} \cdot \frac{\left(-\frac{1}{2\cdot 3} V'''\right)^{\nu}}{\nu!} \cdots$$

that equation (20) with $u = 0$ results in an uncancellable fraction whose denominator contains the factor $1 - z$ to a power not exceeding

$$j + 1 - \lambda,$$

since with $u = 0$ the function $V'$ contains a factor $1 - z$, while the function of $V$ is impossible to divide by $(1 - z)^2$.

On the other hand, from the conditions limiting

$$j, \lambda, \mu, \nu, \ldots,$$

it is not difficult to derive the inequality

$$\lambda \geq 2j - i, \tag{22}$$

which limits the value of $\lambda$ when $2j > i$.

For $2j < i$ and with $\lambda \geq 0$, the difference

$$j + 1 - \lambda$$

remains, obviously, smaller than $i/2 + 1$; due to inequality (22), the difference

$$j + 1 - \lambda$$

remains smaller than $i/2 + 1$ for $2j > i$, and only with $j = i/2$, if such a value is possible, and with $\lambda = 0$ can it reach the size $i/2 + 1$.

Continuing with the supposition

$$i = 2j, \qquad \lambda = 0,$$

which is possible only with $i$ even, we find that this supposition corresponds to only one member of the equation (21),

$$\frac{1, 2, 3, \ldots, i}{2^{i/2}} \frac{(-V'')^{i/2}}{V^{i/2+1}},$$

since with $j = i/2$ and $\lambda = 0$, equation (21) and the associated inequalities imply

$$\mu = \frac{i}{2}.$$

Consequently, with $i$ odd, none of the fractions obtained from the derivatives

$$\frac{d^i}{du^i}\left(\frac{1}{V}\right), \qquad \frac{d^{i-1}}{du^{i-1}}\left(\frac{1}{V}\right), \qquad \frac{d^{i-2}}{du^{i-2}}\left(\frac{1}{V}\right), \ldots,$$

with $u = 0$ can have a factor $1 - z$ in its denominator (after proper cancellation, of course) to a power higher than $(i + 1)/2$, and only the first of these fractions obtained from the derivative

$$\frac{d^i}{du^i}\left(\frac{1}{V}\right),$$

could contain in its denominator the factor $1 - z$ to the power $i/2 + 1$. If, however, we subtract

$$\frac{1, 2, 3, \ldots, i}{2^{i/2}} \cdot \frac{(-V'')^{i/2}}{V^{i/2+1}}$$

from this first fraction, with $u = 0$, then the difference upon proper cancellation leads to a fraction with a denominator containing the factor $1 - z$ to a power smaller than $i/2 + 1$.

Comparing this result with equation (19), and keeping in mind the definitions (18), we conclude that with $i$ odd the expression

$$\frac{d^i}{du^i}\left(\frac{f(e^u, ze^{-au})}{F(e^u, ze^{-au})}\right)_{u=0}$$

is reduced to a rational fractional function in $z$, the denominator of which contains the factor $1 - z$ to a power smaller than $i/2 + 1$.

Thus, on the basis of the conclusions of the investigations §§3 and 4, we find that when $i$ is odd

$$\textit{Math. Expectation}\,\frac{(x_1 + x_2 + \cdots + x_n - na)^i}{n^{i/2}}$$

must approach the limit 0, as $n$ grows to infinity.

It is not difficult to find a simple fraction with the denominator $(1 - z)^{i/2+1}$ that must contain only lower powers of $1 - z$.

Indeed this fraction must be the one whose removal from the expression

$$\frac{1, 2, 3, \ldots, i}{2^{i/2}} \left(\frac{U}{V}\right)_{u=0} \left(\frac{-V''}{V}\right)_{u=0}^{i/2}$$

creates a fraction without the factor $1 - z$ to a power higher than $i/2 - 1$ in its denominator.

On this basis, recalling the relation

$$\left(\frac{U}{V}\right)_{u=0} = \frac{1}{1 - z}, \tag{10}$$

established in §2, we find that the fraction in question could be represented as

$$\frac{1, 2, 3, \ldots, i}{2^i} \cdot \frac{C^{i/2}}{(1 - z)^{i/2+1}},$$

where the number $C$ is defined by

$$\frac{1}{2} C = \frac{F''_{u=0}(e^u, e^{-au})}{F'_{z=1}(1, z)}. \tag{23}$$

Having arrived at this conclusion, by reviewing the investigations of §§3 and 4 we are convinced without a great deal of trouble that the expression

$$\textit{Math. Expectation}\,\frac{(x_1 + x_2 + \cdots + x_n - na)^i}{n^{i/2}},$$

with $i$ even tends toward the limit

$$\frac{1, 2, 3, \ldots, i}{2^i, 1, 2, \ldots, i/2} C^{i/2},$$

as $n$ grows to infinity.

It remains to note that the integral

$$\frac{1}{\sqrt{\pi}} \int_{-\infty}^{+\infty} t^i e^{-t^2}\, dt$$

with $i$ odd is equal to zero and with $i$ even is equal to

$$\frac{1, 2, 3, \ldots, i}{2^i, 1, 2, \ldots, i/2}.$$

Then we arrive at the final conclusion.

If a number $a$ is defined by equation (16), and a number $C$ by equation (23), then

$$\lim_{n \to \infty} \textit{Math. Expectation} \left(\frac{x_1 + x_2 + \cdots + x_n - na}{\sqrt{n}}\right)^i = \frac{1}{\sqrt{\pi}} C^{i/2} \int_{-\infty}^{+\infty} t^i e^{-t^2}\, dt,$$

with $i$ being either odd or even. Obviously, with unlimited growth of the number $n$, the probability of satisfaction of the inequalities

$$t_1 \sqrt{Cn} < x_1 + x_2 + \cdots + x_n - na < t_2 \sqrt{Cn},$$

where $t_1$ and $t_2$ are any given numbers, $t_2 > t_i$, must approach a limit equal to

$$\frac{1}{\sqrt{\pi}} \int_{t_1}^{t_2} e^{-t^2}\, dt.$$

§6. Having investigated one case with numbers connected into a chain and having noted that it does not possess any special characteristics but differs from others only in the simplicity of data, we are able in a few words to extend our conclusions to the general case of numbers connected into a chain established in the above-mentioned paper, "Extension of the Law of Large Numbers to the Dependent Case."

We assume that the numbers in the chain

$$x_1, x_2, \ldots, x_k, x_{k+1}, \ldots$$

can take on the values

$$\alpha, \beta, \gamma, \ldots \tag{24}$$

and that the system

$$\begin{matrix} p_{\alpha,\alpha}, p_{\alpha,\beta}, p_{\alpha,\gamma}, \ldots \\ p_{\beta,\alpha}, p_{\beta,\beta}, p_{\beta,\gamma}, \ldots \\ p_{\gamma,\alpha}, p_{\gamma,\beta}, p_{\gamma,\gamma}, \ldots \\ \cdots\cdots\cdots\cdots\cdots \end{matrix} \tag{25}$$

represents the probabilities that $x_{k+1}$ will attain any value given the value of $x_k$. The first subscript of $p$ indicates the given size of $x_k$, the second the proposed size of $x_{k+1}$.

The numbers (24) and (25) define our final conclusions, but in order to establish a completely definite problem we must introduce also the numbers

$$p_\alpha', p_\beta', p_\gamma', \ldots,$$

which like the numbers $P$, $Q$, $R$ of the special case vanish from the final solution and correspondingly give the probabilities of the equalities

$$x_1 = \alpha, \qquad x_1 = \beta, \qquad x_1 = \gamma, \ldots,$$

when all numbers of our chain

$$x_1, x_2, \ldots, x_n, \ldots$$

remain uncertain.

To begin investigation of the probability distribution of the sum

$$x_1 + x_2 + \cdots + x_n,$$

let us define the probability of the equality

$$x_1 + x_2 + \cdots + x_n = m$$

by the symbol $P_{m,n}$ and introduce the function $\Phi_n$ of $t$ defined by the equation

$$\Phi_n = \sum P_{m,n} t^m. \tag{26}$$

Proceeding then in the same fashion as in the special case, we divide $P_{m,n}$ into additive terms

$$P_{m,n} = \overset{\alpha}{P}_{m,n} + \overset{\beta}{P}_{m,n} + \overset{\gamma}{P}_{m,n} + \cdots \tag{27}$$

and introduce

$$\overset{\alpha}{\Phi}_n = \sum \overset{\alpha}{P}_{m,n} t^m, \qquad \overset{\beta}{\Phi}_n = \sum \overset{\beta}{P}_{m,n} t^m, \ldots, \tag{28}$$

where the symbols

$$\overset{\alpha}{P}_{m,n}, \overset{\beta}{P}_{m,n}, \overset{\gamma}{P}_{m,n}, \ldots$$

are joint probabilities of the satisfaction of the same equality

$$x_1 + x_2 + \cdots + x_n = m$$

and of an additional condition expressed by one of the equalities

$$x_n = \alpha, \qquad x_n = \beta, \qquad x_n = \gamma, \ldots.$$

With these definitions it is not difficult to establish the equations

$$\overset{\alpha}{P}_{m,n} = \overset{\alpha}{P}_{m-\alpha,n-1} p_{\alpha,\alpha} + \overset{\beta}{P}_{m-\alpha,n-1} p_{\beta,\alpha} + \cdots,$$
$$\overset{\beta}{P}_{m,n} = \overset{\alpha}{P}_{m-\beta,n-1} p_{\alpha,\beta} + \overset{\beta}{P}_{m-\beta,n-1} p_{\beta,\beta} + \cdots,$$
$$\cdots\cdots\cdots\cdots\cdots\cdots\cdots\cdots,$$

and from there to go on to equations

$$
\begin{aligned}
\overset{\alpha}{\Phi}_n t^{-\alpha} &= p_{\alpha,\alpha}\overset{\alpha}{\Phi}_{n-1} + p_{\beta,\alpha}\overset{\beta}{\Phi}_{n-1} + p_{\gamma,\alpha}\overset{\gamma}{\Phi}_{n-1} + \cdots,\\
\overset{\beta}{\Phi}_n t^{-\beta} &= p_{\alpha,\beta}\overset{\alpha}{\Phi}_{n-1} + p_{\beta,\beta}\overset{\beta}{\Phi}_{n-1} + p_{\gamma,\beta}\overset{\gamma}{\Phi}_{n-1} + \cdots,\\
\overset{\gamma}{\Phi}_n t^{-\gamma} &= p_{\alpha,\gamma}\overset{\alpha}{\Phi}_{n-1} + p_{\beta,\gamma}\overset{\beta}{\Phi}_{n-1} + p_{\gamma,\gamma}\overset{\gamma}{\Phi}_{n-1} + \cdots,\\
&\cdots\cdots\cdots\cdots\cdots\cdots\cdots\cdots
\end{aligned}
\tag{29}
$$

From equation (29) we can derive all functions

$$\overset{\alpha}{\Phi}_n,\ \overset{\beta}{\Phi}_n,\ \overset{\gamma}{\Phi}_n,\ \ldots$$

and then determine $\Phi_n$ from

$$\Phi_n = \overset{\alpha}{\Phi}_n + \overset{\beta}{\Phi}_n + \overset{\gamma}{\Phi}_n + \cdots.$$

Equations (29) can be represented in the simple symbolic form

$$
\begin{vmatrix}
p_{\alpha,\alpha} - t^{-\alpha}\Phi, & p_{\beta,\alpha}, & p_{\gamma,\alpha}, & \cdots\\
p_{\alpha,\beta}, & p_{\beta,\beta} - t^{-\beta}\Phi, & p_{\gamma,\beta}, & \cdots\\
p_{\alpha,\gamma}, & p_{\beta,\gamma}, & p_{\gamma,\gamma} - t^{-\gamma}\Phi, & \cdots\\
\cdots & \cdots & \cdots & \cdots
\end{vmatrix}\Phi^n = 0,
\tag{30}
$$

where after calculation we must replace

$$\Phi^n,\ \Phi^{n+1},\ \Phi^{n+2},\ \ldots$$

by

$$\Phi_n,\ \Phi_{n+1},\ \Phi_{n+2},\ \ldots.$$

In view of this equation all functions

$$\Phi_0,\ \Phi_1,\ \Phi_2,\ \ldots,\ \Phi_n,\ \ldots$$

could be determined as coefficients in the power series in the auxiliary variable $z$ of the rational function

$$\frac{f(t, z)}{F(t, z)},$$

the denominator of which is defined by

$$
F(t, z) = \begin{vmatrix}
p_{\alpha,\alpha}z - t^{-\alpha}, & p_{\beta,\alpha}z, & p_{\gamma,\alpha}z, & \cdots\\
p_{\alpha,\beta}z, & p_{\beta,\beta}z - t^{-\beta}, & p_{\gamma,\beta}z, & \cdots\\
p_{\alpha,\gamma}z, & p_{\beta,\gamma}z, & p_{\gamma,\gamma}z - t^{-\gamma}, & \cdots\\
\cdots & \cdots & \cdots & \cdots
\end{vmatrix}.
\tag{31}
$$

Before proceeding to further conclusions, it is necessary to note that we are investigating only chains

$$x_1, x_2, \ldots, x_n, \ldots,$$

where the appearance of some of the numbers

$$\alpha, \beta, \gamma, \ldots$$

does not preclude the appearance of others. This important condition could be expressed by means of determinants in the following manner: The determinant

$$\begin{vmatrix} u, & p_{\beta,\alpha}, & p_{\gamma,\alpha}, & \cdots \\ p_{\alpha,\beta}, & v, & p_{\gamma,\beta}, & \cdots \\ p_{\alpha,\gamma}, & p_{\beta,\gamma}, & w, & \cdots \\ \cdots & \cdots & \cdots & \cdots \end{vmatrix}$$

with arbitrary elements

$$u, v, w, \ldots$$

does not reduce to a product of several determinants of the same type.

This condition is not enough, however, for our purpose and thus we must assume† that the determinant we have chosen does not reduce to the product of several determinants with

$$u = p_{\alpha,\alpha}, \qquad v = p_{\beta,\beta}, \qquad w = p_{\gamma,\gamma}, \ldots.$$

Thus, in the special as well as the general case, our deductions are concerned with the mathematical expectation of

$$(x_1 + x_2 + \cdots + x_n - na)^i,$$

where the number $a$ we define under the condition $i = 1$ so as to assure that the mathematical expectation will not increase infinitely with $n$.

Having established that the mathematical expectation under investigation could be expressed as the coefficient of $z^n$ in the power series in $z$, the value of the derivative

$$\frac{d^i}{du^i}\left\{\frac{f(e^u, ze^{-au})}{F(e^u, ze^{-au})}\right\}$$

at $u = 0$, we note that, if the results of the specific case are to apply in general, we have to look at the decomposition of the function $F(1, z)$ into simple factors

$$F(1, z) = \pm(1 - z)(1 - y_1 z)(1 - y_2 z)\cdots$$

† Our conclusions may be extended to many cases we have excluded.

and prove that the factor $1 - z$ appears only once in this transformation and that the numbers

$$y_1, y_2, \ldots$$

are less than one in magnitude.

In other words we must be convinced that 1 serves as a simple and not as a multiple root of the equation

$$\begin{vmatrix} p_{\alpha,\alpha} - y, & p_{\beta,\alpha}, & p_{\gamma,\alpha}, & \cdots \\ p_{\alpha,\beta}, & p_{\beta,\beta} - y, & p_{\gamma,\beta}, & \cdots \\ p_{\alpha,\gamma}, & p_{\beta,\gamma}, & p_{\gamma,\gamma} - y, & \cdots \\ \cdots & \cdots & \cdots & \cdots \end{vmatrix} = 0 \qquad (32)$$

and that the rest of the roots

$$y_1, y_2, \ldots$$

of the same equation are less than 1 in magnitude.

For the proof that 1 is a simple and not a multiple root of equation (32), the following proposition may serve in reference to the determinant.

If all of the elements of the determinant

$$\begin{vmatrix} u, & -b_1, & -c_1, & \cdots \\ -a_1, & v, & -c_2, & \cdots \\ -a_2, & -b_2, & w, & \cdots \\ \cdots & \cdots & \cdots & \cdots \end{vmatrix}$$

satisfy the inequalities

$$a_k \geq 0, \qquad b_k \geq 0, \qquad c_k \geq 0, \ldots \qquad (*)$$

and

$$u \geq a_1 + a_2 + \cdots, \quad v \geq b_1 + b_2 + \cdots, \quad w \geq c_1 + c_2 + \cdots, \quad \cdots; \qquad (**)$$

then it cannot be a negative quantity and might be 0 only in the extreme cases when all the inequalities (**) are equalities, or when it is reducible because some of the inequalities (*) are converted into equalities and it becomes a product of several determinants of the same type and among these there is a certain determinant for which all the inequalities (**) are turned into equalities.†

† A similar proposition is mentioned in the remarks of Herman Minkowsky "Zur Theorie der Einheiten in den algebraischen Zahlkörpern" (Nach. v.d. Kon. Gesel. der Wiss. ze Göttingen a.d. J. 1900).

We are convinced in the certainty of this important proposition by considering that $u, v, w, \ldots$ are variables and by noting that the derivatives of our determinant with respect to $u, v, w, \ldots$ are expressed by similar determinants of lower order; this observation gives us the possibility of gradually extending the theorem to determinants of second order, third order, fourth order, and so forth.

It follows then that having proven the proposition in accordance with our stated conditions, the derivative with respect to $y$ of the left-hand side of equation (32) does not become 0 with $y = 1$, because this derivative, being multiplied by $\pm 1$, could be represented as a sum of determinants

$$\begin{vmatrix} 1 - p_{\beta,\beta}, & -p_{\gamma,\beta}, & \cdots \\ -p_{\beta,\gamma}, & 1 - p_{\gamma,\gamma}, & \cdots \\ \cdots & \cdots & \cdots \end{vmatrix} + \begin{vmatrix} 1 - p_{\alpha,\alpha}, & -p_{\gamma,\alpha}, & \cdots \\ -p_{\alpha,\gamma}, & 1 - p_{\gamma,\gamma}, & \cdots \\ \cdots & \cdots & \cdots \end{vmatrix} + \cdots,$$

which satisfy the conditions of the theorem and are not acceptable only in extreme cases.

Thus the quantity of 1, which in accordance with conditions

$$p_{\alpha,\alpha} + p_{\alpha,\beta} + p_{\alpha,\gamma} + \cdots = 1,$$
$$p_{\beta,\alpha} + p_{\beta,\beta} + p_{\beta,\gamma} + \cdots = 1,$$
$$\cdots\cdots\cdots\cdots$$

must satisfy equation (32), could not be a multiple root of this equation.

Turning to the other roots of the equation (32), let us suppose that $y$ is any one of them.

We may consider a system of numbers

$$\alpha', \beta', \gamma', \ldots$$

not consisting only of zeros and satisfying the equations

$$\alpha' y = p_{\alpha,\alpha}\alpha' + p_{\alpha,\beta}\beta' + p_{\alpha,\gamma}\gamma' + \cdots,$$
$$\beta' y = p_{\beta,\alpha}\alpha' + p_{\beta,\beta}\beta' + p_{\beta,\gamma}\gamma' + \cdots,$$
$$\gamma' y = p_{\gamma,\alpha}\alpha' + p_{\gamma,\beta}\beta' + p_{\gamma,\gamma}\gamma' + \cdots,$$
$$\cdots\cdots\cdots\cdots$$

The numbers $\alpha', \beta', \gamma', \ldots$ could not have the same value because, if they satisfied the system of equalities

$$\alpha' = \beta' = \gamma' = \cdots,$$

their defining equations would require that

$$y = 1.$$

Noting this, let us assume to start with that $\alpha', \beta', \gamma', \ldots$ do not have the same magnitude. With this assumption and with the conditions imposed on $\alpha', \beta', \gamma', \ldots,$ there must be at least one equation where the multiplier of $y$ on the left-hand side could be any of the numbers

$$\alpha', \beta', \gamma', \ldots$$

with largest absolute value, while the right-hand side has coefficients different from 0 and contains numbers

$$\alpha', \beta', \gamma', \ldots,$$

less than the largest in absolute value. From this it follows that

$$|y| < 1,$$

because the sum of the coefficients of

$$\alpha', \beta', \gamma', \ldots$$

in the right-hand side of any equation is equal to 1.

Let us assume next that all the numbers

$$\alpha', \beta', \gamma', \ldots$$

have the same absolute value.

We know, however, that they are not equal to one another. Thus they can be divided into two groups: One group of numbers equals $\alpha'$ and another is unequal to $\alpha'$, being different from $\alpha'$ in sign.

On the other side, having in mind the basic conditions, the set of sums

$$\begin{gathered} p_{\alpha,\alpha}\alpha' + p_{\alpha,\beta}\beta' + p_{\alpha,\gamma}\gamma' + \cdots, \\ p_{\beta,\alpha}\alpha' + p_{\beta,\beta}\beta' + p_{\beta,\gamma}\gamma' + \cdots, \\ \cdots\cdots\cdots\cdots\cdots\cdots\cdots\cdots \end{gathered}$$

cannot be divided into two sets so that from all numbers

$$\alpha', \beta', \gamma', \ldots$$

the sums of the first set would contain, with coefficients† different from zero, only terms equal to $\alpha'$, the sums of the second set, only the ones not equal to $\alpha'$.

Consequently one of these sums certainly contains, with coefficients different from zero, both numbers equal to $\alpha'$ and numbers not equal to $\alpha'$; thus the magnitude of the sum equals the product of the magnitudes of $y$ and $\alpha'$ and must be smaller than the magnitude of $\alpha'$, because with the number of distinct arguments

† In the abstract "Extension of the Law of Large Numbers to the Dependent Case," I made the simple assumption that among these cofficients there are none equal to zero.

the magnitude of their sum is less than the sum of their magnitudes and not equal to it.

From this the following inequality is derived

$$|y| < 1,$$

which had to be proven.

We have now shown that in the expansion of the function $F(1, z)$ into simple factors

$$F(1, z) = \pm(1 - z)(1 - y_1 z)(1 - y_2 z)\cdots$$

the factor $1 - z$ enters to first degree only, while the magnitudes of the coefficients

$$y_1, y_2, \ldots$$

are less than one. All further results for the specific case can be generalized without changes.

In view of this generality we shall not repeat the arguments, but will present only the final conclusion.

Under our assumptions, if the numbers $a$ and $C$ determined from

$$a = \frac{F'_{t=1}(t, 1)}{F'_{z=1}(1, z)}, \qquad \frac{1}{2}C = \frac{F''_{u=0}(e^u, e^{-au})}{F'_{z=1}(1, z)}, \tag{33}$$

do not lead to either impossible or indefinite results, then

$$\lim_{n\to\infty} \textit{Math. Expectation} \left(\frac{x_1 + x_2 + \cdots + x_n - na}{\sqrt{n}}\right)^i = \frac{1}{\sqrt{\pi}} C^{i/2} \int_{-\infty}^{+\infty} t^i e^{-t^2}\, dt$$

for any positive integer $i$; consequently, with limitless increase of $n$, the probability of satisfaction of the inequality

$$t_1 \sqrt{Cn} < x_1 + x_2 + \cdots + x_n - na < t_2 \sqrt{Cn},$$

where $t_1$ and $t_2$ are any given numbers with $t_2 > t_1$, must approach the limit

$$\frac{1}{\sqrt{\pi}} \int_{t_1}^{t_2} e^{-t^2}\, dt.$$

Regarding numbers $a$ and $C$, we may also note that they are correspondingly equal to the limits

$$\textit{Math. Expectation}\, \frac{x_1 + x_2 + \cdots + x_n}{n}$$

and

$$Math.\ Expectation \left(\frac{x_1 + x_2 + \cdots + x_n - na}{\sqrt{n}}\right)^2,$$

as $n$ is increased to infinity.

This observation permits us to extend our conclusions to cases where the set of numbers

$$\alpha, \beta, \gamma, \ldots$$

is not finite.

Our deductions could also be applied to complicated chains where each number is directly connected not with one, but with several preceding numbers.

# REFERENCES


R. Bellman, *Dynamic Programming*, Princeton University Press, Princeton, New Jersey, 1957, Chapter XI.

R. Bellman and S. Dreyfus, *Applied Dynamic Programming*, Princeton University Press, Princeton, New Jersey, 1962.

A. T. Bharucha-Reid, *Elements of the Theory of Markov Processes and Their Applications*, McGraw-Hill Book Company, Inc., New York.

E. B. Dynkin, *Theory of Markov Processes*, Prentice-Hall, Inc., Englewood Cliffs, New Jersey, 1961.

W. Feller, *An Introduction to Probability Theory and Its Applications*, Vol. I, 2nd Ed., John Wiley & Sons, Inc., 1962.

Ronald A. Howard, *Dynamic Programming and Markov Processes*, The M.I.T. Press, Cambridge, 1960.

W. H. Huggins, "Signal-Flow Graphs and Random Signals," *Proceedings of I.R.E.*, 45, 74 (1957).

J. G. Kemeny and J. L. Snell, *Finite Markov Chains*, D. Van Nostrand Company, Princeton, 1960.

S. J. Mason and H. J. Zimmermann, *Electronic Circuits, Signals, and Systems*, John Wiley & Sons, Inc., New York, 1960.

Operations Research Center, M.I.T., *Notes on Operations Research 1959*, Technology Press, Cambridge, 1969, Chapters 3, 5, and 7.

E. Parzen, *Stochastic Processes*, Holden-Day, Inc., San Francisco, 1962.

R. W. Sittler, "Systems Analysis of Discrete Markov Processes," I.R.E. *Transactions on Circuit Theory*, CT-3, No. 1, 257 (1956).

# INDEX